Lecture Notes in Computer Science　　　7489

Commenced Publication in 1973
Founding and Former Series Editors:
Gerhard Goos, Juris Hartmanis, and Jan van Leeuwen

Panayiotis Zaphiris   George Buchanan
Edie Rasmussen   Fernando Loizides (Eds.)

Theory and Practice of Digital Libraries

Second International Conference, TPDL 2012
Paphos, Cyprus, September 23-27, 2012
Proceedings

Volume Editors

Panayiotis Zaphiris
Fernando Loizides
Cyprus University of Technology
Department of Multimedia and Graphic Arts
3603 Limassol, Cyprus
E-mail: panayiotis.zaphiris@cut.ac.cy, fernando.loizides@gmail.com

George Buchanan
City University of London
School of Informatics
Northampton Square, London, EC1V 0HB, UK
E-mail: george.buchanan.1@city.ac.uk

Edie Rasmussen
The University of British Columbia
School of Library, Archival and Information Studies
Irving K. Barber Learning Centre, Vancouver, BC, V6T 1Z1, Canada
E-mail: edie.rasmussen@ubc.ca

ISSN 0302-9743   e-ISSN 1611-3349
ISBN 978-3-642-33289-0   e-ISBN 978-3-642-33290-6
DOI 10.1007/978-3-642-33290-6
Springer Heidelberg Dordrecht London New York

Library of Congress Control Number: 2012946348

CR Subject Classification (1998): H.4, H.2, H.3, H.5, J.1, H.2.8

LNCS Sublibrary: SL 3 – Information Systems and Application, incl. Internet/Web
and HCI

Typesetting: Camera-ready by author, data conversion by Scientific Publishing Services, Chennai, India

Printed on acid-free paper

Springer is part of Springer Science+Business Media (www.springer.com)

Preface

We are delighted to present the proceeding of the 16th European Conference on Research and Advanced Technology for Digital Libraries, running for the second year as "Theory and Practice in Digital Libraries", which ran in Paphos, Cyprus during September 23–27, 2012.

Following the inaugural ECDL in Pisa in 1997, the conference has become established as a major international venue for leading researchers in the field of digital libraries. The global economic turmoil of the last few years has not dented the continued health of the conference. We continue to receive and accept submissions for publication from countries that have not been represented at TPDL before, as an example our host country of Cyprus is well represented with new contributors.

For TPDL 2012, we focussed on attracting contributions on four major themes:

- Applications and User Experience
- Supporting Discovery
- Digital Humanities
- Research Data

These themes are the same as those of TPDL 2011, and while the core computer-science origins of TPDL continue to mature, the approaches of interdisciplinary methods, and the interests of humanists and research scientists in systematic management of their research materials, are widening the range of challenges addressed by digital library researchers.

Our overseas keynote speakers this year represented two contrasting approaches to supporting information seeking. Mounia Lalmas is well known as a global leader in information retrieval research, and represents the technical origins and ideals of many in the DL community. Mounia's passion for excellence and robust methodology has inspired many of her peers across the world. In contrast, Cathy Marshall is immediately familiar to digital library researchers as a leading light in the more human-centred approaches to DL work, and her work has been repeatedly recognised by best-paper awards at the leading DL conferences.

We are also fortunate to have a local keynote speaker, Andreas Lanitis, whose interests are particularly relevant to the contemporary theme of preserving digital heritage. The problems in that domain have been of increasing pertinence to DLs in recent years, as the transience of digital material and the desire to disseminate ancient heritage combine to seek out the means by which long term, remote access to new and old materials can be sustained.

The programme began with the doctoral consortium and our tutorial programme of five varied tutorials, on themes from building digital libraries through to preservation processes.

Following the main conference, four workshops concluded the programme, including the 11th European Networked Knowledge Organization Systems (NKOS) workshop and the 2nd Semantic Digital Archive workshop.

For TPDL 2012, we ran a single programme committee for all submissions, and furthermore combined the call for short papers and posters into a single call for work-in-progress contributions. We continued to use the two-tier model of TPDL 2011 and ECDL 2010, with a main programme committee supervised by a metareview committee. Each contribution was reviewed independently by three or more members of the main committee, and the subsequent discussion and reflection was moderated by one member of the metareview committee.

The submissions received by the conference totalled 139 contributions (129 full papers and work-in-progress, plus and 10 demonstrations). From this, we selected 23 full papers (17.8%), 19 short paper presentations (15%), 14 poster submissions (11%), and 5 demonstrations. Twenty-six countries contributed works for review, the same as last year's conference in Berlin.

The programme of TPDL 2012 proved particularly challenging for the programme committee to finalise. The support of our metareview committee, continuing the practice of recent years at TPDL to use a second level of review, was critical in balancing the merits of the competing submissions. Our programme committee put in many hours of review and discussion, which greatly assisted us as programme chairs, and ensured that our decisions were robust and well-informed.

Fabrizio Sebastiani, our workshop chair, and Christos Papatheodorou, serving as tutorial chair, put in particularly praiseworthy efforts to help in the success of the conference. Our other supporting chairs also provided invaluable assistance in finalising the programme.

The success of TPDL 2012 is also due to our various supporters, but in particular the Coalition for Networked Information (CNI).

September 2012

George Buchanan
Edie Rasmussen
Panayiotis Zaphiris
Fernando Loizides

Organisation

TPDL 2012 was organised by the Cyprus University of Technology, in association with the University of Cyprus and City University London.

Organising Committee

Conference Chair	Panayiotis Zaphiris, Cyprus University of Technology, Cyprus
Organising Chair	Fernando Loizides, Cyprus University of Technology, Cyprus
Proceedings Chair	Fernando Loizides, Cyprus University of Technology, Cyprus
Publicity Chair	Maria Poveda, University of Cyprus, Cyprus

Programme Committee

Programme Chair	George Buchanan, City University London, UK
	Edie Rasmussen, University of British Columbia, Canada
Tutorials	Christos Papatheodorou, Ionian University, Greece
Workshops	Fabrizio Sebastiani, ISTI/CNR, Italy
Posters	Jennifer Pearson, Swansea University, UK
Panels	Rudi Schmeide, TU Darmstadt, Germany
	Andri Ioannou, Cyprus University of Technology, Cyprus
Demonstrations	Trond Aalberg, Norwegian University of Science and Technology, Norway
Doctoral Consortium	Stefan Gradmann, Humboldt University Berlin, Germany
	Birger Larsen, Royal School of Library and Information Science, Denmark

Metareview Committee

David Bainbridge	University of Waikato, New Zealand
José Borbinha	IST, Portugal
Donatella Castelli	ISTI-CNR, Italy
Gobindha Chowdhury	University of Technology, Sydney, Australia
Fabio Crestani	University of Lugano, Switzerland
J. Stephen Downie	University of Illinois Urbana-Champaign, USA
Schubert Foo	Nanyang Technical University, Singapore

Edward A. Fox	Virginia Tech, USA
Ingo Frommholz	University of Bedfordshire, UK
Norbert Fuhr	University of Duisberg-Essen, Germany
Annika Hinze	University of Waikato, New Zealand
Claus-Peter Klas	Fern Universität Hagen, Germany
Carlo Meghini	Consiglio Nazionale delle Ricerche, Italy
Michael Nelson	Old Dominion University, USA
Christos Papatheodorou	Ionian University, Greece
Andreas Rauber	Vienna University of Technology, Austria
Susan Schreibmnan	Trinity College Dublin, Ireland
Hussein Suleman	University of Cape Town, South Africa
Elaine Toms	Sheffield University, UK
Pertti Vakkari	Tampere University, Finland

Programme Committee

Trond Aalberg	Norwegian University of Science and Technology, Norway
Anne Adams	Open University, UK
Robert Allen	Victoria University Wellington, New Zealand
Simon Attfield	Middlesex University, UK
Wolf-Tilo Balke	University of Hannover, Germany
Alejandro Bia	Miguel Hernández University, Spain
Christine Borgman	University of California Los Angeles, USA
Leonardo Candela	ISTI-CNR, Italy
Les Carr	Southampton University, UK
Lillian Cassel	Villanova University, USA
Tinni Choudhury	Middlesex University, UK
Panos Constantopoulos	Athens University of Economics and Business, Greece
Gregory Crane	Tufts University, USA
Pierre Cubaud	CNAM, France
Sally Jo Cunningham	University of Waikato, New Zealand
Theodore Dalamagas	National Technical University of Athens, Greece
Nicola Ferro	University of Padua, Italy
Luis Francisco-Revilla	University of Texas Austin, USA
Nuno Freire	The European Library, Portugal
Luanne Freund	University of British Columbia, Canada
Pablo de la Fuente	University of Valladolid, Spain
Richard Furuta	Texas A&M University, USA
C. Lee Giles	Pennsylvania State University, USA
Marcos Goncalves	Federal University of Minas Gerais, Brazil
Juilo Gonzalo	UNED, Spain
Stefan Gradmann	Humboldt University Berlin, Germany
Kai Grønbaek	Aarhus University, Denmark

Doug Tudhope   Glamorgan University, UK
Shalini Urs   University of Mysore, India
Nicholas Vanderschantz   University of Waikato, New Zealand
Felisa Verdejo   UNED, Spain
Stina Westman   University of Aalto, Finland
Uffe Wiil   Syddansk University, Denmark

Reviewers

Massimiliano Assante   ISTI-CNR, Italy
Vassilis Christophides   Foundation for Research and Technology
                              (FORTH), Greece
Gianpaolo Coro   ISTI-CNR, Italy
Sujatha Das   Pennsylvania State University, USA
Helen Dodd   Swansea University, UK
Martin Doerr   Foundation for Research and Technology -
                              Hellas, Greece
Laurent D'Orazio   Clermont Université - Université Blaise Pascal,
                              France
Giorgos Giannopoulos   National Technical University of Athens,
                              Greece
Madian Khabsa   Pennsylvania State University, USA
Ivano Masiero   University of Padua, Italy
Simone Peruzzo   University of Padua, Italy
Hany Salaheldeen   Old Dominion University, USA
Sven Schlarb   Austrian National Library, Austria
Gianmaria Silvello   University of Padua, Italy
Zhaohui Wu   Pennsylvania State University, USA

Workshops Committee

Fabrizio Sebastiani (Chair)   ISTI/CNR, Italy
Nicola Ferro   University of Padua, Italy
Ed Fox   Virginia Tech, USA
Marcos Goncalves   Federal University of Minas Gerais, Brazil
Vivien Petras   Humboldt University Berlin, Germany
Andreas Rauber   Vienna University of Technology, Austria

Tutorials Committee

Christos Papatheodorou (chair) Ionian University, Greece
Trond Aalberg   Norwegian University of Science and
                              Technology, Norway
Birger Larsen   Royal School of Library and Information
                              Science, Denmark
Ron Larsen   University of Pittsburgh, USA

Table of Contents

Preservation

Linked Data

Analysing and Enriching Documents

Content and Metadata Quality

Folksonomy and Ontology

Information Retrieval

Organising Collections

Extracting and Indexing

Poster Papers

Demonstration Papers

What Would 'Google' Do? Users' Mental Models of a Digital Library Search Engine

Michael Khoo and Catherine Hall

The iSchool, Drexel University, 3141 Chestnut Street, Philadelphia, PA 19104, USA
{khoo,catherine.e.hall}@drexel.edu

Abstract. A mental model is a model that people have of themselves, others, the environment, and the things with which they interact, such as technologies. Mental models can support the user-centered development of digital libraries: if we can understand how users perceive digital libraries, we can design interfaces that take these perceptions into account. In this paper, we describe a novel method for eliciting a generic mental model from users, in this case of a digital library's search engine. The method is based on a content analysis of users' mental representations of the system's usability, which they generated in heuristic evaluations. The content analysis elicited features that the evaluators thought important for the search engine. The resulting mental model represents a generic model of the search engine, rather than a clustering of individuals' mental models of the same search engine. The model includes a number of references to Web search engines as ideal models, but these references are idealistic rather than realistic. We conclude that users' mental models of Web search engines should not be taken at face value. The implications of this finding for digital library development and design are discussed.

Keywords: human-computer interaction, human factors, mental models, search, search engine, users, user-centered design.

1 Introduction

Mental models shape how users perceive and interact with technologies [19, 22]. Understanding users' mental models supports user-centered design, and the creation of technologies and interfaces that take users' expectations into account. This is relevant for the user-centered design of digital libraries [4, 14, 29].

Since the concept was first proposed [6], a range of theoretical approaches to mental models has been proposed [14, 23, 28]. These approaches can be mutually exclusive [23]. In this paper we adopt the definition of Norman, who connects mental models directly with the ways in which users perceive and evaluate the usability of technologies [14, 19, 23]. According to Norman, a mental model is a model that "people have of themselves, others, the environment, and the things with which they interact. People form mental models through experience, training, and instruction. The mental model of a device is formed largely by interpreting its perceived actions and its visible structure" [19]. Norman cites Kempton's study [11] of 'folk theories'

P. Zaphiris et al. (Eds.): TPDL 2012, LNCS 7489, pp. 1–12, 2012.

of residential thermostats. Some users erroneously believe that if a higher temperature is dialed, then more heat is let into a room; such users set a thermostat to a higher temperature in order to heat a room 'faster,' and then dial the thermostat back once the desired temperature is reached. Kempton called this folk theory the 'valve' model. However, a room will heat up at the same rate, regardless of the thermostat setting, until the desired temperature is reached, at which point the switch cuts off the heat. Kempton called this correct theory the 'feedback' theory.

Incorrect and folk mental models can lead to erroneous user interactions. System interfaces should therefore be designed to make it easy for a user to determine what correct actions are possible at any moment, and to evaluate the outcome of user actions. Norman models this as (a) the designer's image ('the conceptualization that the designer has in mind'), and (b) the user's image ('the model … the user develops to explain the operation of the system'), which intersect in (c) the 'system image.' The system image (such as a thermostat dial) is the evidence available to the user, which points to the underlying function of the system (see Figure 1). It represents the system components and their relationships to the user, so that the user can easily interact with the system; it therefore acts as a translational hinge between the designer and the user. When a system image accurately represents the underlying system model, then a device is easy to use. However, when a system image is "incoherent or inappropriate … incomplete or contradictory" then usability problems can arise. In Kempton's study, a rotary thermostat dial presents an *ambiguous* system image. The dial rotates like a volume knob, suggesting that rotation increases the amount of *something*; however, the supply of heat is actually constant, and the point at which the rotation stops actually sets the limit at which the thermostat cuts off the heat supply.)

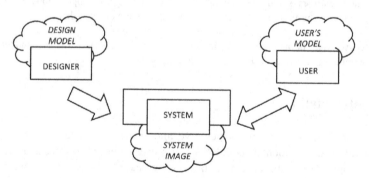

Fig. 1. Norman's model of the role of the system image

2   Background and Literature Review

In this paper, we describe the mental model that a group of digital library users had of that digital library's search engine. There are a number of previous studies of the mental models of users searching libraries and digital libraries. Borgman [3] studied users performing Boolean search operations, and found that even when users had acquired some kind of mental model of how Boolean logic worked, they were unable to articulate it in a coherent fashion. Dimitroff [7] studied student OPAC users, and found a positive relationship between a user's ability to articulate a mental model of search, and their success in searching. Griffiths & Brophy [9] found that Google was the first place most university students looked for information, and held it in high regard ("Google is very straight forward. You put in your word and it searches"). In contrast, few students (with the exception of LIS students) were likely to turn first to library OPACs or academic resources, which were seen as being difficult to understand. Makri et al. [14] observed and interviewed eight users engaged in a variety of information retrieval tasks. They inferred mental models through coding, and identified nine overall themes, including search technologies. Some participants assumed that the search engines of Web sites, e-commerce sites, and digital libraries all worked in a similar manner, while other participants recognized differences but were confused as to their technical nature. They suggest that library and digital library interfaces should provide support for users to make sense of interfaces. Murumatsu and Pratt [16] looked at users' mental models of query transformations by Web search engines (such as behind-the-scenes parsing into Boolean queries) and found that while users recognized that different search engines parsed queries in different ways, they were unable to come up with a reasonable model of how this worked in practice, and there was "a substantial mismatch between most users' mental models of web search engine operation and their actual operation." Nielsen [11] has described a users model of Web search as including a search box for query entry, a search button to run the search, and a list of results "that's linear, prioritized, and appears on a new page." "Deviating from this expected design," notes Nielsen, "almost always causes usability problems." Slone studied public library uses of library catalogs and online search engines, and found that the mental models of all these were often vague and almost 'magical' [25; c.f. 27]. Zhang [29] explored forty-four undergraduate students' mental models of the World Wide Web, using questionnaires, semi-structured interviews, sketches, and observations of search tasks. The mental model identified included a 'search mechanism,' and here students expressed a preference for using Google, although they framed this from utilitarian rather than a technical perspective.

Further relevant research has looked at users' attitudes towards library OPACs, digital libraries, and Web search engines, without necessarily building explicit mental models. Bawdlin and Vilar [1] note a tendency for OPAC and digital library users to expect search engines to perform like Google, and also like e-commerce sites such as Amazon. Becker [2] interviewed students and found that while they could articulate the importance of source evaluation, and how to achieve this with OPACs, in practice they followed the 'path of least resistance,' and used Google. Connaway et al.'s [5] multifaceted study of 'convenience' as a motivating factor in the information seeking behaviors of students and faculty included the finding that subjects (particularly undergraduates) saw Google as a convenient and easy-to-use environment for information seeking, which libraries should seek to replicate. Fast and Campbell [8] found that university students often saw Web search engines (such as Google) as offering relatively simple ease-of-use compared with OPACs ('you can type pretty basic things into the Internet and get a ton of results'), with this perceived ease-of-use generally compensating for any post-session filtering that users might have to engage in. Martzoukou's [15] study of postgraduate students assessed their self-satisfaction with their information seeking skills, and found that while they were experienced Web searchers, and also had an awareness of the limitations of this approach, they had 'minimal motivation to change habitual behavioural patterns' and develop more complex strategies. Finally, Ponsford and vanDuinkerken's [24] user tests of an academic library's new federated search engine, found that while some were willing to spend time to learn to use the new system, many (especially the younger 'Google generation') wanted a 'point-and-click' Google simplicity.

A general theme of many of these studies is that users' expectations of a system can be used to generate requirements for improving that system. In the cases cited above, for instance, Web search engines are seen as forms of requirements for improving OPACs and digital libraries. Just how useful is such an approach, however? In the rest of this paper, we describe how users' reported expectations of Web search engines are often idealistic rather than realistic. We use a content analysis of student heuristic evaluation assignments of ipl2, to identify a generic mental model of a Web search engine in the student evaluators' discourse, which (as we shall show) has a number of gaps with regard to current Web search engine technology.

3   Methods

The Internet Public Library was founded in 1995 as on online reference service. It subsequently developed its own reviewed collections of web sites. Beginning in 2008, the IPL was re-launched as ipl2. The Web site was redesigned, and the catalog crosswalked to Dublin Core metadata, now stored in a FEDORA database [12]. ipl2 is now maintained as a largely volunteer project that provides online reference, reviewed collections, and other services. It has approximately 40,000 catalog records, including 12,500 internal web pages, and links to external web sites. In 2011, ipl2 received 7,842,351 visits (including 5,856,289 unique visitors), and 21,317,480 page views. Recent project work has included building a new metadata tool, evaluating the

Fig. 2. The ipl2 front page and search box

usability of the search interface, and search error messages [21], and understanding users' mental models of ipl2.

Mental models, as abstract entities, can be difficult to observe [14, 29]. Often, researchers will collect data with qualitative instruments (interviews, think-alouds, etc.), and analyze and code these data to identify mental model themes. This study utilizes a hitherto unreported method, based on an analysis of assignments written by student teams in online Masters' "Introduction to HCI" classes over several years. Most students had no prior background in HCI, although many worked with libraries or information systems. The assignment required the students to carry out a heuristic evaluation of ipl2. While the original aim of the assignment was not to generate mental models, it was subsequently realized that the reports could be analyzed for evidence of mental models; for instance, if a report contained the finding that "the ipl2 search engine fails to prioritize search results," it could be inferred that prioritized search results are part of a mental model of what the ipl2 search engine should provide. Similar reasoning can be applied to the other findings in the reports.

The students in the courses read Norman's *The Design of Everyday Things* [19], and studied HCI methods. For a final project, they were asked to carry out a heuristic evaluation [17] of ipl2. ipl2 was chosen to give the students a chance to apply HCI techniques in the context of a realistic client and design brief. Heuristic evaluation was chosen as a technique that works well with distributed virtual teams. The teams interviewed potential ipl2 users (teachers, students, parents, etc.), developed appropriate personas and scenarios, and then carried out the heuristic evaluations. They analyzed and prioritized their evaluation data, and wrote reports detailing their findings and design recommendations.

These reports are a rich source of data regarding interface issues with ipl2. A previous study coded these reports for high-level usability issues [13]. Major themes found included:

- Search and browse issues (38% of issues) related to problems with the search engine, difficulties with browsing, confusing subject categories, poor search results (empty results, etc.), lack of advanced search, lack of result refinement, etc.
- Navigation issues (27% of issues) related to broken links within the site or to external resources, confusion when leaving the site to visit an external resource, lack of navigation back to the home page, and poor quality 'breadcrumb' trails.
- Interface issues (18% of issues) related to a site design that was at times considered to be cluttered, inconsistent, and confusing.

A number of remarks in the reports showed that the student teams took Web search engines as one benchmark for evaluating ipl2. That is, they evaluated the performance of ipl2 against their existing experience of Web search engines, and then suggested that Web search engines were superior, and that ipl2 should implement various aspects of Web search engine technology. The original research question for this paper was therefore to understand further the ways in which the students' mental models valorized Web search engines; that is, which features of Web search engines did users see as desirable, and which could be incorporated into ipl2? However, this relationship turned out to be quite complex, in that while the reports did compare ipl2 to Google, they did so in terms of a low- rather than a high-fidelity model of Google. In the rest of this paper, we will take a closer look at this phenomenon.

The current analysis analyzes 33 de-identified reports produced by 194 students in 33 teams. Both authors iteratively reviewed a complete set of reports for descriptions of the ipl2 search engine. Once identified, relevant passages were cut from the reports, generating a total of 415 'snippets' for the two authors. The snippets ranged in size from one-line bullet points, to detailed page-length descriptions of particular search engine issues, and included both direct and indirect references to Web search engines. Using a card sort methodology, the authors identified an initial set of themes, and then allocated the snippets to these themes. Many of the snippets did not fit tidily, and there were a number of discussions as to the 'best fit.' As the analysis progressed, themes were discarded, modified, and consolidated, until a final set of themes was arrived at. While there is software that can perform somewhat similar analyses, this collaborative hand-sorting has advantages, with the ability to discuss easily any of the snippets leading to contextualised meaning-making of the data.

4   Results

Four high-level themes of users' mental models of the ipl2 search engine were identified in the analysis:

- *Search page issues*. The search page should include tools such as auto-correct, an easy to find 'help' page, etc.
- *Search results page issues*. The search result page should make better sense of the large numbers of search results, address link rot, and distinguish between internal and external Web site links.
- *Proposed design solutions*. Suggestions for improving ipl2 include developing an advanced search page, providing better 'help' pages, saving search histories, and supporting what many of the teams called 'natural language' queries.
- *Recommendations that ipl2 adopt various aspects of Web search technology*. Particularly, the teams recommended that ipl2 become more 'like Google.'

Each theme is discussed in this section, based on snippets allocated in the analysis.

4.1   Search Page Interface Issues

The first theme of the mental model is associated with the usability of the search bar. A number of interface issues were identified, including lack of spelling auto-correct features, and lack of error-correct features. A feature many felt was lacking was a "smart search technology that can predict a user's intended search query when he misspells something, like the Google search engine's 'Did you mean?'", along with predictive text and auto-complete capabilities. This issue was deemed particularly important given ipl2's user population, and one team reported that "the inability of the system to recover from and correct this particular type of error could create an insurmountable barrier for school age children..." Several teams noted problems with the design of the search bar, and a drop-down option that allowed localized searching of some ipl2 collections. Overall, these and other missing features were seen as standard problems that had been solved in many Web browsers, search engines, and e-commerce sites, and as such their absence in the ipl2 search engine was noticeable. As one group put it, adding such features would "drastically improve the flexibility of the search functionality, prevent errors like misspellings, and improve standards that other search engines follow." (How the latter, in particular, would be achieved, was not explained).

4.2   Search Results Page Issues

The second theme concerned the search results page. The presence of dead or broken links negatively affected opinion about ipl2, and it was felt that more maintenance was required, as well as easier mechanisms for users to report bad or broken links. As one report summarized, although "the links that didn't work were to external sites that the ipl2 doesn't control...a broken link will degrade the overall user experience regardless of whose fault it technically is." Several groups also noted that ipl2 did not do enough in terms of providing context and provenance to distinguish between the results for internal ipl2 collections, and those for external web pages. One team reported that "the website gives no indication of which links are located within the ipl2 website and which lead to an external website (other than by looking at the actual link address, which might be too sophisticated for some users to interpret)"; and many felt that when selecting a link to a third-party resource, users should receive some form of alert or warning. Both of these suggestions showed an awareness of the 'library' aspect of ipl2 – that sites were curated and consequently ipl2 had a responsibility to its users that is not typically expected of commercial search engines.

Teams were concerned that large numbers of search results that appeared to lack any apparent prioritization. One group noted that 49 results were returned for the search term "steel," a number considered to be "daunting to the user" (this observation is discussed further below). Lack of obvious indicators of ordering, such as number ranking, clustering, or alphabetization, was thought to hinder a user's ability to find information. Teams also criticized the lack of post-search refinement options. As one team noted, "the modern user expects their search results to have an apparent hierarchy ... there is also the expectation that a user will be able to sort results by simple sets of criteria." Such a lack of transparency in how the results were ranked, along with an inability to refine a results set, negatively affected the "trustfulness" of ipl2 and could lead users "to question the information's relevance and validity." It was therefore suggested that ipl2 provide a clear indication of how search results were ranked. Interestingly, most groups seemed comfortable accepting the implicit ranking of Google and other search engines, but they assumed that ipl2 search results were not ordered in anyway. Post-search refinement options were therefore seen as an expected feature of the ipl2 search engine, and the teams expected to be able to filter search results by criteria such as date, subject, keyword frequency, and most popular sites.

4.3   Proposed Design Solutions

The third theme included solutions that the teams proposed for the problems they had identified, such as an advanced search tool. "Users are not able to enter several search terms, narrow results, or refine existing results," noted one report. The possible underlying technology of an advanced search tool was not however elucidated. The teams tended to think of such a tool as an interface design issue, maybe in terms of adding another Web page, rather than something that would depend on manipulating the underlying ipl2 metadata. (In general, metadata was very rarely discussed in any of the reports). A similar set of solutions called for support for users to reformulate and refine queries, and user accounts for users to share queries and results with each other. This solution was again seen mainly in terms of how such tools might appear in an interface, with the underlying architecture – user accounts, secure logins, etc. – largely ignored.

A further set of solutions asked for easy-to-understand help guides. "The vast majority of the problems stemmed from the lack of help, hints, or guidance from the site," claimed one report. The current help guide was hard to locate, and too technical for most users: "Many users, particularly younger and/or inexperienced people, will find that the wording and complexity of this function is more confusing than helpful," said one team. However, there was also a tension between making the guide simple, and the complexity of the concepts that the teams thought should be included. For instance, as the ipl2 search engine permits Boolean operators, several teams thought that these needed to be explained clearly, especially to younger users, but without an actual technical discussion of what Boolean operators actually were [c.f. 3, 16].

A final set of solutions concerned what the teams called 'natural language' queries. While rarely defined, it was implied that natural language searching would lead to better results and usability; as one report said, "Users also may assume that such a search tool would understand semi 'natural language' searches. Due to the popularity of the search engines Google, Yahoo, and Bing, these assumptions are engrained in users' minds as the way things should work."

4.4   Benchmarking against Web Search Engines

The fourth and final theme consisted of teams benchmarking ipl2 against Web search engines. There were many suggestions that the ipl2 search engine replicate the features and interaction characteristics of existing Web search engines (Google, Yahoo!, and Bing were frequently mentioned). Features such as query auto-complete, referred to as "smart search" technologies, and spelling correction, similar to Google's 'Did you mean ...' feature, were popular. One team observed that some of the health topics might be unsuitable for young children, and so settings "similar to Google's 'Safesearch' can be used to keep certain kinds of content away from prying eyes." Other desired features included an easy-to-use advanced search page; user accounts that allow users to login, save and share searches; an easy to understand help page; and an ability to handle 'natural language' queries.

There were a number of comments that ipl2 could be difficult to use because it was *not* like a Web search engine; as one group commented, "Most personas were particularly challenged by the behavior of ipl2's search function because it did not work the same way as Google's search engine." Several teams suggested that ipl2 just use Google as the default site search (although this does miss the point of what to do about searching through the third party resources in the ipl2 catalog).

4.5   Summary

In summary, the users' mental model of the ipl2 search engine, inferred in the foregoing analysis, emphasizes the return of useful results, and includes:

- a 'smart search' feature (auto-correct, auto-complete, 'did you mean,' etc.)
- an age appropriate 'safe search' feature
- an advanced search page that permits search refinement
- better navigation across multiple results and multiple search pages
- prioritization of search results
- refinement of search queries and filtering of search results
- curation of broken links
- warnings when moving 'off site'
- user accounts for saving and sharing searches
- an easy to understand help page
- an ability to handle 'natural language' queries
- a resemblance to Web search engines that users are already familiar with.

5   Discussion: Like Google, or Google-Like?

Variations on many of the usability issues identified in the analysis have previously been described in the digital library literature, including the fourth issue, in which the teams referred to Google as a search engine that was doing things 'right,' and recommended that the ipl2 search engine be modeled on Google. Identifying and describing this theme was an original intent of the analysis, and if the analysis had stopped here, it would have been tempting to conclude that the users' generic mental model of the ipl2 search engine was indeed based on their experience of Google. However, upon closer examination, a number of dimensions of this similarity are questionable. The users often did not have a firm mental model of what it was that Google actually did, and there were several disconnects between the portrayal of Google in the teams' reports, and Google itself.

In one disconnect, the teams suggested a number of 'Google-like' improvements to ipl2, which are not in fact found in Google. For instance, it was suggested that ipl2 provide user accounts for sharing search results, but Google does not do this (Google does however provide user accounts that include personalized 'as-you-type' search predictions, results and recommendations tailored to users' preferences, etc.). There were suggestions that ipl2 provide a way to refine search results, and allow users to revisit and refine queries, in a form of faceted filtering. Again, Google does not really provide such a service, at least for subjects (although at the time of writing there are various links placed around the edge of Google search results pages that allow refining by date, file type, etc.). There were also requests for alphabetized results, which Google does not provide (perhaps the students were thinking here of OPACs).

In a second disconnect, the ipl2 search engine was critiqued for returning too many irrelevant results. However, Google returns many orders of magnitude more results, for the same queries. A specific example (mentioned above) noted that a search for the word 'steel' returned 49 results in ipl2, which was considered too high. The report commented: "The results certainly are imprecise. The first two results provide relevant sources in this case, but it becomes less useful beyond that. The sixth result, 'Brad's Page of Steel,' concerns acoustic and electric lap steel guitars, which has nothing to do with the evaluator's topic of interest beyond the shared word of 'steel.'" In comparison, however, the same search on Google returns approximately 1.4 billion results, displayed on a page dominated by a local map and links to local businesses.

A third disconnect included recommendations that ipl2 adopt Google's search technology. However, these reports did not acknowledge the different underlying technologies of ipl2 and Google. ipl2 uses the Lucene search engine to search across Dublin Core metadata in a Fedora database. A key component of Google's search technology is PageRank [20], which predicts the quality of a Web page by calculating an index from the inbound links to that Web page, on the assumption that the number of such links represents a quality judgment by other Web users. In practice, it is not possible for ipl2 to have a PageRank-based engine, as it does not have access to an index of the Web upon which to base such calculations.

Overall, the teams often compared ipl2 unfavorably with, and also suggested that ipl2 be more like, Google. However, their descriptions of what it was that Google actually did, and which ipl2 should therefore copy, were imprecise. They often seemed to be not of Google *per se*, but rather of an idealized search engine, to which users attached the label 'Google,' as this was the closest real-life approximation that they could think of (c.f. a similar disconnect in the case of mental models of mobile phones [27]). In other words, the teams did not want the ipl2 search engine to be *like* Google; rather they wanted it to be *Google-like*, in the sense that Google was their preferred search engine. This was reflected in the teams' overall visions of a digital library in which it was possible to seamlessly and effortlessly access and share relevant educational resources.

This finding – that the teams saw Google as an exemplar for ipl2 to follow, without knowing in detail how Google worked – has important implications for the use of mental models in the user-centered design of digital libraries. Digital library users may refer to Web search engines as ideals; this does not mean that digital library designers should take Web search engines as rigorous models for the development of digital libraries (c.f. [26]). To do so would fail to take into account all of the characteristics that users ascribe to Web search engines, but which do not actually exist in these search engines; users' mental models of Web search engines are richly detailed, but they are framed as folk models, rather than technical models [11, 19]. When building mental models of digital library search engines, references to Web search engines should not therefore be taken at 'face value' data, but investigated further to find out exactly what it is that users *do* mean by (for instance) 'being like Google.' This complicates the relationship between the design model, the user's model, and the system image, in Norman's original model. It increases the work that a system image has to do in order to represent a digital library system to users in a facile way, as the system image has take into account users' *folk* models of Web search engines, rather than assuming that they are referring to the characteristics of actual Web search engines.

Finally, there are several limitations to this study. One limitation is that the sample population consisted of Masters students, who are not necessarily typical ipl2 users; however, in the assignment, the use of personas and scenarios helped to address this limitation. A second linked limitation is that the students were also predominantly LIS and MLIS students, which may have biased some of their proposed solutions towards more technical solutions (c.f. Griffith and Brophy's discussion of the differences between library and non-library students [9]). A third limitation is that the assignments were not directed at mental model building, and there was no opportunity to ask the students how they actually thought Web search engines such as Google might work. However, this allowed us to treat the data as 'naturally occurring' discourse, from which mental models could then be inferred. Third, it can be argued that the students were primed to look for problems by the assignment, and thus that the findings are in some sense artificial. However, in general, the consistency of the findings across the teams suggests that this analysis has identified a real phenomenon.

6   Conclusion

Digital libraries provide filtered access to high quality, richly described, expert-evaluated content. At the same time, the resource discovery process in digital libraries can be frustrating, and digital library researchers have known informally for a while that users can benchmark digital libraries unfavorably against Web search engines. This paper has described a method for eliciting a generic mental model from users of ipl2, in order to identify areas for improving the ipl2 search engine. While the initial expectation that the users would preferentially reference Web search engines was confirmed, these references were often found to consist of folk models rather than technical models. The analysis shows that designers, when engaged in user-centered design, should not take users' claims regarding Web search engines at face value, as users often do not know in detail how these search engines work. This finding has significant implications for user-centered design work with digital libraries. The elicitation method itself is also useful, and has interesting potential for studying generic users' mental models of information systems in general.

References

1. Bawdon, D., Vilar, P.: Digital libraries: To meet or manage user expectations. ASLIB Proceedings: New Information Perspectives 58(4), 346–354 (2006)
2. Becker, N.: Google in perspective: Understanding and enhancing student search skills. New Review of Academic Librarianship 9, 84–100 (2003)
3. Borgman, C.: The user's mental model of an information retrieval system. In: Procs. 8th ACM SIGIR Conference, pp. 268–273 (1985)
4. Borgman, C.: Designing digital libraries for usability. In: Bishop, A., Van House, N., Buttenfield, B. (eds.) Digital Library Use, pp. 85–118. The MIT Press, Cambridge (2003)
5. Connaway, L.S., Dickey, T., Radford, M.: "If it is too inconvenient I'm not going to use it:" Convenience as a critical factor in information-seeking behavior. Library and Information Science Research 33, 179–190 (2011)
6. Craik, K.: The Nature of Explanation. Cambridge University Press, Cambridge (1943)

7. Dimitroff, A.: Mental models theory and search outcome in a bibliographic retrieval system. LISR 14, 141–156 (1992)

8. Fast, K., Campbell, D.G.: I still like Google: University Student Perceptions of Searching OPACs and the Web. In: Procs. 67th ASIS&T Annual Meeting, vol. 41, pp. 138–146 (2004)

9. Griffiths, J., Brophy, P.: Student searching behavior and the Web: Use of academic resources and Google. Library Trends 53(4), 539–554 (2005)

10. He, W., Erdelez, S., Wang, F., Shyu, C.: The effects of conceptual description and search practice on users' mental models and information seeking in case-based reasoning retrieval system. Information Processing and Management 44, 294–300 (2008)

11. Kempton, W.: Two theories of home heat control. Cognitive Science 10, 75–90 (1986)

12. Khoo, M., Hall, C.: Merging Metadata: A Sociotechnical Study of Crosswalking and Interoperability. In: JCDL 2010, Brisbane, Australia, June 21-25, pp. 361–364 (2010)

13. Khoo, M., Kusunoki, D., MacDonald, C.: Finding Problems: When Digital Library Users Act as Usability Evaluators. In: 45th Hawaii International Conference on System Sciences (HICSS), Maui, Hawaii, January 4-7 (2012)

14. Makri, S., Blandford, A., Gow, J., Rimmer, J., Warwick, C., Buchanan, G.: A Library or Just Another Information Resource? A Case Study of Users' Mental Models of Traditional and Digital Libraries. JASIST 58(3), 433–445 (2007)

15. Martzoukou, K.: Students' attitudes towards Web search engines – increasing appreciation of sophisticated search strategies. Libri 58, 182–201 (2008)

16. Murumatsu, J., Pratt, W.: Transparent queries: Investigating Users' Mental Models of Search Engines. In: SIGIR 2001, New Orleans, LA, September 9-12, pp. 217–224 (2001)

17. Nielsen, J.: Heuristic Evaluation (2005a),
http://www.useit.com/papers/heuristic/

18. Nielsen, J.: Mental models for search are getting firmer (2005b),
http://www.useit.com/alertbox/20050509.html

19. Norman, D.: The Design of Everyday Things. Basic Books, New York (2002)

20. Page, L., Brin, S., Motwani, R., Winograd, T.: The PageRank Citation Ranking: Bringing Order to the Web. Technical Report: Stanford InfoLab (1999)

21. Park, S.J., MacDonald, C.M., Khoo, M.: Do you care if a computer says sorry? User experience design through affective messages. In: 2012 Conference on Designing Interactive Systems (DIS 2012), Newcastle, UK, June 11-15 (2012)

22. Payne, S.J.: Users' mental models: the very ideas. In: Carroll, J.M. (ed.) HCI Models, Theories, and Frameworks: Toward a Multidisciplinary Science, pp. 135–156. Morgan Kaufmann, Amsterdam (2003)

23. Pisanski, J., Zumer, M.: Mental models of the bibliographic universe. Part 1: mental models of descriptions. Journal of Documentation 66(5), 643–667 (2010)

24. Ponsford, B.C., van Duinkerken, W.: User expectations in the time of Google: Usability testing of federated searching. Internet Reference Services Quarterly 12(1/2), 159–178 (2007)

25. Slone, D.: The influence of mental models and goals on search patterns during Web interaction. JASIST 53(13), 1152–1169 (2002)

26. Swanson, T., Green, J.: Why we are not Google: Lessons from a library Web site usability study. The Journal of Academic Librarianship 37(3), 222–229 (2011)

27. Turner, P., Sobolewska, E.: Mental models, magical thinking, and individual differences. Human Technology 5(1), 90–113 (2009)

28. Westbrook, L.: Mental models: a theoretical overview and preliminary study. Journal of Information Science 30(6), 563–579 (2006)

29. Zhang, Y.: Undergraduate Students' Mental Models of the Web as an Information Retrieval System. JASIST 59(13), 2087–2098 (2008)

An Exploration of ebook Selection Behavior in Academic Library Collections

Dana McKay[1], Annika Hinze[2], Ralf Heese[3], Nicholas Vanderschantz[2], Claire Timpany[2], and Sally Jo Cunningham[2]

[1] Library, Swinburne Institute, PO Box 218 John Street, Hawthorn VIC 3122, Australia
dmckay@swin.edu.au
[2] Dept. of Computer Science, University of Waikato, Private Bag 3105, Hamilton New Zealand
{hinze,vtwoz,ctimpanysallyjo}@waikato.ac.nz
[3] Institute of Computer Science, Freie Universität Berlin, Königin-Luise-Straße 24/26, Berlin
heese@inf.fu-berlin.de

Abstract. Academic libraries have offered ebooks for some time, however little is known about how readers interact with them while making relevance decisions. In this paper we seek to address that gap by analyzing ebook transaction logs for books in a university library.

Keywords: ebooks, log analysis, book selection, HCI, information behavior.

1   Introduction

Consider the process of borrowing a book from a library (digital or physical): the reader searches or browses the collection to identify candidate books that are potentially relevant; they each candidate book to assess its actual relevance; and they borrow those books deemed relevant to explore them more depth at a later date. Of course, finding a useful book is rarely so straightforward: these stages may be sequential or interleaved.

In this paper, we consider a specific aspect of this process: the examination of a candidate book after the candidate has been identified in the library collection; we focus specifically on ebooks. How a reader explores physical books when making relevance decisions—which parts of books are viewed, how quickly, in what order—has been relatively neglected [8], likely because such a study in a physical domain would be intrusive and 'creepy' [16]. To our knowledge this aspect of ebook selection has also not been explored, surprisingly, given that the availability of ebook transaction logs allows such a study to be conducted post hoc so as to avoid disturbing the experience of those using an ebook collection. This paper describes exactly such a study based on the transaction logs of a university library ebook collection.

In Section 2 of this paper we explore the previous research on ebook usage and book selection (both physical and electronic); in Section 3 we describe the ebook collection on which our study is based and gives details of the log sample we analyzed. Section 4 presents the results of our analysis, including which parts of the

P. Zaphiris et al. (Eds.): TPDL 2012, LNCS 7489, pp. 13–24, 2012.
© Springer-Verlag Berlin Heidelberg 2012

books readers examined, and readers' interaction patterns with the ebooks; Section 5 discusses the relationship of these results to the existing understanding of document navigation and selection; and Section 6 presents our conclusions.

2   Related Work

The work related to our study falls into two categories: the use and usability of ebooks (Section 2.1) and the literature on book selection (Section 2.2).

2.1   Use and Usability of ebooks

The use of ebooks for recreational reading is increasing: in 2011 purchases of ebooks on Amazon surpassed those of print books [6]. In the academic sphere—where our study takes place—some disciplines demonstrated higher use of ebooks than print books since as early as 2002 [4], and a slight institution-wide predominance was seen in one study in 2004 [11]. Nearly all studies of ebook use, though, show that uptake varies by discipline (it is more common in technical disciplines) and reader circumstances (for example reader location).

The online nature of ebooks accounts for many of the pros and cons given by readers: readers liked the searchability and currency of ebooks, and the ability to access them anywhere [9, 14], however they were frustrated by restrictions on printing and copying, and by DRM [22]. Users' other frustrations include poor ebook annotation capabilities [16] and in-book browsing facilities [2, 10]. Ebooks were perceived to be more comfortable read on dedicated reading devices than on computer screens [9], but even these devices have problems [15, 19], and a study of students given Amazon Kindles to trial showed the majority did not use them for study-related reading [25].

Despite these problems, however, the largest and most recent survey of academic ebook usage showed over half of all respondents had used at least one ebook [9].

2.2   Book Selection

Rowlands *et al.* noted in 2007 [21] that book selection was a surprisingly unstudied part of the book use process, an assessment that remains true today.

Studies of children's book selection practices are more common than those of adults: Reutzel and Gali [20] followed a number of children through the process of checking out library books, and noted that children were more likely to choose books from eye-level, tended to make judgments based on color, and that while they occasionally flipped through books, they rarely made decisions on the basis of content. Moore [18] witnessed children choosing the first book on any shelf related to their topic, as opposed to browsing to select the most appropriate volume for their needs.

Stelmaszweska *et al.*'s work [23] following computer scientists is the earliest published work based on observation of adults selecting books in a library. This study demonstrated that while these readers did use book content to guide their decision to some extent, they also looked at the book covers to determine books' ages and assessed the amount of dust on books to determine whether they had been recently used. In a physical library, of course, availability is also an issue. Our own previous work on adults in an academic found many of the same things, for example that covers are important and readers make decisions based on perceived book age [8]. Buchanan's bookshop study [3] provides some examples of people choosing between books, and notes that most book shoppers did not, in fact, open the books when making their decisions; however these examples all describe shoppers buying for someone else.

At a more fine-grained level, Stieve *et al.* [24] asked university students to choose between two similar books to meet a pre-defined information need. In this study researchers observed and students reported using the Table of Contents (ToC) and the organization of books to make decisions. The importance of ToC, especially in ebooks, is reinforced by other studies [2, 10]. This finding is also reflected in our companion work on book selection practices in physical libraries [8].

Despite the broad number of studies, there is little work on the actual process readers use to select books of interest or reject those that are not interesting, nor on the data they use to make these decisions. From the studies described above we might guess that choices are based on some combination of book-cover, non-metadata information (such as dust) and the table of contents: our study investigates this question further for academic ebooks.

3   Methodology

Our study used data gathered at Swinburne University of Technology in Melbourne, Australia. Swinburne is a small, dual-sector (university and polytechnic), research-active institution. It has approximately 27,000 students, 770 research postgraduates, and 1270 academic staff. This section describes the ebook collection the data came from and the method used in this study.

3.1   The ebook Collection

The ebooks studied in this paper are provided by EBL (http://www.eblib.com), a large ebook provider in Australia. Swinburne makes nearly 20,000 books from the EBL collection available via the library catalogue; some of these owned by the library, but the majority are available on an ad-hoc basis. The difference between owned and unowned ebooks is transparent to readers; Swinburne uses a patron-driven demand model [7] for ebooks. EBL allows readers five minutes of browsing within books that are not owned by the institution, and ten minutes in books that are. If readers wish to continue to use books beyond this time, or if they wish to copy or print any content the system presents a dialog box inviting users to create a loan; to continue using the book users must click 'yes' on this dialog box, thus providing an affirmative expression of interest. This is the only means by which readers borrow books—they cannot actively decide to check out an ebook and click a button, for example.

EBL books are presented in a web-browser via the EBL interface (Fig. 1); they are not downloadable. The interface allows navigation through ebooks in a variety of ways: the use of a right-hand scrollbar, a left-hand navigation menu based on the table of contents, and a paging navigation interface above the book.

Fig. 1. The EBL interface

3.2   Ebook Log Analysis

The data on which this study was based was collected over the period 1 September to 30 September 2011; this period falls completely during term time at Swinburne, meaning the data is likely to represent a range of academic users, rather than just academic staff and research students. During the study period 9506 people accessed ebooks, of whom 3799 went on to create a loan.

EBL logs anonymised use by individual readers, and captures quite detailed statistics including which books have been used, which pages were examined in each book, the number of minutes spent looking at a book, whether the book was owned by Swinburne, and bibliographic details about the book. When scrolling, only those pages on which a user pauses are logged in the usage stats; pages they scroll past are not.

As we were interested in how readers select (or do not select) ebooks, we created two samples: one of data about books readers went on to borrow, and one of books that readers only browsed. Note that a book not having been borrowed does not mean it was not relevant to the reader's information need, for example they may have found the information they were looking for during the available browsing time or they may have found another more useful book.

The sample on which our study is based is data from 100 randomly selected browsed books, and 100 loaned books, each accessed by a different user. All data is from the period while readers were making a decision, i.e. before a loan was created, and includes all the pages readers viewed to a maximum of 19 pages. We classified the content of these pages using a classification scheme based partly on Reutzel and Gali's observations of children [20], and partly on McKay's earlier work on ebooks [14]. We then examined these results for patterns, and compared loaned books to browsed books.

4   Results

The results of this study can be broadly classified into three categories: the book features readers viewed, the length of time readers spent with books, and how readers examined books and their features.

4.1   Parts of Books Readers Viewed

Our analysis discovered that the five most commonly viewed parts of the book were front matter, chapter headings, table of contents, the first page of content, and the introduction (see Fig. 2). Most commonly used was the front matter; 93% of loaned books and 89% of browsed books showed some use of this content, (this may be an artifact of the EBL system, which displays front matter first by default). 52% (loaned books) and 56% (browsed books) of readers viewed chapter headings. Surprisingly, given that an interactive version of the ToC is available to readers in the left-hand navigation, 41% and 43% of readers viewed the non-interactive printed in-book ToC at some stage, perhaps to investigate page numbers (which are not usually present in the left-hand ToC). Very few readers (2 who borrowed books and 1 who merely browsed) used the index, 3 and 1 respectively used the bibliography, and only one reader—who borrowed the book—viewed the conclusions section. Readers also viewed the introduction (25% loaned / 24% browsed), first page of content (23% / 33%), blank pages (33% / 21%), images (22% / 21%) and tables (13% / 12%).

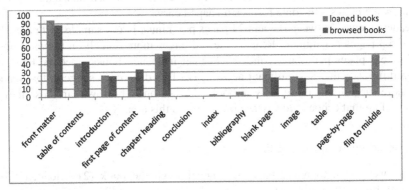

Fig. 2. Parts of books viewed by readers (loaned and browsed books)

4.2   First Three Items Viewed

Almost all readers reviewed the front matter of a book before initiating a more thorough investigation of the book, however this behavior is not predictive of the use of other in-book metadata (for example ToC). Figs. 3 and 4 show the most commonly viewed first, second and third items. Readers typically began moving through a book in a page-by-page manner, either by using the top navigation buttons or by scrolling (see Fig. 1). In 160 books (76 browsed, 84 loaned) this sequential examination of the front matter lasted three pages or more.

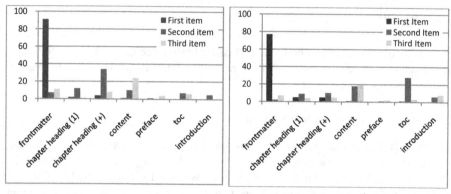

Fig. 3. 1st three parts viewed of 'loaned' books   **Fig. 4.** 1st three parts viewed of 'browsed' books

4.3   Time Taken to Make a Decision

Readers typically either made a decision about a book very quickly or used all of their available browsing time (5 minutes for unowned ebooks, 10 minutes for owned ebooks). In 36% of browsed books and 26% of loaned books readers made a decision within the first minute, rising to 53% and 40% respectively at two minutes. Conversely in 30% of read books and 13% of browsed books users browsed for the available browsing time before making a decision. In all cases—browsed and loaned, owned and unowned books—the data forms a u-shaped curve.

Readers viewed between 0.3 and 24 pages per minute in our sample, with a median rate of 5.3 pages per minute. Readers viewed pages more quickly in browsed books (median 6.9 pages per minute, mean 7.1) than in loaned books (median 4.5, mean 5.7), a difference that is significant at $p=0.017$. This difference suggests that readers read more carefully or thoroughly in books they go on to borrow than those they do not, a finding that echoes McKay's earlier work in this area [14].

4.4   Reading Sequences of Pages

Readers were seen to move page by page through the book (21% loaned / 14% browsed), flip to the middle of a section within the content (49% / 29%), and directly navigate to a chapter heading (51% / 55%). Based on the page visited before and after a page being coded we are able to make assumptions about the ways in which users navigated books. Readers appeared to use the left-hand ToC navigation more often than they entered a page number into the top navigation: 63 of 200 readers in our study apparently used the left-hand navigation while only 14 of 200 entered a specific page number in the top navigation section.

The majority of readers viewed sequences of 3 or more consecutive pages at some stage during their interactions. Borrowers were more likely than browsers to look at more one sequence of 3 or more consecutive pages. Of those that borrowed the books, 35 looked at 1 sequence of three or more pages, 37 looked at 2 sequences, 17 at 3 sequences and only 6 readers interacted with 4 sequences. Of those who only browsed, 44 looked at 1 sequence, 29 looked at 2 sequences and 12 looked at 3 sequences. No reader viewed more than four sequences, and only those who borrowed books viewed more than three.

Many of the readers who looked at just one sequence of consecutive pages did so at the beginning of their interaction. After the first three pages, most readers (60% in loaned books, 43% in browsed books) navigated either to another part of the book (50% loaned, 30% browsed) or away from the book entirely, ending their exploration at this point (10% loaned, 13% browsed). A very small number of readers (4% loaned, 1% browsed) viewed all 19 sample pages consecutively.

4.5   Navigation within a Book

We looked at the sequences of pages accessed in order to explore the ways in which readers navigated books. Readers generally began by paging through a book, either by scrolling or using the paging navigation buttons. (see Fig. 1). This behaviour typically continued through the first three pages of the front matter of a book before readers discovered the left-hand ToC (see Fig. 1), and used this to continue their search. This pattern was seen in 126 of 200 readers. Readers were seen to move page-by-page through books (21% in loaned books, 14% in browsed books) and flip to the middle of a section (49% / 29%) and directly navigate to a chapter heading (51% / 55%).

4.6   Interaction Patterns

To investigate reading patterns we encoded readers' moves between pages as either skips (1 page) or jumps (more than 2 pages, and usually more than 10 pages) forward or back. From this encoding, we identified three patterns: linear progression, contextual confirmation, and exploratory assessment, described further below.

Linear progression (shown in Fig. 5) involves readers paging through the initial parts of the book (we assume using the paging navigation buttons or scrolling), before using the left-hand ToC navigation to jump forward in the book to the start of a chapter. This pattern reflects behaviour observed in our earlier work in physical libraries [8], where readers flipped through books and reading parts of relevant chapters.

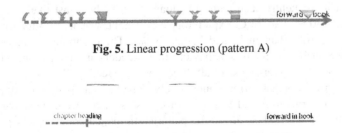

Fig. 5. Linear progression (pattern A)

Fig. 6. Contextual confirmation (pattern B)

The second interaction pattern (Fig. 6) represents a readers jumping to the first page of a chapter (most likely using the left hand ToC navigation) before paging forward two to three pages then jumping backward in the book to the final few pages of the previous chapter. From there they read to their previous reading position and then read further forward in the book. We assume that the reader is reviewing the end of the previous chapter for context to assist with the reading within the current chapter. This interaction pattern was not observed in readers in physical stacks [8].

The third interaction pattern (Fig. 7) shows the reader jumping back and forward throughout the book, seldom looking at more than one page. This interaction is most likely a result of resulted from scrolling (only those pages that completely load are noted in logs), but may also be a result of searching or entering page numbers in the paging navigation. Without knowing which method of interaction readers were using to create this pattern it is difficult to speculate on their intentions.

Fig. 7. Exploratory assessment (pattern C)

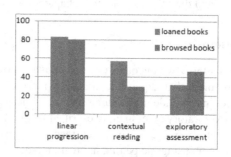

Fig. 8. Distribution of interaction patterns

The patterns were not evenly distributed (see Fig. 8) Most readers used pattern A, linear progression, at some point during their reading—both those who borrowed (83%) and those who browsed (80%) books. Pattern B, contextual confirmation, was the second most frequently seen pattern in loaned books but least likely to seen in browsed books (30%). Pattern C, exploratory assessment, was more often evident in browsed books (46%) than in loaned books (32%). These patterns were relatively clearly distinguishable in reading behavior; most readers used only one or (more commonly) two of the patterns; use of all three was rare (see Fig. 9)

Pattern A (linear progression) was most likely to be the only pattern used; patterns B (contextual confirmation) and C (exploratory assessment) were most commonly used in conjunction with other interaction patterns. In loaned books, a stronger focus on extended reading (pattern AB) is evident even in the first 20 pages, whereas browsed-only books are often approached in a more exploratory manner (AC).

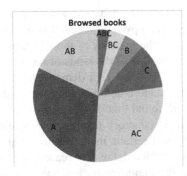 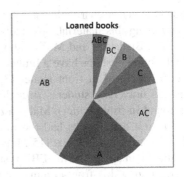

Fig. 9. Proportion of reading patterns used by readers

5   Discussion

There are two major areas of work related to our findings: document navigation, and document selection literature.

5.1   Document Navigation

The navigation patterns seen in Section 4.6 are very interesting: they show a considerable divergence from our work in physical environments, where readers do not typically flip backwards, and especially not in large chunks.

What these patterns do show is consistency with the interaction patterns discovered in the same collection in [14], and with the patterns seen in [13], which examines digital reading of journal articles. The differences in interaction between the ebook collection examined here and our earlier work in physical environments [8], and the backward flipping seen in common with digital article reading in [13], tends to suggest that there is something about digital environments that facilitates or encourages significant backwards movement in text. This conclusion may be supported by comparing [17] which shows some small backward movement in print periodicals to [12] which shows frequent large jumps backward in electronic articles.

If readers are only flipping backward because they can, and not because they must (either to gain context [17], or, as suggested by some commentators—for example [5]—because it is not possible to read deeply online), then digital books are supporting users somehow in a way in which physical books do not. If, conversely, users flip backwards in digital books because they are in some way lost (and users do frequently get lost in digital books, as seen in [10, 15]), then DLs must understand and address the cause of this disorientation to best support users' reading activities. It is as yet an open question whether the large jumps back seen in this and other studies on ebooks are due to disorientation or inclination; further study is needed to clarify this issue.

5.2   Document Selection

The common investigation of front matter seen in our study is unusual among book selection studies. While our own earlier work saw some readers checking for edition information [8], and a charitable interpretation of this behavior would assume that readers were using this material—particularly the copyright page—it is unlikely that this is the case. Buchanan and McKay show that the majority of those seeking books even in high end bookshops have a limited knowledge of the kinds of metadata (such as publisher) contained in front matter [3], and [1] and [26] demonstrate that both academics and university students struggle to create citations (using the kind of information found in front matter). Malama *et al.* found that, at least for fiction books, many readers felt that ebooks had too much introductory material and would have preferred to go straight to the text [15]. It is almost certain, then, that this frequent use of front matter is an artifact of the EBL system (which defaults to readers seeing front matter first), though this front-dominance in the digital realm has been seen in one other study comparing electronic articles to print [12]. It is likely that some other information than front matter is more useful in decision-making, and so DL designers should provide this information to readers first and foremost. What information that might be, however, remains an open question, though some of the major candidates are discussed below.

The EBL system provides readers with instant access to a navigation panel to the left of the book that, for most books, displays some kind of ToC. The ToC is seen to be useful in decision-making in print books [2, 8, 10, 23, 25], print articles [17] and ebooks [2, 10, 15]. Readers in our study used the interactive ToC in the left-hand panel relatively frequently (62% of all readers), suggesting that many at the very least saw this panel. It is surprising, then, how many readers (~40%) nonetheless viewed the in-book version of the ToC, which is not interactive. Either these two groups are mutually exclusive, or the print ToC offers something that the interactive one does not: one possible candidate for this information is page numbers, which are usually available only in the in-book version of the ToC. Page numbers could conceivably be used to determine how much of a book is dedicated to a topic of interest, and therefore should probably be available from any version of a ToC.

Earlier studies show the frequent use of index for information seeking and decision-making within books in both physical [8] and digital formats [2]: a sharp contrast to the data seen here where only a few readers accessed the index. The reason for this discrepancy remains to be discovered.

Images are seen as a source of decision-making data in our own earlier work on print books [8], in children's book decision making [20], and in studies of triage both in physical periodical materials [17] and electronic articles [13]. The lack of use of images in the work presented here is likely due to the long load time of many pages during scrolling, meaning that images are not visible to readers flipping past them in EBL in the same way that they are in print books or pre-loaded electronic documents.

Finally, that so few readers consulted the conclusions section of books in decision-making, even though this section was seen as very important for triaging electronic articles [12, 13] and it is readily available using the left-hand navigation, is perplexing. It may be that conclusions are not as useful in decision-making in long-form documents such as books as they are in short-form documents, or perhaps navigating to conclusions over a long space in a digital environment is prohibitively difficult. This remains a question for future work, but has profound implications for DL design.

6   Conclusions and Future Work

This study raises nearly as many questions as it answers in terms of how readers make decisions about ebooks: ToC is frequently used, but the reasons for the tensions between print and interactive versions of the ToC are not clear. Conclusions and index seem to be less useful than in previous studies, though the reasons are for this are not readily apparent. Front matter is frequently used, though it looks as though this may be an artifact of the EBL system: which information would be most useful presented first is a subject for future study. Finally, images are infrequently used in EBL, almost certainly due to the system failing to present them to readers; DL designers would do well to incorporate image presentation as a key feature of ebook display environments

We do not yet have a clear understanding of whether readers' needs are being met by the current EBL interface. The current interface provides three main methods of navigation for the reader, the left-hand ToC, the top navigation and scrolling; however, there may be different interaction methods that would better support the sampling activities readers want to conduct. This research could be extended to gain understanding of whether current interaction tools adequately facilitate the identified reading patterns and whether other interface features would better support readers' sampling activities.

Acknowledgements. We acknowledge with thanks the kind support of Alison Morin and EBL in facilitating access to the EBL collection. Thanks also to Tony Davies, Nguyen Ly and Fiona Campbell for their technical support for this research.

References

1. Aronsky, D., Ransom, J., et al.: Accuracy of References in Five Biomedical Informatics Journals. J. Am Med. Inform. Assn. 12(2), 225–228 (2005)
2. Berg, S.A., Hoffmann, K., et al.: Not on the Same Page: Undergraduates' Information Retrieval in Electronic and Print Books. J. Acad. Libr. 36(6), 518–525 (2010)
3. Buchanan, G., McKay, D.: In the Bookshop: Examining Popular Search Strategies. In: JCDL 2011. ACM, Ottawa (2011)
4. Christianson, M., Aucoin, M.: Electronic or print books: Which are used? Libr. Collect. Acquis. 29(1), 71–81 (2005)
5. Cull, B.W.: Reading revolutions: online digital text and implications for reading in academia. First Monday 16(6) (2011)
6. Hamblen: Amazon: E-books now outsell print books. Computerworld (2011),
   http://www.computerworld.com/s/article/9216869/Amazon_E_books_now_outsell_print_books2011, (last accessed April 11, 2012)
7. Hardy, G., Davies, T.: Letting the patrons choose: using EBL as a method for unmediated acquisition of ebook materials. In: Information Online 2007, ALIA, Sydney (2007)
8. Hinze, A., McKay, D., Timpany, C., Vanderschantz, N., Cunningham, S.J.: Selection behaviour in the physical library. In: JCDL 2012, Baltimore, MD (to appear, 2012)
9. Li, C., Poe, F., Potter, M., Quigley, B., Wilson, J.: UC Libraries Academic e-Book Usage Survey. California Digital Libraries. Springer, Oakland (2011)

10. Liesaputra, V., Witten, I.H.: Seeking information in realistic books: a user study. In: JCDL 2008, pp. 29–38. ACM, Pittsburgh (2008)

11. Littman, J., Connaway, L.: A Circulation Analysis of Print Books and E-Books in an Academic Research Library. Library Resour. Tech. Ser. 48, 256–262 (2004)

12. Buchanan, G., Loizides, F.: Investigating Document Triage on Paper and Electronic Media. In: Kovács, L., Fuhr, N., Meghini, C. (eds.) ECDL 2007. LNCS, vol. 4675, pp. 416–427. Springer, Heidelberg (2007)

13. Loizides, F., Buchanan, G.: An Empirical Study of User Navigation during Document Triage. In: Agosti, M., Borbinha, J., Kapidakis, S., Papatheodorou, C., Tsakonas, G. (eds.) ECDL 2009. LNCS, vol. 5714, pp. 138–149. Springer, Heidelberg (2009)

14. McKay, D.: A jump to the left (and then a step to the right): reading practices within academic ebooks. In: OZCHI 2011, pp. 202–210. ACM, Canberra (2011)

15. Malama, C., Landoni, M., Wilson, R.: Fiction Electronic Books: A Usability Study. In: Heery, R., Lyon, L. (eds.) ECDL 2004. LNCS, vol. 3232, pp. 69–79. Springer, Heidelberg (2004)

16. Marshall, C.C.: Reading and Writing the Electronic Book. Morgan & Claypool, Chapel Hill (2010)

17. Marshall, C.C., Bly, S.: Turning the page on navigation. In: JCDL 2005, pp. 225–234. ACM, Denver (2005)

18. Moore, P.: Information Problem Solving: A Wider View of Library Skills. Contemp. Educ. Psychol. 20(1), 1–31 (1995)

19. Pearson, J., Buchanan, G., et al.: HCI design principles for ereaders. In: Booksonline 2010. ACM, Toronto (2010)

20. Reutzel, D.R., Gali, K.: The art of children's book selection: A labyrinth unexplored. Reading Psychology 19(1), 3–50 (1998)

21. Rowlands, I., Nicholas, D., et al.: What do faculty and students really think about e-books? Aslib. Proc. 59(6), 489–511 (2007)

22. Shelburne, W.A.: E-book usage in an academic library: User attitudes and behaviors. Libr. Collect. Acquis. 33(2-3), 59–72 (2009)

23. Stelmaszewska, H., Blandford, A.: From physical to digital: a case study of computer scientists' behaviour in physical libraries. JoDL 4(2), 82–92 (2004)

24. Stieve, T., Schoen, D.: Undergraduate Students' Book Selection: A Study of Factors in the Decision-Making Process. J. Acad. Libr. 32(6), 599–608 (2006)

25. Thayer, A., Lee, C.P., et al.: The imposition and superimposition of digital reading technology: the academic potential of e-readers. In: CHI 2011, Vancouver, BC, Canada (2011)

26. Yu, F., Sullivan, J., et al.: What Can Students' Bibliographies Tell Us?- Evidence Based Information Skills Teaching for Engineering Students. Evidence Based Library and Information Practice 1(2) (2006)

Information Seekers' Visual Focus during Time Constraint Document Triage

Fernando Loizides

Cyprus University of Technology
fernando.loizides@cut.ac.cy

Abstract. Time-constraints are a commonly accepted limitation to a user's information seeking process. Physical time constraints can cause users to have a low tolerance of time consuming information seeking tasks. This paper examines the effects of time constraints on the document triage process in an eye-tracked lab-based study. The visual attention of three time constraints are reported on. Similarities and differences to previous triage data are also reported on, contributing to an ongoing research investigation of the general document triage process.

Keywords: Document Triage, Information Seeking, Relevance Decisions.

1 Introduction and Motivation

Document triage is the critical point in the information seeking process when the user first decides the relevance of a document to their information need. Most document triage activities are, by definition, a fast-paced part of the information seeking process [1]. Information seekers will triage documents by skimming their content for relevant information. In-depth reading is minimal and occurs only after a document is perceived to be relevant to the user's information need. During this fast scanning activity, users have a low tolerance for searching through information[7]. We also know that information seekers can be biased against long documents [1]. It is hypothesised that all these factors contribute to users making incorrect judgements on document relevance. The literature also identifies "limited support for the idea that, when there was insufficient time to read all the text available, readers could effectively focus on the most important parts of the text" [2]. This paper investigates the possible impact of time constraints to information seekers performing document triage. By conceptualising information seekers triage behaviour during their fast decision making process, we can identify potential misleading visual areas. In this paper we investigate visual attention relating to document features, when the user needs to sacrifice non-important parts of the document due to time. Using an eye-tracking based study, we report on what parts of a document information seekers visually inspect during document triage given limited time.

P. Zaphiris et al. (Eds.): TPDL 2012, LNCS 7489, pp. 25–31, 2012.

2   Study Description

18 participants were chosen, 7 female and 11 male, for the study. All participants were at postgraduate level or above in a technology related discipline. Their ages ranged from 21 to 50 years of age. The criteria for the selection of appropriate participants was their familiarity with academic documents; namely to have read academic documents before and are therefore accustomed to the general format. Special care was taken however, in order for the participants not to be familiar with the specific documents presented to them in the study. The study design mirrors previous approaches for consistency and comparison purposes [4]. Participants were given two information needs and a set of documents to triage and rate based on an information need and give a score of one to ten (ten being extremely relevant). Instead of an open time-frame, information seekers had limited time to triage each document, a technique employed to give us an indication of the most visually scrutinised features of each document. For this reason we limit the time on each document rather than the overall time. In order to vary the times 6 participants were chosen to triage the documents using each time constraint condition (30, 60 and 90 seconds per document). The documents were presented one at a time on a 19 inch screen, in Portable Document Format (PDF) and the participants' eye gaze was tracked using the non-intrusive Tobii x50 eye-tracker. Following the main study a semi-structured interview was performed in order to gain qualitative feedback from the participants.

3   Results and Discussion

3.1   Subjective Feature Importance and Navigational Patterns

The main title is subjectively considered to be the most important feature of the document. The features which occur on the initial page are also rated highly and headings are ranked within the four most important features. Taking the average ratings given to the document parts we found the two most highly rated as important to be *Main Title* (7.6/10) and the *Abstract* (7.1/10). *Images, headings and tables* received the lowest score (5/10), lower than *plain text* (5.3/10). The *introduction* (6.6/10), *conclusion*(6.7/10), *emphasised text*(6.05/10) and the *headings*(6.85/10) received intermediate average scores. All the categories had a standard deviation between 1.5 and 2.5.

When users perform document triage, they will navigate through a document following a specific pattern [4]. For example, a linear skimming through the document while pausing at points of interest is dubbed the 'step up'When users triage while under the pressure of time constraints, they are found to replicate these patterns. What changes, is the proportion of times each individual pattern is replicated. The most common navigational behaviour across all the groups as well as in each individual group was the "step up navigation". In total each group performed 96 triage instances. The **30 second group** had 4% of the instances performed using a begin and end pattern, 8% of the instances performed using the flatline pattern and 88% of the times using the dominant step pattern. The

60 second group had 3% of the instances performed using the begin and end pattern, 7% of the instances using the flatline pattern and 90% of the instances using the step navigation pattern. The **90 second group** displayed the least amount of variation in terms of patterns. Only one of the triage instances (1%) was performed using the begin and end pattern while there were no instances from the 90 second group which undertool flatline pattern behaviour. 99% of the triage instances where executed with the step navigationl pattern. We can hypothesise that the triage performed by the participants was more opportunistic, in that the users would rely on locating relevant pieces of information, as they skimmed through the entirety of the document at a fast pace.

3.2   User Attention

Participants were allowed to examine every document for a limited amount of time; up to 5 seconds extra was permitted after the limit was reached for the decision to be orally given to the investigator. Participants with a 30 second limit, 60 second limit and 90 second limit are referred to as Group 30, Group 60 and Group 90 respectively. The experimental setup used a Latin square approach to reduce effects such as fatigue on the behaviour of the participants. We calculate visual processing values [3]. An overview of the findings can be seen in Figure 1.

Table 1. Area Occupied and Time Spent Scrutinising Each Document Part

	Area	30 (%)	30 (S.D)	60 (%)	60 (S.D)	90 (%)	90 (S.D)
Initial Page	10.80	42.70	49.28	46.70	33.39	27.00	9.09
Main Title	1.10	10.80	15.44	5.50	9.28	9.50	9.01
Abstract	1.20	16.5	19.66	15.1	7.88	21.1	8.64
Introduction	2.30	8.10	5.91	5.90	7.86	7.10	8.64
Headings	2.10	15.30	6.47	8.10	6.47	8.20	6.47
Emph. Text	1.90	2.20	1.51	2.80	1.84	4.60	3.09
Images	6.20	4.90	8.87	6.80	8.94	10.50	7.74
Diag./Tables	2.30	12.50	5.94	4.60	4.90	6.30	4.40
Conclusion	1.60	6.10	6.88	5.00	9.14	7.50	5.80

Total Time. Participants in Group 30 took an average of 33 seconds to complete triage on every document (SD: 11.8 secs), thus using their whole time on average to the full and had to be asked to give a rating. Participants in Group 60, took on average 45 seconds to complete triage per document (SD: 15.2 secs) and felt comfortable giving an answer. This is an average of 15 seconds less than the mentioned limited time. Group 90 Participants completed triage on average at 61 seconds per document (SD: 18.8 secs), an average of 29 seconds less than the limit time. An ANOVA test produces significant results between the average times taken and the amount of time given to the participants at the $p < 0.05$ level for the three conditions [$F(2,285) = 140.88$, $p = 0.286^{-42}$].

Post-hoc Tukey's HSD tests showed statistical significance between all groups at the 0.01 level. Participants were asked to comment on when the relevance decision was made. The feedback presented the initial page as the point at which the decision was made, whereas the remaining triage was performed in order to verify their decision. When questioned directly about why they did not, therefore, spend all their time on the initial page, since the time constraint was strict, participants from all three groups suggested that they "just wanted to be sure that the document was what I think it is". In other words, they wanted to make sure that the abstract and main title were not misleading. We observe that being able to scrutinise the entire document rather than only a part of the document, is also considered a benefit for the users.

Initial Page. The initial page of a document most often includes the main title, abstract and part, if not all, of the introduction; key features in relevance decision making [5]. We are not certain of the reason why there is a drop in visual attention on the initial page by the 60 second group. We eliminated the effects of bias from previous triage operations by alternating the order in which the the individual time constraints are assigned to the participants. Although further testing will be needed to identify the cause for this phenomenon, we hypothesise a reason. We are reminded that participants follow a largely 'step up' approach in triaging documents when there is a time limitation. It may be the case that participants in Group 60 did not have the time to return to the initial page after scanning the remainder of the document. An ANOVA test revealed a statistical significance at the $p < 0.05$ level for the three conditions [$F(2,285) = 13.96$, $p = 0.16$ x 10^{-5}]. Upon further scrutiny Tukey's HSD tests showed statistical significance between the 30 and 90 groups and the 60 and 90 at the 0.01 level. There was no statistical difference between the 30 and 60 second groups at the 0.05 level.

In general, the initial page is scrutinised extensively compared to the remaining features, and this is reflected in the reduced attention of the remaining features. The increased attention of the initial page; specifically the abstract and introduction, creates a domino effect of reduced attention on the remaining features. The statistical relevance between the times, suggests that the 60 and 90 second groups are more inclined, and able, to scrutinise the remaining features, whereas the 30 second group mostly choose to focus on the initial page features, yet skim rapidly to verify their decisions on the rest of the document.

Main Title. The main title, which was subjectively rated as the most important feature to assist in document triage on a document, was viewed by all our participants. A substantial amount of visual attention is given to the main title, making it the second most visually important document feature across all three time groups. No significant differences were uncovered using an ANOVA at the $p < 0.05$ level [$F(2,285) = 2.06$, $p = 0.13$] between the three groups.

Abstract. The abstract is the most visually processed feature on all three time constraints. An ANOVA test revealed a statistical significance at the $p < 0.05$ level for the three conditions [$F(2,285) = 7.85$, $p = 0.0005$]. A Tukey's HSD test reveals statistical differences between the 30 and 90 groups and between the 60 and 90 groups at the 0.05 level.

Introduction. That the introduction is often located on, or partly located on, the first page of the document. The initial page is weighted high in viewing time to viewable area ratio. Within all three groups, the introduction rates moderately when compared to all the document features. The visual processing ratio however (3.52, 2.56 and 3.08 respectively), testifies to the fact that there is attention given to the introduction by all three groups. Using an ANOVA test we find strong significance between groups at the $p < 0.05$ level $[F(2,285) = 12.25, p = 0.03 \times 10^{-4}]$. Upon further scrutiny, we find statistical significance between all groups at the 0.05 level using Tukey's HSD.

Headings. Headings have been commented on as having "a modest impact on users' navigational behaviour" during document triage [4]. When comparing the ratios during fast triage there is an exception. Time to area ratios were 7.28, 3.85 and 4.68 respectively for the three groups. These findings testify to a very high level of attraction by headings while an information seeker is performing document triage with a limited time. Using an ANOVA test we find strong significance between groups at the $p < 0.05$ level $[F(2,285) = 42.25319907, p = 0.7 \times 10^{-10}]$. Using Tukey's HSD tests we find statistical significance between all groups at the 0.05 level. Users with time constraints produced area to time ratios of 7.29 (Group 30), 3.86 (Group 60) and 3.90 (Group 90).

Emphasised Text. This shows that, on average, emphasised text receives slightly elevated scrutiny during fast triage compared to previous studies [4]. Using an ANOVA test we find strong significance between groups at the $p < 0.05$ level $[F(2,285) = 6.76, p = 0.001]$. Using Tukey's HSD tests for multiple comparisons, the 30 and 90 groups and the 60 and 90 groups have statistical significance at the 0.05 level.

Images. Images were considered the least visually important feature across all groups. Using an ANOVA test we find strong significance between groups at the $p < 0.05$ level $[F(2,285) = 4.91, p = 0.008]$. A Tukey's HSD post hoc test reveals a statistical difference only between the 30 and 60 groups at the 0.05 level.

Diagrams and Tables. The time to area ratios are 5.43, 2 and 2.73 respectively for the 30, 60 and 90 second groups. Using an ANOVA test we find significance between groups at the $p < 0.05$ level $[F(2,285) = 3.47, p = 0.03]$. Post hoc testing finds statistical significance at the 0.05 level between the 30 and 90 second groups, using Tukey's HSD.

Conclusions Section. The conclusions section is considered a highly influential feature regarding users' visual focus during normal document triage [4]. We hypothesise that since the conclusion is located towards the end of a document, this makes it harder to find. We also know that the attention of participants falls inversely to the page number of the document [4]. The conclusion therefore, being close to the end can also be a contributing factor to the reduced attention of the information seekers. Using an ANOVA test we find significance between groups at the $p < 0.05$ level $[F(2,285) = 4.38, p = 0.01]$. Statistical differences, using a Tukey's HSD post hoc test, exists between the 30 and 60 second groups at the 0.05 level.

3.3   Document Relevance Ratings

Figure 2 shows the average rating and standard deviation for the three groups, as well as the average rating from findings in [4]. A Mann-Whitney test for between subjects studies with non-parametric data gives us no statistical significance between the relevance ratings of the three groups. It is commonly accepted that information seekers are prone to making several erroneous relevance decisions. Our findings reveal that, statistically, this error rating did not decrease and reducing time in this scenario did not reduce the time efficiency [6] of the triage task. Further testing is required to investigate the effects of time on relevance decisions on a larger scale.

Table 2. Average ratings and standard deviation for Group 30, Group 60, Group 90 and Group ∞ (open time frame)

	Group 30 (%)	Group 60 (%)	Group 90 (%)	Group ∞ (%)
Average Score	5.10	5.82	6.02	5.99
St. Deviation	2.41	2.42	2.15	2.20

4   Summary

In this paper the effects of limited time on document triage in a process we label *rapid triage* are reported on. Several findings correspond well with previous research data and we can verify their validity during rapid triage also. These include the subjective ratings of the participants on the document features and the extensive visual scrutiny of the initial page and features such as the abstract and main title. We are able to make further reports also on new evidence which contributes to our understanding on document triage. Although the 'step up' pattern is the most popular navigation method for document triage [4], we see an increased use of this pattern in this specific study. Given a limited time frame, information seekers will choose to navigate using a 'step up' behavioural pattern more frequently than without the constraint of time (Significant to the $p < 0.05\%$ level using a t-test, $N = 416$). We have reported that the initial page is considered by the participants, both subjectively and from the observational data, to be the most visually attractive page of a document for relevance decisions. We also know from previous experiments that users claim to make a relevance decision during the initial part of their triage process [4] and most often from the content of a document's first page features. Even given a limited time however, information seekers will most probably choose to navigate beyond the initial page in order to "verify their decision" using the rest of the document.

References

1. Buchanan, G., Loizides, F.: Investigating document triage on paper and electronic media. In: Procs. of the European Conf. on Reasearch and Advanced Technology for Digital Libraries, vol. (35), pp. 416–427 (2007)
2. Duggan, G., Payne, S.: Skim reading by satisficing: Evidence from eye tracking. Journal of Experimental Psychology: Applied, 228–242 (2009)
3. Just, M.A., Carpenter, P.A.: Eye fixations and cognitive processes. Cognitive Psychology, 441–480 (1976)
4. Loizides, F., Buchanan, G.: An Empirical Study of User Navigation during Document Triage. In: Agosti, M., Borbinha, J., Kapidakis, S., Papatheodorou, C., Tsakonas, G. (eds.) ECDL 2009. LNCS, vol. 5714, pp. 138–149. Springer, Heidelberg (2009)
5. Saracevic, T.: Comparative effects of titles, abstracts and full text on relevance judgments. Journal of the American Society for Inf. Science (22), 126–139 (1969)
6. Shneiderman, B.: Designing the User Interface: Strategies for Effective Human-Computer Interaction, 3rd edn. Addison-Wesley Longman Publishing Co., Inc., Boston (1997)
7. Spink, A., Jansen, B.J., Wolfram, D., Saracevic, T.: From e-sex to e-commerce: Web search changes. Computer 35, 107–109 (2002)

Which Words Do You Remember?
Temporal Properties of Language Use in Digital Archives*

Nina Tahmasebi, Gerhard Gossen, and Thomas Risse

L3S Research Center, Appelstr. 9a, Hannover, Germany
{tahmasebi,gossen,risse}@L3S.de

Abstract. Knowing the behavior of terms in written texts can help us tailor fit models, algorithms and resources to improve access to digital libraries and help us answer information needs in longer spanning archives. In this paper we investigate the behavior of English written text in blogs in comparison to traditional texts from the New York Times, The Times Archive, and the British National Corpus. We show that user generated content, similar to spoken content, differs in characteristics from 'professionally' written text and experiences a more dynamic behavior.

1 Introduction

The rise of the Web has allowed more people to publish texts by removing barriers that are technical but also social such as the editorial controls that exist in traditional media. The resulting language tends to be more like spoken language because people adapt their use to the medium [10]. We can see evidence of this adaption in the vocabulary: Authors on the Web use more new or non-traditional terms. This results in a very dynamic language. However, the terms used to refer to a concept can change between the contexts of the document author and the user of an archive. This makes finding relevant documents in an archive harder.

Knowing about the change rate of modern language we can target algorithms towards capturing these changes in long term archives. In this paper we complement our earlier work on term evolution by analyzing the dynamics of languages itself. We compare traditionally written texts (news corpora, written part of British National Corpus (BNC)) and spoken language and the language the Web (spoken part of BNC, two TREC blog crawls). Using the BNC as a ground truth, we investigate if language in user generated text behaves like spoken language and is thus more dynamic than language used in traditional text.

2 Related Work

Language evolution has been a research topic for a long time and gained increasing interest [9]. A good overview of the field can be found in [4]. This research

---

* This work is partly funded by the European Commission under ARCOMEM (ICT 270239).

P. Zaphiris et al. (Eds.): TPDL 2012, LNCS 7489, pp. 32–37, 2012.

focuses on the origins and development of languages and involves a wide range of scientific areas. Some work has been done in the areas of information retrieval for the access of content. Abecker et al. [1] showed how medical vocabulary evolved. Kanhabua et al. [6] proposed an algorithm for detecting time-based synonyms and showed that using these synonyms for query expansion greatly improved retrieval effectiveness in a digital archive. A special case of evolution, outdated spellings of the same term, has been addressed in [5] where a rule based method is used for deriving spelling variations that are later used for information retrieval. Bamman et al. [2] propose a method for automatically identifying word sense variation in a collection of historical books. They automatically classify the word senses in the corpus and track the rise and fall of those senses over a span of two thousand years. The work in this paper contributes to the state of the art by providing more statistical insights into the dynamics of Web language.

3 Hypothesis

Our hypothesis is that the language used in user generated text from the Web (called *Web language*) behaves like spoken language and is thus much more dynamic than the language used in traditionally written texts (called edited text). To support our hypothesis we use three main measures described below.

The first important property is the deviation from standard vocabulary which we measure using the *dictionary recognition rate* (DRR). We define the DRR as the proportion of terms recognized by current dictionaries (WordNet[1] [8] and Aspell[2], see also [11]). Unrecognized terms are for example proper names, domain specific terms, spelling errors and new terms. We use the DRR of the edited datasets as an estimate for the number of domain specific and new terms.

The second property is the change rate in the vocabulary from one year to the next. We measure this using the *vocabulary overlap* between the *dataset dictionaries* for two years. While the DRR captures general language properties, this measure describes the language dynamics in a specific dataset. As we expect different behavior for high frequency compared to infrequent terms, we split the dataset dictionaries by frequency into three parts using a 25%/50%/25% split[3]. For each adjacent year we measure the part overlaps as well as the total overlap between all terms. We expect a decrease in overlap from the terms of highest to those of lowest frequency because the frequent ones form the stable core of the language whereas the most infrequent ones are typically misspellings.

The third and final property is the change in frequency of individual terms in adjacent years. We define K_{diff} *terms* as the set of terms that have a change in frequency greater than 50%, measured as (for term t and year i):

$$K_{\mathrm{diff}}(t) = abs\left(\frac{fr_i(t) - fr_{i+1}(t)}{max(fr_i(t), fr_{i+1}(t))} \right)$$

---

[1] We use the stemmer from the MIT Java WordNet interface JWI(http://projects. csail.mit.edu/jwi/) as well as the WordNet "exception entries" for irregular words.

[2] http://aspell.net/

[3] The 25%/50%/25% split was empirically determined.

Table 1. Used datasets

	type	timespan	docs
NYTimes	newspaper	1987–2007	1.8 mio
Times	newspaper	1960–1985	2.0 mio
BNCwr	misc. written	1986–1993	2628
BNCsp	misc. spoken	1991–1994	613
Blogs	blogs	2005–2008	14.9 mio

Table 2. Dictionary recognition rates for WordNet and Aspell

	WN DRR	Aspell DRR
NYTimes	94.6%	96.5%
Times	92.0%	94.0%
BNCwr	95.6%	97.1%
BNCsp	92.6%	98.4%
Blogs	87.7%	89.5%

where $fr_i(t)$ is the normalized frequency for term t in the dataset dictionary corresponding to year i. A high amount of K_{diff} terms shows a highly dynamic language behavior. By considering the dataset dictionary overlap together with K_{diff} terms we can measure how many terms stay between two adjacent years and among those, how many change substantially in frequency.

4   Experiments

We use four datasets (see Table 1) which we divide into two different classes: traditional written text (New York Times, Times, written part of BNC (BNCwr) [3]) and Web language and spoken language (TREC-BLOG and spoken part of BNC (BNCsp), from here on described as *user generated content*).

All datasets are pre-processed to filter out noise. We process the texts using TreeTagger[4] to lemmatize terms and extract nouns and Lingua::EN::Tagger[5] to extract noun phrases. We process each dataset in yearly collections and create a *dataset dictionary* with the identified nouns and noun phrases. The Blogs dataset is a combination of the two TREC-Blog datasets Blogs06 [7] and a subset of Blogs08 [12]. We filter out non-English documents using a simple heuristic.

4.1   Analysis Results

We split the datasets into yearly collections and show the yearly average for each experiment. However, in the experiments regarding DRR we process BNCwr and BNCsp as a whole because of the limited size of the datasets.

Dictionary Recognition Rate: Table 2 shows the DRR of all datasets. For NYTimes and BNCwr the DRR is 94.6%–97.1%. The unrecognized terms are most likely proper names because the Aspell DRR is consistently higher than of WordNet (Aspell contains many person names, WordNet almost none). Among the user generated datasets the DRR is consistently lower than for the written datasets. An exception is the Aspell DRR for the BNCsp (98.4%) which is high because of the person names used in conversations that are recognized by Aspell.

The DRR of Blogs is lower than the DRR of NYTimes. To some extend this is due to non-English blog entries being included in the dataset. However, a

---

[4] http://www.ims.uni-stuttgart.de/projekte/corplex/TreeTagger/

[5] http://search.cpan.org/dist/Lingua-EN-Tagger/

Table 3. Dictionary Overlaps: Dictionaries divided into top 25%, middle 50%, and bottom 25% and compared year by year. Average overlap for all years is shown here.

	Top	Middle	Bottom	Total
NYTimes	76.8%	41.3%	9.3%	60.8%
Times	75.1%	42.3%	9.6%	60.5%
BNCwr	70.4%	30.4%	7.25%	55.2%
BNCsp	50.0%	16.6%	4.2%	43.6%
Blogs	54.8%	14.9%	6.0%	38.9%

Table 4. K_{diff} measure for all datasets. A value of 0.5 means an increase or decrease of frequency $\geq 50\%$ between two consecutive years. The values shown in this table are an average for all years in each dataset.

	0.5	0.55	0.6	0.65	0.7	0.75
NYT	13.0%	9.8%	7.3%	5.1%	3.3%	2.0%
Times	14.0%	10.8%	8.1%	5.8%	3.9%	2.5%
BNCwr	11.2%	10.1%	8.2%	6.7%	5.4%	3.6%
BNCsp	19.6%	18.8%	16.7%	14.3%	13.1%	10.9%
Blogs	17.7%	15.9%	13.8%	11.7%	9.5%	7.4%

subsequent test on a sample of the dataset showed that approximately 96.2% of all documents were English. If we correct for this error by adding 3.8% to the DRR of blogs and compare it to NYTimes we find that the first is significantly lower at 99% significance. As NYTimes contains edited texts and has no errors we can use it as an upper bound on the possible DRR of a dataset with a large number of person names. The significantly lower DRR of Blogs shows that Web language deviates more from a standard dictionary than the language used in the NYTimes.

Dataset Dictionary Overlap: In Table 3 we can see the overlaps of nouns and noun phrases in yearly collections (dataset dictionaries) of consecutive years for all datasets. We use this measure to indicate the temporal dynamics of each dataset. For written datasets the average top overlap is 70%–77% which is significantly higher than all other values. Intuitively this makes sense as the most common nouns and noun phrases are unlikely to disappear from one year to another. For example, *health, service* and *government* will remain frequent topics of discussion so we consider them *temporally stable*. Other terms are *locally frequent* or *temporally unstable*: They are in the top parts only for a single year and disappear after that. These terms are highly affected by events or technical innovations, e.g., *iranian movie, olympic gold, groovebox, payroll disparity*. For the user generated datasets we see a much lower top overlap (50%–55%). The rate with which terms are exchanged from the vocabulary is much higher in user generated datasets. This is in particular true for the Blogs dataset as terms have be used by many different authors to be considered as frequent. For all datasets, the bottom overlap is very low (less than 10%). This also makes intuitively sense as this sets contains many person names or misspellings that are less stable.

The total overlap also differs between the two types of datasets. Among the written datasets around 55%-60% of all terms will also appear the next year. However, among the user generated datasets this overlap is between 39%-44%. In our news datasets the total overlap is around 60% even though the very aim of a newspaper is to cover many different events. Thus 60% overlap indicates a stable language but varying events. For Blogs the total overlap is 39%. The big difference is most likely a product of topic variations, temporally unstable terms as well as misspellings. However, misspellings of common terms are likely to follow a pattern across all datasets and thus play a smaller role in the lower overlap. Hence the low overlap of user generated datasets is a sign of high dynamics.

K_{diff} **Terms:** As a final measure we consider the K_{diff} of terms as the relative frequency of terms between consecutive years. Thus this is also a measure of the popularity of a term. For our experiments we only consider differences in frequency larger than 50% ($K_{diff} \geq 0.5$). In Table 4 we see the average amount of terms in each dataset that have a K_{diff} of at least 0.5. A high K_{diff} indicates a more dynamic language as more terms increase or decrease in popularity. In Table 4 we see that that there is an average K_{diff} of 11%–14% for the traditionally written datasets compared to 18%–20% in user generated datasets. The difference between the groups is statistically significance at 95% level. We also see that as the K_{diff} threshold increases we see a larger difference between the two different types of datasets. There are 2%–4% terms with $K_{diff} \geq 0.75$ for traditionally written in comparison to 7%–11% K_{diff} terms in the user generated datasets.

5   Discussion

Overall our experiments show that language from user generated text (speech or blogs) is more dynamic than traditionally written language like that in newspapers. Due to the relatively small size of the BNC, we choose to compare the written and spoken parts of the BNC to each other and the remaining datasets separately. Because of the high quality of the BNC, we use the relationship between its parts as a ground truth to compare against the other datasets.

The spoken part of BNC is more dynamic than the written parts: The DRR is lower for BNCsp, indicating the use of more non-standard terms and spellings and the dictionary overlap for nouns and noun phrases (Table 3) is consistently lower for BNCsp, just as the K_{diff} terms (Table 4) are higher by 7%–9%.

The relation between spoken and written language from the BNC also holds for Web language and written language. In general, the news datasets behave similarly while the Blogs dataset is markedly different. This holds for the DRR where Blogs has consistenly lower values for both dictionaries, the dictionary overlap (20% lower for Blogs), as well as the K_{diff} terms (5%–6% larger for Blogs). The difference between the news datasets is quite small and very likely the consequence of the presence of OCR errors in Times but not NYTimes.

In general, we find that the relationship between spoken and written language is similar to that between Web language and written language. In conclusion, we find that Web language behaves more dynamically with a higher change rate.

6   Conclusions and Future Work

Our experiments showed that language from user generated content like conversations or blog content behaves more dynamically than language used in traditionally written texts. To measure dynamics we used the number of terms that appear in or disappear from the vocabulary as well as the number of terms that have radical changes in their frequency. As ground truth we used the written and spoken parts of the British National Corpus, as representatives of real life datasets we used datasets such as the New York Times and the Times Archive (traditionally written texts) as well as the TREC-BLOG dataset (user generated texts). The relationship between written and spoken texts in the BNC is mimicked by the relationship between traditionally written text and user generated text.

However, in our experiments we have seen results that we partly attribute to a high amount of person names and proper nouns. It remains future work to investigate to what extent these play a role and determine their role properly. It would also be interesting to investigate how variations spread and to see which types of variations make it to a wider audience and become established.

The TREC-BLOG dataset we used in our experiments has some issues: The topic distribution is unknown and the number of entries per year varies heavily. We propose to create a dataset that follows a set of topics over a longer time to encourage and simplify research on temporal language evolution.

Acknowledgments. We would like to thank Times Newspapers Limited for providing the archive of The Times and our colleague Nattiya Kanhabua for constructive discussions.

References

1. Abecker, A., Stojanovic, L.: Ontology evolution: Medline case study. In: Wirtschaftsinformatik: eEconomy, eGovernment, eSociety, pp. 1291–1308 (2005)
2. Bamman, D., Crane, G.: Measuring historical word sense variation. In: JCDL, pp. 1–10 (2011)
3. The British National Corpus, version 3, BNC Consortium (2007)
4. Christiansen, M., Kirby, S.: Language evolution. Studies in the evolution of language. Oxford University Press (2003)
5. Ernst-Gerlach, A., Fuhr, N.: Retrieval in text collections with historic spelling using linguistic and spelling variants. In: JCDL, pp. 333–341 (2007)
6. Kanhabua, N., Nørvåg, K.: Exploiting time-based synonyms in searching document archives. In: JCDL, pp. 79–88 (2010)
7. Macdonald, C., Ounis, I.: The TREC Blogs06 Collection: Creating and Analysing a Blog Test Collection. DCS Technical Report Series (2006)
8. Miller, G.A.: WordNet: A Lexical Database for English. Communications of the ACM 38, 39–41 (1995)
9. Pinker, S., Bloom, P.: Natural selection and natural language. Behavioral and Brain Sciences 13(4), 707–784 (1990)
10. Segerstad, Y.: Use and adaptation of written language to the conditions of computer-mediated communication. Ph.D. thesis, Göteborg University (2002)
11. Tahmasebi, N., Niklas, K., Theuerkauf, T., Risse, T.: Using Word Sense Discrimination on Historic Document Collections. In: JCDL, pp. 89–98 (2010)
12. TREC-BLOG (2012), http://ir.dcs.gla.ac.uk/wiki/TREC-BLOG

Toward Mobile-Friendly Libraries: The Status Quo

Dongwon Lee*

The Pennsylvania State University, University Park, PA 16802, USA
dongwon@psu.edu

Abstract. As the number of users accessing web sites from their mobile devices rapidly increases, it becomes increasingly important for libraries to make their homepages "mobile-friendly." However, to our best knowledge, there has been little attempt to *survey* how ready existing libraries are towards this upcoming mobile era and to quantitatively *analyze* the findings via data exploration methods. In this paper, using the W3C's tool, `mobileOK`, we characterize the mobile-friendliness of comprehensive set of more than 400 libraries with respect to locations (e.g., world-wide vs. US vs. EU) and types (e.g., desktop vs. mobile). Based on our findings, we conclude that majority of current libraries (regardless of locations and types) be *not* mobile-friendly at all (with low mobile-friendliness scores of 0.16–0.21). Using mobilization tools, in addition, we demonstrate that the mobile-friendliness of library homepages can be improved significantly (i.e., 67%–82%). As such, much more efforts to make library homepages more mobile-friendly are greatly needed.

1 Introduction

To accommodate today's library users, most libraries currently offer well-designed homepages on the Web where users can find basic information such as opening hours, direction to branches, contact information, etc. Digital libraries in addition offer advanced browsing and searching capabilities on their catalogs, books, and medias. Traditional users have accessed such libraries' homepages from their home desktop machines. However, this trend is changing rapidly. According to Morgan Stanley's forecast in 2009[1], for instance, mobile usage will be at least double that of the desktop/laptop within the next 5 years. Interestingly, recently, Duke university library reported[2] that mobile access to their libraries has tripled from fall 2010 to fall 2011. Therefore, to be able to accommodate such mobile users, libraries have to prepare to make their homepages *mobile-friendly*. First, let us use the following informal definition:

Definition 1 (Mobile-Friendly Web Page). *A web page that can be rendered well in mobile devices is called as the mobile-friendly web page.* □

---

* In part supported by NSF DUE-0817376 and DUE-0937891 awards.

[1] http://www.morganstanley.com/institutional/techresearch/
mobile_internet_report122009.html

[2] http://prezi.com/a1fmjdk-mjhb/mobile-web-stats-fall-2011/

P. Zaphiris et al. (Eds.): TPDL 2012, LNCS 7489, pp. 38–50, 2012.
© Springer-Verlag Berlin Heidelberg 2012

Note that the definition of mobile-friendliness is vague and subjective at best. Often, whether a web page is going to be displayed well in a mobile browser or not depends on many factors. W3C's Mobile Web Best Practices [1], for instance, suggests that mobile-friendliness of a page be related with: (1) the types of *content* involved (e.g., narrow image vs. flash-based animation), (2) the *capabilities* of mobile devices and networks used (e.g., basic cellular radio access vs. 4G LTE), and (3) the *context* in which the content is received by the user (e.g., sitting at a desk vs. sitting on a subway). Our focus, in this paper, is more pertinent to "traditional" web browsing of libraries' homepages, excluding other content presentation options or emerging multimodal technologies (e.g., multimedia messaging, and podcasts) on diverse mobile platforms (e.g., iPad).

Regardless of the precise definition of mobile-friendliness, libraries today deal with the issue differently. Some libraries make their homepages (and the whole site) to render well for both desktop and mobile devices. For instance, using feature like CSS3 Media Query [2], one can build a single web page design that renders well across multiple devices. Other libraries make and maintain two separate contents (with separate URLs to such contents): one for desktop and the other for mobile devices. Conventionally, main web site URLs such as "www.foo.com" are used for desktop contents, while special URLs such as "m.foo.com" or "foo.com/m/" are reserved for mobile contents. However, as we will show in this paper, many libraries world wide currently do not have home-pages designed for mobile users in mind. Often, a site maintains two separate URLs and contents for desktop and mobile users. To differentiate these two types of homepages, in this paper, we use the following terms:

Definition 2 (Desktop/Mobile Homepage). *When a library homepage is made for user agents from fixed devices (e.g., desktop, laptop, tablet), the home-page, denoted as HP_d, is called a* **desktop homepage**. *Similarly, when made for users from hand-held devices (e.g., cell phone, smartphone, mp3 player), the homepage, denoted as HP_m, is called a* **mobile homepage**. □

Despite the importance of mobile-friendliness of library homepages, in general, very little is known about the current status of mobile-ready libraries and their characteristics. In recent years, in library communities, a few attempts to conduct a limited scale of survey have been made (e.g., [3]). However, none of them studied the mobile-friendliness of library pages using measurable metrics in a substantially large scale. In this paper, therefore, we present a comprehensive study to measure various types of library homepages world wide, identify several important characteristics of both desktop and mobile pages of libraries, and their "mobile-friendliness" characteristics in detail.

2   Related Work

Mobile-Friendly Libraries/Museums: One of the recent popular topics in library and museum communities is the migration of existing library and museum sites to support mobile device users. For instance, [4] describes key design and development strategies on how to create mobile-friendly library sites using the case of Oregon State University (OSU) libraries as an example. [5] reviews an array of mobile applications appropriate in museum settings. Unlike ours, their focus is on visitors who visit museum physically (as opposed to visiting web sites). [6] focuses on the issue of mobile learning of nomadic learners in libraries and examines various standards and technologies by IETF or W3C that enable such mobile learning. [7] examines the mobile web landscape such as mobile devices, mobile web apps, mobile library initiatives. It also presents how to create a mobile experience and get started using the mobile Web. Finally, [3] conducts case studies of four selected institutions and university libraries. and concludes that "offering mobile access to digital collections is still a relatively new endeavor for libraries and museums."

Mobile-Friendly Web: There exist tools to check mobile-friendliness of a web page according to common practices (e.g., W3C `mobileOK` Basic Tests 1.0 [8], W3C Mobile Web Best Practices 1.0 [1], mobiReady[3]). Since there are no agreed-upon standards for mobile-friendliness, however, it is not uncommon to have disagreeing result across different tools. More importantly, there is currently no fundamental understanding as to features of web pages affecting their mobile-friendliness. For instance, the Google Mobilizer[4] and the GOMO initiative[5] aim to help generate mobile-friendly version of a web page. While they generate a decent quality output, often, for a page with complex internal structure/graphics, all they do is to strip off textual contents from the page. Therefore, commercial advertisers who want to keep the graphical design aspect of web pages will not find them useful. Using CSS3 Media Query [2], a designer is able to extend a single page design across desktop, tablet, and mobile devices. However, it helps little in rewriting existing mobile-unfriendly web pages into mobile-friendly ones. Commercial tools (e.g., WireNode, Mobify, bMobilized, Onbile) are in abundance to help advertisers create mobile-friendly web sites. However, many of them focus on creating mobile web sites from the scratch (as opposed to rewriting existing ones), or require intensive labor by web designers. Furthermore, none provides an objective measure to indicate how good the rewriting is.

In academic literature, there have been a few attempts to adapt existing web pages for mobile devices (e.g., [9,10,11,12]). Although useful, none of them provides quantitative scores w.r.t. how similar a page is (before and after the rewriting). Because of the limitations in mobile handheld devices, including small screen size, narrow network bandwidth, low memory capacity, and limited computing power and resources, researchers explored different methods to improve loading and visualizing large documents on handheld devices. For instance, [13] discussed how to avoid distorting web pages in mobile devices using segmentation of contents and ranking therein. [14] argued that changing the layouts of web pages would simplify or delete contents of the pages, leading to

---

[3] http://ready.mobi/

[4] http://www.google.com/gwt/n

[5] http://www.howtogomo.com/

undesired misunderstanding. Instead, the authors designed systems to facilitate users' browsing experience. **Our Contribution:** Compared to aforementioned works, ours is unique in that: (1) Using W3C's mobileOK, we quantify the mobile-friendliness of library home pages; (2) Compared to previous surveys of small scales where 10-30 libraries (and their mobile web support) are evaluated qualitatively, we investigate a much substantial number of libraries (> 400) in a systematic fashion; and (3) Using various data exploration methods, we analyze the mobile-friendliness results of those libraries and unearth several interesting findings.

3   Design of the Study

3.1   Testing Mobile-Friendliness with mobileOK

The mobile-friendliness of a given web page (i.e., whether or not the page can be rendered well on a mobile device) may change dramatically, depending on the contents, capabilities, and contexts of the evaluation. For instance, the same web page may be displayed well in the latest iPhone with 4G but not so in a barebone cell phone with poor network bandwidth. Therefore, inherently, it is challenging to test whether a given web page is mobile-friendly or not. There are several tools to test mobile-friendliness of web pages or sites (e.g., W3C mobileOK, mobiReady, Gomez, iPhoney). In this paper, among these, we decided to use the W3C mobileOK checker that supports programmable APIs (as opposed to web interface) and focuses on the mobile-friendliness of basic mobile devices based on several W3C's recommendations. This checker performs various tests defined in the mobileOK Basic Tests 1.0 specification [8], which is based upon a limited subset of the Mobile Web Best Practices [1]. The mobileOK Basic Tests 1.0 includes 25 tests, each of which in turn includes multiple sub-tests, yielding a total of 100 sub-tests. In our experiments, we use mobileOK checker library v 1.4.2. The score returned by mobileOK checker is computed based on the number and severity of failures of 25 tests carried out over a web page. Each failure can be diagnosed in a severity level between 1 (low) and 6 (critical). More severe failures will cause more penalty in the score evaluation. Utilizing the scores returned from mobileOK, now, we define the following:

Definition 3 (MF-score). *MF-score (Mobile-Friendliness-score) of a web page p refers to a score that* mobileOK *assigns to p (after scaled down to 0–1 range).*□

If a web page gets an MF-score of 1, it implies that the page is likely to be laid out well in a barebone cell phone. Conversely, the MF-score of 0 means that most mobile devices will not be able to render (part of) the page or will not be able to render the page in a reasonable time frame. It is important to note that the MF-score given by mobileOK is not determined linearly. For instance, the MF-score of 0.5 does not imply the passing of half of 25 tests in mobileOK nor suggest 100% more mobile-friendly than that of 0.25.

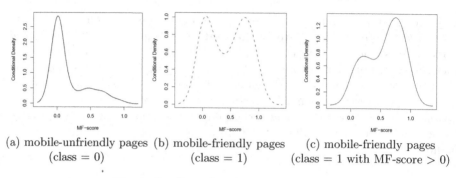

(a) mobile-unfriendly pages (b) mobile-friendly pages   (c) mobile-friendly pages
      (class = 0)   (class = 1)   (class = 1 with MF-score > 0)

Fig. 1. Conditional probability distribution

3.2   Setting the Ground Rules

In this section, using the top-500 sites of US from Alexa[6], we attempt to simulate the "ground truth" data set based on *human judges* and *heuristics*, and validate the effectiveness of `mobileOK` accordingly.

Heuristics-Based Ground Truth. First, we accessed the top-500 sites from mobile phone emulators, and identified URLs to their counterpart mobile version by following the re-direction of HTTP responses. At the time of repeated experiments, on average, 465 out of 500 top sites were accessible via HTTP requests, and 44.7% (208 out 465) of top sites turned out to maintain separate URLs/contents for mobile device users. For those 208 top sites with two explicitly made URLs/contents for desktop and mobile users, we consider those homepages made for mobile users (e.g., "`google.com/m`") as the candidate "ground truth" mobile-friendly homepages. The reason is that these homepages are likely to be explicitly designed for mobile device users by human experts or in-house production system in large companies with sufficient resources.

Figure 2 shows average MF-scores of those 208 top sites. The first two bars show that between two counterpart homepages for the same company, mobile versions have the higher average MF-score than desktop versions have, showing 133% improvement. Considering mobile versions are specifically made for mobile devices, their

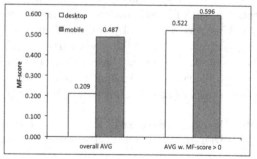

Fig. 2. Average MF-scores of 208 HP_d and HP_m selected from Alexa's top sites

higher MF-scores are expected. However, on average, overall MF-score of all mobile versions is only 0.487, showing that there exists much room for

---

[6] http://www.alexa.com/topsites/countries/US

improvement. That is, although those 208 top sites have made their homepages specially designed for mobile device users, with respect to W3C's mobility standard tests, their overall quality is still rather *poor*. Furthermore, next two bars on the right show the refined average of MF-scores of those sites whose MF-score > 0. As illustrated in Figures 1, the interpretation of the homepage p with MF-score=0 is bimodal (i.e., p is mobile-unfriendly or `mobileOK` incorrectly assigns p's MF-score as 0 even if p is in fact mobile-friendly). Therefore, here, by eliminating those sites that got 0 as MF-score, we tried to mitigate the outlier effect on the calculation of MF-scores. Overall, the average MF-score of mobile homepages rises to 0.596, showing 14% improvement of MF-score from their counterpart desktop homepages.

Human-Judged Ground Truth. Second, using only top-200 sites, three human judges visited each site using two mobile phones (i.e., Apple iPhone 4, HTC Inspire 4G, Samsung Galaxy S simulator), and judged the site as either mobile-unfriendly (class=0) or mobile-friendly (class 1). Note that judges did *not* have a prior agreement on the definition of mobile-friendliness, simulating normal users' *ambiguous* perception. Using the majority voting scheme, at the end, a site with at least two "1"s is labeled as class=1 and with at least two "0"s as class=0. Next, after removing those 10 sites that were inaccessible, we obtained a total of 190 data points (i.e., 71 sites in class=0 and 119 sites in class=1). We first measured how much agreement among judges there is on the mobile-friendliness judgements using the Cohen's *Kappa* measure from the social sciences: $Kappa = \frac{P(A)-P(E)}{1-P(E)}$, where $P(A)$ is the proportion of the observed agreement between two judges, and $P(E)$ is the proportion of the times two judges would agree by accident. Since there are three judges, we computed three pair-wise Kappa measures and used their average. In addition, since class (i.e., 0 or 1) distribution is skewed, we used the marginal statistics to calculate $P(E)$, and obtained a final Kappa value of 0.7126. In general, a Kappa value between 0.67 and 0.8 is regarded as fair agreement between judges. Therefore, we can conclude that our human-judged ground truth is in a fair agreement. Figures 1(a) and (b) show kernel density estimators of MF-score distributions using data in class=0 and class=1, respectively. Observe that the high densities around 0 in Figure 1(a) and the bimodal shape around both 0 and 0.75 in Figure 1(b). This implies that when a page p gets a high MF-score, p's probability to be labeled as mobile-friendly by human judges is very high. However, when p's MF-score is close to 0, it could be either of two reasons: (1) MF-score does not well reflect the human perception on mobile-friendliness (since p that human judges viewed as mobile-friendly got the MF-score close to 0), or (2) p is simply poorly designed and mobile-unfriendly. In other words, *the interpretation of cases with MF-score close to 0 should be made with care* (since it could mean one of two reasons). To further validate this implication, we removed 26 "contradicting" data points in class=1 whose MF-score is 0, and got Figure 1(c). Now, observe the unimodal distribution with high densities around 0.75. Figure 1(c) can be considered as an increasing function, implying that a page with a "higher" MF-score be "more" likely to be mobile-friendly.

4 Experimental Results

4.1 Set-Up

For our study, we prepared the URLs of library homepages of various types from the following lists:

- L_{world}: A list of 167 world-wide national libraries (one national library per country)[7].
- L_{us}: A list of top-100 largest (with respect to volumes held) US public libraries[8].
- L_{eu}: A list of 46 European national libraries in "The European Library" project[9].
- L_{mobile}: A list of 105 mobile homepages (HP_m) of world-wide public libraries[10] (e.g., "m.psu.edu/library/"). These libraries maintain specially-made homepages for mobile device users exclusively.
- $L_{desktop}$: A list of 105 desktop homepages (HP_d), a counterpart to L_{mobile} (e.g., "www.libraries.psu.edu").

All lists (except L_{mobile}) contain the URLs of conventional library homepages, made mainly for desktop users–i.e., HP_d. Using W3C's mobileOK checker library v 1.4.2, we first set our program's user-agent setting as "Mozilla/5.0 (iPhone; U; CPU iPhone OS 4_3_3 like Mac OS X; en-us) AppleWebKit/533.17.9 (KH TML, like Gecko) Version/5.0.2 Mobile/8J2 Safari/6533.18.5" so that the web server of a target library would think that the request comes from a mobile device, iPhone. If the target library detects our user-agent type correctly and forwards the request to their specially-made mobile homepages (HP_m), however, then we treat this mobile homepage as the main URL of the target library and test it accordingly. In obtaining the MF-score of a given library homepage, ideally, one has to fetch all web pages of the library site and use the average MF-score of all pages. However, this approach was abandoned due to the following reasons: (1) The typical number of web pages per library site varies greatly. However, in general, for the large libraries that we investigate in this paper, such number ranges from thousands to tens of thousands, without even including dynamically generated pages. Therefore, we need to reduce the number of pages to experiment; and (2) When a user browses a library's homepage using her mobile device, it is likely that she would bounce back and exit the site immediately if the first encountered URL of the site does not give satisfactory mobile experience. Therefore, in this experiment, we focus on the main URL of the homepage, and ignore all subsequent internally linked pages.

All subsequent experiments have been carried from Nov. 2011 to Feb. 2012. Each target library has been tested for three times, each test two weeks apart, and the average MF-score of three measurements was used for analysis. If the web

---

[7] http://en.wikipedia.org/wiki/List_of_national_libraries
[8] http://www.ala.org/tools/libfactsheets/alalibraryfactsheet22
[9] http://www.theeuropeanlibrary.org/
[10] http://libsuccess.org/index.php?title=M-Libraries

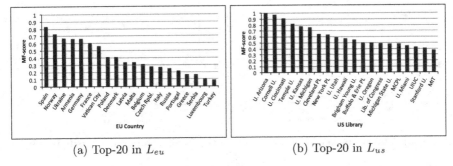

(a) Top-20 in L_{eu} (b) Top-20 in L_{us}

Fig. 3. Top-20 MF-scores of homepages of libraries in different regions of the world

server of a target library did not respond (within 60 sec) for all three attempts, we removed the library from the analysis.

4.2   Location Based Analysis

First, we compared the MF-scores of three lists (L_{eu}, L_{us}, and L_{world}). In general, substantial percentages of libraries failed a large number of sub-tests of mobileOK, yielding the MF-score of 0. At the end, 47.7%, 55.2%, and 44.3% of the lists (L_{eu}, L_{us}, and L_{world}) garnered the MF-score of 0, respectively, indicating that the majority of library homepages are currently *"mobile-unfriendly."* This is a bit surprising result, considering that libraries in the three lists are either the biggest national library of countries or one of the largest public ones in US (with possibly more sufficient resources for funding and IT than smaller libraries). Top-20 ranks (with respect to MF-score) of two lists are shown in Figure 3. In summary, as shown in Figure 4, all three lists show similar patterns–i.e., rather low MF-scores (with median: 0.31, 0.32, 0.27, mean: 0.185, 0.16, 0.22, and std. dev.: 0.25, 0.25, 0.29 for L_{eu}, L_{us}, and L_{world}, respectively), and similar inter-quartile range (IQR), a bit skewed toward to the left (i.e., low mobile-friendliness).

Next, we examined to see if there is any correlation between MF-score of libraries and other factors. However, using L_{us}, first, we found no significant correlation between MF-scores and the ranks of US libraries (with respect to volumes) or state populations where libraries are located, etc. Using L_{world} data, next, we examined the correlation between the MF-score of national libraries and how active

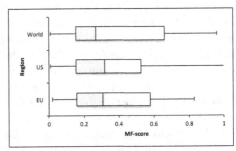

Fig. 4. A box-and-whisker plot of libraries (w. MF-score > 0) using L_{eu}, L_{us}, and L_{world}

people use mobile web with their mobile devices. That is, our conjecture was that if a country has more active mobile web usage, it is more likely that web

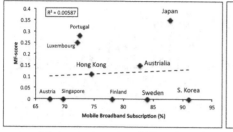 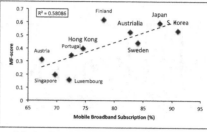

      (a) Against national libraries   (b) Against top-5 most visited sites

Fig. 5. Linear regression between mobile broadband subscription percentage (X-axis) and MF-scores of 10 countries with most active mobile web access (Y-axis) using: (a) national libraries, and (b) top-5 most visited sites

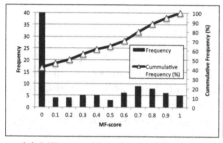 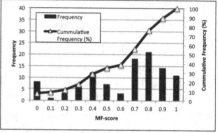

    (a) MF-score histogram of $L_{desktop}$ (b) MF-score histogram of L_{mobile}

Fig. 6. MF-score frequency histograms of $L_{desktop}$ (i.e., HP_d) and L_{mobile} (i.e., HP_m)

sites of the country (including library homepages) are designed to be mobile-friendly. From the Global mobile statistics 2012 data set[11], we first obtained top-10 countries with active mobile usages. Then, using Alexa.com's top sites data, we also obtained the top-5 most-visited commercial sites of those 10 countries. When the regression is made against the MF-score of national library of those top-10 countries, as shown in Figure 5(a), a very low coefficient of determination (i.e., $R^2 = 0.00587$) indicates that X value of the graph (i.e., active mobile web usage of a country) cannot predict well Y value of the graph (i.e., MF-score of its national library). However, when we instead used the average MF-score of top-5 most-visited commercial sites of those 10 countries and correlated it against the mobile broadband subscription percentage, as shown in Figure 5(b), some correlation is found. An R^2 of 0.58086 implies that 58% of the variance in MF-score is predictable from the mobile broadband subscription percentage. In order words, in those countries with heavy mobile web usage, top companies with high traffic are already well aware of and prepared for mobile web browsing. However, unfortunately, this is currently not the case for national libraries in those countries.

---

[11] http://mobithinking.com/mobile-marketing-tools/latest-mobile-stats

4.3 Mobile-Oriented vs. Desktop-Oriented

In this section, we investigated the difference of mobile-friendliness between HP_d and HP_m. The list L_{mobile} contains URLs of homepages that are specially made for mobile device users. Such URLs typically have formats like "m.foo.com" or "www.foo.com/m/". From these HP_m, we first reverse guessed their original HP_d such as "www.foo.com", and manually corrected any remaining errors. Then, by comparing these HP_d and HP_m w.r.t mobile-fiendliness, we can see the impact of having mobile-friendly library homepages. Overall, the average MF-score of HP_m, 0.586 (std. dev.=0.282), is significantly higher than that of HP_d, 0.318 (std. dev.=0.34). Since HP_m are explicitly designed for mobile browsing, such a result is expected.

In zooming in for more details, Figure 6 shows MF-score frequency histograms and cummulative frequency (%) of both HP_d and HP_m, respectively. In Figure 6(a), note that there are 40 libraries whose MF-score came out as 0, implying poor mobile-friendliness of their homepages. However, in Figure 6(b), this number is reduced to only 7. Furthermore, for those bins of higher MF-score in Figure 6(b), their frequencies are much bigger than those in Figure 6(a), implying that many libraries in L_{mobile} have been designed to be more mobile-friendly.

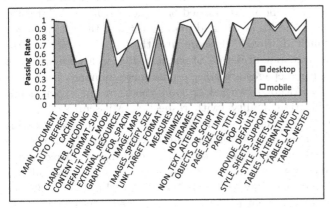

Figure 7 shows a stacked graph of *passing rates* (i.e., the fraction of libraries that passed a particular sub-test in mobileOK) for $L_{desktop}$ and L_{mobile} (X axis includes 25 sub-tests of W3C's mobileOK). Note that passing

Fig. 7. Comparison of passing rates of 25 sub-tests in mobileOK for $L_{desktop}$ and L_{mobile}

rates of L_{mobile} are usually higher than those of $L_{desktop}$. A few sub-tests are noticeable here (e.g., IMAGE_MAPS, IMAGES_SPECIFY_SIZE, NO_FRAMES, OBJECTS_OR_SCRIPT, POP_UPS, PAGE_SIZE_LIMIT, and TABLES_LAYOUT). Homepages of $L_{desktop}$ failed more on sub-tests than those of L_{mobile}. Overall, those poorly designed library homepages suffer from a large page size and loading time (with many embedded resources), a poor layout design, and lack of support for mobile access.

4.4 Mobilizing Library Homepages

In this section, we report our small pilot study to see if one can quickly improve the mobile-friendliness of library pages. From the L_{us} list, first, we randomly

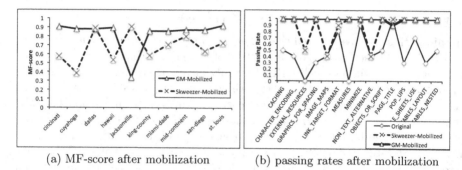

(a) MF-score after mobilization   (b) passing rates after mobilization

Fig. 8. Changes of MF-scores and passing rates after mobilization

picked 10 public libraries all of which had the MF-score of 0 and caused a lot of critical/severe violations according to W3C `mobileOK` Basic Tests 1.0 [8] and W3C Mobile Web Best Practices 1.0 [1]. Then, from available tools that can instantly transcode the given web page into mobile-friendly one, we chose two well-known ones: Google Mobilizer[12] and Skweezer[13], and converted main homepages of 10 libraries into mobile-optimized ones (i.e., *mobilized*). After the mobilization, we measured the MF-score and passing rates of their homepages. Figure 8(a) shows the much improved MF-score after the mobilization (from the original score of 0). When we look at details of passing rates, in Figure 8(b), it is also clear that mobilized homepages are able to pass more number of sub-tests of `mobileOK` successfully. Although varied in some cases, in our experiments, we note that the mobilization quality (with respect to MF-score) of Google Mobilizer is in general better than that of Skweezer. Note that since existing mobilizer software such as Google Mobilizer and Skweezer largely transcodes pages by stripping off contents from pages, often, the resulting mobilized design contains dull text-based contents and links. Therefore, such a mobilizing software is to be used as a quick-and-dirty way to test the mobile-friendliness of mobilized pages, and more human involvement is likely to be expected in practice.

5 Limitations and Future Plan

We recognize the limitations of the current study and plan to conduct the following research in future: (1) While we validated the effectiveness of `mobileOK` in quantifying the mobile-friendliness of a web page in Section 3.2, a more extensive user study would be able to provide stronger grounding. For instance, using Amazon's Mechanical Turk system, one could do crowd-source based human evaluation of library pages and compare the results with that by `mobileOK`; (2) Current focus of `mobileOK` scoring is on the general "browsing" task of library homepages. Since users often access library homepages with specific tasks in mind

---

[12] http://www.google.com/gwt/n
[13] http://www.skweezer.com

(e.g., finding direction to libraries, searching book information), a more fine-grained task-specific study along with mobile-friendliness of homepages would be interesting; and (3) Since "tablets" users are increasing rapidly, we also plan to extend the current study to include tablets as mobile platforms.

6 Conclusion

In this paper, we have analyzed homepages of more than 400 libraries world wide with respect to their mobile-friendliness. From our study, we have found: (1) W3C's `mobileOK` can be an effective tool to estimate the mobile-friendliness of a web page; (2) According to `mobileOK`, majority of 400+ library homepages (regardless of their locations and types) have very poor MF-scores (in the range of 0.16–0.21); (3) Library homepages explicitly designed for mobile devices (i.e., HP_m from L_{mobile}) have significantly higher MF-scores than those made for desktop viewing (i.e., HP_d from $L_{desktop}$); and (4) Using mobilization tools, the mobile-friendliness of ten US public library homepages can be improved significantly (i.e., 67%–82%).

Acknowledgement and Availability: Author thanks Haibin Liu for providing script codes and Woo-Cheol Kim for helpful discussion on Section 3.2. All final implementation codes and data sets to run the presented experiments are publicly available at: `http://pike.psu.edu/download/tpdl12/`.

References

1. Rabin, J., McCathieNevile, C.: Mobile Web Best Practices 1.0. Technical report, W3C (July 2008), `http://www.w3.org/TR/mobile-bp/`
2. Lie, H.W., Celik, T., Glazman, D., van Kesteren, A.: Media Queries. Technical report, W3C (July 2010), `http://www.w3.org/TR/css3-mediaqueries/`
3. Mitchell, C., Suchy, D.: Developing Mobile Access to Digital Collections. D-Lib. Magazine 18 (January 2012)
4. Griggs, K., Bridges, L.M., Rempel, H.G.: library/mobile: Tips on Designing and Developing Mobile Web Sites. Code4lib Journal 8 (November 2009)
5. Raptis, D., Tselios, N., Avouris, N.: Context-based Design of Mobile Applications for Museums: A Survey of Existing Practices. In: Int'l Conf. on HCI with Mobile Devices & Services (2005)
6. Saravani, S.J.: Standards Informing Design of Library Service Delivery to Mobile Devices and Nomadic Learners. In: Conf. and Exhibition, VALA (2010)
7. Kroski, E.: On the Move with the Mobile Web: Libraries and Mobile Technologies. Library Technology Reports 44 (July 2008)
8. Owen, S., Rabin, J.: W3C mobileOK Basic Tests 1.0. Technical report, W3C (December 2008), `http://www.w3.org/TR/mobileOK-basic10-tests/`
9. Chen, Y., Xie, X., Ma, W.Y., Zhang, H.J.: Adapting Web Pages for Small-Screen Devices. IEEE Internet Comp. 9, 50–56 (2005)
10. Erol, B., Berkner, K., Joshi, S.: Multimedia Clip Generation From Documents for Browsing on Mobile Devices. IEEE Trans. on MultiMedia (TMM) 10, 711–723 (2008)

11. Xiao, X., Luo, Q., Hong, D., Fu, H., Xie, X., Ma, W.Y.: Browsing on Small Displays by Transforming Web Pages into Hierarchically Structured Subpages. ACM Trans. on the Web (TWEB) 3 (January 2009)
12. Zhang, D.: Web Content Adaptation for Mobile Handheld Devices. ACM Comm. of the ACM (CACM) 50 (February 2007)
13. Hu, W., Kaabouch, N., Yang, H., Hu, W.: Technologies and Systems for Web Content Adaptation. In: Handheld Computing for Mobile Commerce: Applications, Concepts and Technologies, pp. 263–277 (2010)
14. Arase, Y., Hara, T., Nishio, S.: Web Page Adaptation and Presentation for Mobile Phones. In: Handheld Computing for Mobile Commerce: Applications, Concepts and Technologies, pp. 240–262 (2010)

Listen to Tipple: Creating a Mobile Digital Library with Location-Triggered Audio Books

Annika Hinze and David Bainbridge

Dept. of Computer Science, University of Waikato,
Private Bag 3105, Hamilton New Zealand
{hinze,davidb}@waikato.ac.nz

Abstract. This paper explores the role of audio as a means to access ebooks while the user is at the locations that are referred to in the books. The books are sourced from a digital library and can either be accompanied by pre-recorded audio or synthesized using text-to-speech. The paper discusses the implications of audio access for ebook with particular reference to HCI challenges.

1 Introduction

This paper proposes a new method for finding and reading ebooks – via location-based audio. It is our assumption that most ebooks are bought for recreational reading. Though ebook acquisition is, naturally, restricted to places with access to the electronic web, the (physical) locations of ebook consumption may vary widely. With the introduction of electronic readers, ebook consumption is no longer limited to the home or office, but is also potentially mobile (that is, able to be consumed while moving).

Geographic positioning is increasingly a significant contextual component for the users of mobile applications. For example, many new mobile apps on smart phones offer location-aware information. So far, ebook access has not made use of this contextual information. We here suggest a new form of ebook consumption: a location-based audio reader. The paper explores the challenges and opportunities for such a system and reports experiences in its development and explorative use.

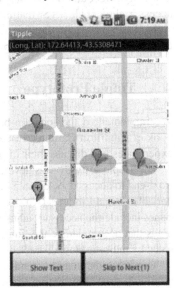

Fig. 1. Audio at location Worcester Street

2 Scenario

As a scenario, imagine a visitor to Christchurch, New Zealand. Installed on their phone is the location-based audio reader, Tipple, loaded with a pro-

P. Zaphiris et al. (Eds.): TPDL 2012, LNCS 7489, pp. 51–56, 2012.

file of their interests, which includes the English author Rose Tremain. Walking from Christchurch Cathedral towards Latimer Square, their phone makes a "chirp chirp" noise alerting them to an area of interest nearby.

Worcester Street is a location where Harriet Blackstone, the protagonist of Tremain's book *The Colour* buys a book before heading out into the New Zealand bush. Using the map displayed on-screen, our visitor decides to divert to this location (visitor location shown in orange in Figure 1).

Once there, they select the opening chapter of *The Colour*, one of several chapters listed on their phone as being set at this location, and start listening to it as it is read aloud through their headphones. Other locations nearby that also appear in the book are also shown on the map (book locations shown in red).

3   Related Work

No comparable study of ebook access has been previously undertaken, even though in a 2003 study the top ebook functions preferred by people were named as audio, bookmarks & dictionary (19%) and mobility (30%) [8]. Spoken audio in ebook access has so far only been considered for desktop-based digital library systems. Typical examples are reading learning tools and text-to-speech for easy access – such as systems for visually impaired or dyslexic people, see, *e.g.*, [2,7,12].

Some ebook readers have Text-to-Speech features (TTS) built into them, including the Amazon Kindle device, the BeBook Mini V5 and the iPad. Some rights holders have disabled the text-to-speech on their ebooks as they are concerned that it will adversely affect audio book sales. Audio access to ebooks also creates concerns about access rights [11]. TTS features are predominantly targeted at blind or partially sighted persons; it is often named together with large print and Braille [6]. None of the available ebook readers supports location-based access.

4   Tipple Prototype

We implemented the location-based audio reader Tipple and explored its use for a book of Hamilton Gardens. In the remaining sections of this paper we summarise our experiences and discuss their implications for ebook acquisition and consumption.

The Tipple software was developed not for single ebook access but as a location-based front-end to access ebooks via a digital library. Previously we have studied self-contained portable Digital Libraries (without location reference or audio) on mobile devices [3]. HCI issues include the necessary adjustments for displaying library content on smart phones or iPads that accommodate the limitations of screen size. Access situations are often dominated by multi-tasking, distracted attention, and limited input options [10].

Tipple builds on and extends concepts from two systems: TIP and Greenstone. The mobile Tourist Information Provider (TIP) combines an event notification service and a location-based service to alert its users to attractions in the vicinity [9]. The open source Greenstone software was chosen for the digital library component for this project [4].

Greenstone documents are prepared in advance (when a digital library collection is formed), with the installation configured for location-based search using a place name recognition package, which adds location mark-up to the documents. For Tipple, we needed to provide audio information to the user even though the books in the digital library may not be available with pre-recorded audio. For those cases, a Text-to-Speech function was used.

Access to the books uses the chapter-level metaphor for narrative books (order of chapters predefined) and an article metaphor for reference books (independent of ordering). For audio-books we required a Greenstone location-search at the chapter level and we changed the indexing in Greenstone to operate at the chapter level also. The system selects audio books that are relevant to the user's location. Again, their relevance is determined at a chapter level: if any chapter of a book is related to a given location, the book will be selected. All relevant audio books will be listed for the user, with highlighted chapters referring to the current location. The user can choose to listen to the audio content or to view the text retrieved from the digital library. We also offer the option to play these chapters by travel-route sequence.

5   Usage Experience Discussion

To evaluate Tipple we undertook a field trial of the software. One of the difficulties of studying location-based audio access to ebooks is the need to provide users with sufficient example data. For our study we decided first to explore the use of a guide book, as no suitable works of fiction were available for the region we live in. The ebook selected was one describing the local council's public gardens (Hamilton Gardens). The text of the book was manually prepared by enriching it with location data (GPS areas for each chapter). While this ebook has a defined sequential order, it can essentially be accessed non-sequentially, given its structure, and still make sense. While not an exact match for the type of ebook content we ultimately wish to explore, this simpler format still allowed insight into issues created by audio access. Here we focus on those results that pertain to the audio access of ebooks.

Participant Demographics. The sixteen participants who took part were students at the Computer Science Department (14 participants aged 20 to 35, two aged 35 to 50; three female and 13 male). An extended study with members of the general public is currently ongoing. Most participants had previously visited the Hamilton Gardens, which allowed us to compare visits with and without the audio book system. The participants' familiarity with and use of books and audio books is divided into two groups: most of them read books often or always (50%), the others read rarely. The majority would not take works of fiction on travels (median of 'rarely'), even fewer would take travel guides (median at 1.5; mode is 'never'). Audio books or guides were almost completely unused.

Satisfaction. Four people used a text-based access to the ebook and 12 people the audio-based access. All four readers of the digital text were satisfied with the interaction (see Figure 2); 8 of the 11 participants with audio were satisfied, with one not answering. Participants P1 and P11 observed that the Text-to-Speech audio became boring after some time, as it was rather monotonous. For the audio access, it was suggested that we create smaller items with an option to continue instead of offering all available information for each location at once.

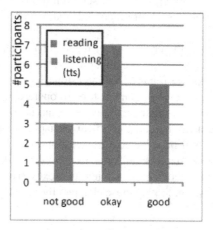

Fig. 2. Access type satisfaction

Access Preference. Figure 3 summarises the access type preferences of the participants for our study. All audio participants preferred the audio presentation over text; none of them asked for text presentation. The reason most often given was the advantage of being able to look at the surroundings and not having to read from a small screen.

As a possible additional use of text they mentioned the advantage of skimming and quick decision making. Two of the participants using the text display preferred text over audio; P13 felt it would have 'looked crazy' if they had been walking around with an audio guide and they preferred the option of quick skimming. P14 was a non-native speaker and preferred to read the text in their own time.

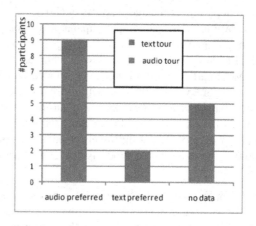

Fig. 3. Access type Preferences

Shared Audio. Our study participants went alone into the gardens to listen to the ebook. In an informal pilot to our study, two participants went together and expressed the desire to listen to the audio book together. This would constitute a new access pattern to ebooks as users do not typically search and read ebooks in groups. One study explored the aspect of reading ebooks together [13]. They observed that the "act of reading is typically solitary" and identified as an exception the reading in (study) groups. Audio access to ebooks is more naturally shared among a group of people, but has not been studied so far.

Audio interaction and collaboration between users has been evaluated by Aoki *et al.* [1], who developed an electronic guide that allowed museum visitors to explicitly share audio information (pre-recorded and spoken). They found that most couples used shared audio as a conversational resource. Evjemo *et al.* [5] also report that nearly 40% of their study participants explicitly asked for the opportunity to share audio content in a museum and this included nearly 90% of those who usually had company when visiting museums.

Language Issues. Several participants noted the limited quality of the text-to-speech function. They reported that listening to the very monotonous voice becomes "boring" and thus hard to follow. The researchers observed an additional issue when developing the chapters referring to the Māori garden. No text-to-speech implementation is available for Māori; moreover, the spoken text was to be delivered in English and a mixing of languages is not supported. As a work-around we had to duplicate each chapter that contained Māori names: one for displaying the text and one in which all Māori words were transcribed into phonetic English. Culturally this is not a viable option nor is it a manageable approach for larger bodies of text.

6   Conclusion

Emphasis is placed here on the role of audio access as this is a little-understood area for ebooks. Our field study is the first one of its kind, exploring the use of audio access to ebooks on location. The book selected for our study was a reference-style book without ongoing narrative. Further issues are to be expected in books with ongoing narratives, which are currently explored in further studies.

For this study, the location mark-up was inserted manually into the metadata of each book chapter in the digital library. For semantic enrichment of whole books (with possible fine-granular mark-up) automatisms need to be used, such as using Greenstone's text-mining capability for locating place names. A simple text-analysis with a gazetteer will not be sufficient.

Finally, discussions of ebook acquisition have so far not considered audio support or location mark-up in metadata.

Acknowledgements. We would like to thank the management of Hamilton Gardens for their support of this study.

Reference

1. Aioki, P.M., Grinter, R.E., Hurst, A., Szymanski, M.H., Thornton, J.D., Woodruff, A.: Sotto Voce: Exploring the Interplay of Conversation and Mobile Audio Spaces. In: CHI 2002, pp. 431–438 (2001)
2. Bae, K.-J., Jeong, Y.-S., Shim, W.-S., Kwak, S.-J.: The ubiquitous library for the blind and physically handicapped: a case study of the LG Sangnam Library, Korea. IFLA Journal 33(3), 210–219 (2007)
3. Bainbridge, D., Jones, S., McIntosh, S., Jones, M., Witten, I.H.: Running Greenstone on an iPod. JCDL, 419–419 (2008)
4. Bainbridge, D., Witten, I.H.: Greenstone digital library software: current research. In: JCDL 2004: Joint Conference on Digital Libraries, pp. 416–416 (2004)
5. Evjemo, B., Akselsen, S., Schormann, A.: User acceptance of digital tourist guides lessons learnt from two field studies. In: HCI, pp. 746–755 (2005)
6. Garrod, P., Weller, J.: Ebooks in UK Public Libraries: where we are now and the way ahead, Networked Services Policy Task Group, Issue Paper No. 2 (January 2005)
7. Goh, D.H.-L., Sepoetro, L.L., Qi, M., Ramakhrisnan, R., Theng, Y.-L., Puspitasari, F., Lim, E.-p.: Mobile Tagging and Accessibility Information Sharing Using a Geospatial Digital Library. In: Goh, D.H.-L., Cao, T.H., Sølvberg, I.T., Rasmussen, E. (eds.) ICADL 2007. LNCS, vol. 4822, pp. 287–296. Springer, Heidelberg (2007)
8. Golub, K.: Digital libraries and the blind and visually impaired. In: CARNet Users Conference, pp. 25–27 (2002)
9. Henke, H.: Consumer Survey on eBooks by OeBF, http://ns2.idpf.org/doc_library/surveys/IDPF_Consumer_Survey_2003.pdf (last accessed September 2011)
10. Hinze, A., Voisard, A., Buchanan, G.: TIP: Personalizing information delivery in a tourist information system. Journal of Information Technology and Tourism 11(4) (2009)
11. Jones, M., Marsden, G.: Mobile Interaction Design. John Wiley & Sons (2005)
12. Kerscher, G., Fruchterman, J.: The Soundproof Book: Exploration of Rights Conflict and Access to Commercial EBooks for People with Disabilities. First Monday 7(6-3) (June 2002)
13. Prahallad, K., Black, A.: A text to speech interface for Universal Digital Library. Journal of Zhejiang University SCIENCE 6A(11), 1229–1234 (2005)
14. Pearson, J., Buchanan, G.: CloudBooks: An Infrastructure for Reading on Multiple Devices. In: Gradmann, S., Borri, F., Meghini, C., Schuldt, H. (eds.) TPDL 2011. LNCS, vol. 6966, pp. 488–492. Springer, Heidelberg (2011)

Re-finding Physical Documents: Extending a Digital Library into a Human-Centred Workplace

Annika Hinze and Amay Dighe

Dept. of Computer Science, University of Waikato,
Private Bag 3105, Hamilton New Zealand
{hinze,aad11}@waikato.ac.nz

Abstract. It is often difficult for busy people to keep track of or re-find documents in their own workplace. Very few methods have been developed for finding a physical object's location in an office. Most of the existing methods require that some kind of structured approach be followed by the user. We created a "Human-Centred Workplace" system that does not require orderly users. The system embeds passive tags in documents and uses cameras in the office to track changes in the documents' locations. This paper introduces the design and implementation of the system, explores its use in an office environment and gives a initial evaluation of our prototypical implementation.

1 Introduction

Most internet searches are searches to re-find information [7,13]. Similarly, people's workplaces are filled with documents, which often need to be re-found in the physical environment. The problem of digital re-finding and its physical equivalent are often related as previously digital information is printed, annotated and then placed somewhere in the office. Librarians and archivists have long developed methods for organizing physical documents. However, people rarely want to follow such a structured approach because of the considerable discipline it requires.

Our approach aims to reflect real-life situations in which people do not fill in metadata nor place documents in the appropriate folders. Instead of trying to identify the exact position of a physical document we aim to trigger a user's memory of when and where they last used a document. We do this by presenting them with an image of the document and contextual information about its intended use, as well as feedback about the observed last location of the document. Repeated searches fail to capture previous access context and previously established meaning. Retrieval of information requires the blurring of workspace boundaries [1,16]; the result is a human-centred workspace that merges the digital, physical and conceptual world. Our system constitutes a first step towards building such a Human-Centred Workspace (HCW).

The remainder of this paper is structured as follows: Section 2 provides an overview of related work that inspired this project. Section 3 introduces the HCW system architecture and Section 4 gives details about the implementation. Section 5 illustrates

P. Zaphiris et al. (Eds.): TPDL 2012, LNCS 7489, pp. 57–63, 2012.

the system's use cases by way of a scenario. Advantages and shortcomings of the current implementation are discussed in Section 6. The paper concludes with an overview of planned extensions and further steps in our research.

2   Related Work

SOPHYA is a system to digitally manage physical documents [14]. It uses a physical object container (a shelf) in which each document is placed into a container unit (a folder) with an attached RFID tag. SOPHYA stores information about the documents. The user needs to fill in the meta-data and place the document in the right container. This system, the first of its kind, lacks the freedom to place documents anywhere in the office. Furthermore, meta-data needs to be entered and maintained manually.

The Fused Library uses RFID tags for object localisation [1]. Queries could be formulated by moving tagged objects within the physical space of a library. The focus here is on information seeking and access. Re-finding of particular objects is not directly supported. Other systems provide physical interfaces for searching (e.g., navigational blocks [3]) or ambient information display (phicons to hold references to digital information [10]). None of these directly address the problem of locating physical documents. Re-finding of documents and artefacts could be supported by using navigational blocks as an interface to "probe" the environment. Phicons [10] could be a way of alerting a user to a document's location.

Andrew Walsh used Quick Response (QR) codes and mobile phones in a library environment [19]. He made locations of physical objects accessible and inspired our use of QR codes in a system to monitor and re-find printed documents in offices.

3   Human-Centred Workplace (HCW)

The HCW system automatically identifies documents as they are printed and monitors the print-outs' movements through the office. We use passive tags (QR codes) when printing documents, and cameras to follow these codes. This allows for accurate document identification, retrieval of digital information about the documents as well as physical retrieval of documents. Furthermore, it enables the users to monitor their own document-printing behaviour and provides feedback about information previously accessed.

Figure 1 shows our HCW system architecture. The user interaction with the system is as follows: (1) User starts to print a document. (2) Document is automatically tagged with a QR code passive tag on the front page and then printed. (3) The user places the document in the office. (4) The camera continually monitors the document in the office workplace and reads the QR code. (5) User searches for the document via the HCW. (6) The system identifies the document's location (via QR code and image comparison) and reports it back to the user.

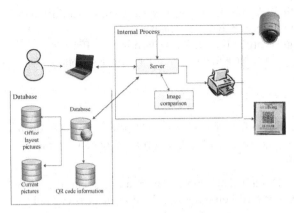

A camera periodically records the office situation. As the user prints the document, an entry with the print details is entered in the database. When the user places a printed document in the office, the camera will capture the image and, with the help of a QR reader, will decode the QR tags on the documents. In the next step, this information is

Fig. 1. Architecture of the Human-Centred Workplace

used to search for a possible match in the document database. Both document information and captured image are sent to the server gateway, where image comparison is used to further identify the location of the document within the image (e.g., on the desk). Finally, the user receives detailed information about the document's current location (context of use and location). The digital documents are thenplaced in a digital library for easy access. The easy access to information about the physical location of a document's hardcopy is not yet integrated into the digital library environment.

3.1   Implementation Details

The system is divided into three phases: (1) Generating QR code, (2) Reading QR code, and (3) Document localization.

Phase 1: Generating the QR Code. When a document is printed, an electronic identifier is embedded in the printout for later re-finding of the physical copy. We store information related to the document in the database and encode a reference to this information in the form of a QR code as identifier. In our prototype implementation, we automatically convert all documents into Word format. Native insertion of QR codes into PDF documents could be implemented using itext [11]. For Word documents, we implemented an Active X control to generate the QR code (using the StrokeScribe tool [18]). The first five lines of a document are encoded; additional use of a hash code and timestamps for the document are also supported. After the automatic QR code encoding, the system prints the document with the code attached. Alternatively one could explore using Xerox® Barcode Printing [20].

Phase 2: Reading the QR Code. The system continually monitors and locates the QR code in the room (via web-cam). It decodes the information from the tag (via reader software) and stores the new document location together with the current image in the database for further document localization when requested by the user. We used a number of readers: Kaywa [14], desktop reader [4], and Fun2D [7] (see Sect. 6).

Phase 3: Document localization. When a user tries to locate a physical document, the images collected in Phase 2 are further analysed by comparing [15] the images taken over time with images of the original office layout. Each place in the office is given a semantic mark-up, which can be reported back to the user as location description.

3.2   Use Cases

We now illustrate the HCW system with four use cases: (1) printing of new document, (2) re-finding a (forgotten) document, (3) locating a printed document, and (4) attempted location of a document.

Printing a New Document. When a user attempts to print a document, the HCW system checks whether it has been printed before. When the document is entered for the first time, the user is asked to insert some notes on the intended use of the paper (e.g., "related work for paper *xyz*" or "copy for person *a*"). This step is not mandatory but may help the user later to remember the context in which they printed the paper. The document is then printed with the embedded QR code linking to the database entry.

Re-Finding a Document. When a user attempts to print a document previously entered into the system (i.e. it was printed earlier), the HCW system warns that a physical copy of the document exists and asks if the user wishes to proceed with printing. The system gives further details about the previous printout of the document to help in remembering the context. The system identifies the location of the physical document (e.g. on the book shelf next to the desk). This step may avoid duplicate printing and may allow the user to access prior comments on the printout.

Locating a Printed Document. The user may remember a previously used paper and search for it using the HCW interface. A list of papers they previously read (and printed) is returned. The user may then decide whether or not to search for the printouts.

Attempted Location of a Document. If a user (mistakenly) searches for a document they did not previously print, the HCW system fails to find the document. The user may then print the document, knowing that they do not yet have a physical copy in their office. The same information as in the first use case is collected by the system.

4   Discussion

The purpose of our HCW system is to make the management of physical documents in a workplace easier by using an open approach that does not require the user to maintain orderly document keeping. We here discuss the design implications of our initial observations of the HCW prototype.

- We generated QR codes by first converting PDF documents into Microsoft Word™ format and then embedding the QR tag in the document to be printed. Inserting a QR code into the Word document's main body has the potential to change the layout of the document. Inclusion of the code into the document's margins is therefore preferable. Moreover, a native support for both PDF and Word documents is desired and planned for the next version.

- Reading and identifying QR code tags in real-time is one of the main features of the software. We worked with a number of cameras placed around the workplace. Within a short distance from the camera (up to three feet), the QR code reading results were acceptable and allowed us to correctly identify the code and monitor its movements. When we tested QR code reading by increasing the distances between camera and document to more than three feet, the application failed to read the QR code. Thus, the system can be reliably used with a single camera within the "personal distance" [8] of up to 2.5 feet. Within the "social distance" of about 4 to 12 feet, several cameras are needed within the environment. Other options are the enlarging of the QR code up to one foot [6].
- We tested the Kaywa reader [14], the desktop QR code reader [4], and Fun2D reader [7] wih a fixed-mount security camera, a web camera and hand-held phone cameras. The results of our testing were mostly positive, but we encountered difficulties in repeat-readings of documents placed directly in front of the camera. Sometimes repeated attempts to read the same QR code from different directions were needed. This may be acceptable for a typical application on a mobile phone, where a user can re-try to read a QR code. However, for repeated automatic QR code detection, better reliability of QR code detection is needed.
- A seamless integration into the digital library environment has currently not yet been achieved. Both the search for digital documents and for their physical representation is supported but not from a common interface. Implementation of a simple extension is planned once the initial HCW system tests are successful. Such integration would then allow the exploration of further uses cases such as full-text search and tracking of material to be read.

5   Conclusions

Here we briefly summarise our HCW project to date and discuss directions of future work. In comparison to other approaches, our project aims to provide a system that would support users in their established workflows without the need to change engrained working patterns (e.g. by asking them to sort their papers into predefined slots or dedicated physical places). We decided to use QR codes instead of the more typical RFID tags as these can be printed "on the fly" and would be available to the users without interrupting their work. After generating QR codes for documents, the physical location of papers needs to be monitored within the office. Most problems were caused by the inadequate performance of QR code readers, making it necessary to place a number of cameras within an office environment (as only readings at less than 3 feet are reliable). Using image comparison for localization achieved positive results and allowed for document localisation with user-defined semantics. The HCW system not only helps users to find and track their documents but also monitors a user's document printing habits and thus limits the possibility of document duplication in the office. Overall, the system design has been successful in principle as we were able to track and re-find documents. However, limitations of cameras and QR reader applications need to be addressed.

We are planning to further integrate the HCW system into a personal digital library to allow seamless access to all digital and physical documents a user has previously encountered. The system has only been trialled with a small sample of users and documents. Long-term usage still needs to be explored for both performance and user acceptance.

References

1. Buchanan, G.R.: The fused library: Integrating digital and physical libraries with location-aware sensors. In: 10th Annual Joint Conference on Digital Libraries (JCDL 2010), pp. 273–282. ACM, New York (2010)
2. Bush, V.: As We Think. Atlantic Monthly (July 1945)
3. Camarata, K., Yi-Luen Do, E., Johnson, B.R., Gross, M.D.: Navigational blocks: Navigating information space with tangible Media. In: 7th International Conference on Intelligent User Interfaces (IUI 2002), pp. 31–38. ACM, New York (2002)
4. Dansl. Desktop QR code. Available online at dansl: Interactive creativity: (2010), http://www.dansl.net/ (entry on March 8 2010), (retrieved September 2011)
5. Davies, S.: Still building memex. Communication of the ACM 54(2), 80–88 (2011)
6. Eismann, O.: QR Codes Scanning Distance (2012), http://qrworld.wordpress.com/2011/07/16/qr-codes-scanning-distance/ (entry July 16, 2011) (last accessed April 10, 2012)
7. Fun 2D. Retrieved from Nationwide Barcode (2012), http://www.nationwidebarcode.com/fun-with-qr-codes/ (last accessed April 10, 2012)
8. Hall, E.T.: The Hidden Dimensions. Anchor Books, New York (1990)
9. Hoven, E.V., Eggen, B.: The effect of cue media on recollection. Human Technology: An Interdisciplinary Journal on Humans in ICT Enviroments 5(1), 47–67 (2009)
10. Ishii, H.: Tangible Bits: designing the seamless interface between people, bits and atoms. In: 8th International Conference on Intelligent User Interfaces (IUI 2003). ACM Press, New York (2003)
11. Itext. IText library for creating and manipulating PDF documents (2012), http://itextpdf.com (last accessed April 10, 2012)
12. Jervis, M.G., Masoodian, M.: SOPHYA: A system for digital management of ordered physical document collection. TEI, 33–40 (2010)
13. Jones, W., Teevan, J.: Personal Information Management. University of Washington Press, Seattle (2007)
14. Kaywa (2011) Kaywa QR code, http://reader.kaywa.com/ (retrieved December 06, 2011)
15. Lemkin, P.F.: 2DWG meta-database of 2D electrophoretic gel images on the Internet. Electrophoresis 18, 2759–2773 (1997)
16. Sellen, A.J., Harper, R.H.: The Myth of the Paperless Office. MIT Press, Cambridge (2003)

17. Tyler, S.K., Teevan, J.: Large scale query log analysis of re-finding. In: Web Search and Data Mining, New York (February 2010)
18. StrokeScribe. StrokeScribe: Tool for barcode integration (2012),
    `http://strokescribe.com/en/qr-code.html` (last accessed April 10, 2012)
19. Walsh, A.: Quick response codes and libraries. University of Huddersfield, Computer Science. Library of Hi Tech News, Huddersfield (2009)
20. Xerox® Intelligent Barcode Printing Solutions, information,
    `http://www.xeroxofficesapsolutions.com/index.cfm#barcode`

User Needs for Enhanced Engagement with Cultural Heritage Collections

Mark S. Sweetnam[1], Maristella Agosti[2], Nicola Orio[3], Chiara Ponchia[3],
Christina M. Steiner[4], Eva-Catherine Hillemann[4],
Micheál Ó Siochrú[1], and Séamus Lawless[5]

[1] Department of History, Trinity College Dublin, Ireland
{sweetnam,osiochrm}@tcd.ie
[2] Department of Information Engineering, University of Padua, Italy
maristella.agosti@unipd.it
[3] Department of Cultural Heritage, University of Padua, Italy
nicola.orio@unipd.it, chiara.ponchia.1@studenti.unipd.it
[4] Knowledge Management Institute, Graz University of Technology, Austria
{christina.steiner,eva.hillemann}@tugraz.at
[5] Knowledge and Data Engineering Group, Trinity College Dublin, Ireland
seamus.lawless@scss.tcd.ie

Abstract. This paper presents research carried out in order to elicit user
needs for the design and development of a digital library and research
platform intended to enhance user engagement with cultural heritage
collections. It outlines a range of user constituencies for this digital library. The paper outlines a taxonomy of intended users for this system
and describes in detail the characteristics and requirements of these users
for the facilitation and enhancement of their engagement with and use
of textual and visual cultural artefacts.

1 Introduction

In recent decades, considerable investment has been made in the digitization
of cultural heritage collections. This has involved the creation of digital representations of cultural artefacts, and of associated metadata describing those
artefacts. As a result, a continually expanding volume of content is now available
to humanities scholars. However, these collections are typically monolithic and
can be difficult to search and navigate. Existing platforms for digital archives
tend either to provide basic tools for interacting with a range of collections, or
more complex tools that are tied to a very specific type of collection, or even
to a single archive. By contrast, the research presented in this paper is aimed
at delivering the next generation of cultural experience, by guiding, assisting
and empowering every user's interaction with a digital cultural collection. It is
intended not merely to provide a set of tools that are optimized to exploit a single cultural resource, or type of resource, but to deliver a fully featured digital
library and research platform that is "corpus-agnostic". This research has been

P. Zaphiris et al. (Eds.): TPDL 2012, LNCS 7489, pp. 64–75, 2012.

Fig. 1. 1641 and IPSA Collections accessible through the CULTURA interface

conducted as part of the CULTURA project[1] (Cultivating Understanding and Research through Adaptivity).

The design of this platform has required extensive engagement between humanities, information engineering and computer science research groups. The humanities researchers, based in Trinity College Dublin (TCDH) and the University of Padua (UNIPD), are both engaged with very different types of collection. TCDH have a historically significant textual corpus (the 1641 Depositions[2]), while UNIPD are providing a visual corpus of illuminated medieval astrological and herbal manuscript codices (the *Imaginum Patavinae Scientiae Archivum*, or IPSA[3]). Both the 1641 Depositions and IPSA collections are being made accessible via the CULTURA research platform, as depicted in Figure 1.

Designing the structure of a digital library that serves the needs of the users of these disparate collections has required scholars to gain an understanding of the contrasting methodologies employed by colleagues across different disciplines within the humanities. The object of this is not to arrive at a platonic ideal of "the humanities researcher" but to identify and exploit the commonalities in methodology and process that do exist while also supporting the different and distinctive desires of each discipline.

---

[1] CULTURA is co-funded under the 7th Framework Programme of the European Commission, URL: http://www.cultura-strep.eu/
[2] 1641 Depositions Website, URL: http://1641.tcd.ie/about.php
[3] IPSA Website, URL http://www.ipsa-project.org/

It is vital that the needs and requirements of the humanities researcher and community are addressed in the design of digital libraries. Rather than information engineers or computer scientists driving the design and development of a system that is subsequently subjected to user trials, effective design must elicit and incorporate user input at every stage of design and implementation. Successfully accomplishing this requires effective collaboration between researchers in the humanities and information engineering and computer science throughout the design, implementation and testing of a research environment.

This paper describes the outcomes of a process of investigating and identifying the commonalities in methodology and process across the two humanities research communities described above. From this process, a series of user community characteristics and requirements are derived. These characteristics and requirements form the basis of the design of a digital library platform which supports the distinctive desires of each community. Section 2 details related work and literature. Section 3 describes the approach taken during this research. This is followed by Section 4 which describes the community characteristics and Section 5 which reports on identified user needs. Section 6 reports a summary of the outcomes of this research and a discussion of some future work.

2   Background Literature

In addition to direct engagement with users, the literature has been consulted in order to identify accepted requirements in digital libraries and collections. Besides explicit research studies on understanding user requirements [1] research on user acceptance in the context of digital libraries can provide useful information [16]. This literature review confirmed the importance of a user-centered approach [6] to the development of the CULTURA environment, in order to ensure the system gains acceptance and delivers maximum benefit.

Key requirements for digital libraries that are consistently outlined in the literature are that the system should: (i) be easy to learn and easy to use; and (ii) deliver reliable search results [9, 10]. The design of the system, as well as the terminology used, should be clear, consistent and easy to understand [15]. Moreover, there exists a preference for visual-based interfaces that support users in finding information [9]. In particular, map-based visualisations have demonstrated value [12]. Constructing complex searches involving multiple terms has been identified as a key challange [4], and the use of visual interfaces to enable faceted searching of collections has been identified as an effective way of addressing this challange [14].

It is clear from the literature that different user groups will likely have different requirements for a digital library [4, 11, 17]. In particular, differences in the level of knowledge and experience with search functionality, and information communication technologies in general, have proven to be factors which influence users' needs, and the acceptance of a system [5, 7, 8]. As a result, a digital library should optimally cater for differences in user characteristics; for instance it should adapt to the different degrees of expertise and experience of users. Direct connections to other resources or digital libraries have also been found to be a feature desired by users [9]. The integration of heterogeneous information resources, as well as the development of services fostering and allowing cooperation among users has increasingly become a key factor [2]. Moreover, the consideration of previous user habits and practices is important for the design of a digital library or web portal [1] and familiar aspects of a new system should be highlighted in order to support users' acceptance [13].

Overall, the information that was derived from the literature on relevant aspects of user requirements provided a useful basis and complement for the engagement with user constituencies that was conducted as part of this research. The literature actually demonstrated a high level of correspondence with the results gained from the empirical work with users, as outlined in the following.

3   Research Approach

The aim of the research conducted by the CULTURA project is to stimulate and support the communities of interest which form around digital humanities collections. Based on interactions with users of the artifact collections, the following taxonomy was developed to describe the types of users who form these communities of interest:

- Professional researchers: established academics, experienced in the general area covered by the resource, but not necessarily with the specific content of the resource.
- Apprentice investigators: students at advanced undergraduate and postgraduate level. Some knowledge of the historical period and/or cultural context addressed by the resource.
- Informed users: researchers who are not professional academics but have knowledge of some aspects addressed by the resource
- General public: adults and children.

In order to characterize users and identify and address their needs, researchers used a combination of surveys and interviews, along with more sustained interactions with users of the resource. Surveys were carried out with users from both TCDH and UNIPD. Twenty-five students were surveyed by TCDH, 15 at undergraduate level, and 10 at postgraduate. At UNIPD, 21 undergraduate students and 49 postgraduate students were surveyed. Undergraduate students were taking courses specific to the disciplines of history and art history. Postgraduate users included representatives of these disciplines, as well as students who were completing programmes in digital humanities, and cultural heritage. These surveys were supplemented by detailed focus-group type interactions. In addition, user needs were discussed in detail with 16 users from the professional researcher category (UNIPD 6, TCDH 10).

Users were encouraged to focus on their needs, and on the benefits they would hope to gain from a well-designed and effectively implemented digital research environment. Initially this feedback was based on pre-existing web interfaces, but subsequently a prototype implementation of CULTURA was evaluated with users. Detailed engagement with users was carried out using a co-operative inquiry methodology to elicit and refine user needs. This action research methodology focuses on performing research with users rather than on users and is based on close and sustained cooperation between researchers and users (co-researchers) to develop a framework of inquiry and to explore their experiences and impressions of a research topic. In addition to eliciting the needs and desired benefits of users in each category, interactions with users enabled the description of a model research process for researchers using both of the content collections. This allowed a provisional set of use cases to be developed. The definition of these use cases is the beginning of an ongoing and dynamic process.

In the context of the corpus-agnostic approach adopted by the project, it is noteworthy that these use cases manifest a great deal of common ground. The key areas of interest – people, places, and events – can very readily be conceptualized in a way that ensures their relevance to the work of researchers across various disciplines. Naturally, not all use cases overlap, and it would be a poor service to humanities researchers if the resulting environment were to become the digital equivalent of a methodological Procrustean bed. Thus, in addition to their overlap, the use cases defined reflect the distinctive and specific needs of each discipline. Extensive work on the elicitation of user needs is essential for providing robust and useful ways of engaging with cultural heritage collections.

The following sections describe the findings of the user study in more detail, with a specific focus on the characteristics and requirements of each of the user constituencies described above.

4   Community Characteristics

Professional Researchers: These users are experienced researchers, with a strong publication record. They are the most demanding of the users, and are ruthlessly interested in tools that assist in the production and publication of research. They have little time for technological novelty unless it provides a demonstrable benefit to their work. In addition, these users are researchers who have developed particular methodologies, practices, and workflows that they are comfortable with. In many cases, these practices have been established over decades. This means that there is a significant degree of inertia to be overcome before these approaches and practices will be altered. Adaptability and customizability are key requirements for this group. A resource that has the flexibility to accommodate and enhance existing workflow practices is more likely to be used than one that requires a degree of "re-tooling" in order to make use of it. Those users know their data, and are accustomed to the task of synthesizing, analyzing, and exploiting it. As established scholars, they have a very clear sense of the importance of unmediated access to primary sources. For this reason, they tend to be skeptical about technological intervention, and are troubled by the idea that their view of the sources is being filtered or distorted in any way. For this reason, control and scrutability of any personalization and adaptation are crucial. By the same token, however, these researchers are well placed to appreciate the value added to a cultural heritage collection by tools that enable new ways of looking at that collection, or that make it possible to perceive patterns and features not previously recognized. Thus, these researchers expressed a keen appreciation of the potential of visualization techniques for opening up new avenues of inquiry. This constituency's familiarity with the source materials means that they are excellently placed to evaluate the usefulness and accuracy of the impression given by these visualization techniques.

The professional researchers interviewed also tended to have a clear and relatively inelastic view of research collaboration. This tended to be modeled on existing practice, and may include collaboration (or the sharing of resources and annotations) with students. In general, they demonstrated some ambivalence to a wider and less formalized collaborative framework.

Apprentice Investigators: These researchers are students who are beginning to acquire familiarity both with the research process and with the content of the heritage collections. They place high importance on tools that allow them to gain this familiarity as rapidly as possible. In particular, the ability to visualize collections was identified as an especially useful way for students to gain an overall understanding of the structure of the depositions themselves and of the events that they describe. Map-based visualization was singled out as especially useful. At the same time, these students indicated a detailed grasp of the potential benefits of other visualizations, and of the possible pitfalls associated with their use. The representatives of this constituency provided an excellent snapshot of the historian's dilemma about the use of personalization and adaptivity as a guide to their sources. Their low level of familiarity with the sources means that they are very well placed to appreciate the value of the sort of guidance that an adaptive resource would provide. However, some participants expressed concern that this guidance would effectively distort their understanding of their sources, and thus of the stream of historical events. All participants could see the potential value in this guidance, but expressed the need for building trust in the operation of the system. Control, scrutability, and transparency feature highly as essentials in developing user confidence in adaptivity and thus fostering its use.

These users place high value on tools that assist in general explorations of the collection. At the same time, they are as conscious as the professional researchers of the need to produce concrete research outputs. For this reason, they identified the ability to track interactions with the research interface and consistently to replicate and cite the results of these interactions as an essential element in building their confidence in using the system. Such tools would free them to explore the data without having to worry about noting the route that they are taking. The apprentice investigators interviewed evidence a commendable appreciation of the priority of primary sources. This understanding is exemplified in a number of ways. In the context of the 1641 Depositions collection, students felt it important that they have access to the digitized originals as well as the transcription. Additionally, users expressed this priority in their requirement for a robust and effective search mechanism that would allow them to reliably and quickly identify the items relevant to their interest at a particular moment. Users also expressed the need to be able to "drill down" from visualization to underlying source as readily as possible. In keeping with existing research, these apprentice investigators proved more enthusiastic about the possibilities of collaboration than their senior colleagues. [3,4] Some of the students interviewed have recent experience of carrying out a collaborative research project. They were able to identify ways in which CULTURA could have assisted that collaboration. They identified choice and control as the crucial elements of a collaborative process. They conceptualized possible methodologies for collaboration in terms of "following" as with Twitter[4], or the sharing facilities present in Diigo[5].

While these users have not yet developed a workflow to the extent that the professional researchers have, their interests, personality, and preferred work process are already being expressed. The choice of digital tools is an important part of their workflow. They expressed a strong preference that the project environment, in addition to providing tools for analysis and organization internally, should feature a high degree of interoperability. In particular, the ability to export annotations or aggregated sets of data based on the results of search queries was consistently identified as a highly desirable feature.

Informed Users: This constituency includes users belonging to relevant societies or interest groups, and cultural institutions. These informed users are, in some senses, a sub-set of the general public. Their knowledge of, and sometimes their interest in, the sources contained in the project environment is generally quite specific.

General Public: Characterizing this constituency presents a considerable challenge. It is large and diverse, and users will bring a very wide range of interests, technical abilities, and contextual and/or historical awareness to collections of cultural artefacts. In addition, attempts to get members of this group to imagine a resource, as distinct to responding to one that already exists, are fraught with challenges. For these reasons, we will be in a position to interact more usefully with this constituency when a more complete version of the project platform exists and can be tested.

School Children: School children are a particularly challenging constituency to address, but successfully supporting their engagement with historical sources offers the prospect of very rewarding gains. In the case of the 1641 Depositions it has been possible to engage directly with pupils and history teachers. Teachers were particularly enthused about the access provided to primary sources. This has the potential not only to educate students about the specific historical

---

[4] URL http://twitter.com/
[5] URL http://www.diigo.com/

context of the artefacts, but also to teach them about historical methodologies, and especially the use of primary sources. The language of the Depositions was identified as a serious obstacle to student engagement. Teachers – who are not generally specialists in this period – identified the need for supporting material that might guide them through the Depositions. They identified visualization techniques as being particularly useful in offering post-primary students a way into the Depositions.

5   User Needs

5.1   Common Results

Unsurprisingly, there is considerable overlap between the user groups surveyed. A number of user requirements were so basic as to be common to each of the constituencies surveyed, and others were shared by a subset of those constituencies. For all users, accurate search was identified as the one "make or break" feature for engaging with a cultural heritage collection. In the case of the 1641Depositions, particularly, the frustrations of searching over unnormalized data mean that users with experience of the website all clamored for the ability to search over a normalized corpus. In addition to early modern spelling variation, the metadata for the IPSA collection features a number of languages: Italian, Latin, and a variety of dialects. The ability to search for all variants of person and place names was identified as particularly important, the need for reliable full text search also had a high priority for all users. In addition, all users highlighted the usefulness of faceted search, that provided the ability to filter their content dynamically across all available categories of metadata. Users also emphasized the importance of being able to bookmark the results of this search and to retrieve or recreate a given filter configuration easily.

While direct searching is an important way of allowing users to engage with collections, it is only one of a number of methods identified as useful by users. Users from all groups identified visualizations as a potentially useful way of interacting with the collections. Map-based visualization was identified as being especially useful. Users displayed openness to other forms of visualization strategy, but highlighted the need for clear support material and instructions to extract the maximum possible benefit from such visualizations. This finding represents a shift from the attitude to visualizations noted in earlier research [17] and may be indicative of a wider disciplinary shift.

Professional researchers and apprentice investigators both highlighted the importance of being able to add in-line annotation to collection items. Users of the Depositions highlighted the importance of being able to select and annotate chunks of text, while IPSA users wanted to be able to annotate sections of images. Users also highlighted the usefulness of being able to link annotations together, both within a single document, and across documents. These annotations were identified as one of the most important sources of added value for these users. This finding goes beyond the focus on information finding that characterizes much research on user requirements [4, 11], and addresses the ways in which users can make the collection their own, enriching and enhancing their developing engagement with its contents. These categories of user are also aware of the value of undirected investigation of a source, of exploring (playing with) the available material, using searches and visualizations to explore its content. However, they are also familiar with the situation of not being able to recover a previously noticed fact or idea that is urgently required to complete a paper. For this reason, they emphasized the value of a log of their interactions with a resource within the research environment. They also felt that there would be considerable value in using this log along with a recommendations feature, to render its operations more transparent and accountable. Those users also highlighted the importance of being able to export annotations, aggregated depositions and other data. This requirement is motivated by considerations of data security and of workflow integration. Users, both academics and students, stressed the importance of being able to use the project environment not just as a one-stop-shop, but as part of a wider research process, the ability to export data for analysis elsewhere is seen as a key requirement. In addition, users of the 1641 Depositions expressed a wish to be able to export sub-sets of depositions (the results of a search or query, for example) as text files, for additional analysis, or simply for off-line reading. The same two groups of users stressed that it is essential for a research platform to make a contribution to producing published research. This concern means that they need easy access to the data underlying any of the visualizations generated by within the environment, and the ability to export that data for publication or further analysis.

Professional researchers, apprentice investigators and interested users emphasized the importance of the project environment supporting a variety of project structures. At the most basic level, all users from these groups wanted to be able to organize their bookmarks and annotations into separate folders. Members of the apprentice investigators group, reflecting on their experience with other sorts of technology, suggested that a tagging system would be a useful adjunct to a more rigid system of bookmarking. To support the variety of ways in which projects evolve, all users require the ability to copy existing bookmarks and annotations into new folders as well as being able to create a new project from scratch. These users all highlighted the organic way in which many academic projects develop, and the concomitant importance of being able to reconfigure the research environment accordingly.

The requirements outlined above are common to two or more of the user groups surveyed. In the following, the requirements specific to individual groups are outlined.

5.2   User Group-Specific Results

Professional Researchers: In addition to needs shared with other user groups, professional researchers outlined a number of specific requirements. These involved interoperability, collaboration, and, in the context of IPSA, tools to facilitate research on the relationship between images across time. For these researchers the resources initially contained in a given research environment are only part of a larger picture. Thus, they place a high premium on the environment's ability to accommodate and allow interoperation with other collections. These users have a specific understanding of how collaboration would best assist their work. Essentially, they see value in two types of collaborative research. They require a research environment to support collaboration with other researchers on a specific topic or project. Such collaboration is limited to that project, and is probably curtailed to a specific period of time. Secondly, they see the value of collaboration as a pedagogic tool, and believe that this potential would be best supported by the ability to share a specific sub-set of their annotations with students, and to see annotations by students that those students have chosen to share with them.

For art historians, there are two types of historical-artistic search that are especially valuable. Firstly, it is useful to follow the development of the illustration system of a specific text. It has to be possible to follow the evolution of a single image through the centuries. Iconography gradually changed over the centuries and it is highly desirable that researchers to be assisted in tracing this process of evolution. Secondly, an environment for users of these image collections should allow the researcher to provide information to describe the relations between images, possibly through contextual annotations. Ideally, these annotations could be used in the visualization of the links also for use of the other categories of users. The environment should also facilitate the generation of an image stemma codicum (a family-tree representation of different chains of derivation starting from a given image). These should visualize the connections between images and annotations. The stemma codicum should dynamically reflect the addition of new information by the professional researcher. A graphical way of making explicit the lack of a relation between images should be found, because stating that there is no relation between two images (even if they are visually similar) is another important scientific result. Those users would benefit from the possibility of recalling all the similar images that are present in the archive. It is important to note that, for this particular case, they are more interested in search functionalities that are based on textual descriptors, rather than on content-based similarity searches between images.

Apprentice Investigators: Controllable collaboration is of particular importance for apprentice investigators. These users have an understanding of peer-to-peer collaboration very similar to that expressed by users in the professional researcher constituency. However, they also envisaged considerable potential benefit from a solution that allowed them to "follow" more senior academics and to see such annotations as they choose to make public.

Informed Users: These users constitute a heterogeneous group, with a range of requirements. In general, however, they tend to approach such collections with quite specific queries in mind, which are less intensive than those that professional researchers might pursue, but, at the same time, are more detailed than would be typical of the general public. Often, these queries will be a part of a wider investigation, and speed, accuracy and straightforward accessibility are, therefore, particularly important. These users are informed in quite a different sense. They have limited interest in the specifics of the collections, but are keen to identify solutions that would assist in the showcasing and exploitation of their own content bases. Given the limited resources available to many of these institutions in the present environment, it is important to foresee ingestion and preparation of additional content bases, and the reuse of existing metadata as straightforward and efficient as possible.

General Public: These users typically have very little contextual information about the collections. They identified the need for accessible introductions to the collections, explaining the material they contain, and its historical context.

6   Conclusions and Further Work

User characterisation and the development of use cases is an ongoing process of refinement and enhancement. A co-operative inquiry methodology is central to the design and implementation of digital libraries that house humanities collections, and that make them available to the researchers that use them. Thus, as prototypes of the CULTURA research platform continue to be developed, researchers will work closely with end users to evaluate the usefulness of the environment. This process is crucial to the effective design of a research environment that will be genuinely and robustly useful for a wide range of users who come to the resource, and the collections that it makes available, with a range of experience, interests, and needs.

The research reported in this paper demonstrates the value of a design approach that begins with users. The user taxonomy that it presents provides a very useful model for digital libraries and research platforms that attempt to address the needs of a wide range of users. This is particularly relevant in relation to making cultural artefacts widely accessible. Tis research has generated an authentic set of general and user-group specific design requirements that could help inform the design of a range of resources for digital humanists and for collections of cultural artefacts.

Acknowledgements. The work reported has been partially supported by the CULTURA project, as part of the Seventh Framework Programme of the European Commission, Area "Digital Libraries and Digital Preservation" (ICT-2009.4.1), grant agreement no. 269973. The authors would like to thank all the CULTURA partners for the useful discussions on many aspects related to the user engagement with cultural heritage collections.

The authors are grateful to the anonymous referees for their very helpful comments.

References

1. Agosti, M., Crivellari, F., Di Nunzio, G., Gabrielli, S.: Understanding user requirements and preferences for a digital library web portal. Int. Jour. on Digital Libraries 11, 225–238 (2010)
2. Agosti, M.: Digital libraries. In: Melucci, M., Baeza-Yates, R. (eds.) Advanced Topics in Information Retrieval. The Information Retrieval Series, vol. 33, pp. 1–26. Springer, Heidelberg (2011)
3. Barrett, A.: The information-seeking habits of graduate student researchers in the humanities. The Journal of Academic Librarianship 31(4), 324–331 (2005)
4. Buchanan, G., Cunningham, S.J., Blandford, A., Rimmer, J., Warwick, C.: Information Seeking by Humanities Scholars. In: Rauber, A., Christodoulakis, S., Tjoa, A.M. (eds.) ECDL 2005. LNCS, vol. 3652, pp. 218–229. Springer, Heidelberg (2005)
5. Fields, B., Keith, S., Blandford, A.: Designing for expert information finding strategies. In: Fincher, S., Markopoulos, P., Moore, D., Ruddle, R. (eds.) People and Computers XVIII - Design for Life, pp. 89–102. Springer, London (2005)
6. Fox, E.A., Hix, D., Nowell, L.T., Brueni, D.J., Wake, W.C., Heath, L.S., Rao, D.: Users, user interfaces, and objects: Envision, a digital library. JASIS 44(8), 480–491 (1993)
7. Hong, W., Thong, J., Wong, W.M., Tam, K.T.: Determinants of user acceptance of digital libraries: An empirical examination of individual differences and system characteristics. Jour. of Management Information Systems 18, 97–124 (2002)
8. Hsieh-Yee, I.: Effects of search experience and subject knowledge on the search tactics of novice and experienced searchers. JASIS 44, 161–174 (1993)
9. Kani-Zabihi, E., Ghinea, G., Chen, S.: Digital libraries: What do user want? Online Information Review, pp. 395–412 (2006)
10. Kimani, S., Panizzi, E., Catarci, T., Antona, M.: Digital library requirements: A questionnaire-based study. In: Theng, Y.L., et al. (eds.) Handbook of Research on Digital Libraries, pp. 287–297. IGI Global, Hershey (2009)
11. Loizides, F., Buchanan, G.: What patrons want: supporting interaction for novice information seeking scholars. In: JCDL, pp. 427–428 (2009)
12. McIntosh, S.J., Bainbridge, D.: An integrated interactive and persistent map-based digital library interface. In: ICADL, pp. 321–330 (2011)
13. Nov, O., Ye, C.: Resistance to change and the adoption of digital libraries: An integrative model. JASIST 60, 1702–1708 (2009)
14. Suominen, O., Viljanen, K., Hyvönen, E.: User-Centric Faceted Search for Semantic Portals. In: Franconi, E., Kifer, M., May, W. (eds.) ESWC 2007. LNCS, vol. 4519, pp. 356–370. Springer, Heidelberg (2007)
15. Thong, J., Hong, W., Tam, K.Y.: Understanding user acceptance of digital libraries: What are the roles of interface characteristics, organizational context, and individual differences? Int. Jour. of Human-Computer Studies 57, 215–242 (2002)
16. Thong, J., Hong, W., Tam, K.: What leads to user acceptance of digital libraries? Communications of the ACM 47, 79–83 (2004)
17. Watson-Boone, R.: The information needs and habits of humanities scholars. RQ 34(2), 203–216 (1994)

Digital Library Sustainability and Design Processes

Anne Adams and Pauline Ngimwa

Institute of Educational Technology (IET)
Open University, Milton Keynes, UK
{a.adams,p.g.ngimwa}@open.ac.uk

Abstract. This paper highlights the importance of sustainability in digital library design processes and frames these arguments within current digital library forums and literature. Sustainability of digital libraries is analysed through an empirical study of 10 best practice digital library projects across three African countries (Uganda, South Africa, Kenya). Through a retrospective review of the projects design processes the paper focuses on the role of technologies / platforms (bespoke, open source, proprietary, web 2 and mobile) in sustainability of these systems. In-depth interviews from 38 stakeholders were triangulated against a documentary analysis and observational data and the findings integrated through a grounded theory analysis. The results identify the importance of flexibility in technologies that enable customization of educational digital resources to meet specific institutional and subject discipline needs. Comparative Evidence is presented that highlights poor sustainability when inflexible systems do not consider scalability or maintenance issues.

Keywords: Sustainability, design processes, flexibility, African HE, case studies.

1   Introduction

The issues of digital library (DL) sustainability are essential to ensure that advances in DL design and development have a greater impact upon practitioners and current practices. However, reviewing this concept is problematic since it relates to a broad range of issues from digital library infrastructure to interface as well as those related to social, organizational and policy issues. A good starting point might be to historically place the concept within this digital library forum in relation to theory and practice.

TPDL and its predecessor ECDL have always presented technical and theoretical advances in the digital library field. However, this forum also highlights the value of advancing our understanding of practice for practitioners. This requires a careful merging of different perspectives, disciplines and methods as well as taking a reflexive account of our approaches to research. Traditionally digital library interface and usability studies have tended to take a practice based approach tending to focus on applications in relation to user experience, tasks and context. In comparison technical research has tended to focus on the issues of architecture and infrastructure often presenting novel solutions to problems in information access, organisation and management. Over recent years the fields of digital humanities and research data have merged these two approaches in

P. Zaphiris et al. (Eds.): TPDL 2012, LNCS 7489, pp. 76–88, 2012.

reference to their particular fields of study. It is within these two fields that we see a growing emphasis on sustainability in both the development of innovative technologies and their application within practice based contexts [1, 2, 3, 4].

When applying innovative technical developments within a practice based context, issues of scalability and sustainability are paramount in the design process. Of particular importance are the issues of DL applications, platforms and infrastructure. It is in these decisions between developers, user groups and organisational stakeholders that the cross-over between theory and practice can be made through the application of innovation. As an introduction to this paper's empirical research, we present a preliminary content analysis (See Table 1) of applications / platforms used in papers published between 2007 and 2011 within the ACM JCDL and the TPDL/ECDL forums. It should be noted that many papers focusing on DL practice based contexts without referencing specific applications / platforms were not considered in this analysis thus skewing the data towards technical papers. Also papers reviewing several technologies were placed according to the category they first fell into working top down according to the order in table 1. Although skewed, this analysis does present an interesting overview of digital library technologies in design processes within which to place this research.

Table 1. Literary analysis of DL applications appearing in JCDL & TPDL/ECDL full papers between 2007 and 2011

Applications / Platforms	JCDL	TPDL/ECDL
Bespoke	67	73
Open source	9	14
Proprietary	2	7
Web 2.0/virtual world e.g. second life	3	6
Mobile technology	1	2

As we can see from table 1 there is a far greater emphasis on bespoke technologies (e.g. CSM, SIMDL, MHMpara, TIME) than open source systems (e.g. greenstone, Dspace, Fedora). Although skewed towards technical papers the analysis shows that there are less papers detailing how to adapt open source systems for particular practice based contexts. This could highlight a gap in the knowledge base since system adaptations for practice based contexts are important for sustainability [2].

Ultimately, within the current economic climate effective design processes producing sustainable digital libraries in practice is paramount. Digital resources improve the quality of teaching and learning [5] and within African HE enable greater economic revitalisation. However there are problems with the sustainability, innovation and leadership in the development of these resources [6, 7, 8]. For these reasons this paper presents results from several case studies reviewing digital library design processes across three African countries.

2   Related Work

Sustainability of designs especially within interaction technologies has been an area of interest for sometime. Blevis [3] for example has discussed sustainable interaction design

(SID) where sustainability is presented in terms of design values, methods and reasoning. Sustainability in this context is linked with the concept of a viable future (i.e. in environment, public health, social justice and equality) where Blevis argues that in order to promote sustainability, we need to link invention and disposal, promote renewal and reuse, promote quality and equality, and use natural models among other strategies. DiSalvo et al. [4] share a similar perception in their paper of sustainable HCI. For example they lay a lot of focus on technology solutions to problems of sustainability (e.g. by improving the material design of technologies, encouraging sustainable behaviour or improving the efficiency of our everyday activities). They argue that we can achieve sustainability in design (i.e. in mitigating material effects of software/hardware) and through design (i.e. by influencing sustainable lifestyles or decision-making).

Existing work related to sustainability in digital libraries tends to concentrate on economic and social resources; preservation of digital content; and collection development. It appears like little attention is given to the issue of sustainability from a digital library design perspective. Hamilton [2] for instance considers the relevance of the preservation of hardware, software and the operating systems on which the software is run in order to ensure sustainability of digital materials. He further notes that economic sustainability can be ensured through factors such as quality of the product, digital library project branding and having champions who have influence on budget holders as well as having funding options. While all these factors contribute to overall digital library sustainability, the one issue closest to design considerations is the quality of the product. In discussing product quality, Hamilton takes a usability perspective to technical issues such as extensive testing and piloting to ensure users are not distracted from product value by poorly designed products.

McArthur et al. [1] have also considered sustainability issues within the context of the NSF-supported National Science Digital Library (NSDL) program and have highlighted two issues: availability and preservation of digital content for users; and financial and social resources required for long-term sustainability. As in the case of Hamilton, this work highlights the importance of sustainability issues in digital library projects, however, there is little in this paper to inform us on the practicalities of sustainable design for such projects.

Ashraf and Gulati [9] have examined sustainability of digital libraries from preservation and collection development where they argue that a sustainable digitization program needs to be fully integrated into traditional collection development strategies. They too consider factors related to financial sustainability.

The above highlighted studies provide broader insights of how we can ensure sustainability of digital library projects. This suggests existence of a gap in knowledge of how we can we can ensure sustainability through the design process for these projects. However one study conducted by Norberg et al. [10] came close to suggesting ways of achieve sustainability through an iterative design process. The purpose of this study was to demonstrate the value of usability studies following an iterative design process in order to create more sustainable and user centered digital libraries. They found that serving needs of multiple users is an iterative process involving an on-going dialog with users. The focus on the user-centeredness in this study is important to note as it alludes to the notion of designs centering on values.

According to Cockton [11], the concept of value is a unifying concept for design whereby the intended value of a digital artefact provides a focus for research, design and evaluation, and a common ground for all stakeholders including users of the artefact. He argues that this conceptualisation of values needs to be reflected in the development framework from the initial identification of the product opportunities to the installation and operation of digital products and services.

As we can see from this research literature there is a growing importance in understanding the sustainability issue within the digital library design process. Despite all the reviewed studies having examined broad issues of sustainability, none of them focused specifically on gaining a rich understanding of the phenomenon itself and what is needed to promote sustainability. To effectively understand the multifaceted nature of sustainability and how to achieve it, there is a need to review these issues across different discipline, institutional and national contexts.

3   Digital Library Case Studies

This paper presents evidence from three case studies of universities in Uganda, South Africa and Kenya focusing on a retrospective review of ten digital library project design processes. These projects were determined as 'best practice' projects, based on criteria that considered the presence of: (i) technology innovation in the library and learning programs, and (ii) policies linked to the operations of the university. We later added a comparable project which represented weaker presence of innovative technologies. Table 2 presents a summary of the 10 projects studied and the distribution and roles of the 38 participants. Within this study there are no bespoke or proprietary digital library examples as within these institutions these projects did not present any of the elements highlighted for 'best practice' identified.

Table 2. Summary of case study projects and participants

		S. Africa	Kenya	Uganda
Technology projects	DSpace	1	0	4
	Web-based	2	1	0
	Mobile technology	0	1	1
	Total	**3**	**2**	**5**
Key stakeholder participants	Digital librarians	4	1	6
	E-learning technologies	1	1	1
	Academics	4	1	6
	Students	3	1	7
	Community projects staff	0	1	1
	Total	**12**	**5**	**21**

3.1   Uganda Case Study Projects

This case study had five projects. In addition to the presence of innovative technologies, there were policies that supported the design process of these digital library projects. Below is a brief description of each of these projects.

i) Clinical mobile digital library project: This was using Personal Digital Assistants (PDAs) designed for rural clinicians by university academics, librarians, Ministry of Health and project sponsors. This DL contained digital resources from the university library including international peer-reviewed journal articles which were relevant to the rural clinicians to support their continuing education and clinical practice.

ii) College knowledge management system: This project was collaboratively designed by academics and digital librarians. This involved the use of DSpace application and librarians' knowledge management skills to create a repository of well organized and accessible college knowledge output while creating a virtual presence for the college.

iii) Problem Based Learning (PBL) digital resource support system: This was specially developed to support students in health related disciplines. This comprised of a digital collection partly supported by DSpace application, appropriate technology infrastructure and a dedicated information intermediary to facilitate the access and utilization of this service. Figure 1 below shows PBL students using this service:

Fig. 1. Students using PBL supported facility

iv) Digitized music collection: This project was collaboratively designed and developed by library digitization experts and music academics. This was housed in the library and was intended for the Music department students and academics. In addition, it was aimed at preserving Ugandan music and open to members of the public. Digitized files were hosted in the Institutional repository (described below) which meant that they were organized using DSpace application. DSpace supports a range of digital content including music. It also supports open access to resources.

v) Institutional Repository: This was developed by the library to host institution's academic and research output (e.g. research papers, theses and dissertations) and learning resources i.e. the music files described earlier. DSpace application was used to organize and manage the collection. The project was a response to the Open Access Movement which encourages free access and sharing of resources.

3.2 South Africa Case Study Projects

There were three projects in this case study. Just like in Uganda, the design process in the projects was supported by innovative technologies as well as policies. Below is a brief description of these projects:

i) Digital library supported by Web 2.0 applications. This was developed by digital librarians. They designed it around web 2.0 applications in order to reach their younger student population who were no longer using library resources but were very active in the virtual world space. They also used a virtual game to design an information literacy program in order to make it more appealing to this student population (figure 2).

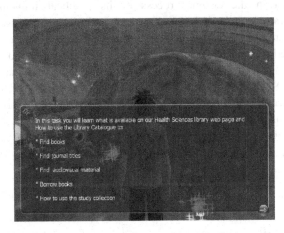

Fig. 2. Literacy program game in library's innovative technologies project (source: [12])

ii) A comprehensive virtual learning environment (VLE) with digital library resources seamlessly integrated into the system. This was developed jointly by librarians, academics and e-learning technologists.

iii) Institutional repository: this was similar to the Ugandan one described earlier.

3.3 Kenya Case Study Projects

This case study had two projects. They both had presence of human-related factors but unlike the other two case studies, we found a weak policy framework which affected the impact and sustainability of these projects.

i) Educational digital library: This was a learner centered digital library whose success depended on the collaboration between digital librarians, academics and e-learning experts in the design process.

ii) Community based agricultural knowledge management system. This project involved researchers developing an agricultural knowledge management system for two rural farming communities, using a participatory design approach and mobile devices. The team consisted of UK experts in telecommunications, renewable energy sources, sensor technology, education and design, working with local experts at the case study university and local communities.

4   Data Collection and Analysis

This paper presents 38 in-depth interviews (audio recorded for 40-60 minutes) over a period of four months (September-December, 2009) with academics, information professionals, library designers, e-learning technologists, community project staff and students involved in the design process or its practical implementation. This qualitative data was triangulated with a document analysis, and observational studies.

Participants were purposively sampled according to their participation in the projects. Data analysis used a grounded theory approach [13] selectively following Charmaz's [14] simplified version of open and selective coding. Transcribed data coded line-by-line (with NVivo 8) identified open followed by selective coding which was facilitated by collaborative coding sessions, supporting inter-rate reliability, clarifying emerging categories and reducing subjectivity. The findings have points illustrated with verbatim extracts from participants identified by roles not as individuals. Attempts to anonymise individual, social groupings and institutions were made to reduce the potential for privacy invasion.

5   Results

The findings identified several aspects of the DL design process that support or inhibit sustainability (e.g. stakeholder and user engagement, policies, usability). The important role of policies in supporting this sustainability has been discussed in a separate paper [15]. However, one factor that is often overlooked is the issue of technology application / platform, which will be the focus of our analysis for this paper. The findings revealed that two types of application / platforms (i.e. Open Source Software and Web 2.0) supported sustainability of digital library projects across the three case studies. Conversely, where these technologies were absent, the projects ended up being unsustainable. The main reason why these technologies supported sustainability was identified as there flexibility enabling systems to meet the intended practice based projects' needs. Below these findings illustrate how flexibility ensured projects' needs were met leading to sustainability.

5.1   DSpace Application

One of the characteristics of DSpace Open Source Software is that it allows customization. It also supports any form of digital content including text, images, moving images, mpegs and data sets[1]. This flexibility enables it to accommodate specific needs and requirements.

In this research, five projects were identified as utilizing DSpace either to meet the broader institutional needs or to meet specific learners' needs. Two institutional repositories (i.e. in Uganda, 3.1 v, and in South Africa, 3.2 iii, case studies) and a College knowledge management system (3.1 ii) were effectively tailored, using DSpace, to meet specific institutional needs. For example, both institutional repositories depended on academics and students, without prior knowledge of content management methods, to be able to deposit their academic output in the system. The flexibility in DSpace allowed them to do this.

> *"We have got the students doing it themselves now, and getting the materials and uploading them to the UPSpace. So they have a collection of architects they are working on, they get the stuff digitised and uploaded on to the UPSpace".* (Digital Librarian 4-SA)

Because DSpace is open source, it supported the Open Access aspect of these institutional repositories which could be accessed without security restrictions.

> *"...it is a project where we are compiling research output from the lecturers and students at Makerere University into an online database [DSpace] that can be accessed even by the general public."* (Librarian 3-UG)

Other projects also highlighted how these digital library systems were flexible to the learners' needs and could also be tailored to meet the specific discipline needs of music and health students. These included the Digitization of music collections (3.1 iii) and Problem Based Learning (PBL) digital resource support system (3.1 iv) for music and health students respectively. DSpace was used to make the projects relevant to the learning needs of these students. For example the music project required an application that had the capability for managing audio files which were necessary for the students' music analysis class. DSpace is flexible enough to handle such audio files and therefore was used to meet this need:

> *"...we are planning to put the audio files in the DSpace, because it has the capability of holding audio files."* (Librarian 4-UG)

The PBL project supported PBL students studying within local communities in a program called Community Based Education and Service (COBES). Students depended on local knowledge to support their learning needs and required access to a digital collection that integrated different local content. DSpace was used to organize this collection within the institutional repository to meet these specific learning needs:

---

[1] http://www.dspace.org

"...resources have been very helpful in supporting PBL ... when the lecturer gives them a topic... the first place to go to is the computer lab, because they know there are digital resources ..." (Librarian 1 -UG)

5.2   Web 2.0 Applications

Web 2.0 applications similarly create this flexibility and leverage collaboration between people where even non-technical users can participate in systems' customization to perform specific tasks [16, 17]. Web 2.0 applications were again used to tailor resource to meet user needs in the Web 2.0 Digital library project in South Africa (3.1 i). Digital librarians in this project noticed a problem with their younger student population who were utilizing library resources less often than expected. Almost all of them were identified as being actively involved with social networking technologies to acquire and share knowledge. Consequently, the library positioned itself to reach out to these students by utilizing a range of Web 2.0 applications, i.e. facebook, blogs, U-tube and flickr. They then integrated the functionality of recommendations and collaboration into the digital resources.

"We felt that many of our library users are involved in all these web 2.0 applications, that's what they are using, they are using less and less the library databases, ... but they are googling, and they are using all these tools, and we said we have got to reach them. Take the library databases, the library articles, library tools to them by using these web 2.0 tools and that's what motivated us to do that." (Librarian 1-SA)

They also developed an information literacy program adapted from a virtual game within a Web2.0 context (see figure 2) and piloted it among the students. The game, raised students' engagement and re-visits to the library resources:

"...they did a pilot with a number of undergrad students at the main campus and at the medical library, they developed a game just at a pilot to train them in information literacy, and the students could play and while playing this game they could learn. ... That was very successful, we got very good feedback from the students ... the students were very very excited about this..." (Librarian 1-SA)

The librarians also created blogs to engage directly with these users. They encouraged students to post queries about their assignments difficulties and put up physical notices in the library (see Figure 3) encouraging them to use the blog:

5.3   Absence of Flexible Technologies in the Design Process

The research findings also established that the absence of these technologies contributes to unsustainable digital library projects. This was especially apparent in the Kenyan case study. The educational library project particularly highlighted a

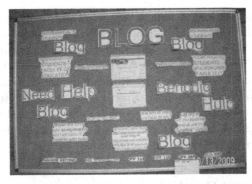

Fig. 3. A blog's notice in the library (Source: Research fieldwork)

disconnection between technology innovation, user centeredness and sustainability of the digital resources developed. The system was designed around web-base resources of international databases and was intended to meet the needs of students and academics. However, research findings established that this system avoided use of flexible technologies. This in turn meant that individual departmental and individual users' needs were not always met. For example, academics and students needed personalized digital resources to support their teaching and learning needs, but were not using this generic library as it was not meeting their needs.

In order to personalize their information needs, these users were designing their own personal systems. For example, academics created personal portals where they uploaded relevant digital resources to support students' needs:

> "... A few lecturers out of their personal initiatives have their personal portals..." (Students -KEN)

Another example is where some students wanted local information which they could share. As this was not provided for, they got together with their lecturer and created a web resource which used a Web 2.0 application allowing customization of local information and knowledge sharing:

> "It is the students who came up with this idea and said "why don't we build our own site?" So we constructed our own site with these students who are actually agriculturists. .. It's called Try-African-Food." (Academic 1 -KEN)

According to the students, developing such personalized portals enabled them to customize agricultural research information so that others would benefit:

> "... So we are making it a lot easier for people to access the same information ... repackaging to reach the clientele we want to reach. Lets take an example of our local food... this kind of food has some nutrition benefits, and you repackage this into various information i.e. how to cook it etc." (Student -KEN)

This personalization happened without support of the library systems or the skills of the librarians. Although the institution had skilled librarians and academics, they were unable to collaborate and come to a joint understanding about the needs to be met by the system. The resultant situation was that users found their own ways of dealing with information sharing needs. However, these approaches often lacked wider university impact and were unsustainable due to limited resources (i.e. only a few individual students) to maintain them. For example, the student's web resource has since become deactivated after these students, who were in their final year, had left leaving no one to take up the portal maintenance.

6   Discussion and Conclusion

Borgman et al. [5] have emphasized that educational digital resources are critical in improving the quality of teaching and learning. The findings from this research identified the importance of flexible technologies that enable customization of educational digital resource projects to meet specific institutional and subject discipline needs. In particular the results highlight the importance of Web 2.0 applications and DSpace Open Source Software in practice based application of digital libraries. Although, we are aware of limitations of systems such as Web 2.0 it must be noted that the best practice projects identified across Africa adopted flexible technologies because they found them to be institutionally sustainable and providing the desired impact (i.e. user-centeredness). It is this flexibility that was crucial in the customization process. Bardzell [16] uses the concept of a 'quality of pluralism' to describe certain design artefacts which *"resist any single, totalizing or universal point of view"* [16 p.1305]. The flexible applications detailed in these projects appear to have some resemblance with Bardzell's description. By being flexible enough to customize the projects to intended needs, they dissolved any universality or generalization of previously existing systems such as web-based digital databases. In comparison the Kenyan digital library project showed how inflexible academic intuitional systems will be replaced by home grown departmental resources. However, this then leaves these bespoke systems poorly maintained as students and academics leave institutions. Although these systems may appear to be adapted to user needs there scalability and thus sustainability is limited. The implication here is that designers of such systems interested in value-centered designs [11] should consider using these technologies.

Sustainability has always been a crucial issue for Africa, as it is growing in importance for the rest of the world. Because of the developing and under-resourced nature of African HE, digital libraries present not only an enabler for greater economic revitalization but also, through limited resources, encounter many barriers with connectivity and sustainability. Unfortunately, libraries in African HE have been criticized for not showing appropriate leadership and commitment in adopting innovation [6]. In order for these libraries to make a positive and sustainable contribution to teaching and learning, in African HE as well as globally, they must be designed innovatively by integrating the instructional functionality in the design process [7]. The design process must also involve key players (i.e. academics, librarians and students) and their practice based needs and contexts [8]. This is an environment that has for a long time been concerned with sustainability of projects which are often funded by external sources and are thus expected to find sustainability solutions long after the seed funding has been exhausted. This research highlights how flexible systems can economically support effective and innovative design processes and produce sustainable digital libraries that fit practice and user needs. Ultimately, within the current economic climate effective design processes producing sustainable digital libraries in practice is paramount.

This study reviewed the issue of sustainability using examples from Africa. It would be interesting to see how this compares with DL projects in other regions. The brief analysis of studies outlined in the introduction of this paper reflects a scarcity of DL projects using adaptable/flexible systems (e.g. Open Source and Web 2.0). From our studies we identified these systems were used to increase sustainability. However, further research is required into the life of DL projects using bespoke applications to clearly ascertain the sustainability of these systems.

Acknowledgments. This research was supported by an Open University studentship. We would like to thank our participants in South Africa, Kenya and Uganda.

References

1. McArthur, D.J., Giersch, S., Burrows, H.: Sustainability issues for the NSDL. In: Proceedings of the 2003 Joint Conference on Digital Libraries (JCDL 2003), p. 39. ACM, Houston (2003)
2. Hamilton, V.: Sustainability for digital libraries. Library Review 53(8), 392–395 (2004)
3. Blevis, E.: Sustainable interaction design: invention & disposal, renewal & reuse. In: Proceedings of the SIGCHI Conference on Human Factors in Computing Systems (CHI 2007), pp. 503–512. ACM, New York (2007)
4. DiSalvo, C., Sengers, P., Brynjarsdóttir, H.: Mapping the landscape of sustainable HCI. In: Proceedings of the SIGCHI Conference on Human Factors in Computing Systems (CHI 2010), pp. 1975–1984. ACM, Atlanta (2010)
5. Borgman, C.L., Gilliland-Swetland, A.J., Leazer, G.L., Mayer, R., Gwynn, D., Gazan, R., Mautone, P.: Evaluating digital libraries for teaching and learning: a case study of the Alexandria Digital Earth Prototype (ADEPT). Library Trends, Special Issue on Assessing and Evaluating Library. Services 49, 228–250 (2000)
6. Manda, P.A.: Access to electronic library resources and services in academic and research institutions in Tanzania. In: Rosenberg, D. (ed.) Evaluating Electronic Resource Programs and Provision: Case Studies from Africa, and Asia, International Network for the Availability of Scientific Publications (INASP), Oxford (2008)
7. Dong, A., Agogino, A.: Design principles for the information architecture of a SMET education digital library. In: Proceedings of the JCDL 2001, pp. 314–321. ACM Press, Roanoke (2001)
8. Marchionini, G., Plaisant, C., Komladi, A.: The people in digital libraries: Multifaceted approaches to assessing needs and impact. In: Bishop, A.P., House, N.A.V., Buttenfield, B.P. (eds.) Digital Library Use. MIT Press (2003)

9. Tariq, A., Anand, G.P.: Digital Libraries: A Sustainable Approach. In: Ashraf, T., Sharma, J., Gulati, P.A. (eds.) Developing Sustainable Digital Libraries: Socio-Technical Perspectives, pp. 1–18 (2010)
10. Norberg, L.R., Vassiliadis, K., Ferguson, J., Smith, N.: Sustainable design for multiple audiences: The usability study and iterative redesign of the Documenting the American South digital library. OCLC Systems & Services 21(4), 285–299 (2005)
11. Cockton, G.: Designing worth is worth designing. In: Proceedings of the 4th Nordic Conference on Human-Computer Interaction: Changing Roles, NordiCHI 2006 (2006)
12. Van Wyk, J.: UP Library Service's Web 2.0 Journey. Knowledge Management Practitioners Group of Pretoria Meeting. CSIR Knowledge Commons, Pretoria (2009)
13. Strauss, A., Corbin, J.: Basics of Qualitative Research: grounded theory procedures and techniques. Sage Publications, London (1990)
14. Charmaz, K.: Grounded theory. In: Lewis-Beck, M.S., Bryman, A., Liao, T.F. (eds.) The Sage Encyclopedia of Social Science Research Methods, pp. 440–444. Sage Publications, London (2004)
15. Ngimwa, P., Adams, A.: Role of policies in collaborative design process for digital libraries within African higher education. Library Hi. Tech. 29(4), 678–696 (2011)
16. Bardzell, S.: Feminist HCI: taking stock and outlining an agenda for design. In: Proceedings of the SIGCHI Conference on Human Factors in Computing Systems (CHI 2010), pp. 1301–1310. ACM, Atlanta (2010)

Creation of Textual Versions of Historical Documents from Polish Digital Libraries*

Adam Dudczak, Miłosz Kmieciak, and Marcin Werla

Poznań Supercomputing and Networking Center,
ul Z. Noskowskiego 12/14, 61-704 Poznań, Poland
{maneo,milosz,mwerla}@man.poznan.pl
http://dl.psnc.pl

Abstract. This paper describes the results of initial work aimed at increasing the number and improving the quality of textual versions of the historical documents available in Polish digital libraries. Digital libraries community is missing tools that integrate existing digitisation workflow with customizable OCR engine and crowd–based text correction, this paper describes work on providing such a solution. Apart from today's state of the art in this field, this paper includes a description of the Virtual Transcription Laboratory (VTL) prototype, a crowdsourcing platform that utilize the Tesseract OCR engine. The last chapter outlines results of the prototype's evaluation on real life dataset of historical documents from the IMPACT project. Results prove the applicability of the proposed solution as an enhancement of the digitisation workflow.

1 Introduction

Recent years witnessed a very intensive development in the field of digital libraries in Poland. Users have gained access to various kinds of resources including historical and modern books, manuscripts, articles and newspapers as well as iconographic materials. In order to facilitate the use of those resources Poznań Supercomputing and Networking Center has created the Polish Digital Libraries Federation[1] (DLF) portal. The portal offers an unified access to information about digital objects, gathered via the OAI–PMH protocol from more than 90 existing digital libraries in Poland [1]. As the searching mechanism in DLF is based on metadata, they offer only a selective view on the digital objects' content. In order to increase the retrieval capabilities, various solutions have been proposed, e.g. aimed at normalization and semantic enrichment of metadata records, as described in [2]. This work explores a parallel approach that facilitates the creation of textual versions of the historical documents available in Polish digital libraries in order to increase the number and improve the quality of textual versions. Full–text version can increase the retrieval precision and

---

* Presented results were developed as a part of PSNC activities within the scope of the SYNAT project (http://www.synat.pl) funded by the Polish National Center for Research and Development (grant no SP/I/1/77065/10).

[1] http://fbc.pionier.net.pl

P. Zaphiris et al. (Eds.): TPDL 2012, LNCS 7489, pp. 89–94, 2012.

recall, but it can also be used as a base for various digital humanities research including linguistic analysis, or creation of digital editions. Whether it will be a full–text search engine or the creation of transcript of a historical document, one needs to assure that full–text version of the object exists and is of a sufficient quality.

This paper is organised as follows. First chapter identifies existing solutions including commercial and open source OCR software and crowdsourcing text corrections platforms. In the next chapter the prototype of the Virtual Transcription Laboratory is described. The last chapter of this paper describes results of an experiment conducted on top of ground truth data prepared by PSNC in the IMPACT[2] project.

2   Description of Existing Solutions

The results of the survey from [3] provide information about the digitisation process in the year 2010 conducted by 26 institutions that constitute the Polish digital libraries network. As reported, only around 41.6% of digitised resources that can be considered as text had been processed with OCR engine. In most cases, the full text retrieved from OCR is randomly sampled and evaluated by librarians. If the recognition quality is poor, two correction scenarios are possible: customisation of recognition engine's parameters or manual correction. However, as the survey's results show, the correction is performed very rarely due to the lack of technical knowledge regarding OCR engines or the high cost of the manual correction process. Therefore, new tools and methods are required in order to enhance the digitisation workflow, e.g. include custom OCR software training and crowd–based text correction.

The most common OCR software used by Polish librarians (as stated in [3]) is ABBYY FineReader[3] which is perceived as an industry standard in this field. FineReader can be customised e.g. in order to train the recognition engine for a particular type of text in two ways. In the first case, customisation is conducted interactively by the user, so it cannot be automated. In the second case, the core of the recognition engine is trained by the ABBYY company itself on the basis of provided documents. Although this results in a better quality, it requires paying additional fee and applies only to a particular software version and setup.

In order to get better results and reduce customisation costs, different institutions that digitise similar historical documents might share the effort of the OCR software customisation. Although use of existing commercial software for this purpose might be hard, this is not the case for the open source projects like Tesseract, OCRopus and Gamera. Each offers layout analysis, training capabilities and rich output formats (e.g. hOCR), as well as customisability. Among these three, Tesseract is still gaining popularity, mainly because of its maturity and support for most of the European languages. It is also used as the recognition engine in this work.

---

[2] IMProving ACCess to Text, http://digitisation.eu

[3] http://www.abbyy.com

Successful crowdsourcing projects prove that correction can be easily handled by a group of volunteers, in particular researchers, hobbyists and scholars that are interested in work with particular manuscripts or printed documents. Examples of such projects are Transcribe Bentham, Trove Newspapers, Distributed Proofreaders and IBM Concert [4]. However, among these projects only IBM Concert makes a direct use of the OCR software. It guides users through the correction process of a OCR output, passing results back to the recognition engine as training data. Note that in case of the biggest projects like Trove Newspapers and Distributed Proofreaders, documents need to be processed locally prior to the submission to the service.

3   Prototype of Virtual Transcription Laboratory

On top of the analysis of existing OCR software and crowdsourcing tools, a prototype of Virtual Transcription Laboratory (VTL) has been created. The proposed system enables users to create an annotated full–text version of historical documents. In order to reduce costs of such work, VTL gives access to the OCR service with tools allowing to create custom recognition profiles and a Web portal which simplifies correction and creation of the document's rich transcript. In current setup, the VTL application bases on Tesseract, but its architectural design allows to extend it by other software, either open source or commercial one.

Customisation of Tesseract consists of creating new recognition profiles specific to a particular language, its characters set and font. Preparation of data for a new recognition profile has to be done manually. In order to ease the profile creation process, additional tools have been developed: the Cutouts web application and the Page Generator toolkit. The former application allows to create a mapping between glyphs and the character set for a particular font and language. Given an image of a scanned page and the recognised glyphs from the Tesseract's output, the user iterates over glyphs and corrects the initial recognition result. It may contain false positive and false negative errors, as well as inaccurate pattern matching error in terms of bounding area or corresponding character. Corrected data is stored on the server side, it contains glyph patterns, character mappings, binarization thresholds and coordinates of the bounding areas, composing ground truth datasets for given pages. Datasets are processed later on with the Page Generator toolkit in order to create pairs of denoised images and expected correct recognition results. These pairs can correspond to the original pages or can be artificially created basing on provided text, in order to control the vocabulary or distribution of characters that are present in the document. Eventually, this data feeds the recognition profile.

The interface of the transcript editor from the VTL Web portal is presented in Figure 1. Users transcribe images from projects they have created or projects published by other users of the portal. Projects are managed by their owners who can add or remove images. For each image, transcription is made from scratch or, optionally, images can be processed with a custom OCR profile. This

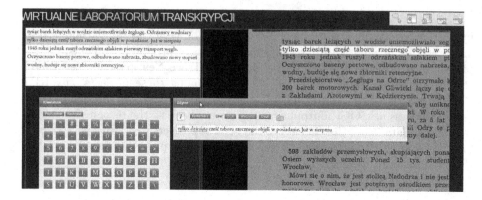

Fig. 1. Interface of Virtual Transcription Laboratory

results in content and coordinates which can be selectively imported as long as the recognition quality is sufficient. Alternatively, an existing transcript can be imported into project. The owner may decide to make their projects publicly available for the purpose of crowd–based correction. As all changes made to the transcripts are tracked, selected modifications can be discarded, if the project's owner finds them inappropriate. Results of the transcription can be exported from VTL portal in hOCR format at any time.

4   Verification and Results

The proposed solution has been verified in order to prove its applicability to full–text content creation for the historical printed documents. The verification scenario consisted of the following steps:

1. Choosing a dataset from a particular historical period consisting of ground truth and images available in Polish digital libraries;
2. Selecting and retrieving training data from the gathered ground truth;
3. Creating the custom recognition profile related to datasets' historical period;
4. Importing the OCR results into VTL in order to create transcripts;
5. Transcript's accuracy evaluation based on the gathered ground truth.

In the first step, a dataset was selected from the IMPACT data[4]. It contained 471 images and corresponding ground truth of 23 Polish documents created between 16th and 18th century. Pages were scanned at resolution of 300dpi and stored using the lossless TIFF file format in grayscale or RGB color space. Ground truth included information on full text content of the image (with expected characters accuracy of 99.95%) as well as coordinates of paragraph regions. In addition, the ground truth included glyphs coordinates and font type indication for 288 images, which can be used as training data.

---

[4] http://dl.psnc.pl/activities/projekty/impact/results/

The given dataset can be easily transformed into the form of Cutouts output and subsequently processed with the Page Generator toolkit. Based on the provided information, a dataset of 288 pages was filtered out by the font type (fraktur vs. antiqua) and the layout was simplified to contain only text of the main paragraphs (e.g. marginalia were skipped). As a result, almost 400 thousands separate fraktur glyphs were collected of 164 distinct characters, with 35 characters represented by at most three different glyphs (0,15‰ of all glyphs). During the next step, the collected glyphs were used to generate 288 denoised original images along with their textual content, which were passed to the training process. As a result, fraktur recognition profile was created and used for importing two sets of original images. The first set, referred to as *real*, is composed of 288 historical pages, the same that glyphs dataset was retrieved from[5]. The second one contained extra 183 images of lower quality, it contained damaged pages or pages partially printed using antiqua fonts. Therefore, the second set called *noise* can be perceived as much harder to create full text content of a high accuracy.

In the last step, the results of the recognition process were compared with the ground truth data on the level of characters, based on the character accuracy measure ACC_0 proposed in [5]. This measure bases on the *edit distance d* between two texts and the total number of characters in the ground truth text referred to as n. As defined, the maximum accuracy equals one, but when $d > n$ the accuracy value is negative and unbounded, unbalancing its arithmetic mean. Therefore, the accuracy measure ACC is redefined as follows:

$$ACC = \max\{ACC_0, 0\} = \max\left\{\frac{n-d}{n}, 0\right\}. \tag{1}$$

Table 1 shows the mean and median value of recognition accuracy for the two considered datasets and their sum. As expected, the best results were achieved for the *real* dataset. As noised data negatively influenced the OCR results, the mean accuracy value decreased by 6.9 percentage points in case of the summed datasets and by 17.9 percentage points for the *noise* dataset. In all three cases, the accuracy distribution has a negative skew, e.g. in case of *real* dataset, the accuracy is lower than 40% for only 7.29% of images. Moreover, 61.81% of results for images in this set are evaluated at characters accuracy between 60% and 80%. The accuracy distribution for two considered sets is demonstrated by histogram from Table 2.

5 Summary

This paper shows that creation of custom recognition profiles could be a viable enhancement for institutions interested in digitisation of historical documents. Thanks to the open nature of the proposed approach, recognition profiles can be

---

[5] Note, that original images contain much more noise than the training data, and hence compose an independent verification set.

Table 1. The mean and median value of recognition accuracy for two considered sets and their sum

	Real	Both	Noise
Avg(ACC)	62.30%	55.36%	44.44%
Median(ACC)	65.29%	59.41%	48.43%

Table 2. The accuracy distribution for the two considered sets

	10%	20%	30%	40%	50%	60%	70%	80%	90%	100%	SUM
Real	4	5	6	6	22	53	98	80	14	0	288
Noise	11	13	16	25	33	45	35	5	0	0	183

developed jointly by interested institutions. The proposed crowdsourcing platform is a great complement to a dedicated OCR service as it allows the correction of the recognition results by volunteers or professionals. Together they significantly simplify the creation of full–text versions of existing historical documents. Moreover, an experimental evaluation of created recognition profile on real Polish historical documents shows that more than 60% of good quality scans can be recognised at accuracy level between 60% and 80%, although they contain fraktur fonts, complicated layout as well as historical sets of characters.

Further work will include evaluation of the VTL main portal as a crowdsourcing platform, including its usability and the quality of corrections produced by real users. The evaluation will be extended by addressing word accuracy rates and identifying the most common types of OCR error. Possibilities related to additional tools accelerating process of text correction and OCR training will also be investigated.

References

1. Lewandowska, A., Werla, M.: Pionier network digital libraries federation interoperability of advanced network services implemented on a country scale. Computational Methods in Science and Technology, 119–124 (2010)
2. Mazurek, C., Sielski, K., Stroiński, M., Walkowska, J., Werla, M., Węglarz, J.: Transforming a Flat Metadata Schema to a Semantic Web Ontology: The Polish Digital Libraries Federation and CIDOC CRM Case Study. In: Bembenik, R., Skonieczny, L., Rybiński, H., Niezgodka, M. (eds.) Intelligent Tools for Building a Scient. Info. Plat. SCI, vol. 390, pp. 153–177. Springer, Heidelberg (2012)
3. Dudczak, A., Kmieciak, M., Werla, M.: Country scale infrastructure for creation of full text versions of historical documents from Polish Digital Libraries. Presented at Interedition Symposium: Scholarly Digital Editions, Tools and Infrastructure, The Hague, Netherlands (2012)
4. Neudecker, C., Tzadok, A.: User Collaboration for Improving Access to Historical Texts. Liber Quarterly 20(1), 119–128 (2010)
5. Alexandrov, V.: Error evaluation and applicability of ocr systems. In: Proceedings of the 4th International Conference Conference on Computer Systems and Technologies: e-Learning. CompSysTech 2003, pp. 308–313. ACM, New York (2003)

Increasing Recall for Text Re-use in Historical Documents to Support Research in the Humanities

Marco Büchler[1], Gregory Crane[2], Maria Moritz[1], and Alison Babeu[2]

[1] Institute for Computer Science, Leipzig University, Germany
{mbuechler,mmoritz}@e-humanities.net
[2] Department of Classics, Tufts University, Boston, USA
{gregory.crane,alison.jones}@tufts.edu

Abstract. High precision text re-use detection allows humanists to discover where and how particular authors are quoted (e.g., the different sections of Plato's work that come in and out of vogue). This paper reports on on-going work to provide the high recall text re-use detection that humanists often demand. Using an edition of one Greek work that marked quotations and paraphrases from the Homeric epics as our testbed, we were able to achieve a recall of at least 94% while maintaining a precision of 73%. This particular study is part of a larger effort to detect text re-use across 15 million words of Greek and 10 million words of Latin available or under development as openly licensed TEI XML.

Keywords: historical text re-use, hypertextuality, Homer, Athenaeus.

1 Introduction

The ability to detect where one text uses portions of another has emerged as a distinct task within information retrieval, with prominent applications including the detection of plagiarism (places where one text takes credit for the contents of another text) and duplicate documents (e.g., multiple postings of the same entry). The detection of where one text makes use of another has also long been a core task for students of the textual record. Humanists have, for centuries, manually scoured key texts for their sources, seeking not only explicit quotations but summaries, paraphrases and allusions, often seeking to identify textual re-use where the original no longer survives except insofar as other sources quote or paraphrase it. We report here on the results of work from the eTRACES project[1] that made use of textual data from the Perseus Digital Library(PDL)[2] and that focused upon replicating the high recall that humanists have achieved through intensive manual analysis.

Figure 1 shows part of the textual notes for Axel Ahlberg's 1913 edition of Sallust's *Catilinarian Conspiracy*, a first century BCE account of an attempted

---

[1] http://etraces.e-humanities.net
[2] http://www.perseus.tufts.edu/hopper/

P. Zaphiris et al. (Eds.): TPDL 2012, LNCS 7489, pp. 95–100, 2012.

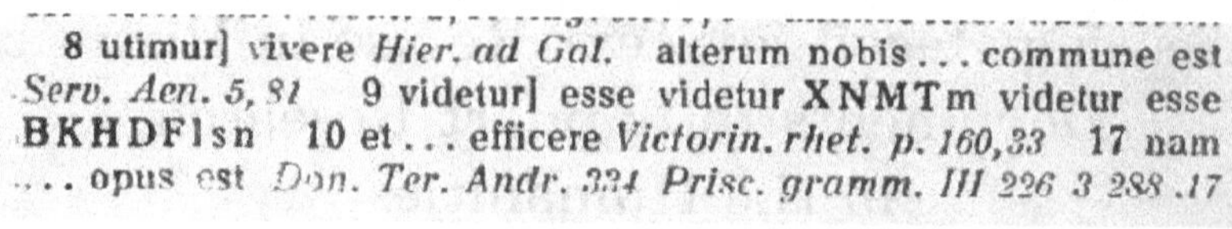

> 8 utimur] vivere *Hier. ad Gal.*   alterum nobis ... commune est
> *Serv. Aen. 5, 81*   9 videtur] esse videtur **XNMTm** videtur esse
> **BKHDFlsn**   10 et ... efficere *Victorin. rhet. p. 160,33*   17 nam
> ... opus est *Don. Ter. Andr. 324 Prisc. gramm. III 226 3 288 .17*

Fig. 1. Manually produced record of text re-use

coup in Rome. This apparatus contains the textual variants familiar to humanists (e.g., manuscripts *XNMTm* read *esse videtur* rather than simply *videtur*) but most of the apparatus criticus details places where subsequent texts quoted from the opening of Sallust's history – with more than twenty citations to texts that quote this passage. All of the work in the figure above was conducted manually by scholars who worked over generations identifying and publishing this data. It is important to emphasize the collaborative nature of such text re-use detection – Figure 1 illustrates an example of crowd-sourced work, with a crowd that is widely separated in time rather than space.

When we compared our initial methods to an instance where an edition of one text (Athenaeus' *Deipnosophistai'*, a Greek work written in the 3rd century CE) had extensively labeled quotations of two other texts (the *Iliad* and *Odyssey* of Homer), we found that we were only achieving a recall of approximately 50%. We consequently report here on our efforts to improve recall so that it is comparable to that of traditional, labor-intensive, but effective, methods.

The rest of this paper is organized as follows: section 2 surveys the state of the art and previous work, section 3 describes the mining of a traditional print publication for labeled instances of text re-use that were then used to evaluate, and improve upon, recall, and section 4, details the results of using those high recall methods to detect text re-use between Homer's *Iliad* and *Odyssey* and the texts of Athenaeus.

2   Related Work

Text re-use has a number of applications including restatement retrieval [1], near duplicate detection [2,3], and automatic plagiarism detection [4,5].

In addition, the use of computational methods to track text re-use within classical and historical texts has been explored by a number of different researchers. John Lee has examined textual re-use between the Synoptic Gospels of the Greek New Testament, with a particular focus on sentence alignment [6]. [7] discussed how a textual aligner they developed to highlight variants between different manuscript versions of an author's text might also be used to explore text re-use between historical documents.

The field of text re-use is also closely related to the study of intertextuality, or how a text's meaning has been influenced and shaped by other texts. [8] proposed a model that distinguishes various types of intertextuality including different types of text re-use such as quotation and allusion in their development of the HyperHamlet Corpus. Similarly, the Tesserae Project, based at the State University of New York-Buffalo has created a freely available tool on the Web[3] that can be used for analyzing text re-use (intertextuality) in Latin poetry. Their tool automatically identifies matching two-word phrases or bigrams in different Latin poets using two different search algorithms [9] and was recently utilized in a comparison of two Latin poets [10].

Other related research in the study of intertextuality has been presented by Kane et al. [11], who have described the development of the Electronic Manipulus Florum project[4], a digitized collection of Latin quotations, as well as the Janus search engine that finds overlap between user query text and the Florum quotations (despite the existence of complex variants). They noted that the Janus search engine could also be used to find textual overlaps between other random texts as well.

3   Maximizing Recall

In this section, we describe our experiences tracking textual re-use in the 10 million words of Classical Greek and 7 million words of Classical Latin found in the source texts released by the PDL between June 2011 and March 2012. While we focused initially upon maximizing precision, the small number of results caused us to gradually adjust our settings to increase recall. This was not done by decreasing thresholds but instead by changing the language model. In our revised approach we made three fundamental changes that are outlined below.

3.1   Bigram vs. Unigram Fingerprinting

Initially, we used *bigram shingling* to fingerprint our texts, a method that identified 553,285 links between different sentences in the PDL. The vast majority of these links (99.7%) were not between different works, however, but were instead internal links found within 18 of 20 of the most frequently re-used works. The strongest re-use of a work within itself was observed for Euclid's *Elements*, a work that contains many repetitive mathematical expressions (e.g., line segments AB) thus making many text passages similar to each other. Furthermore, works of Plato and Homer were also very prominent in this top 20, a factor caused by either author or genre specific expressions or discourses. To address these and other issues, we modified the pre-processing, removed diacritics and made all letters lower-case, changes that resulted in an increase of relevant results of almost 19%.

3.2   Normalized Weighting vs. Absolute Weighting of the Overlap

In the initial setting we used Broder's resemblance [12] with a threshold of 0.7. This measure normalizes the overlap between 0 and 1 of two pairwise compared

---

[3] http://tesserae.caset.buffalo.edu/

[4] http://manipulusflorum.com/

sentences. While this measure works very well in plagiarism detection or in studying text re-use in the Bible, it proved to be less sufficient for the literary texts found within the PDL, where more than 80% of text re-use examples involve quotations of 5 words or less within larger sentences. Sentence to sentence comparison thus missed most of these smaller yet critical quotations. As a result, we shifted our approach from one where weighting was normalized according to whole sentences to one that instead searched for distinctive words and phrases within each sentence. In addition to using the *tf.idf* measure, we focused on weighting by type of word class, where *nouns* and *verbs* were ranked higher and various kinds of function words were ranked lower. Furthermore, we also conducted different experiments investigating the role of verbs in text re-use. One of the most challenging tasks that emerged from utilizing the different weighting strategy was the need to be able to distinguish between multi-word units such as complex names like 'King Alexander the Great' and real examples of text re-use. In our evaluation data, we also discovered that slightly more than 90% of all textual re-uses that were found contained at least one verb.

3.3   Sentence Based, Non Overlapping Segmentation of Re-use Units vs. Fixed-Size, Moving Window

Sentences as they appear in modern editions of ancient sources were rarely found in the manuscripts that serve as our basis for these ancient texts, but were instead developed by modern editors in order to make print editions easier to read – different editors often added their own unique punctuation that was variant from other editors. Given this fact and our observation that most text re-use involves just a few key words, we decided to focus upon the *locality* of text re-use. A window size of 5 provided the best trade-off between flexibility in re-writing and overlaps that were too dense.

In our most recent settings, texts from the PDL are preprocessed by a lemmatisation step, followed by removing diacritics. Potential re-use is observed by a moving window of size 5 (cf. section 3.3). Every re-use unit is fingerprinted by words and not by bigrams (cf. section 3.1). Finally, linked re-use candidates are scored by a word class heuristic to support the detection of nouns and verbs (cf. section 3.2).

4   Finding Homer in Athenaeus

To measure our ability to detect instances of text re-use, we chose to use as our testbed a digital edition of Athenaeus' *Deipnosophistai* where editors had explicitly marked 353 instances where this text quoted or paraphrased the Homeric epics. Using this digital edition, we were able to distinguish three cases of TEI-based inline annotations (cf. figure 2).

All three different kinds of annotation (cf. figure 2) were used for evaluation in a separate way due to the degree of certainty as to where the text re-use actually started and ended. The *bibl* tag simply indicates a re-use, however,

Cit-quote-bibl	blockquote	bibl without quote

`<cit>`
  `<quote>`du/o ku/nes a)rgoi\ ei
  `(/ponto</quote>`
    `<bibl n="Hom. Od. 2.11">`Od. 2.11`</bibl>`
`</cit>`
	`<quote rend="blockquote">`
	  • `<line>`
	      a)gxou= d' i(stame/nh e)/pea
	      ptero/enta proshu/da
	        `<bibl n="Hom. Il. 4.92">`Il.
	        4.92`</bibl>`
	  `</line>`
	  • `<line>`
	      a)li' a)/ge nu=n ma/stiga kai\ h
	      (ni/a sigalo/enta
	        `<bibl n="Hom. Il. 5.226">`Il.
	        5.226`</bibl>`
	  `</line>`
	`</quote>`
		`<p>`
		[...]a)nti\ tou= proe/pinon. kuri/ws
		ga/r e)sti tou=to propi/nein, to\ e
		(te/rw| pro\ e(autou= dou=nai piei=n.
		kai\ o(*)odusseu\s de\ para\ tw=| *
		(omh/rw|
		  `<bibl n="Hom. Od. 13.57">`Od.
		  13.57`</bibl>`
		  [...]
		`</p>`

Fig. 2. Three different annotation styles of quotation depending on certainty of re-use boundaries

neither the boundaries nor the position in relation to the re-use. In addition, the *bibl* tag can stand both before and after re-use in scenario 3 (cf. right column of Figure 2).

		Odyssey	Iliad	
found	Cit-quote-bibl	84	80	
	blockquote	34	50	
	bibl without quote	40	43	331
not found	Cit-quote-bibl	1	1	
	blockquote	11	7	
	bibl without quote	2	0	22
		172	181	353

Fig. 3. Results of text re-use detection between Homer's *Odyssey* and *Iliad* in all books of Athenaeus

Using recently developed techniques (cf. section 3), we were able to find 331 of the 353 annotated references (with a final recall of 0.938 and an average precision of 0.73). In detail, the proposed method can reach a precision of 1.0 for five or more overlapping words, while if three or four words are overlapping then precisions of 0.7 and 0.8 can be obtained respectively. Two word overlaps were difficult to evaluate, however, since it was quite often not possible to determine what was a bigram or a co-occurrence and what was a two word text re-use. Twenty two links were missed by our experiments (cf. figure 3) and five of them represented one word text re-uses. We do assume, however, that text re-use is not only about single words or concepts, but a minimum overlap of two words between two text passages makes it nearly impossible to find those one-word re-uses. Finally, two links were not found since they are highly paraphrased.

5   Conclusion

While our initial methods for text re-use detection provided high precision, when recall was measured against an existing gold standard the results proved to be too low to meet the needs of many humanists. We found that we were able to achieve much higher recall (94%) with reasonable precision (73%), identifying in particular many instances where text re-use of three to five words occurred in much longer sentences. Our next goal is to create a comprehensive map of text re-use in and across Greek and Latin literature, with the Perseus Digital Library CC licensed library as the starting point for this work.

References

1. Balasubramanian, N., Allan, J.: Syntactic Query Models for Restatement Retrieval. In: Karlgren, J., Tarhio, J., Hyyrö, H. (eds.) SPIRE 2009. LNCS, vol. 5721, pp. 143–155. Springer, Heidelberg (2009)
2. Potthast, M., Stein, B.: New Issues in Near-duplicate Detection Data Analysis, Machine Learning and Applications. In: Studies in Classification, Data Analysis, and Knowledge Organization, pp. 601–609. Springer, Heidelberg (2008)
3. Wang, J.H., Chang, H.C.: Exploiting Sentence-Level Features for Near-Duplicate Document Detection. In: Lee, G.G., Song, D., Lin, C.-Y., Aizawa, A., Kuriyama, K., Yoshioka, M., Sakai, T. (eds.) AIRS 2009. LNCS, vol. 5839, pp. 205–217. Springer, Heidelberg (2009)
4. Hoad, T.C., Zobel, J.: Methods for identifying versioned and plagiarized documents. J. Am. Soc. Inf. Sci. Technol. 54(3), 203–215 (2003)
5. Alzahrani, S., Salim, N., Abraham, A.: Understanding plagiarism linguistic patterns, textual features, and detection methods. IEEE Transactions on Systems, Man, and Cybernetics, Part C 42(2), 133–149 (2012)
6. Lee, J.: A computational model of text reuse in ancient literary texts. In: Proceedings of the 45th Annual Meeting of the Association of Computational Linguistics, Prague, Czech Republic, Association for Computational Linguistics, pp. 472–479 (June 2007)
7. Bourdaillet, J., Ganascia, J.G., Pierre, U., Curie, M.: J.g: Alignment of noisy unstructured text data. In: Proc. of the IJCAI Workshop on Analytics for Noisy Unstructured Text Data (AND 2007) of the 20th International Joint Conference on Artificial Intelligence (IJCAI), pp. 139–146 (2007)
8. Trillini, R.H., Quassdorf, S.: A 'key to all quotations'? a corpus-based parameter model of intertextuality. LLC 25(3), 269–286 (2010)
9. Coffee, N., Koenig, J.P., Poornim, S., Forstall, C., Ossewaarde, R., Jacobson, S.: The tesserae project: Intertextual analysis of latin poetry (2011), http://dh2011abstracts.stanford.edu/xtf/ view?docId=tei/ab-215.xml;query=;brand=default (last accessed February 14, 2012)
10. Forstall, C.W., Jacobson, S.L., Scheirer, W.J.: Evidence of intertextuality: investigating paul the deacon's angustae vitae. Literary and Linguistic Computing 26(3), 285–296 (2011)
11. Kane, A., Tompa, F.W.: Janus: the intertextuality search engine for the electronic manipulus florum project. Literary and Linguistic Computing 26(4), 407–415 (2011)
12. Broder, A.Z.: On the resemblance and containment of documents. In: Compression and Complexity of Sequences (SEQUENCES 1997), pp. 21–29. IEEE Computer Society (1997)

PrEV: Preservation Explorer and Vault for Web 2.0 User-Generated Content*

Anqi Cui[1], Liner Yang[1], Dejun Hou[2],
Min-Yen Kan[2], Yiqun Liu[1], Min Zhang[1], and Shaoping Ma[1]

[1] State Key Laboratory of Intelligent Technology and Systems,
Tsinghua National Laboratory for Information Science and Technology,
Department of Computer Science and Technology,
Tsinghua University, Beijing 100084, China
{cuianqi,lineryang}@gmail.com, {yiqunliu,z-m,msp}@tsinghua.edu.cn
[2] School of Computing, National University of Singapore, Singapore
houdejun214@gmail.com, kanmy@comp.nus.edu.sg

Abstract. We present the **P**reservation **E**xplorer and **V**ault (*PrEV*) system, a city-centric multilingual digital library that archives and makes available Web 2.0 resources, and aims to store a comprehensive record of what urban lifestyle is like. To match the current state of the digital environment, a key architectural design choice in *PrEV* is to archive not only Web 1.0 web pages, but also Web 2.0 multilingual resources that include multimedia, real-time microblog content, as well as mobile application descriptions (e.g., iPhone app) in a collaborative manner. *PrEV* performs the preservation of such resources for posterity, and makes them available for programmatic retrieval by third party agents, and for exploration by scholars with its user interface.

Keywords: Preservation, Archive Visualization, API, Web 2.0, User-Generated Content, NExT, PrEV.

1 Introduction

Not long ago, the Web was a largely homogenous digital environment, with web servers serving static authored web pages, enriched by embedded image, audio and video resources. We consumed these resources as readers, and our indexers and archivers did the same. Archiving such content, while difficult due to scale and the necessary curation, was otherwise technically feasible. This type of initiative is exemplified by the Internet Archive's Wayback Machine[1], which provides a URL-based navigation on web pages. Through a calendar interface,

---

* This work was supported by Natural Science Foundation (60903107, 61073071), National High Technology Research and Development (863) Program (2011AA01A207) and the Research Fund for the Doctoral Program of Higher Education of China (20090002120005). This work has been done at the NUS–Tsinghua EXtreme search centre (NExT).

[1] http://www.archive.org/web/web.php

P. Zaphiris et al. (Eds.): TPDL 2012, LNCS 7489, pp. 101–112, 2012.

a scholar can view many archived instances of specific web pages, reaching back as far as the mid 1990s.

Fast forward to today's Web, a web that is centered on a new form of content: User-Generated Content (UGC). Such content is tacked on at the end of Web 1.0 pages (e.g., news articles with commenting) or at centralized (e.g., restaurant reviews on Yelp) or decentralized (e.g., personal blog sites running WordPress) Web sites. It comprises people's opinions and comments and is updated more frequently than the near-static Web 1.0 pages. Today's web is now much more interactive than before, better catered to the spectrum of devices that we use to consume digital resources now.

This interactivity has come with a cost: the Web has become more fragmented and harder to archive. Popular social media sites must restrict access to sensitive personal content. Many pages are dynamically created via push technologies, making simple crawed versions of pages largely devoid of content. Smartphone and tablet applications ("app") make up a large percentage of consumed bandwidth, but accessing information about these apps is restricted to their proprietary devices and their communication protocols. With the dynamicism of Web 2.0, we may only be able to capture a particular user experience – as how a website looks might change from user to user, instant to instant[2]. Clearly, to archive the spectrum of UGC that collectively represent today's Web is more challenging.

However, it is a challenge that we must rise to, in order to paint a holistic picture of life today for future generations to appreciate. A key observation we make is that such an archiving initiative must present these myriad resources in a unified manner. If we only archive piecemeal, the future scholar may make erroneous conclusions based on his incomplete picture of our lives. To surmount the challenge of the archiving the Web 2.0 spectrum, we scale back the scope of the resources that we archive. Our project, the **P**reservation **E**xplorer and **V**ault, hereafter *PrEV*[3] is currently fielded to archive only content related to the two capital cities of Beijing and Singapore, in both Chinese and English.

In preserving content for posterity, we must amass a large focused collection of resources that are valuable and of potential interest to developers looking to harvest and extract information from today's web. For this reason, we make another key decision to opportunistically support third-party programming data access.

We use two running scenarios to motivate the approach and architecture taken with *PrEV*:

(1) Suppose many years later, Ryan, a secondary student is doing his term project about Singaporean hawker center (cooked food center) history, He aims to obtain some historical pictures of people and food at hawker centers, but finds only secondary texts that mention what it was like to eat a meal at the

---

[2] http://blog.dshr.org/2012/05/harvesting-and-preserving-future-web.html

[3] So named to contrast with its umbrella project name – NExT: the NUS–Tsinghua EXtreme search centre.

hawker center. They contain digested accounts, and do not have the raw data and primary source, first-hand data that Ryan needs for his project.

PrEV addresses this problem by archiving pictures of Singapore (including hawker center experiences), associating them with the time and place where they were taken. In encountering *PrEV*, Ryan issues the simple query "hawker", which lists all the relevant records in the history (Fig. 1, left). Ryan picks out the picture records in different time points. Ryan uses the word cloud summary feature, (Fig. 1, right) highlighting the most frequent terms found in the records relevant to the query. Browsing the data, he generates several more word clouds for different periods in the collection, discovering salient aspects of hawker centers, including locations (*bedok, toa payoh*), and cuisine (*chicken rice*). From these records, he issues subsequent queries to *PrEV*, browsing through to find related blog, forum and Twitter posts that give Ryan a feel for the old-style dishes and environment that made up the hawker center environment of 2012.

Fig. 1. *PrEV* records with respect to "hawker". Left: collection statistics with respect to the query; Right: a (stemmed) word cloud summary restricted to microblog data from Twitter and Weibo (a Chinese microblogging service), and adding Flickr records.

(2) A startup company, Blueberry, aims to develop their "Follow Me" iPhone travel guide application for Beijing and Singapore. Before they start, the design team wants to review the existing competitor apps to understand their weaknesses through the user's comments. However, they have to read through competitor app descriptions and reviews manually, as the data in the Apple App Store can only be accessed through an iPhone or iPad and not by any server based system.

Later, the team comes across *PrEV*, which archives this data of interest; in particular, descriptions and reviews of mobile apps about both cities.. They contact *PrEV*'s administrators and obtain an API key for retrieving the large amount of archived data programmatically. They query *PrEV* using a REST-compliant syntax and receive the results in JSON format, and are able to see how the application's descriptions and popularity changed over time.

Both scenarios are largely supported by our current implemented version of *PrEV*. Underlying the scenarios is our assumption that users want to retrieve the

{ "total":86, "count":10, "totalpage":9, "page":0, "data":[
  { "crawlresource":"twitter", "encoding":"en", "tweetcreatedat":"Sun Oct 02 19:56:51 +0800 2011",
    "url":"http://twitter.com/#!/ChristianLeeVO/status/120467276104339457",
    "maincontent":"Check this video I shot of Pulau Ubin for our Travel Now Singapore webseries and iphone
app http://t.co/mvfgJDqP via @youtube" },
  { "crawlresource":"weibo", "encoding":"zh", "weibocreatedat":"Sat Jan 14 19:13:27 +0800 2012",
    "url":"http://www.weibo.com/1910529591/y0Lf5mFAk",
    "maincontent":"#App推荐# 旅程规划：Routes. Planning your journeys【出行必备】, iPad/ iPhone通用。这款应用
可让你规划旅游景点，像是到某景点去拜访或是去某家大卖场购物等等。它会算出需要多远的距离以及所需的时间，就像...
http://t.cn/z0gtfEr （分享自 @App每日推送）" },
  { "crawlresource":"sgbjapps", "encoding": "others", "crawltime":"Wed Dec 23 00:00:00 +0800 2009",
    "maincontent":"Do You Love Travel ? If Yes, You Should Not Miss This App. Updated For Now! Download
this app to your iPhone to enjoy these beautiful scenery anywhere you go! These pictures are HD Photo You
can download the image to your iPhone or iPod and make it to wallpaper. No Ad No Wifi!",
    "name":"A Tourist Paradise <Singapore>" }, ...] }

Fig. 2. Some actual multilingual *PrEV* data relevant to the travel application domain. Results subsampled to highlight the variety of sources (Twitter, Weibo and App Store).

historical records on some topic (e.g. an entity, a product), filtering the content not only by keyword but by other facets such as time and data source.

We present *PrEV*'s architecture and implementation in the remainder of this paper. As the scenarios illustated, *PrEV* has three main missions: 1) to archive the myriad Web 2.0 about cities, 2) provide them in an exploratory browsing interface, as well as 3) providing them in a programmatic interface. To achieve this, *PrEV* uses a three-layer framework: 1) a preservation layer to store the different types of records in a unified structure, 2) an indexing layer that allows fast retrieval on different facets, and 3) an interface for presenting the data to both browsing users and programmatic agents. The loosely coupled design makes it possible to preserve both text and multimedia records together, as well as to provide retrieval and visualization from different views.

2 Related Work

There is much work relating to the wider aspects of digital preservation. We limit our discussion to the most relevant work on archiving Web data and user-generated content. We also briefly review the user interfaces and visualization techniques that have been used to explore such archives.

Web Archiving. The preservation of the Web has been an issue of interest early, as the Web became the method of choice to disseminate information. Research topics include crawling methodologies, version control, recovery of broken links, among other topics. A seminal system that continues to operate today is the Internet Archive's Wayback Machine [14], which takes a broad approach to archive historical web pages by URL, collecting multiple snapshots of websites. Their effort has been a reference for many succeeding studies, including country-restricted web archives in Norway [2] and China [24], among other national initiatives. The International Internet Preservation Consortium (IIPC) [8] associates libraries in different countries to preserve the Interent contents (mainly web pages) crawled by each library themselves. While some relevant work describes how to make curation decisions in Web archiving to focus web crawling efforts, in *PrEV* we take a broad approach, collecting and storing any data provided by trusted third parties.

Some of the studies on web preservation face some technical challenges of data format standard, storage safety, scaling issues or selection priorities [3,10,19,20]. In contrast, our focus differs in our system objective, i.e. to provide access to Web data, including but not limited to traditional Web pages. Our challenges center around the organization of the variegated data types that we collect, and ensuring usable access to the data in a unified interface.

Another web archiving research focus is to discover and restore access to deleted or missing pages [2,5,11,17,22]. To support this, up-to-date crawls and rate of change estimation are a necessity, as pages change constantly. This phenomonon is exacerbated in the scope of *PrEV*'s Web 2.0 data, in which UGC are often short but updated frequently [3].

Separately, proper data access is also a concern in shared archives that span users from multiple institutions or organizations [4]. Such archives may have to meet different requirements in data access and sharing [9]. Specific infrastructures have been designed for such multi-level access control. Some studies design a multi-layer architecture, with different layers geared towards handling data preservation, indexing or search access [11]. Since our system is a public data archive, we have fewer restrictions on our data, but we ingest data from certain sources that have restrictions (e.g., Twitter). We adopt a similar multi-layer architecture in *PrEV*, loosely coupling different functionalities as serial layers.

Web 2.0 Archiving. Web 2.0 is about user-generated content. This bottom-up flow – from users to website – brings more challenges for preservation [7,23]. The multimedia, real-time and streaming UGC are usually difficult to crawl with traditional techniques. Many Web 2.0 pages' ultimate appearance within a user's web browser conditions not on just the web page but external cascaded style sheets, embedded applets, scripts, and more recently iframe contents and dynamic content written by push technologies (i.e., AJAX/XHR). More efficient headless browsers that simulate the actual rendered page and execute the embedded scripts are needed to crawl and preserve such contents.

While the call to arms to preserve UGC is widely known, UGC are individually the focus of different research groups. Twitter is perhaps the UGC source with the most active archiving movement. [15,18,25] archive Twitter data for their own analyses. Existing commercial websites also provide access to the resyndicated Twitter archive, such as *Topsy*[4], *TwimeMachine*[5] and *indextank*[6]. They offer keyword-based retrieval with different facets (time, language, etc.) on the Twitter messages. Similarly, researchers working on analyzing Flickr images, YouTube videos, Yelp reviews are all crawling these sites individually. Currently, there is no unified platform for researchers or users to obtain a holistic view that cuts across UGC sources, to search through all these UGCs, even for limited topics.

---

[4] http://www.topsy.com/

[5] http://www.twimemachine.com/

[6] http://www.indextank.com/

Within the realm of private social network data, less work has been done. A notable exception, is McGown and Nelson's work [16] which developed a browser extension to back up a user's Facebook presence. Greplin[7] takes a similar approach, emcompassing multiple social network sites. Both embody personal backup solutions, but not large-scale preservation.

In contrast to all of these works, *PrEV* aims to un-silo these Web 2.0 resources and archive them together.

Web History Visualization. Once the Web data is crawled and archived, how can it be effectively presented to the end user? Two web page preservation systems, the Internet Archive and the Memento Project [21] present the historical page, given a manually-specific date. While fine for traditional Web 1.0 web pages, it may not make sense for UGC since they are often spread over multiple 2.0 sites. On the other hand, changes between different versions of a page are not presented directly. In most studies, changes are measured with respect to text (keywords, tags, etc.) [1,6,12,13]. In a typical change visualization, different versions of the page or relevant change areas are listed in columns, then related terms in the adjacent versions are connected with lines to show the change flow. Currently in *PrEV*, we are working towards defining what constitutes change. We present a stationary summary of a set of resources, via a *word cloud* generated from different resources, instead of simply comparing different versions.

3   System Architecture

The *Preservation Explorer and Vault* involves multiple data contributors, while serving the data to end users through both Web and API service endpoints. To achieve a loosely coupled system, we divide the system into three layers, as shown in Fig. 3. The three layers operated independently from each other, although some layers operate concurrently on individual machines.

1. *The Preservation Layer* interacts with the crawlers and handles long term storage. It reads the raw, crawled data and archives them in the file system or database permanently.
2. *The Indexing Layer* enables the necessary retrieval functionality of frontend service endpoints. Each archived record is processed into an indexable version, so that users can retrieve them by both content (keyword) and different facets (metadata). Multiple levels of indexing and processing can be run to generate different levels of automated analyses on the raw archived data.
3. *The Interface Layer* serves the data to end users. The interface modality varies based on the user requirements: e.g., content-based queries, visualizations or command-line access.

While the multi-layer design nicely modularizes the responsibilities at each layer, the interface between layers is the challenge. Issues that have been addressed include specifying data formats between layers, and lowering the latency between the initial crawl and eventual availability in the system. We now give a detailed view of each layer.

---

[7] http://www.greplin.com/

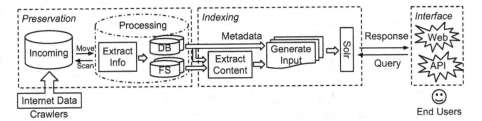

Fig. 3. *PrEV*'s overall three-layer system architecture

3.1   The Preservation Layer

At this first layer, *PrEV* collects data from different crawlers.

PrEV is a central repository project that binds several city-centric search research projects together. Staff and students spread across two institutions, Tsinghua University in China and the National University of Singapore in Singapore, are involved, and many of them run their own crawlers to fetch Web 2.0 data sources for their individual projects. Some of them crawl resources with respect to their own city, while some others crawl global resources. Our project mandates that they share their trusted crawled data with our central project. Up till now, we have been collecting multilingual data of city lifestyles from crawlers covering more than 300 million records (shown in Table 1). Though most of the resources are static (such as microblogs or news articles), dynamic resources (forum posts, product reviews, app data etc.) are gradually becoming a larger percentage as these 2.0 data sources are re-crawled more frequently.

A series of steps from collecting to storing are demonstrated below.

Incoming Data Detection. Our preservation process is architected as a federation of independent crawlers who push data to the central repository at their convenience. The crawlers, run by individual researchers, use the *PrEV* master server as the single point of contact to upload data to. To be clear, our system does not crawl resources but collects data from the crawlers.

Each crawler is registered as a user on the *PrEV* server. The master server also has a *PrEV* user that houses a writable directory, in which other users can write to. Crawlers can copy their crawled data to the server at any time. Crawlers also create a zero-sized *unlock* file, after a transfer is finished, to denote the associated file was successfully stored on the server. The *PrEV* user periodically scans the common directories for new files and their associated unlock files, and moves the files to a staging area for processing. In this collaborative way, different resources are united into the central server.

Sources from the individual crawlers are independent between each other. Our central repository benefits from this strategy, that requests from different locations may provide different views (responses) of a same resource at one time. Therefore, duplicate resources from different crawlers are all kept in the system; duplicate handling is the province of the indexing and presentation layers.

Table 1. City lifestyle data resources from multiple crawlers (as of May 2012)

Data Type	Resource	No. of Records
Microblog messages	Twitter	229 M
	Weibo	139 M
Photos with texts	Flickr, Panoramio	2 M
Food forum restaurants	27 Singapore sites	6 M (pages)
	Fantong, Dianping (Chinese)	78 K
Public forum posts	4 Chinese forums	1 M (approx.)
Product review products	7 e-commerce sites	70 K
News articles	Sina News	224 K
	Guardian, Channel NewsAsia, Skysports, CNN, Economist, FoxNews, NewYorkTimes, StraitsTimes	59 K
Wiki articles	Hudong (Chinese)	1 M
Traffic records	Singapore	24 K
	Beijing	19 K
Question Answering articles	Baidu Zhidao (Chinese)	33 K
	Yahoo! Answers, WikiAnswers	52 K
Mobile Apps	US App Store	617 K
	Android Market	345 K
	Blackberry, Windows	162 K

Data Format Recognition. Due to the diversity of data that *PrEV* archives, we allow several different formats for crawler submission. A microblog post only consists of a short text string, while a photo incorporates megabytes of binary data and text for its description and comments. We use three simple submission data formats to reduce the overhead for the staff maintaining the crawlers.

1. *Short text* data (e.g., microblog entries) are stored in a text file, i.e. each line of the file is considered as a single record. The meta-information is provided within the file.
2. *Single record files* (e.g., image files) are a raw, binary source of the record. Any record metadata is provided in a separate description file that accompanies the submission.
3. *Multiple record files* (e.g., web page archive) allow crawlers to compress and store multiple single raw files together, In this case, we consider the file as a concatenation of the records. A crawler must also provide the offset and size of each record, for extracting the corresponding record from the file, in a separate description file.

In the latter two cases, the raw and README files are zipped into a single file for upload. The zipped file is later extracted for content analysis and retrieval. This strategy also makes it possible to consider multiple files as one single record. For example, the source code (text) of a web page and its associated images and CSS style files are considered as a complete snapshot of the web page. Note that some crawlers only crawl texts or even ignore the header information of a web page. For this reason, we do not explicitly force crawlers to archive in well-known formats such as the Web Archiving File Format (WARC), although they may.

Record Storage and Backup. After extracting each record from the uploaded file, *PrEV* stores each record's metadata information in the database for management; any raw files are stored as simple files on a distributed file system.

The database is periodically backed up in a RAID 1 system (fully redundant mirror), while the raw files are stored on a RAID 6 file systems, able to withstand two simultaneous disk failures.

3.2   The Indexing Layer

A separate indexing process enables faceted navigation on the metadata stored in the databases. We use a single installment of Apache Solr[8] as our indexing service. It provides an HTTP-based method for data injection, in which the data to be indexed are submitted to a web service. In this way, we implement incremental indexing (while not particularly efficient). The workflow of the indexing system is shown in Fig. 3:

1. Fetch new records from the preservation layer.
2. For each record, extract the facets to be indexed based on its type. For example, *PrEV* extracts the text content of the microblog messages, descriptions of the photos, body text of the web pages. Other facets include the resource, data format, crawled time, record ID, etc. Facets are defined per-resource type.
3. The extracted fields are submitted for injection into the Solr server. After indexing, the corresponding record in the preservation layer is marked as "indexed", to prevent multiple indexing instances.

We use dynamic fields (supported by Solr and its underlying Lucene search core) to handle specific facets associated with certain data types. These fields are used for specialized query and presentation. For example, the *author* of a microblog post may be helpful to end users in defining their search scope, but authors are not generally attributed with all records that *PrEV* indexes. Therefore, for the microblog data type, we define the author's screen name and profile image as two additional dynamic fields, which can be used to retrieve a certain user's posts. The Solr `traverse` process is executed periodically to add the latest records to the index.

3.3   The Interface Layer

The interface layer currently has two service endpoints, which we described through the earlier scenarios. These are a web frontend for individual users and an API frontend for enterprise-level use.

Web Frontend. In the first scenario, Ryan asks *PrEV* to provide him the statistics on relevant records, as well as a summary of the text contents (Fig. 1).

With the help of the indexing layer, the web frontend issues a number of database queries to provide a calendar view of the number of matching records. The calendar view is hierarchical, allowing results to be drilled down from a year view to months, and to individual days, providing both macro and micro views

---

[8] http://lucene.apache.org/solr/

of the data. We show the proportions of different data types as dynamically-generated pie charts drawn via the Google Chart API. The page is implemented with AJAX to improve the user experience.

Results at any level can be used to form a word cloud, to get a feel for the individual resources at each level. The word cloud is dynamically generated, based on a random sample of records in the relevant range (usually a keyword + time range).

API Frontend. Enterprise-level users, as in the second scenario, usually process a much larger number of records at once, requiring a batch mechanism to retrieve data. *PrEV* provides a RESTful API service that implements an API which includes user-level authentication to achieve this. The functions implement facet search, specified in parameters, to access the data in a flexible manner. For example, users may choose the data containing some query from certain resources within a specified time range.

On the *PrEV* website, we created a forum that combines user management and API registration. In addition to the standard troubleshooting and broadcast use of such a forum, a forum user – with appropriate permission – is issued a standard API key for authenticated API requests. The API key is assigned after the user registration is approved in the forum.

The API uses user-level authentication and performs two services: rate limiting and data access management. Each API call needs to provide the API key in the request. The rate limiting ensures that users can only send up to their allowed quota of requests per hour. The data access management ensures that a user can only access the types of data he has been authorized for. We also created an API sandbox, to help familarize our users with the functionalities we provide in the programmatic API.

The workflow below demonstrates the steps from reading the request to generating the response:

1. The web service receive a URI as the request from the user. The URI must contain a parameter as the API key.
2. The system checks if the API key is valid. If so, it finds the corresponding user information, including his rate limiting level and data access level. Otherwise, the request is rejected via generation of an HTTP error.
3. The system checks the rate limiting counts. If the user has exceeded his current use quota, the request is rejected.
4. The system generates a Solr query based on the request type, user's data access level and the request parameters, and sends this to the Solr server.
5. The system reads the query results from the Solr server, then transforms it to the response format, and returns it to the end user. The response header contains rate limiting information, while the body contains the data.

The rate limiting counts of each user is reset per cycle. This strategy is used by most RESTful API websites (such as Twitter). Besides the header of the response, the user can access one certain API to query his rate limit status, or request for their accessible resources.

4   Conclusion and Future Work

We have presented *PrEV*, the Preservation Explorer and Vault. *PrEV* is a city-centric archiving system, modeled to archive and unify multilingual data as represented by the current Web 2.0 paradigm. Our indexing layer implements faceted search, allowing users to access the data in a flexible manner. This includes internal natural language processing engines, which freely access the raw and previously-processed archives to process and deposit back annotations on the material.

While still under development, we argue that our user-oriented interfaces and APIs already provide flexibility for both individual scholars looking to browse the archival data and enterprise-level automation that seek to programmatically access a large amount of the crawled data. In addition, we plan to continuously involve more resources such as geo-location based contents, personalized pages in different languages.

In future work, we plan to continue to improve system performance, and support more community standards (web archive access via Memento [21]) and conduct formal evaluations, while enhancing the user interfaces to support better visualization of the changes in the collection. We plan to implement comparative visualization that will complement the faceted aspects of the current *PrEV* collection.

Acknowledgments. We are indebted to the many students and staff who provided the crawlers that supply *PrEV* with its data. The NExT Search Centre is supported by the Singapore National Research Foundation and Interactive Digital Media R&D Program Office, MDA under research grant (WBS: R-252-300-001-490).

References

1. Adar, E., Dontcheva, M., Fogarty, J., Weld, D.: Zoetrope: Interacting with the ephemeral web. In: Proceedings of the 21st Annual ACM Symposium on User Interface Software and Technology, pp. 239–248. ACM (2008)
2. Albertsen, K.: The paradigma web harvesting environment. In: Proceedings of the 3rd Workshop on Web Archives, pp. 49–62 (August 2003)
3. Ball, A.: Web archiving. Tech. rep., Digital Curation Centre, UKOLN, University of Bath (March 2010)
4. Campbell, L.E.: Recollection: Integrating Data through Access. In: Agosti, M., Borbinha, J., Kapidakis, S., Papatheodorou, C., Tsakonas, G. (eds.) ECDL 2009. LNCS, vol. 5714, pp. 396–397. Springer, Heidelberg (2009)
5. Chang, H.: Enriched Content: Concept, Architecture, Implementation, and Applications. Ph.D. thesis, New York University (2003)
6. Collins, C., Viegas, F., Wattenberg, M.: Parallel tag clouds to explore and analyze faceted text corpora. In: IEEE Symposium on Visual Analytics Science and Technology, VAST 2009, pp. 91–98. IEEE (2009)
7. Dougherty, M., Meyer, E., Madsen, C., Van den Heuvel, C., Thomas, A., Wyatt, S.: Researcher engagement with web archives: State of the art (2010)

8. Hallgrímsson, T.: The International Internet Preservation Consortium (IIPC). In: Conference of Directors of National Libraries (CDNL 2005), Oslo, Norway, pp. 14–18 (2005)

9. Hockx-Yu, H.: The past issue of the web. In: Proceedings of the ACM WebSci Conference 2011, pp. 1–8 (2011)

10. Hodge, G.: An information life-cycle approach: Best practices for digital archiving. Journal of Electronic Publishing 5(4) (2000)

11. JaJa, J., Song, S.: Robust tools and services for long-term preservation of digital information. Library Trends 57(3) (2009)

12. Jatowt, A., Kawai, Y., Tanaka, K.: Visualizing historical content of web pages. In: Proceedings of the 17th International Conference on World Wide Web, pp. 1221–1222. ACM (2008)

13. Jatowt, A., Kawai, Y., Tanaka, K.: Page history explorer: Visualizing and comparing page histories. IEICE Transactions on Information and Systems 94(3), 564 (2011)

14. Kahle, B.: Preserving the Internet. Scientific American 276(3), 82–83 (1997)

15. Kwak, H., Lee, C., Park, H., Moon, S.: What is Twitter, a social network or a news media? In: Proceedings of the 19th International Conference on World Wide Web, pp. 591–600. ACM (2010)

16. McCown, F., Nelson, M.: What happens when facebook is gone? In: Proceedings of the 9th ACM/IEEE-CS Joint Conference on Digital Libraries, pp. 251–254. ACM (2009)

17. Nelson, M., McCown, F., Smith, J., Klein, M.: Using the web infrastructure to preserve web pages. International Journal on Digital Libraries 6(4), 327–349 (2007)

18. Petrovic, S., Osborne, M., Lavrenko, V.: The Edinburgh Twitter corpus. In: Proceedings of the NAACL HLT 2010 Workshop on Computational Linguistics in a World of Social Media, pp. 25–26 (2010)

19. Ronald Jantz, M., Mlis, M.: Digital archiving and preservation: Technologies and processes for a trusted repository. Journal of Archival Organization 4(1-2), 193–213 (2007)

20. Seadle, M.: Selection for digital preservation. Library Hi Tech. 22(2), 119–121 (2004)

21. Van de Sompel, H., Nelson, M., Sanderson, R., Balakireva, L., Ainsworth, S., Shankar, H.: Memento: Time travel for the web. Arxiv preprint arxiv: 0911.1112 (2009)

22. Song, S.: Long-term information preservation and access. Ph.D. thesis, University of Maryland, College Park (2011)

23. Thomas, A., Meyer, E., Dougherty, M., Van den Heuvel, C., Madsen, C., Wyatt, S.: Researcher engagement with web archives: Challenges and opportunities for investment (2010)

24. Yan, H., Huang, L., Chen, C., Xie, Z.: A new data storage and service model of China web infomall. In: 8th European Conference on Research and Advanced Technologies for Digital Libraries The 4th International Web Archiving Workshop (IWAW 2004), Bath, UK (2004)

25. Yang, J., Leskovec, J.: Patterns of temporal variation in online media. In: Proceedings of the fourth ACM International Conference on Web Search and Data Mining, pp. 177–186. ACM (2011)

Preserving Scientific Processes
from Design to Publications

Rudolf Mayer[1], Andreas Rauber[1],
Martin Alexander Neumann[2], John Thomson[3], and Gonçalo Antunes[4]

[1] Secure Business Austria
Favoritenstrasse 16, 1040 Vienna, Austria
{mayer,rauber}@sba-research.at
[2] Karlsruher Institut für Technologie
Kaiserstrasse 12, 76131 Karlsruhe, Germany
mneumann@teco.edu
[3] Caixa Magica Software
Rua Soeiro Pereira Gomes, Lote 1, 8 F, Lisbon, Portugal
john.thomson@caixamagica.pt
[4] INESC ID, Instituto de Engenharia de Sistemas e Computadores, Investigacao e
Desenvolvimento, Rua Alves Redol 9, 1000029 Lisbon, Portugal
goncalo.antunes@ist.utl.pt

Abstract. Digital Preservation has so far focused mainly on digital objects that are static in their nature, such as text and multimedia documents. However, there is an increasing demand to extend the applications towards dynamic objects and whole processes, such as scientific workflows in the domain of E-Science. This calls for a revision and extension of current concepts, methods and practices. Important questions to address are e.g. what needs to be captured at ingest, how do the digital objects need to be described, which preservation actions are applicable and how can the preserved objects be evaluated. In this paper we present a conceptual model for capturing the required information and show how this can be linked to evaluating the re-invocation of a preserved process.

Keywords: Digital Preservation, Context.

1 Introduction

Digital Preservation deals with ensuring the long-term access to digital information objects in the face of changing technologies or designated user communities. So far, the main focus of research in this area has targeted digital objects that are static in their nature, such as text and multimedia documents. There is however the need to extend the research towards dynamic objects (such as interactive art or video games), and beyond to whole processes and workflows. The latter is an emerging topic especially in disciplines such as E-Science, where data-intensive experiments form a core of the research. These experiments and their results need to be verifiable to others in the community. They need to be

P. Zaphiris et al. (Eds.): TPDL 2012, LNCS 7489, pp. 113–124, 2012.

preserved as researchers need to be able to reproduce and build on top of earlier experiments to verify and expand on the results. A recent report underlines the importance of *Big Data*, noting that it emerges as a new paradigm for scientific discovery that reflects the increasing value of observational, experimental and computer-generated data in virtually all domains, from physics to the humanities and social sciences [1], an aspect which has also been emphasised in the so-called fourth paradigm [3].

Business processes frequently also need to be preserved for issues such as liability cases, where e.g. a company needs to prove that it executed its processes correctly, and faults did not occur because of their manufacturing.

Preserving complete processes and other types of non-static digital objects calls for reassessing and extending current methods and practices. Important research questions to be addressed include

- What needs to be captured at ingest? We need to go beyond single files (and their metadata), up to potentially including and exceeding complete computer systems (or at least a description thereof to be able to recreate them at a later point), and any additional documents that might be needed to understand and operate this process. However, many of todays processes are not limited to single systems but make use of remote services enabled by the Internet of Services (IoS) and Software as a Service (SaaS), and these need to be captured as well .
- How do these digital objects need to be described? We need to characterise not only single files, but several different aspects of the process. This starts from a top-level where organisational parameters need to be described, down to the technical description of the systems the process depends on, including hardware, operating systems, software, and third-party libraries and services. We introduce a *process context model* that addresses these two questions, by proposing a set of aspects that need to be captured and means of storing them in a structured way.
- Which preservation actions are applicable? In practice, there will be a need for combining several different preservation actions, such as migration of specifications and documents, code migration/cross-compilation, or emulation of hardware or software utilised in the process.
- How can a preserved process be verified and evaluated? We need to ensure that the execution of the (modified) process at a later stage is equivalent to the original process. We show the applicability of a recent framework developed for comparing original and emulated versions of digital objects [2].

The remainder of this paper is structured as follows. Section 2 introduces a model to capture contextual information of a process. Section 3 then presents a scientific experiment as a case study for a process, and demonstrates how the context model can be applied. In Section 4, we discuss how verification of preserved processes can be achieved. Finally, we present conclusions and an outlook on future work in Section 5.

2 Capturing Process Context

Todays digital preservation approaches focus on preserving (mostly static) digital objects, and additional information about these objects. In terms of the Open Archival Information System (OAIS) [4], this additional information can be denoted as *Representation Information*, i.e. the information needed so that Designated Communities can understand the digital object at a later point in time, as well as Preservation Description Information (PDI), i.e. is the additional metadata needed to manage the preservation of the objects.

The context of information needed for preserving processes is considerably more complex, as it not only requires dealing with the structural properties of information, but also with the dynamic behaviour of processes.

The successful digital preservation of a business process requires capturing sufficient detail of the process, as well as its context, to be able to re-run and verify the original behaviour at a later stage, potentially under changed and evolved conditions. This may include different stakeholders and parties, different or evolved enabling technologies, different system components on both hardware and software levels, changed services (or terms of service) by external service providers (potentially caused as well by a change in technologies or components), and differences in other aspects of the context of the business process.

To enable digital preservation of business processes, it is therefore required to preserve the set of activities, processes and tools, which all together ensure continued access to the services and software which are necessary to reproduce the context within which information can be accessed, properly rendered and validated.

To address these challenges, we have devised a context model to systematically capture aspects of a process that are essential for its preservation and verification upon later re-execution. The model consists of approximately 240 elements, structured in around 25 major groups. It is implemented in the form of an ontology, which on the one hand allows for the hierarchical categorisation of aspects, and on the other hand shall enable reasoning, e.g. over the possibility of certain preservation actions for a specific process instance. The ontology is authored in the Web Ontology Language (OWL). We developed a set of plug-ins for the Protégé ontology editor to support easier working with the model.

The context model corresponds to some degree to the representation information network [5], modelling the relationships between an information object and its related objects, be it documentation of the object, constituent parts and other information required to interpret the object. However, our model extends on this to understand the entire context within which a process, potentially including human actors, is executed, forming a graph of all constituent elements and, recursively, their representation information.

The model was derived from the combination of two approaches. The first approach was a top-down approach, utilising models from enterprise architecture frameworks. Most prominently, we employed the Zachman Framework [11], a schema for classifying artefacts. It consists of a two-dimensional classification

	DATA *What*	FUNCTION *How*	NETWORK *Where*	PEOPLE *Who*	TIME *When*	MOTIVATION *Why*
Objective/Scope (contextual) *Role: Planner*	List of things important in the business	List of Business Processes	List of Business Locations	List of important Organizations	List of Events	List of Business Goal & Strategies
Enterprise Model (conceptual) *Role: Owner*	Conceptual Data/ Object Model	Business Process Model	Business Logistics System	Work Flow Model	Master Schedule	Business Plan
System Model (logical) *Role:Designer*	Logical Data Model	System Architecture Model	Distributed Systems Architecture	Human Interface Architecture	Processing Structure	Business Rule Model
Technology Model (physical) *Role:Builder*	Physical Data/Class Model	Technology Design Model	Technology Architecture	Presentation Architecture	Control Structure	Rule Design
Detailed Reprentation (out of context) *Role: Programmer*	Data Definition	Program	Network Architecture	Security Architecture	Timing Definition	Rule Speculation
Functioning Enterprise *Role: User*	Usable Data	Working Function	Usable Network	Functioning Organization	Implemented Schedule	Working Strategy

Fig. 1. Zachman Enterprise Architecture Framework

matrix, where the columns represent communication questions (why, how, what, who, where, and when), and the rows transformations, as shown in Figure 1.

For a bottom-up approach, existing taxonomies such as the PREMIS data dictionary [9], as well as a scenario based analysis, have been employed. A number of scenarios for business process preservation, from various domains, have been devised, and their relevant aspects been identified. One of these processes, a scientific experiment, will be described as case study in Section 3.

Two sections of this model are depicted in Figure 2. Each item represents a class of aspects, for which a specific instance of the context model then creates concrete members, which are then related to each other with properties.

Figure 2(a) details aspects on software and specifications. Technical dependencies on software and operating systems can be captured and described via CUDF (Common Upgradeability Description Format) [10] for systems which are based on packages, i.e. where there is a package universe (repositories) and a package manager application. Such an approach allows to capture the complete software setup of a specific configuration and store it as an instace of the context model. The setup can then be recreated from this model at a later point.

Related to the software installed, capturing information on the licences associated to them allows for verifying which preservation actions are permissible for a specific scenario. Software and/or its requirements are formally described in specification documents. Specific documents for a process are created as instances of the appropriate class, and related to the software components they describe.

Configuration, also depicted in Figure 2(a), is another important aspect, closely related to software (and hardware). It is fundamental to capture these aspects, as the specific configuration applied can alter the behaviour of an operating system or software component maybe even to a greater extent than the

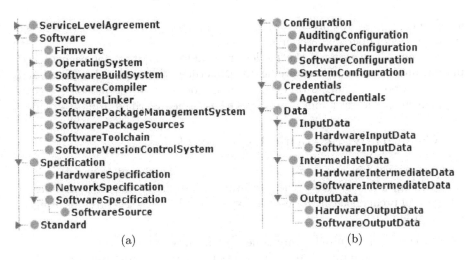

Fig. 2. Sections of the Context Model

specific version of a software utilised might influence the process outcome. Capturing this configuration might not always be easy. Again, in systems that rely on packages for their software, these packages tend to provide information about default locations for configuration files, which might be a start for capturing tools.

Another important aspect of the context model deals with several types of data consumed and created by a process, as seen in a section of Figure 2(b). We distinguish between data that originates from hardware or software, and whether this data is input to or output of the process, or created and consumed inside the process, i.e. output from one process step and input for another. Capturing this data is an important aspect in verifying that a re-execution of a process yields the same results as the original process, as we will detail in Section 3. It may be easily captured if the process is formally defined in a workflow engine. In other cases, it may be more difficult to obtain, e.g. by observing network traffic or system library calls.

Other aspects of the model cover for example human resources (including e.g. required qualifications for a certain role), actors, or legal aspects such as data protection laws. Location and time-based aspects need to be captured for processes where synchronisation between activities is important. Further important aspects are documentation and specifications, on all different levels, from high-level design documents of the process, use-case specifications, down to test documents, etc.

While the model is very extensive and captures a lot of aspects, it should be noted that a number of aspects can be filled automatically. Especially if institutions have well-defined and documented processes, capturing the roles involved, documents created and used, and similar aspects can be captured more easily. The software setup and the licenses associated with these software components are other aspects that are feasible to be automatically captured, at least on certain types of operating systems.

Also, not all sections of the model are equally important for each type of process and preservation requirement – sometimes the technical aspects migh be more important, and in other cases, more focus has to be put on e.g. legal requirements. Therefore, not every aspect has to be described in most detail.

3   Case Study: Scientific Experiment

The process used in our case study is a scientific experiment in the domain of data mining, where the researcher performs an automatic classification of music into a set of predefined categories. The experiment involves several steps, roughly in this order:

- Music data is acquired from online content providers.
- For this music data, genre assignments are obtained from websites such as Musicbrainz.org.
- A web-service is employed to extract numerical features describing certain characteristics of the audio files.
- The numerical description and the genre assignments are combined.
- This forms the basis for learning a machine learning model, which is finally employed to predict genre labels for unknown music.

Besides these steps, several scripts are used to convert data formats and for other similar tasks.

An implementation of this scientific experiment workflow with Taverna[7] is given in Figure 3. Taverna is one of the most prominent scientific workflow management systems (SWMS) existing today. It allows scientists to easily combine services (remote services or programs/scripts) and infrastructure for their research, and have a complete and documented model of their experiment process.

The process depends on several components that are not under direct control of the researcher, most notably the web service used to extract the numeric feature representation from the audio files. Also the software used for machine learning frequently has new releases, and changes in the process outcome might be due to bugs being solved or introduced. Thus there is the risk that that the workflow might lead to different results at a later execution time. In such an event, it is relevant to know that (a) something has changed, and (b) in which component of the workflow it has changed.

In the Taverna implementation, scripts and commands that would have been executed with the shell of the operating system have been migrated to scripts in the Taverna-supported language *beanshell*, based on the Java programming language. Note that this already constitutes some form of migration of the original process, at the same time rendering it more platform independent. Taverna is capable of capturing the data exchanged between the process steps as provenance data, which can be stored in the Taverna-specific format Janus [6] (the also available Open Provenance Model format [8] contains only information of the invoked process steps, but not the actual data).

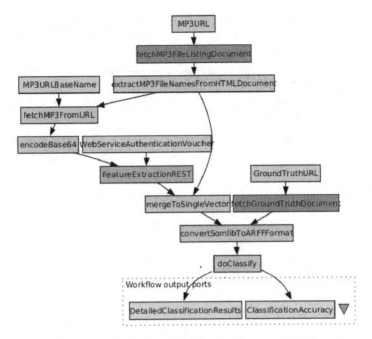

Fig. 3. Scientific workflow modelled in the Taverna Workflow engine

Through a series of iterations, we modelled this scientific experiment in the above presented Context Model. Figure 4 gives an overview on the concrete instances and their relations identified as relevant aspects of the business process context.

As the scientific experiment is a process mostly focusing on data processing, the majority of the identified aspects are in the technical domain – software components, external systems such as the web service to extract the numerical audio features from, or data exchanged and their format and specification. However, also goals and motivations are important aspects, as they might heavily influence the process. As such, the motivation for the providers of the external systems is relevant, as it might determine the future availability of these services. Commercial systems might be more likely to sustain than services operated by a single person for free.

Another important aspect in this process are licences – depending on which licence terms the components of our process are released under, different options of preservation actions might be available or not. For closed-source, proprietary software, migration to a new execution platform might be prohibited.

A central aspect in the scientific process is the AudioFeatureExtractionService, i.e. the remote web-service that provides the numeric representation for audio files. The service needs as input files encoded in the MP3 format (specified by the ISO standard 11172-3). More specifically, as they are binary files, they need to be further encoded with Base64, to allow for a data exchange over the HTTP protocol. The web-service further accepts a number of parameters

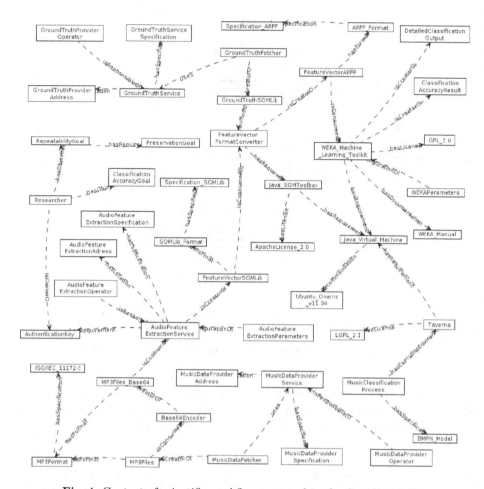

Fig. 4. Context of scientific workflow captured in the Context Model

that control the exact information captured in the numeric representation; they are specified in the AudioFeatureExtractionSpecification, which for example also covers a detailed information on how the extraction works. The service requires an authorisation key. The operator of the web-service provides the service for free, but grants authorisation keys that are non-transferable between different researchers. Finally, the feature extraction service provides the numeric description as ASCII file, following the SOMLib format specification.

As a software component used locally, the WEKA machine learning toolkit requires a Java Virtual Machine (JVM) platform to execute. The JVM in turn is available for many operating systems, but has been specifically tested on a Linux distribution, Ubuntu, version "Oneiric" 11.04. WEKA requires as input a feature vector in the ARFF Format, and a set of parameters controlling the learning algorithm. These parameters are specified in the WEKA documentation. As output result, the numeric performance metric "accuracy" is provided, as well as a textual, detailed description of the result. WEKA is distributed under the terms of the open-source GNU Public License (GPL) 2.0, which allows for source code modifications.

After this experimentation process, a subsequent process of result analysis and distillation is normally performed, taking input from the experiment outcomes, and finally leading to a publication of the research in the form of e.g. a conference or journal paper. This, again, may be modelled either as a single information object (the paper) connected to the process, and thus to all data and processing steps that led to the results published, or as a more complex process in its own, specifically if a paper reports on meta-studies across several experiment runs. This can be modelled via connections between the individual components, e.g. the paper (and its supporting representation information network and entire context model) and the process context model.

4   Evaluation of Preserved Processes

One important aspect in Digital Preservation is the verification that the pre-served object is equivalent to the original one, i.e. that the identified significant properties are equal. This is also valid when preserving business process – fi-nally, after capturing the context of a process, one needs to asses whether two executions of the specific business process (or workflow), are equivalent.

A framework for evaluating whether two versions of a digital object are equiva-lent is presented in [2]. To this end, the authors propose to measure and compare significant properties of the original and modified object. Important steps in the this framework include a (1) description of the original environment, (2) the identification of external events influencing the object's behaviour, (3) the deci-sion on what level to compare the two objects, (4) recreating the environment, (5) applying standardised input to both environments, and finally (6) extracting and (7) comparing the significant properties. Even though [2] focuses mostly on emulation of environments, the principles have also been discussed and are ap-plicable specifically for entire processes, and will work virtually unchanged also for migration approaches, when complex objects are transformed e.g into a new file format version.

Relating to the framework for comparing systems presented above, external events (2) in the scientific experiment can be the external system used – the ones for providing the data, and the ones providing functionality such as the feature extraction web-service. The level to compare two instances of a process (3) is primarily on the interface between the different process steps. Provenance data captured during process execution can be utilised to apply the standardised input (5) to the current instance of the workflow. Significant properties in the process are mostly the data output from the overall process, and from each intermediate step. Provenance data should be able to capture these significant properties (6).

```
<rdf:Description rdf:about="{nsTaverna}/2010/workflow/{idWF}/processor/
    MusicClassificationExperiment/out/ClassificationAccuracy">
  <janus:has_value_binding rdf:resource="{nsTaverna}/2011/
      data/{idDataGrp}/ref/{idDataPort0}"/>
  <rdfs:comment rdf:datatype="{nsW3}/2001/XMLSchema#string">
    ClassificationAccuracy
  </rdfs:comment>
  <janus:is_processor_input rdf:datatype="{nsW3}/2001/XMLSchema#boolean">
    false
  </janus:is_processor_input>
  <janus:has_port_order rdf:datatype="{nsW3}/2001/XMLSchema#long">
    0
  </janus:has_port_order>
  <rdf:type rdf:resource="http://purl.org/net/taverna/janus#port"/>
</rdf:Description>
<rdf:Description rdf:about="{nsTaverna}/2011/data/{idDataGrp}/ref/{idDataPort0}">
  <rdfs:comment rdf:datatype="{nsW3}/2001/XMLSchema#string">
    80.0
  </rdfs:comment>
  <janus:has_port_value_order rdf:datatype="{nsW3}/2001/XMLSchema#long">
    1
  </janus:has_port_value_order>
  <janus:has_iteration rdf:datatype="{nsW3}/2001/XMLSchema#string">
    []
  </janus:has_iteration>
  <rdf:type rdf:resource="http://purl.org/net/taverna/janus#port_value"/>
</rdf:Description>
```

Listing 1.1. Example provenance data of Taverna for the process output *ClassificationAccuracy* (cf. Figure 3). The first RDF Description element defines the output port *ClassificationAccuracy*, the second element contains the actual value of "80.0". Note that some identifiers have been abbreviated, marked by {...}.

```
<rdf:Description rdf:about="{nsTaverna}/2010/workflow/{idWF}/processor/
    MusicClassificationExperiment/out/DetailedClassificationResults">
  <janus:has_value_binding rdf:resource="{nsTaverna}/2011/
      data/{idDataGrp}/ref/{idDataPort1}"/>
  <rdfs:comment rdf:datatype="{nsW3}/2001/XMLSchema#string">
    DetailedClassificationResults
  </rdfs:comment>
  <janus:is_processor_input rdf:datatype="{nsW3}/2001/XMLSchema#boolean">
    false
  </janus:is_processor_input>
  <janus:has_port_order rdf:datatype="{nsW3}/2001/XMLSchema#long">
    0
  </janus:has_port_order>
  <rdf:type rdf:resource="http://purl.org/net/taverna/janus#port"/>
</rdf:Description>

<rdf:Description rdf:about="{nsTaverna}/2011/data/{idDataGrp}/ref/{idDataPort1}">
  <rdfs:comment rdf:datatype="{nsW3}/2001/XMLSchema#string">
    1    2:Hip-Hop    2:Hip-Hop        0.667  (3.359461)
    2    2:Hip-Hop    2:Hip-Hop        0.667  (3.294687)
    3    1:Classica   1:Classica       0.667  (2.032687)
    4          3:Jazz       3:Jazz     0.667  (2.536849)
    5    1:Classica   1:Classica       0.667  (1.31727)
    6    1:Classica         3:Jazz  +  0.667  (3.46771)
    7          3:Jazz 1:Classica    +  0.333  (2.159764)
    8    2:Hip-Hop    2:Hip-Hop        0.667  (3.127645)
    9          3:Jazz       3:Jazz     0.667  (3.010563)
    10   2:Hip-Hop    2:Hip-Hop        0.667  (4.631316)
  </rdfs:comment>
</rdf:Description>
```

Listing 1.2. Example provenance data of Taverna for the process output *DetailedClassificationResults* (cf. Figure 3). The first RDF Description element defines the output port *DetailedClassificationResults*, the second element contains the actual value, one entry for each file tested, with the actual class, the predicted class, and the confidence of the classifier in the prediction. Note that some identifiers have been abbreviated, marked by {...}.

The recorded provenance data can be utilised to verify whether the new version of the process still renders the same results. To this end, as the evaluation framework suggests, one can automatically reapply the inputs and verify the recorded outputs, similar to what would be performed in automated software testing. An example of the provenance data recorded for the two process outputs, the percentage of correctly classified instances, and the detailed classification results, are given in Listings 1.1 and 1.2. Note that some unique identifiers, such as URLs as namespaces, and identifiers for the workflow and specific data elements, have been abbreviated for space reasons.

Each listing contains two RDF *Description* elements, where the first one defines the output port, and contains as a sub-element the identifier of the element containing the actual value, which is the second *Description* element in both listings. With the identifiers used in the *rdf:about* attributes, it is possible to uniquely identify the process step (and iteration, if the step is looped over) the data originates from.

The provenance data can further be used for implementing a watch service for software and external service dependencies, e.g. by periodically executing the process with all historic recordings of previous executions, either as a complete process, or for each process step individually.

5   Conclusions and Future Work

Preservation of business processes is a challenge that has so far not been tackled by Digital Preservation research. However, there is a need to re-run processes at a later time, for example in areas such as E-Science, or for liability reasons, when one needs to show that a process was executed correctly, maybe according to some contract. In this paper, we therefore presented a model to capture and document important aspects of a process and its context, enabling the preservation of the process for later re-execution. We showed how this model can be utilised to describe a process from the E-Science domain, a scientific experiment in the area of machine learning.

Many aspects of the context model can be populated automatically, such as the software dependencies and their licences. This holds especially true if these processes are formally defined and modelled, and executed in a workflow engines. Then, one can automatically record provenance data, i.e. the input and output data of the whole process, as well as data exchanged between different process steps. The evaluation and verification of the preserved process is a challenging task, but it can be easier performed in such fully documented processes. This provenance data can then be employed to verify whether later executions of the process render the same results.

Future work includes development of software components to extract more aspects of the business process in an automatic way, and modules to verify the preserved process to be equivalent to the original instance. One particular research challenge will be the complexity of preserving processes which might be composed of hundreds of parts and components, which all might have dependencies on each other. This will require more refined approaches than traditional Digital Preservation offers today.

Acknowledgments. Part of this work was supported by the project TIMBUS, partially funded by the EU under the FP7 contract 269940.

References

1. Glinos, K.: E-infrastructures for big data: Opportunities and challenges. ERCIM News 2012(89) (2012)
2. Guttenbrunner, M., Rauber, A.: A Measurement Framework for Evaluating Emulators for Digital Preservation. ACM Transactions on Information Systems (TOIS) 30(2) (2012)
3. Hey, T., Tansley, S., Tolle, K. (eds.): The Fourth Paradigm: Data-Intensive Scientific Discovery. Microsoft Research, Redmond, Washington (2009)
4. International Organization For Standardization. OAIS: Open Archival Information System - Reference Model, Ref. No ISO 14721:2003 (2003)
5. Marketakis, Y., Tzitzikas, Y.: Dependency management for digital preservation using semantic web technologies. International Journal on Digital Libraries 10, 159–177 (2009)
6. Missier, P., Sahoo, S.S., Zhao, J., Goble, C., Sheth, A.: *Janus*: From Workflows to Semantic Provenance and Linked Open Data. In: McGuinness, D.L., Michaelis, J.R., Moreau, L. (eds.) IPAW 2010. LNCS, vol. 6378, pp. 129–141. Springer, Heidelberg (2010)
7. Missier, P., Soiland-Reyes, S., Owen, S., Tan, W., Nenadic, A., Dunlop, I., Williams, A., Oinn, T., Goble, C.: Taverna, Reloaded. In: Gertz, M., Ludäscher, B. (eds.) SSDBM 2010. LNCS, vol. 6187, pp. 471–481. Springer, Heidelberg (2010)
8. Moreau, L., Freire, J., Futrelle, J., McGrath, R.E., Myers, J., Paulson, P.: The Open Provenance Model: An Overview. In: Freire, J., Koop, D., Moreau, L. (eds.) IPAW 2008. LNCS, vol. 5272, pp. 323–326. Springer, Heidelberg (2008)
9. PREMIS Editorial Committee. Premis data dictionary for preservation metadata. Technical report (March 2008)
10. Treinen, R., Zacchiroli, S.: Description of the CUDF Format. Technical report (2008), http://arxiv.org/abs/0811.3621
11. Zachman, J.A.: A framework for information systems architecture. IBM Systems Journal 26(3), 276–292 (1987)

Losing My Revolution:
How Many Resources Shared on Social Media Have Been Lost?

Hany M. SalahEldeen and Michael L. Nelson

Old Dominion University, Department of Computer Science
Norfolk VA, 23529, USA
{hany,mln}@cs.odu.edu

Abstract. Social media content has grown exponentially in the recent years and the role of social media has evolved from just narrating life events to actually shaping them. In this paper we explore how many resources shared in social media are still available on the live web or in public web archives. By analyzing six different event-centric datasets of resources shared in social media in the period from June 2009 to March 2012, we found about 11% lost and 20% archived after just a year and an average of 27% lost and 41% archived after two and a half years. Furthermore, we found a nearly linear relationship between time of sharing of the resource and the percentage lost, with a slightly less linear relationship between time of sharing and archiving coverage of the resource. From this model we conclude that after the first year of publishing, nearly 11% of shared resources will be lost and after that we will continue to lose 0.02% per day.

Keywords: Web Archiving, Social Media, Digital Preservation.

1   Introduction

With more than 845 million Facebook users at the end of 2011 [5] and over 140 million tweets sent daily in 2011 [16] users can take photos, videos, post their opinions, and report incidents as they happen. Many of the posts and tweets are about quotidian events and their preservation is debatable. However, some of the posts and events are about culturally important events whose preservation is less controversial. In this paper we shed light on the importance of archiving social media content about these events and estimate how much of this content is archived, still available, or lost with no possibility of recovery.

To emphasize the culturally important commentary and sharing, we collected data about six events in the time period of June 2009 to March 2012: the H1N1 virus outbreak, Michael Jackson's death, the Iranian elections and protests, Barack Obama's Nobel Peace Prize, the Egyptian revolution, and the Syrian uprising.

P. Zaphiris et al. (Eds.): TPDL 2012, LNCS 7489, pp. 125–137, 2012.

2   Related Work

To our knowledge, no prior study has analyzed the amount of shared resources in social media lost through time. There have been many studies analyzing the behavior of users within a social network, how they interact, and what content they share [3, 19, 20, 23]. As for Twitter, Kwak et al. [6] studied its nature and its topological characteristics and found a deviation from known characteristics of human social networks that were analyzed by Newman and Park [10]. Lee analyzed the reasons behind sharing news in social media and found that informativeness was the strongest motivation in predicting news sharing intention, followed by socializing and status seeking [4]. Also shared content in social media like Twitter move and diffuse relatively fast as stated by Yang et al. [22].

Further more, many concerns were raised about the persistence of shared resources and web content in general. Nelson and Allen studied the persistence of objects in a digital library and found that, with just over a year, 3% of the sample they collected have appeared to no longer be available [9]. Sanderson et al. analyzed the persistence and availability of web resources referenced from papers in scholarly repositories using Memento and found that 28% of these resources have been lost [14]. Memento [17] is a collection of HTTP extensions that enables uniform, inter-archive access. Ainsworth et al. [1] examined how much of the web is archived and found it ranges from 16% to 79%, depending on the starting seed URIs. McCown et al. examined the factors affecting reconstructing websites (using caches and archives) and found that PageRank, Age, and the number of hops from the top-level of the site were most influential [8].

3   Data Gathering

We compiled a list of URIs that were shared in social media and correspond to specific culturally important events. In this section we describe the data acquisition and sampling process we performed to extract six different datasets which will be tested and analyzed in the following sections.

3.1   Stanford SNAP Project Dataset

The Stanford Large Network Dataset is a collection of about 50 large network datasets having millions of nodes, edges and tuples. It was collected as a part of the Stanford Network Analysis Platform (SNAP) project [15]. It includes social networks, web graphs, road networks, Internet networks, citation networks, collaboration networks, and communication networks. For the purpose of our investigation, we selected their Twitter posts dataset. This dataset was collected from June 1st, 2009 to December 31st, 2009 and contains nearly 476 million tweets posted by nearly 17 million users. The dataset is estimated to cover 20%-30% of all posts published on Twitter during that time frame [21]. To select which events will be covered in this study, we examined CNN's 2009 events timeline[1].

---

[1] http://www.cnn.com/2009/US/12/16/year.timeline/index.html

We wanted to select a small number of events that were diverse, with limited overlap, and relatively important to a large number of people. Given that, we selected four events: the H1N1 virus outbreak, the Iranian protests and elections, Michael Jackson's death, and Barrack Obama's Nobel Peace Prize award.

Preparation: A tweet is typically composed of text, hashtags, embedded resources or URIs and usertags all spanning a maximum of 140 characters. Here is an example of a tweet record in the SNAP dataset:

```
T    2009-07-31 23:57:18
U    http://Twitter.com/nickgotch
W    RT @rockingjude: December 21, 2009 Depopulation by Food Will Begin http://is.gd/1WMZb
     WHOA..BETTER WATCH RT plz #pwa #tcot
```

The line starting with the letter **T** indicates the date and time of the tweet creation. While the line starting with **U** shows a link to the user who authored this particular tweet. Finally, the line starting with **W** shows the entire tweet including all the user-references "@rockingjude", the embedded URIs "http://is.gd/1WMZb", and hashtags "#pwa #tcot".

Tag Expansion: We wanted to select tweets that we can say with high confidence are about a selected event. In this case, precision is more important than recall as collecting every single tweet published about a certain event is less important than making sure that the selected tweets are definitely about that event. Several studies focused on estimating the aboutness of a certain web page or a resource in general [12, 18]. Fortunately in Twitter, hashtags incorporated within a tweet can help us estimate their "*aboutness*". Users normally add certain hashtags to their tweets to ease the search and discoverability in following a certain topic. These hashtags will be utilized in the event-centric filtration process.

For each event, we selected initial tags that describe it (Table 1). Those initial tags were derived empirically after examining some event-related tweets. Next we extracted all the hashtags that co-occurred with our initial set of hashtags. For example, in class H1N1 we extracted all the other hashtags that appeared along with *#h1n1* within the same tweet and kept count of their frequency. Those extracted hashtags were sorted in descending order of the frequency of their appearance in tweets. We removed all the general scope tags like *#cnn*, *#health*, *#death*, *#war* and others. In regards to aboutness, removing general tags will indeed decrease recall but will increase precision. Finally we picked the top 8-10 hashtags to represent this event-class and be utilized in the filtration process. Table 1 shows the final set of tags selected for each class.

Tweet Filtration: In the previous step we extracted the tags that will help us classify and filter tweets in the dataset according to each event. This filtration process aims to extract a reasonable sized dataset of tweets for each event and to minimize the inter-event overlap. Since the life and persistence of the tweet itself is not the focus of this study but rather the associated resource that appears

Table 1. Twitter hashtags generated for filtering and their frequency of occurring

Event	Initial Hashtags	Top Co-occurring Hashtags
H1N1	'h1n1'	'swine'=61,829 'swineflu'=56,419 'flu'=8,436
Outbreak	=61,351	'pandemic'=6,839 'influenza'=1,725 'grippe'=1,559 'tamiflu'=331
M. Jackson's	'michaeljackson'	'michael'=27,075 'mj'=18,584 'thisisit'8,770 'rip'=3,559 'jacko'=3,325
Death	=22,934	'kingofpop'=2,888 'jackson'=2,559 'thriller'=1,357 'thankyoumichael'=1,050
Iranian	'iranelection'	'iran'949,641 'gr88'=197,113'tehran'=109,006 'freeiran'=13,378
Elections	=911,808	'neda'=191,067 'mousavi'=16,587 'united4iran'=9,198 'iranrevolution'=7,295
Obama's	'obama'=48,161 &	'nobel'=2,261 'obamanobel'=14 'nobelprize' 'nobelpeace'=113
Nobel Prize	'peace'=3,721	'barack'=1292 'nobelpeaceprize'=107

in the tweet (image, video, shortened URI or other embedded resource), we will extract only the tweets that contain an embedded resource. This step resulted in 181 million tweets with embedded resources (http://is.gd/1WMZb in the prior example). These tweets were further filtered to keep only the tweets that have at least one of the expanded tags obtained from Table 1. The number of tweets after this phase reached 1.1 million tweets.

Filtering the tweets based on the occurrence of at least one of the hashtags only is undesirable as it will cause two problems: First, it will introduce possible event overlap due to general tweets talking about two or more topics. Second, is that using only the single occurrence of these tags will yield a huge amount of tweets and we need to reduce this size to reach a more manageable size. Intuitively speaking, strongly related hashtags will co-occur often. For example, a tweet that has #h1n1 along with #swineflu and #pandemic is most likely about the H1N1 outbreak rather than a tweet having just the tag #flu or just #sick. Filtering with this co-occurrence will in turn solve both problems as by increasing relevance to a particular event, general tweets that talk about several events will be filtered out thus diminishing the overlap, and in turn it will reduce the size of the dataset.

Next, we increase the precision of the tweets associated with each event from the set of 1.1 million tweets. In the first iteration we selected the tag that had the highest frequency of co-occurrence in the dataset with the initial tag and added it to a set we will call the selection set. After that we check the co-occurrence of all the remaining extracted tags with the tag in the selection set and record the frequencies of co-occurrence. After sorting the frequencies of co-occurrence with the tag from the selection set, we pick the highest one to keep add it to the selection set. We repeat this step of counting co-occurrences but with all the previously extracted hashtags in the selection set from previous iterations.

To elaborate, for H1N1 assume that the hastag '#h1n1' had the highest frequency of appearance in the dataset so we add it to the selection set. In the next iteration we record the how many times each tag in the list appeared along with '#h1n1' in a same tweet. If we selected '#swine' as the one with the highest frequency of occurrence with the initial tag '#h1n1' we add it to the selection list and in the next iteration we record the frequency of occurrence of the remaining hashtags with both of the extracted tags '#h1n1' and '#swine'. We repeat this step, for each event, to the point where we have a manageable size dataset which we are confident in its 'aboutness' in relation to the event.

Table 2. Tweet Filtration iterations and final tweet collections

Event	Hashtags selected for filteration	Tweets Extracted	Operation Performed	Final Tweets
MJ	michael	27,075		
	michael & michaeljackson	**22,934**	Sample 10%	**2,293**
Iran	iran	949,641		
	iran & iranelection	911,808		
	iran & iranelection & gr88	189,757		
	iran & iranelection & gr88 & neda	91,815		
	iran & iranelection & gr88 & neda & tehran	**34,294**	Sample 10%	**3,429**
H1N1	h1n1	61,351		
	h1n1 & swine	44,972		
	h1n1 & swine & swineflu	42,574		
	h1n1 & swine & swineflu & pandemic	**5,517**	Take All	**5,517**
Obama	obama	48,161		
	obama & nobel	**1,118**	Take All	**1,118**

Two problems appeared from this approach with the Iran and Michael Jackson datasets. In the Iran dataset the number of tweets was in hundreds of thousands and even with 5 tags co-occurrence it was still about 34K+ tweets. To solve this we performed a random sampling from those resulting tweets to take only 10% of them resulting in a smaller manageable dataset. The second problem with the Michael Jackson dataset upon using 5 tags to decrease it to a manageable size we realized there were few unique domains for the embedded resources. A closer look revealed this combination of tags was mostly border-line tweet spam (MJ ringtones). To solve this we used only the two top tags "#michael" and "#michaeljackson", and then we randomly sampled 10% of the resulting tweets to reach the desired dataset size (Table 2).

3.2 Egyptian Revolution Dataset

The one year anniversary of this event was the original motivation for this study [13]. In this case, we started with an event and then tried to get social media content describing it. Despite its ubiquity, gathering social media for a past event is surprisingly hard. We picked the Egyptian revolution due to the role of the social media in curating and driving the incidents that led to the resignation of the president. Several initiatives were commenced to collect and curate the social media content during the revolution like R-sheif.org[2] which specializes in social content analysis of the issues in the Arab world by using aggregate data from Twitter and the Web. We are currently in the process of obtaining the millions of records related to the Arab Spring of 2011. Meanwhile, we decided to build our own dataset manually.

There are several sites that curate resources about the Egyptian Revolution and we want to investigate as many of them as possible. At the same time,

---

[2] http://www.r-shief.org/

we need to diversify our resources and the types of digital artifacts that are embedded in them. Tweets, videos, images, embedded links, entire web pages and books were included in our investigation. For the sake of consistency, we limited our analysis to resources created within the period from the 20th of January 2011 to the 1st of March 2011. In the next subsections we explain each of the resources we utilized in our data acquisition in detail.

Storify: Storify is a website that enables users to create stories by creating collections of URIs (e.g., Tweets, images, videos, links) and arrange them temporally. These entries are posted by reference to their host websites. Thus, adding content to Storify does not necessarily mean it is archived. If a user added a video from YouTube and after a while the publisher of that video decided to remove it from YouTube the user is left with a gap in their Storify entry. For this purpose we gathered all the Storify entries that were created between 20th of January 2011 and the 1st of March 2011, resulting in 219 unique resources.

IAmJan25: Some entire websites were dedicated as a collection hub of media to curate the revolution. Based on public contributions, those websites collect different types of media, classify them, order them chronologically and publish them to the public. We picked a website named IAmJan25.com, as an example of these websites, to analyze and investigate. The administrators of the website received selected videos and images for notable events and actions that happened during the revolution. Those images and videos were selected by users as they vouched for them to be of some importance and they send the resource's URI to the web site administrators. The website itself is divided into two collections: a video collection and an image collection. The video collection had 2387 unique URIs while the image collection had 3525 unique URIs.

Tweets From Tahrir: Several books were published in 2011 documenting the revolution and the Arab Spring. To bridge the gap between books and digital media we analyzed a book entitled *Tweets from Tahrir* [11] which was published on April 21st, 2011. As the name states, this book tells a story formed by tweets of people during the revolution and the clashes with the past regime. We analyzed this book as a collection of tweets that had the luxury of a paperback preservation and focused on the tweeted media, in this case images. The book had a total of 1118 tweets having 23 unique images.

3.3   Syria Dataset

This dataset has been selected to represent a current (March 2012) event. Using the Twitter search API, we followed the same pattern of data acquisition as in section 3.1. We started with one hashtag, #Syria, and expanded it. Table 3 show the tags produced from the tag expansion step. After that each of those tags were input into a process utilizing the Twitter streaming API and produced the first 1000 results matching each tag. From this set, we randomly sampled 10%. As a result, 1955 tweets were extracted each having one or more embedded resources and tags from the expanded tags in Table 3.

Table 3. Twitter #Tags generated for filtering the Syrian uprising

Initial Hashtags	Extracted Hashtags
'Syria'	'Bashar' 'RiseDamascus' 'GenocideInSyria' 'STOPASSAD2012' 'AssadCrimes' 'Assad'

Table 4 shows the resources collected along with the top level domains that those resources belong to for each event.

Table 4. The top level domains found for each event ordered descendingly by the number of resources

Event	Top Domains (number of resources found)
MJ	youtube (110), twitpic (45), latimes (43), cnn (30), amazon (30)
Iran	youtube (385), twitpic (36), blogspot (30), roozonline (29)
H1N1	rhizalabs (676), reuters (17), google (16), flutrackers (16), calgaryherald (11)
Obama	blogspot (16), nytimes (15), wordpress (12), youtube (11), cnn (10)
Egypt	youtube (2414), cloudfront (2303), yfrog (1255), twitpic (114), imageshack.us (20)
Syria	youtube (130), twitter (61), hostpic.biz (9), telegraph.co.uk (5)

4   Uniqueness and Existence

From the previous data gathering step we obtained six different datasets related to six different historic events. For each event we extracted a list of URIs that were shared in tweets or uploaded to sites like Storify or IAmJan25. To answer the question of how much of the social media content is missing we test those URIs for each dataset to eliminate URI aliases in which several URIs identify to the same resource. Upon obtaining those unique URIs we examine how many of which are still available on the live web and how many are available in public web archives.

4.1   Uniqueness

Some URIs, especially those that appear in Twitter, may be aliases for the same resource. For example "http://bit.ly/2EEjBl" and "http://goo.gl/2ViC" both resolve to "http://www.cnn.com". To solve this, we resolved all the URIs following redirects to the final URI. The HTTP response of the last redirect has a field called *location* that contains the original long URI of the resource. This step reduced the total number of URIs in the six datasets from 21,625 to 11,051. Table 5 shows the number of unique resources in every dataset.

Table 5. Percentages of unique resources from all the extracted ones we obtained per event and the percentages of presence of those unique resources on live web and in archives. All resources = 21,625, Unique resources = 11,051.

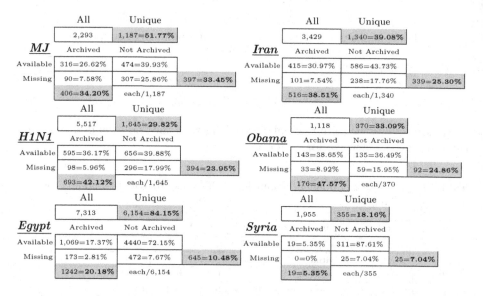

4.2   Existence on the Live-Web

After obtaining the unique URIs from the previous step we resolve all of them and classify them as Success or Failure. The *Success* class includes all the resources that ultimately return a "200 OK" HTTP response. The *Failure* class includes all the resources that return a "4XX" family response like: "404 Not Found", "403 Forbidden" and "410 Gone", the "30X" redirect family while having infinite loop redirects, and server errors with response "50X". To avoid transient errors we repeated the requests, on all datasets, several times for a week to resolve those errors.

We also test for "Soft 404s", which are pages that return "200 OK" response code but are not a representation of the resource, using a technique based on a heuristic for automatically discovering soft 404s from Bar-Yossef et al. [2]. We also include no response from the server, as well as DNS timeouts, as failures. Note that failure means that this resource is *missing* on the live web. Table 5 summarizes, for each dataset, the total percentages of the resources missing from the live web and the number of missing resources divided by the total number of unique resources.

4.3   Existence in the Archives

In the previous step we tested the existence of the unique list of URIs for each event on the live web. Next, we evaluate how many URIs have been archived in public web archives. To check those archives we utilize the Memento framework. If there is a memento for the URI, we download its memento timemap and analyze it. The timemap is a datestamp ordered list of all known archived versions (called "mementos") of a URI. Next, we parse this timemap and extract the number of mementos that point to versions of the resource in the public archives. We declare the resource to be archived if it has at least one memento. This step was also repeated several times to avoid the transient states of the archives before deeming a resource as unarchived. The results of this experiment along with the archive coverage percentage are presented in Table 5.

5   Existence as a Function of Time

Inspecting the results from the previous steps suggests that the number of missing shared resources in social media corresponding to an event is directly proportional with its age. To determine dates for each of the events this we extracted all the creation dates from all the tweet-based datasets and sorted them. For each event, we plotted a graph illustrating the number of tweets per day related to that event as shown in figure 1. Since the dataset is separated temporally into 3 partitions, and in order to display all the events on one graph we reduced the size of the x-axis by removing the time periods not covered in our study.

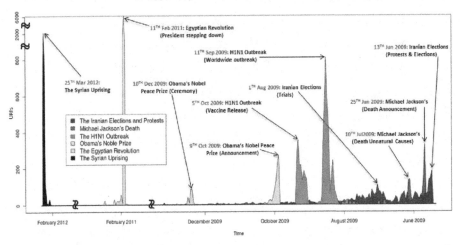

Fig. 1. URIs shared per day corresponding to each event and showing the two peaks in the non-Syrian and non-Egyptian events

Upon examining the graph we found an interesting phenomena in the non-Syrian and non-Egyptian events: each event has two peaks. Upon investigating history timelines we came to conclusion that those peaks reflect a second wave of social media interaction as a result of new incident within the same event after a period of time. For example, in the H1N1 dataset, the first peak illustrates the world-wide outbreak announcement while the second peak denotes the release of the vaccine. In the Iran dataset, the first peak shows the peak of the elections while the second peak pinpoints the Iranian trials. As for the MJ dataset the first peak corresponds to his death and the second peak describes the rumors that Michael Jackson died of unnatural causes and a possible homicide. For the Obama dataset, the first peak reveals the announcement of his winning the prize while the second peak presents the award-giving ceremony in Oslo. For the Egyptian evolution, the resources are all within a small time slot of 2 weeks around the date 11th of February. As for the Syrian event, since the collection was very recent there was no obvious peaks. Those peaks we examined will become temporal centroids of the social content collections (the datasets). MJ (June 25th & July 10th 2009), Iran (June 13th & 1st August 2009), H1N1 (September 11th & 5th October 2009), and Obama (October 9th & December 10th 2009). Egypt was (February 11th 2011) and the Syria dataset also had one centroid on March 27th 2012. We split each event according to the two centroids in each event accordingly. Figure 1 shows those peaks and Table 6 shows the missing content and the archived content percentages corresponding to each centroid.

Table 6. The Split Dataset

	MJ		Iran		H1N1		Obama		Egypt	Syria
% Missing	36.24%	31.62%	26.98%	24.47%	23.49%	25.64%	24.59%	26.15%	10.48%	7.04%
% Archived	39.45%	30.78%	43.08%	36.26%	41.65%	43.87%	47.87%	46.15%	20.18%	5.35%

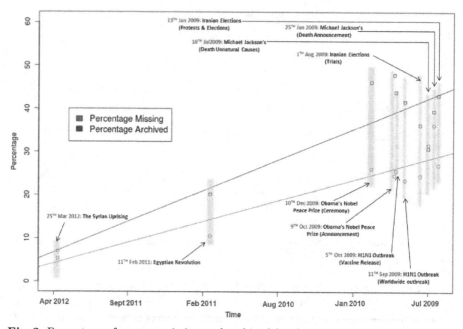

Fig. 2. Percentage of content missing and archived for the events as a function of time

Figure 2 shows the missing and archived values from Table 6 as a function of time since shared. Equation 1 shows the modeled estimate for the percentage of shared resources lost, where *Age* is in days. While there is a less linear relationship between time and being archived, equation 2 shows the modeled estimate for the percentage of shared resources archived in a public archive.

$$Content\ Lost\ Percentage = 0.02(Age\ in\ days) + 4.20 \qquad (1)$$

$$Content\ Archived\ Percentage = 0.04(Age\ in\ days) + 6.74 \qquad (2)$$

Given these observations and our curve fitting we estimate that after a year from publishing about 11% of content shared in social media will be gone. After this point, we are losing roughly 0.02% of this content per day.

6   Conclusions and Future work

We can conclude that there is a nearly linear relationship between time of sharing in the social media and the percentage lost. Although not as linear, there is a similar relationship between the time of sharing and the expected percentage of coverage in the archives. To reach this conclusion, we extracted collections of tweets and other social media content that was posted and shared in relation to six different events that occurred in the time period from June 2009 to March 2012. Next we extracted the embedded resources within this social media content and tested their existence on the live web and in the archives. After analyzing the percentages lost and archived in relation to time and plotting them we used a linear regression model to fit those points. Finally we presented two linear models that can estimate the existence of a resource, that was posted or shared at one point of time in the social media, on the live web and in the archives as a function of age in the social media.

In the next stage of our research we need to expand the datasets and import other similar datasets especially in the uncovered temporal areas (e.g., the year of 2010 and before 2009). Examining more datasets across extended points in time could enable us to better model these two functions of time. Also several other factors beside time would be analyzed to understand their effect on persistence on the live web and archiving coverage like: publishing venue, rate of sharing, popularity of authors and the nature of the related event.

Acknowledgments. This work was supported in part by the Library of Congress and NSF IIS-1009392.

References

1. Ainsworth, S.G., Alsum, A., SalahEldeen, H., Weigle, M.C., Nelson, M.L.: How Much of the Web Is Archived? In: Proceedings of the 11th Annual International ACM/IEEE Joint Conference on Digital Libraries, JCDL 2011, pp. 133–136 (2011)

2. Bar-Yossef, Z., Broder, A.Z., Kumar, R., Tomkins, A.: Sic Transit Gloria Telae: Towards an Understanding of the Web's Decay. In: Proceedings of the 13th International Conference on World Wide Web, WWW 2004, pp. 328–337 (2004)

3. Benevenut, F., Rodrigues, T., Cha, M., Almeida, V.: Characterizing User Behavior in Online Social Networks. In: Proc. of ACM SIGCOMM Internet Measurement Conference, SIGCOMM 2009, pp. 49–62 (2009)

4. Lee, C.S., Ma, L., Goh, D.H.-L.: Why Do People Share News in Social Media? In: Zhong, N., Callaghan, V., Ghorbani, A.A., Hu, B. (eds.) AMT 2011. LNCS, vol. 6890, pp. 129–140. Springer, Heidelberg (2011)

5. Facebook official fact sheet, http://newsroom.fb.com/content/default.aspx?NewsAreaId=22

6. Kwak, H., Lee, C., Park, H., Moon, S.: What is Twitter, a Social Network or a News Media? In: Proceedings of the 19th International Conference on World Wide Web, WWW 2010, pp. 591–600 (2010)

7. Mohr, G., Kimpton, M., Stack, M., Ranitovic, I.: Introduction to Heritrix, an Archival Quality Web Crawler. In: 4th International Web Archiving Workshop, IWAW 2004 (2004)

8. McCown, F., Diawara, N., Nelson, M.L.: Factors Affecting Website Reconstruction from the Web Infrastructure. In: Proceedings of the 7th ACM/IEEE-CS Joint Conference on Digital Libraries, JCDL 2007, pp. 39–48 (2007)

9. Nelson, M.L., Danette Allen, B.: Object Persistence and Availability in Digital Libraries. D-Lib Magazine 8(1) (January 2002)

10. Newman, M.E.J., Park, J.: Why social networks are different from other types of networks. Phys. Rev. E 68(3), 036122 (2003)

11. Nunns, A., Idle, N.: Tweets From Tahrir, ISBN-10: 1935928457

12. Phelps, T.A., Wilensky, R.: Robust Hyperlinks Cost Just Five Words Each. Technical Report, UCB/CSD-00-1091, EECS Department, University of California, Berkeley (2000)

13. SalahEldeen, H.M., Nelson, M.L.: Losing My Revolution: A year after the Egyptian Revolution, 10% of the social media documentation is gone, http://ws-dl.blogspot.com/2012/02/2012-02-11-losing-my-revolution-year.html

14. Sanderson, R., Phillips, M., Van de Sompel, H.: Analyzing the Persistence of Referenced Web Resources with Memento. CoRR, arXiv:1105.3459 (2011)

15. Stanford SNAP Project Dataset, http://snap.stanford.edu/

16. Twitter numbers, http://blog.Twitter.com/2011/03/numbers.html

17. Van de Sompel, H., Nelson, M.L., Sanderson, R., Balakireva, L.L., Ainsworth, S., Shankar, H.: Memento: Time Travel for the Web. Technical Report, arXiv:0911.1112 (November 2009)

18. Wan, X., Yang, J.: Wordrank-based Lexical Signatures for Finding Lost or Related Web Pages. In: Zhou, X., Li, J., Shen, H.T., Kitsuregawa, M., Zhang, Y. (eds.) APWeb 2006. LNCS, vol. 3841, pp. 843–849. Springer, Heidelberg (2006)

19. Wilson, C., Boe, B., Sala, A., Puttaswamy, K.P., Zhao, B.Y.: User Interactions in Social Networks and their Implications. In: Proceedings of the 4th ACM European Conference on Computer Systems, EuroSys 2009, pp. 205–218 (2009)

20. Wu, S., Hofman, J.M., Mason, W.A., Watts, D.J.: Who Says What to Whom on Twitter. In: Proceedings of the 20th International Conference on World Wide Web, WWW 2011, pp. 705–714 (2011)

21. Yang, J., Leskovec, J.: Patterns of Temporal Variation in Online Media. In: ACM International Conference on Web Search and Data Minig, WSDM 2011, pp. 177–186 (2011)
22. Yang, J., Counts, S.: Predicting the Speed, Scale, and Range of Information Diffusion in Twitter. In: 4th International AAAI Conference on Weblogs and Social Media, ICWSM 2010 (May 2010)
23. Zhao, D., Rosson, M.B.: How and Why People Twitter: The Role that Microblogging Plays in Informal Communication at Work. In: Proceedings of the ACM 2009 International Conference on Supporting Group Work, GROUP 2009, pp. 243–252 (2009)

Automatic Vandalism Detection in Wikipedia with Active Associative Classification

Maria Sumbana, Marcos André Gonçalves, Rodrigo Silva,
Jussara Almeida, and Adriano Veloso

Department of Computer Science, Universidade Federal de Minas Gerais, Brazil
{inesumbana,mgoncalv,rmsilva,jussara,adrianov}@dcc.ufmg.br

Abstract. Wikipedia and other free editing services for collaboratively generated content have quickly grown in popularity. However, the lack of editing control has made these services vulnerable to various types of malicious actions such as vandalism. State-of-the-art vandalism detection methods are based on supervised techniques, thus relying on the availability of large and representative training collections. Building such collections, often with the help of crowdsourcing, is very costly due to a natural skew towards very few vandalism examples in the available data as well as dynamic patterns. Aiming at reducing the cost of building such collections, we present a new active sampling technique coupled with an on-demand associative classification algorithm for Wikipedia vandalism detection. We show that our classifier enhanced with a simple undersampling technique for building the training set outperforms state-of-the-art classifiers such as SVMs and kNNs. Furthermore, by applying active sampling, we are able to reduce the need for training in almost 96% with only a small impact on detection results.

1 Introduction

The proliferation of collaboratively generated content has become an increasing phenomenon on the Web 2.0. One example of how communities can produce collaborative content on a large scale is Wikipedia[1], where anyone can edit, modify, or revise articles, given that some copy and modification rights are preserved [2]. This online encyclopedia currently contains more than 17 million articles, written in various languages.

However, this large amount of information, made available to everyone virtually without any control, opens opportunities for "malicious acts" of people trying to exploit the services for their own benefit or to intentionally degrade their integrity and trustworthiness (e.g., by including polluted material such as pornography or incorrect information) [10]. Wikipedia, in particular, has historically suffered from non-cooperative user actions such as vandalism, edit wars, and lobbyism [7]. Vandalism is defined in Wikipedia as "any addition, removal, or change of content in a deliberate attempt to compromise the integrity of the system"[2]. Vandalism detection is currently done mostly manually by volunteers [10], with obvious drawbacks in terms of cost and scalability, given its current size and growth rate. These drawbacks motivated the development of

---

[1] www.wikipedia.com

[2] http://en.wikipedia.org/wiki/Wikipedia:Vandalism

P. Zaphiris et al. (Eds.): TPDL 2012, LNCS 7489, pp. 138–143, 2012.

automatic detection techniques [6]. These techniques rely on specialized large collections, built with the help of crowdsourcing [4], to serve as training data for supervised methods. There is also an annual CLEF competition similar to those of TREC [5,7].

Despite these efforts, there are still challenges to be faced to build effective vandalism detection techniques. For instance, the best reported solutions present precision figures around 77% [7], implying that there is still room for significant improvement in detection effectiveness. Moreover, the most effective solutions available are based on supervised learning techniques, and are thus challenged by the fact that most articles are in fact non-vandalized and, thus, available training collections tend to be very unbalanced. This imbalance may impact the learning process, and ultimately hurt detection effectiveness. Furthermore, since finding vandalized articles for training is hard, building new training collections is costly. This cost becomes a key limiting factor if we consider that, in real-world situations, as vandalism actions evolve and new patterns arise to circumvent existing detection mechanisms or to exploit new features of the application, there is a constant need for rebuilding the training collection with up-to-date examples so that these new patterns are introduced into the detection model.

In this context, we here tackle the problem of Wikipedia vandalism detection focusing particularly on reducing the cost of building training collections. Towards this goal, we employ a novel active sampling technique. We evaluate our strategy, comparing it against two state-of-the-art classifiers, notably SVM and kNN [1]. We first demonstrate that our classifier, coupled with a very simple undersampling technique that uses balanced samples of the training data, produces vandalism detection results that greatly outperform those of SVM and kNN (by as much as 22%, in terms of Macro-F1). Furthermore, we also show that active sampling is capable of reducing the need for training in almost 96% while still producing results that are very close to those when the complete training set is used, thus making our solution very practical for real-case scenarios.

2   Vandalism Detection Method

We address the task of detecting vandalism in Wikipedia as a binary classification problem. Given an edit e introduced by a user in a given article, our goal is to automatically detect whether e is an act of vandalism or not. Formally, the task of detecting vandalism in Wikipedia takes as inputs a training set $\mathcal{D}$ and a test set $\mathcal{T}$. The training set $\mathcal{D}$ consists of records $< e, l_i >$, where e is an edit and $l_i \in \{0, 1\}$ is a label that identifies e's class, i.e., regular edit or vandalism. Moreover, each edit e is represented as a list of m feature-values $\{f_1, f_2, \ldots, f_m\}$. $\mathcal{D}$ is used to learn a model $\mathcal{M}$ that relates features of the edits to their corresponding classes. The test set $\mathcal{T}$ consists of records with unlabeled edits ($< e, ? >$). The model $\mathcal{M}$ is used to predict the class for each edit in $\mathcal{T}$.

2.1   On-Demand Associative Classifier for Vandalism Detection

We here adopt LAC (Lazy Associative Classification [9]) to classify each edit in $\mathcal{T}$ as either vandalism or regular. LAC exploits the fact that often there are strong associations between feature values and classes, and such associations can be used to predict the class of unseen elements (e.g., edits in $\mathcal{T}$). Rules $\mathcal{X} \rightarrow l_i$ indicate the association between a set of feature values $\mathcal{X}$ and a label l_i. LAC learns a model $\mathcal{M}$ composed of association rules extracted from $\mathcal{D}$ in two main phases, namely rule extraction and class prediction.

In order to ensure effectiveness while extracting rules from $\mathcal{D}$, LAC performs a demand-driven rule extraction. That is, the rule extraction process is performed only at classification (i.e., vandalism detection) time. The method projects the search space for rules according to information in edits in $\mathcal{T}$, allowing efficient rule extraction. It projects/filters the training data according to the features in edit $e \in \mathcal{T}$, and extracts rules for this projected training data, which is denoted as $\mathcal{D}^e$. This ensures that only rules that carry information about edit e are extracted from the training data, drastically bounding the number of possible rules. The confidence of a rule, denoted by $\theta(\mathcal{X} \to l_i)$, measures the strength of the association between $\mathcal{X}$ and class l_i, and is estimated by the conditional probability of l_i being the class of edit e, given that $\mathcal{X} \subseteq e$. LAC predicts the class of an edit $e \in \mathcal{T}$ by combining the confidences of all useful rules $\mathcal{X} \to l_i$. More specifically, let $\mathcal{R}_{l_i}^e$ be the set of rules predicting the class of e as l_i. $\mathcal{R}_{l_i}^e$ is interpreted as a poll, in which each rule $\mathcal{X} \to l_i \in \mathcal{R}_{l_i}^e$ is a vote given by features in $\mathcal{X}$ for class l_i. The weight of a vote $\mathcal{X} \to l_i$ depends on the strength of the association between $\mathcal{X}$ and l_i, given by $\theta(\mathcal{X} \to l_i)$. The process of estimating the probability of l_i being the class of an edit e starts by summing weighted votes for l_i and then averaging the sum by the total number of votes for l_i. The respective score function $s(e, l_i)$ thus gives the average confidence of the rules in $\mathcal{R}_{l_i}^e$. We then estimate the probability $p(l_i, e)$ of l_i being the class of e by normalizing $s(e, l_i)$ by the summation of scores obtained by e for both classes. The class of e is set as the one with the highest value of $p(l_i, e)$.

2.2 Fewest Rules First Active Sampling

In this section we present an active sampling strategy [8] to reduce the manual effort involved in building the vandalism detection model $\mathcal{M}$, thus referring to it as ASVD (Active Sampling for Vandalism Detection). The key assumption is that, by carefully choosing a smaller set of training examples, it may still be possible to learn a model $\mathcal{M}$ that is as effective as the model learned from a much larger training set. Consider a large set of unlabeled edits $\mathcal{U} = \{u_1, u_2, \ldots, u_n\}$. The goal is to select a small subset of edits in $\mathcal{U}$ carrying almost the same information of all edits in $\mathcal{U}$. These highly informative edits will compose the training set $\mathcal{D}$, and, ideally, $|\mathcal{D}| \ll |\mathcal{U}|$. Particularly, ASVD exploits the redundancy in feature-space that exists between different edits in $\mathcal{U}$. That is, many edits in $\mathcal{U}$ may share some of their feature-values. ASVD uses this fact to perform an effective sampling strategy, which is based on a "fewest rules first" heuristic.

Intuitively, if an edit $u_k \in \mathcal{U}$ is inserted into $\mathcal{D}$, then the number of rules matching edits in $\mathcal{U}$ and sharing feature-values with u_k will possibly increase. In contrast, the number of rules matching those edits in $\mathcal{U}$ that do not share any feature-value with u_k will remain unchanged. Thus, the number of rules extracted for each edit in $\mathcal{U}$ can be used as an approximation of the amount of redundant information between edits already in $\mathcal{D}$. ASVD exploits this idea by selecting edits that contribute primarily with non-redundant information (i.e., edits with the fewest numbers of rules). Specifically, the sampling function $\delta(\mathcal{U})$ returns an edit in $\mathcal{U}$ according to the following rule: $\delta(\mathcal{U}) = \{u_k \mid \forall u_j : |\mathcal{R}^{u_k}| \leq |\mathcal{R}^{u_j}|\}$, where $|\mathcal{R}^{u_k}|$ is the number of rules matching u_k, i.e., rules $\mathcal{X} \to l_i$ such that u_k contains all features in $\mathcal{X}$. The edit returned by the sampling function is inserted into $\mathcal{D}$, but remains in $\mathcal{U}$. The sampling procedure continues, iteratively, selecting a new edit u_j from $\mathcal{U}$, one at a time. Note that, at each new iteration, the number of rules extracted from $\mathcal{D}$ for each edit in $\mathcal{U}$ is likely to change due to the edit inserted into $\mathcal{D}$ in the previous iteration. The intuition behind always choosing the edit in $\mathcal{U}$ which demands the fewest rules is that such edit should share fewer feature-values with edits already in $\mathcal{D}$, which serves as evidence that $\mathcal{D}$ does not contain edits that are similar to u_k. Thus, the information provided by u_k is not redundant.

This iterative selection process stops when there is no edit in $\mathcal{U}$ that is more informative than any edit already in $\mathcal{D}$. This occurs when ASVD selects an edit which was previously inserted into $\mathcal{D}$ (recall that selected edits are *not* removed from $\mathcal{U}$). At this point, continuing with the selection process would lead to the same edit being picked repeatedly. Thus, the training set $\mathcal{D}$ contains the most informative edits, and can then be used by the classifier (e.g., LAC) to detect vandalism among the edits in $\mathcal{T}$.

3 Evaluation Methodology

Our evaluation is performed on the PAN-WVC-10 dataset [4], consisting of 32,452 English edits in 28,468 articles, out of which 2,391 edits were found to be vandalism. This is a very skewed dataset, with 87.7% of the instances belonging to the regular class. Each edit is represented by a list of m feature-values. We adopt the same features used in [3] (most of them are based on the textual difference between old and new article revisions). As baselines, we use two widely known classifiers, namely SVM and kNN.

For ASVD, recall that the training set is initially empty and grows one element at a time as examples are selected from the unlabeled set and then labeled, stopping when an example that had been previously picked is selected. We here evaluate whether ASVD would perform better if more examples were selected. We do so by defining *selection rounds*. That is, in the first round, ASVD selects examples from the initial set of unlabeled examples as described above. The training set built from these examples is referred to as $\mathcal{D}_1$. Once ASVD stops, we remove the selected examples from the unlabeled set and run it once again. The examples selected by ASVD in this second round are merged with those selected in the first round to compose a new training set $\mathcal{D}_2$. The same procedure is repeated in order to build $\mathcal{D}_3$. Note that $\mathcal{D}_1 \subset \mathcal{D}_2 \subset \mathcal{D}_3$. We refer to the execution of LAC with the training sets $\mathcal{D}_1$, $\mathcal{D}_2$ and $\mathcal{D}_3$ as Round1, Round2 and Round3, respectively. As baseline for comparison, we also evaluate the effectiveness of randomly selecting samples from an unlabeled set to build the training set. Specifically, we randomly select the same number of examples selected by ASVD in each round of experiments, and use LAC to learn the vandalism detection model. We refer to these experiments as Random1, Random2 and Random3.

Our evaluation is performed according to a 5-fold cross validation design. Particularly, parameter selection for each classifier was performed using cross-validation in the training set. All reported results are averages over all 5 test sets, and we compare them using statistical significance tests (Student's t-test) with a 95% confidence level.

4 Experimental Results

We start discussing the results obtained by using the full training sets with no effort to balance class proportions (i.e., without undersampling). Table 1 shows results along with corresponding 95% confidence intervals in terms of precision, recall and Macro-F_1. Clearly no classifier performs particularly well. SVM presents the best Macro-F_1

Table 1. Classifier Performance

		Without UnderSampling				
		Regular Edits		Vandalism		Macro
Classifier		Precision	Recall	Precision	Recall	F_1
SVM		93.86±0.0015	99.3±0.0023	68.61±0.0167	18.48±0.0.0108	62.75.±0.0108
KNN		93,5±0.4421	98,3±0.1905	40,1±3.9141	14,3±1.8145	58.4±1.2518
LAC		92.6±0.0740±	1 ± 0.000	0 ± 0.000	0 ± 0.000	47.8±0.0008
		With UnderSampling				
	Proportion	Regular Edits		Vandalism		Macro
Classifier	p	Precision	Recall	Precision	Recall	F_1
SVM	4.4:1	96.66±0.0018	94.54±0.00158	46.09±0.0069	58.59±0.0231	73.58±0.0.0068
KNN	3.5:1	95.75±1.105	86.77±4.998	26.26±3.1353	55.83±13.5691	62.82±1.6924
LAC	1.5:1	97.9±0.0023	95.5±0.0064	57.3±0.0161	75.1±0.0265	80.9±0.0065

Table 2. Results for Active Sampling with a Negative:Positive Proportion of 1.6:1

	Regular Edits		Vandalism		Macro		
Classifier	Precision	Recall	Precision	Recall	F_1	%train_sel	%train_bal
Round1	96.7±0.0083	95.5±0.0174	53±0.7039	58.77±0.11053	75.22±0.103	1.3	0.4
Round2	96.8±0.0107	95.9±0.0195	56±0.0970	59.8±0.1428	76.1±0.0163	2.68	0.8
Round3	96.7±0.0091	96.4±0.0193	59.4±0.0796	59±0.1208	77.0±0.0060	4.14	1.13
Random 1	94.2±0.002	98.4±0.003	57±0.037	23.8±0.034	62.9±0.0164	1.3	0.4
Random 2	94.4±0,002	97.2±0.014±	62±0.023	26.8±0.360	63.9±0.0169	2.68	0.8
Random 3	94.4±0.0805	98.5±0.114	64±2,167	26.3±2.364	65.5±1.236	4.14	1.13

values, followed by kNN, and the latter presents large variability in results, mainly in the vandalism class. LAC, in turn, has the worst performance in terms of Macro-F_1. Indeed, it was not able to correctly classify any vandalism instance. Although producing better results, both SVM and kNN leave most vandalism instances undetected.

We hipothesize that the poor performance of the classifiers, LAC in particular, could be due to severe data skewness, with only a small fraction of examples being vandalism instances. However, such skew is expected in practice as it is very hard to obtain examples of vandalized articles. To test our hypothesis, we performed experiments using a simple undersampling technique. We balanced the training sets by eliminating instances of the majority class (i.e., regular edits) until we reach a certain proportion p of the number of regular edit examples to the number of vandalism examples. To determine the best value of p, we ran a series of validation experiments with each classifier, using a portion of the training set as validation set. These experiments indicated that the best results were obtained with p around 1.5 for LAC, 3.5 for kNN and 4.4 for SVM.

We next ran a series of new classification experiments with the undersampling technique, using the best value of p for each classifier found in the validation set. Results are also shown in Table 1. Note that, in this setup, LAC outperforms both SVM and kNN, with an average Macro F_1 of 80.9%. LAC's results are particularly high in recall: we are able to retrieve about 75% of all vandalism edits while almost 60% of these predictions are in fact vandalisms. This coupled with the facts that very few regular edits are misclassified and that the number of vandalism edits in practice is much smaller than regular edits, make our strategy very suitable to be used as a filtering system for human editors that can judge if the predicted vandalism edits are in fact malicious acts.

We now turn our attention to the active selection of elements for the training set. Table 2 shows the results for the classification using ASVD in Rounds 1, and 2, and

3 as well as the cumulative percentage of training set that was selected in each round. In these results a similar balancing strategy was applied. Specifically, we split the edits selected by ASVD in two halves, using one for training and the other for validation to determine the best negative:positive proportion (p). Our results are similar as before, with the best value of p equal to 1.6. The final fraction of edits used for training, after balancing the training set, is seen in the final column of the table.

We can see that ASVD selected very few instances: 1.3% of all the available unlabeled set in the first round, 2.68% in the second and 4.14% in the third. This corresponds to labeling roughly 40, 80, and 110 edits. Moreover, the results from Round 3 are statistically tied or only marginally different from those of Rounds 1 and 2. Thus, if the labeling budget is short, one may choose to stop after the second or even the first round, with no significant loss in detection effectiveness. More importantly, the effectiveness with this very small amount of training is very close to the results produced by LAC using the whole training: in Round 3 the Macro F_1 is only 3.9% lower. Finally, our active choices are always better than randomly selecting a sample with the same size, with large performance differences (up to 19.6% in Macro F_1 in Round 1).

5   Conclusion

We proposed an active learning method for Wikipedia vandalism detection which is very competitive against the best existing supervised solutions while using only a very small fraction of the training set used by those methods. This makes our solution very practical for real situations. As future work, we want to explore other features as well as test our proposed methods in face of evolving patterns and within classifier ensembles.

References

1. Baeza-Yates, R.A., Ribeiro-Neto, B.A.: Modern Information Retrieval, 2nd edn. Pearson Education Ltd., Harlow (2011)
2. Belani, A.: Vandalism Detection in Wikipedia: a Bag-of-Words Classifier Approach. CoRR, abs/1001.0700 (2010)
3. Javanmardi, S., McDonald, D.W., Lopes, C.V.: Vandalism Detection in Wikipedia: A High-Performing, Feature-Rich Model and its Reduction Through Lasso. In: Proc. 7th International Symposium on Wikis and Open Collaboration, pp. 82–90 (2011)
4. Potthast, M.: Crowdsourcing a Wikipedia Vandalism Corpus. In: Proc. ACM SIGIR, pp. 789–790 (2010)
5. Potthast, M., Holfeld, T.: Overview of the 2nd International Competition on Wikipedia Vandalism Detection. In: CLEF (Notebook Papers/Labs/Workshop) (2011)
6. Potthast, M., Stein, B., Gerling, R.: Automatic Vandalism Detection in Wikipedia. In: Proc. 30th European Conference on IR Research, pp. 663–668 (2008)
7. Potthast, M., et al.: Overview of the 1st International Competition on Wikipedia Vandalism Detection. In: CLEF (Notebook Papers/Labs/Workshop) (2010)
8. Silva, R., Gonçalves, M.A., Veloso, A.: Rule-Based Active Sampling for Learning to Rank. In: Gunopulos, D., Hofmann, T., Malerba, D., Vazirgiannis, M. (eds.) ECML PKDD 2011, Part III. LNCS, vol. 6913, pp. 240–255. Springer, Heidelberg (2011)
9. Veloso, A., Meira Jr., W., Zaki, M.J.: Lazy Associative Classification. In: Proc. IEEE International Conference on Data Mining, pp. 645–654 (2006)
10. Wikipedia. Vandalism on Wikipedia (2012)

Applying Digital Library Technologies to Nuclear Forensics

Electra Sutton, Chloe Reynolds, Fredric C. Gey, and Ray R. Larson

School of Information and UC Data
University of California, Berkeley
Berkeley, California, USA, 94720-4600
{electra,gey}@berkeley.edu,
{chloe_reynolds,ray}@ischool.berkeley.edu

Abstract. Digital Libraries will enhance the value of forensic endeavors if they provide tools that enable data mining capabilities. In fact, collecting data without such tools can result in investigators becoming overwhelmed. Currently, the quantity of highly dangerous radioactive materials is increasing with the advancement of civilizations' scientific inventions. This creates a demand for an equivalently sophisticated forensics capability that prevents misuse and brings malicious intent to justice. Our forensics approach applies digital library and data mining techniques. Specifically, the forensic investigator will utilize our digital library system which has been enhanced with advanced data mining query tools in order to determine attribution of material to their geographic sources and threat levels, enabling tracing and rating of smuggling activities.

1 Introduction

Radioactive materials can cause harm in many ways, from poisoning to bombs. Today these materials are abundantly applied in the medical, energy, weapons, and space exploration fields. Keeping populations safe from their malicious use is the job of forensic investigators worldwide. Recent records from the Illicit Trafficking Database (ITDB) at the International Atomic Energy Agency (IAEA) show that high-value HEU (Highly Enriched Uranium), Pu, Cs-137, Am-241, Ir-192, Cr-60 and Sr-90 have been traded illegally [1]. Therefore, tracking these activities is urgent.

A literature search [3] has revealed that neither the database research community nor the information retrieval research community are aware of the unique search challenges posed by the nuclear forensics identification problem. Each of these communities has significant resources which could be brought to bear in solving the fundamental problems of large-scale nuclear forensic discovery. Among other goals, a primary focus of this digital library is to interest these communities in the issues of nuclear forensic identification and to supply education. This is the first project of its kind to apply modern search technology to the nuclear forensics matching and identification problems. The methodologies explored (directed graph similarity, automatic classification, and rule-based

P. Zaphiris et al. (Eds.): TPDL 2012, LNCS 7489, pp. 144–149, 2012.

matching) are expected to break new ground in providing rigorously evaluated approaches to nuclear signature matching. Just as traditional forensics (fingerprint matching, DNA matching) benefited from algorithms developed by information search specialists in their area, we expect the area of nuclear forensics to be significantly improved by focused attention by researchers in the search area. In particular, the projects results should help support the work of the Nuclear Smuggling International Technical Working Group.

In the remaining sections of the paper we provide descriptions of the data collections involved. Finally we explain the data mining techniques used and our implementation. In the conclusion we discuss future plans for the system.

2   Data Collections

Since this digital library is about search and search evaluation, it is crucial to have some data upon which to evaluate the effectiveness of our methodologies. We expect to include, among others to be assembled, the following data sets:

1. Radioactive decay chain patterns. The unclassified Nuclear Structure and Decay Data (NuDat) data set [7] will be used to build graphs of radioactive decay. The data set contains over 165 thousand records of isotopes of elements, decay daughters, radiation type, half life and energy levels.
2. Location identification collections
   (a) Uranium Oxide Quality Control data set. This proprietary dataset from Lawrence Livermore National Laboratory contains principal components analysis to infer location from indicator trace elements which are geospecific [6]. The data set has 3,436 sample measurements for 21 sites. The data set has licensing restrictions but we are attempting to obtain access to the data.
   (b) Pit Manufacturing Measurements. Isotopic measurement records of pit manufacturing materials from the Plutonium Metal Standards Exchange Program at Los Alamos National Laboratory. The data consists of analytical measurements within and between worldwide laboratories.
   (c) Actinide measurement test results. This unclassified dataset is a report of all measurements made on actinide samples and contains the assay actinide and its isotopic and trace element amounts. It is produced by researchers from six national laboratories and the Atomic Weapons Establishment.
3. The Nuclear Fuel Cycle and Weapon Development Cycle [8]. The Pacific Northwest National Laboratory created it on behalf of the United States Department of Energy. Nuclear materials are naturally-occurring and can also be manufactured in a lab. Once obtained, either sort goes through any of various multi-stage processes to either become fuel or nuclear weapons material. This figure depicts the life cycle stages of both outcomes: fuel or weapon development. At each stage, the figure also names the processing facility types that convert materials from one stage to the next in the life cycle.

4. Facilities. The International Atomic Energy Agency (IAEA) lists of worldwide reactors.

3   Data Mining Technologies

Our intent is to create a library of empirical radio chemical data with forensics capabilities in order to preserve public safety while accommodating the important utility of radioactive materials for medicine, energy and security. The nuclear materials addressed by our library collection are economically valuable commodities to criminals. As such, great resources are employed to smuggle nuclear materials to illicit customers who may have destructive or catastrophic objectives. To interfere with such aims, we utilize a digital library with advanced search capabilities.

3.1   Forensics Capacity

In order to furnish the forensics community with suitable protective force, the system's design not only incorporates advanced digital library and search capabilities, but also incorporates the practices and protocols of forensics. A forensics operation is distinguished by requirements for provability, traceability and credibility. The results of a query must stand up in court since the goal of a forensics operation is to make a claim during judicial proceedings. Therefore our design focuses on technologies that would survive the scrutiny of trial court requirements and other legal proceedings. Also, our data sources must constitute a reliable, authoritative base. We use collections from certified national laboratories. The following technologies were selected by virtue of their traceability to authoritative sources and for their quantifiable characteristics that will be credible in court.

3.2   Technologies

The data mining techniques employed are association relationships, a graph theory distance algorithm and rule-based classification. Each one was selected because it satisfies forensic practices and protocol requirement of reproducible results to which a subject matter expert witnesses can testify.

1. Relationship associations are utilized to make a conceptual link between the radiochemical properties of a nuclear material sample and where a sample lies in the Nuclear Fuel and Weapon Life Cycle corresponding to a threat level, as well as geographic origin, as discussed in the Data Collections section. These associations are drawn by using the transitive property of relationships using set operations and therefore can be demonstrated to have an irrefutable relationship.

2. The graph theory node distance algorithm identifies the threat level classification for nuclear materials. Node distance metrics determine the distance of a nuclear material's progress along the Nuclear Fuel and Weapon Life Cycle [8]. The lower the distance from current stage to end stage, the higher the threat level.

3. Rule-based data mining for classification of nuclear material signatures. This method is utilized to predict the radio chemical signatures of nuclear materials given temporal changes and the laws of radioactive decay.

4   Implementation

4.1   Attribution

Implementation of the attribution search involves association relationships that are found by joining data sets on common instances with known class labels and with the set operation of transitivity. Attribution will be further explained below.

Given chemical measurements of nuclear material samples, our digital library can return probabilistic geographic origin points and threat levels. Threat level refers to a material's position in the nuclear weapons life cycle, which begins at naturally-occurring Uranium and Thorium and progresses through various processes of enrichment - any of which has multiple stages - eventually becoming weapons grade material.

Chemical measurements of samples reveal trace element signatures which correspond to particular geographic sites or regions. Specific sites hold facilities that are capable of certain material processing steps (mining, enriching, etc.). These processes are associated with specific stages of the weapons life cycle. The closer a material is to the weapons grade end stage of the weapons development life cycle, the higher that material's threat level is. By multiple association relationships and the transitive property then, a user can input a smuggled material's chemical composition as a query. The query result will be (attribution of) the material's probable geographic origin and threat level.

4.2   Threat Level

Implementation of the search for a threat level of a nuclear material involves k-nearest neighbor distance computations [4] using the Nuclear Fuel Cycle and Weapons Development Process chart[8]. Considering each stage in this cycle as a node, node distance metrics can assign a numeric threat level to a given sample. The distance (in number of nodes) from a material's current state to the end state of weapons-grade material, determines its threat level; the smaller the distance, the more dangerous a material is. The distance attribute of this technique provides a compelling way to express viability of a material as a fuel or weapon and its threat level.

As mentioned in the previous section, the result of the relation association search is a point within the Nuclear Fuel Cycle and Weapons Development

Process, e.g., chart position 5.10 "Spent Fuel or Special Irradiated U Target Containing Pu." This point or position identifies the type of facility where a material with particular radio chemical properties would be found due to the manufacturing processes and development state of the material.

The metric aspect is suitable for forensics practices since it makes the classification of a material's threat level measurable and illustratable in a courtroom setting. The threat level will also serve as a metric for the material's value as a commodity (to smugglers).

4.3   Rule-Based Matching

Any given nuclear material has a nuclide signature. This signature is determined by the properties of radioactive decay and fission. The training data set for our signature classifier is a set of measured quantities of isotopes and trace elements for particular nuclear material samples from well-established laboratories. This training set has created class labels for signatures, e.g., Highly Enriched Uranium (HEU) - more than 20 % of Pu-240 - and Low Enriched Uranium (LEU) - less than 4 % of Pu-240.

Laboratory data is usually incomplete and contains anomalies such as disagreeing production date isotopic ratios. For those reasons, we will generate our own set of simulated measurements. In the paper 'Characterization of nuclear fuel using multivariate statistical analysis' [5], the authors state, "It is impractical to obtain such data from real nuclear fuel sample analyses. Instead, the ORIGEN-ARP code package (Bowman, 2000) was used to predict U and Pu isotopic compositions at various burn up values." In the AAAS/APS report on nuclear forensics [2], simulated data was also used to characterize different types of nuclear reactors. This simulated data set was generated by software called MCODE. MCODE is similar in function to ORIGEN-ARP but uses more computing resources because it implements a Monte Carlo method. For this project we expect to generate simulated data using one of the code packages mentioned above. We will build simulated decay chain graphs using the radioactive decay rules inferred from the NuDat database [7]. Our implementation will supplement the incomplete and anomalous laboratory measurements with our own simulated data.

One note in composing classifications for forensic use of raw data is that the modeling process must be sensitive to unexpected values. Unexpected values are important as unique attributes which could be evidence to prove a case for particular events rather than for generalized scientific knowledge. Investigative practices that search for malicious intent are focused on finding singular activities that can be traced to particular perpetrators rather than finding general observations about a phenomenon.

5   Conclusion

In this paper we have described the preliminary work being carried out on this project, and discussed some of design and search issues involved in nuclear forensics search. Our work in data collection, data mining and forensic matching algorithm creation is ongoing. We are in the process of collecting and indexing the various databases discussed in the Data Collections section and are investigating algorithms for graph-based search with time components.

Acknowledgements. The work described in the paper is part of the grant "ARI-MA: Recasting Nuclear Forensics Discovery as a Digital Library Search Problem" which was funded by the U.S. National Science Foundation and Department of Homeland Security (award #ECCS-1140073).

References

1. International Atomic Energy Agency. Illicit Trafficking Database (ITDB) Factsheet (2009), http://www-ns.iaea.org/downloads/security/itdb-fact-sheet.pdf
2. American Physical Society (APS) / American Association for the Advancement of Science (AAAI) Joint Working Group: Nuclear Forensics: Role, State of the Art, Program Needs (2008)
3. Gey, F., Larson, R., Sutton, E.: ARI-MA Recasting nuclear forensics discovery as a digital library search problem, grant proposal to the NSF/Domestic Nuclear Detection Office Academic Research Initiative from UC Berkeley (2012)
4. Lifshits, Y.: Branch and bound algorithms for nearest neighbor search: Lecture 1. Steklov Institute of Mathematics at St. Petersburg, California Institute of Technology, RuSSIR, Russian Summer in Information Retrieval (2007)
5. Robel, M., Kristo, M.J.: Characterization of nuclear fuel using multivariate statistical analysis. Lawrence Livermore National Laboratory report (2007)
6. Robel, M., Kristo, M.J., Heller, M.: Nuclear forensic inferences using iterative multidimensional statistics. In: Institute of Nuclear Materials Management 50th Annual Meeting Tucson, AZ (July 2007)
7. Sonzogni, A.: Nudat (nuclear structure and decay data) database (2010)
8. Willingham, C.E.: Nuclear fuel cycle and weapons development process. Prepared for the United States of America Department of Energy by the Pacific Northwest National Laboratory (2004), http://www.pnl.gov/publications/abstracts.asp?report=228018

Identifying References to Datasets
in Publications

Katarina Boland[1], Dominique Ritze[2], Kai Eckert[2], and Brigitte Mathiak[1]

[1] GESIS - Leibniz Institute for the Social Sciences, Cologne, Germany
{katarina.boland,brigitte.mathiak}@gesis.org
[2] Mannheim University Library, Mannheim, Germany
{dominique.ritze,eckert}@bib.uni-mannheim.de

Abstract. Research data and publications are usually stored in separate and structurally distinct information systems. Often, links between these resources are not explicitly available which complicates the search for previous research. In this paper, we propose a pattern induction method for the detection of study references in full texts. Since these references are not specified in a standardized way and may occur inside a variety of different contexts – i.e., captions, footnotes, or continuous text – our algorithm is required to induce very flexible patterns. To overcome the sparse distribution of training instances, we induce patterns iteratively using a bootstrapping approach. We show that our method achieves promising results for the automatic identification of data references and is a first step towards building an integrated information system.

Keywords: Digital Libraries, Information Extraction, Recognition of Dataset References, Iterative Pattern Induction, Bootstrapping.

1 Introduction

In empirically oriented fields of research such as the social sciences, primary data from surveys, interviews and other studies lay the basis for publications and the continuing research process. Traditionally, primary data and publications are stored in separate systems. Libraries usually concentrate on publications while research institutions mainly focus on research data. Connections between these resources are usually not available. This leads to an unfavorable situation for researchers. They need to seek connections between publications and primary data and query structurally different information systems to gain access to further information.

A typical use case would be a researcher in the field of social sciences who investigates "the opinions of German citizens about the social state" and consults several publications concerned with that topic. These publications, however, may present differing results and draw divergent conclusions. Without knowing which research data they are based on, the results cannot directly be compared and interpreted. Unless the publications and corresponding studies are interlinked, the researcher has to read every document sentence by sentence to identify the

P. Zaphiris et al. (Eds.): TPDL 2012, LNCS 7489, pp. 150–161, 2012.

underlying research data. With existing links, however, the researcher would be able to recognize at first glance which publications base on the same data and which of the results can be compared directly.

Manually establishing links between research data and publications is a nontrivial and time-consuming task which does not fit to the demand of transparency in research. By generating these links automatically, we aim to increase transparency and thereby improve the traceability and reproducibility of research results. We identify connections by applying a bootstrapping algorithm based on pattern induction. Starting with a study name as seed, it generates patterns which detect other references to studies and primary data. These can again be used to generate new patterns. After several iterations, the algorithm returns a list of research data references for each publication. Identifying these links constitutes the first step towards our aim of creating an integrated retrieval system.

In this paper, we first discuss to what extent abstracts and bibliographic metadata for publications and studies can help to identify connections between the data types. We then analyze the specific characteristics of study references in publication full texts. Finally, we introduce and experiment with an iterative pattern induction method to recognize references to research data in publications and give an outlook on future applications of our work.

2   Related Work

2.1   Citation Mining

Several methods have been developed to automatically find connections between publications based on citations. Although finding citations and finding references to datasets appear to be very similar tasks, they differ in some important aspects. In contrast to study references, citations are usually listed in a bibliography which significantly simplifies the linking process. Systems like TIERL [1] do not consider any information except the bibliographies to detect citations. Other systems like ParsCit [2] additionally analyze the contexts of citation to apply machine learning methods. In [3], an approach for computing the similarity of publications based on the proximity of citations is introduced. However, for citations there are standardized specifications, albeit varying, that facilitate recognition and mining. This is not the case for references to research data.

2.2   Named Entity Recognition

A similar task to finding study references is named entity extraction. Within this field of research, texts are analyzed in order to find named entities, e.g. names of persons or cities. In this sense, study names can be seen as a specific group of named entities. Existing approaches try to extract named entities for example from speeches [4]. A training set is used to learn how to identify entities. In our case, such a training set is not available and its generation would be very time-consuming. Another technique [5] uses unsupervised learning in combination with web search engines. This works quite well to extract facts from websites but studies are usually not frequently mentioned in the Web. Based on the specific characteristics of studies and the unavailability of a training set, it is very difficult to apply named entity extraction methods on our task.

2.3 Iterative Pattern Induction

Weakly supervised bootstrapping algorithms for automatic pattern induction have been applied in a variety of different fields. All these methods start with a small set of manually created patterns or training instances and iteratively expand their training sets by labeling subsets of the test data. One area of application is concerned with finding hyponyms [6], part-of relationships [7] or even more domain-oriented relations between concepts [8,9]. Similar to our application, the creation of lexicons can be automated by learning patterns and extracting the appropriate noun phrases. For example, this is used to generate semantic lexicons with different categories [10] or to construct medical treatment lexicons [11]. In contrast to these approaches, prior candidates cannot be easily found in our application scenario. This is because study references often appear in captions or footnotes rather than in complete sentences. This makes syntactic preprocessing difficult. Furthermore, since the contexts of the unstandardized references are very heterogeneous, flexible patterns are needed that may capture mentions in contexts such as footnotes and captions as well as mentions in continuous text. The induction of patterns that, for example, only analyze the two preceding words of a predefined candidate may therefore not be sufficient for our task. Thus, we investigate the use of patterns that do not require the identification of prior candidates but instead enable the algorithm to automatically detect the boundaries of study names.

3 Approach

To find study references in publications, we investigated different types of data: bibliographic metadata, abstracts and full texts. The former two data types are often available even if full texts are not. We therefore started by analyzing whether they alone can be helpful for detecting references. For these preinvestigations, we used studies from da|ra[1], the registration agency for social science research data, and publications from SSOAR[2], the Social Science Open Access Repository. To find references to datasets, we extracted the study titles from the metadata and checked whether they can be found in the abstracts and full texts of publications. Unfortunately, this is rarely the case because the extracted study titles are usually very extensive, for example "ALLBUS/GGSS 1996 (Allgemeine Bevölkerungsumfrage der Sozialwissenschaften/German General Social Survey 1996)." Such detailed names are seldomly referenced. Instead, it is more

---

[1] http://www.gesis.org/dara
[2] http://www.ssoar.info/

erfolgt die Darstellung und Diskussion der empirischen Ergebnisse Hierfür werden
die Daten des Sozio-oekonomischen Panels (SOEP) aus den Jahren 1990 und 2003
verwendet und für beide Zeitpunkte werden die Einflussfaktoren mittels linearer

a) Regressionsmodelle geschätzt.

---

1 Herangezogen wurden außerdem Allbus, Allensbacher Erhebungen, Eurobarometer, International
Social Survey Program, International Social Justice Project, Sozio-ökonomisches Panel, World

b)   Values Survey.

---

*Tabelle 1: Bevölkerungsvorausberechnung für Deutschland nach Altersgruppen - Anteile in
Prozent*

c) *(Datenbasis: 10. Bevölkerungsvorausberechnung des Statistischen Bundesamtes, Variante 5)*

---

*Tabelle 3: Stichprobe der Untersuchung in den Jahren 2003 und 2004 sowie Größe der Stich-
probe, mit gültigen Daten aus beiden Erhebungen*

d) *(Quelle: Ditton u.a. 2005a)*

---

*Grafik 7: Einschätzung der wirtschaftlichen Lage: Einschätzung der eigenen wirtschaftlichen Lage
(in Prozent)*

e) *(Quellen: Allbus/Sozialstaatssurvey)*

Fig. 1. Different referencing styles for datasets

common to use abbreviations like "ALLBUS 96." Without knowledge about abbreviations or synonyms, finding references is difficult. To check whether study
references occur in abstracts at all, we analyzed the ALLBUS bibliography[3].
Only about 700 of 2000 publications which are listed in the bibliography mention
"ALLBUS" or some longer version in combination with a year. Since ALLBUS is
a very common study in Germany, we assume that other studies are referenced
even more rarely.

Based on these preinvestigations, we decide to focus on full texts as they usually contain more references to datasets. Additionally, we refrain from searching
for references based on lists of study names. In the following sections, we investigate the characteristics of references in full texts (Sect. 3.1) and finally introduce
our pattern induction method (Sect. 3.2).

3.1   Characteristics of References to Research Data in Full Texts

Although citation standards have been proposed in the past [12], datasets are
to date not referenced in a standardized way, cf. [13]. We therefore analyzed
a random set of documents from SSOAR to gain insights on how datasets are
actually referenced in full texts. Figure 1 demonstrates a few different reference
styles we encountered. Study names are highlighted.

References to datasets are usually neither listed in a dedicated index nor
included in the bibliography. They almost exclusively appear in the body of a

---

[3] http://www.gesis.org/fileadmin/upload/dienstleistung/
daten/umfragedaten/allbus/Bibliographie/Biblio25_ris.txt

publication, except for some rare mentions in the abstract. Within the full texts, they can be found in captions of figures or tables (Figs. 1c, d and e), in footnotes (Fig. 1b) and in the continuous text (Fig. 1a).

As described before, the mention of study names themselves is not standardized. Studies are referenced using their proper title (Fig. 1b), an abbreviation (Fig. 1a), an alternative name or a different spelling (Figs. 1a and 1b). In some cases, authors cite the primary publication of a study instead of referencing the dataset itself (Fig. 1d). Figures 1b and e illustrate different ways of enumerating multiple datasets that were used.

Due to these various different possibilities to reference datasets, any method for their detection must be very flexible to recognize all types. Additionally, this variance makes manual creation of rules difficult. At the same time, it leads to the sparse data problem for machine learning approaches: as a large number of different reference styles exist, there is only a small number of mentions for each style.

3.2   Method

As our previous experiments with metadata of studies indicate, lists of study names are not helpful to reliably find references. Thus, we take a different approach: we try to identify contexts that typically include references to research data. Since we are not in possession of an annotated corpus to use for supervision and because examples for the different reference style are sparse, we apply an iterative bootstrapping method to overcome these problems.

The algorithm is depicted in Fig. 2. It starts with a study name as seed. This name should refer to an unambiguous dataset or study. It has to be mentioned frequently enough in the test corpus to allow the induction of patterns from its contexts. The selection of the seed is the only supervision our algorithm requires. As a first step, the text corpus is queried for the seed study name. For efficient search, we use a Lucene[4] index. Next, the contexts of all mentions are extracted. Based on this context set, the algorithm seeks to identify patterns that predict the existence of a study reference.

Construction of Patterns. Besides predicting occurrences of a reference, the patterns need to enable the algorithm to detect its boundaries. Patterns must therefore always include the words as well as brackets or punctuations which surround a dataset mention. For example, the whole sequence *(Quelle:Allbus)* would be extracted when the algorithm detects *Allbus* as a dataset mention. In order to avoid the generation of low-precision patterns, we require that each pattern must consist of at least one non-stop word based on a list of stop words consisting of determiners and prepositions. Additionally, at least one of the words must not consist of only punctuation or a single character. To achieve a higher

---

[4] http://lucene.apache.org/

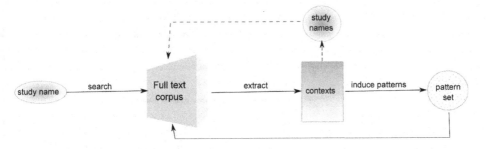

Fig. 2. Overview of the algorithm

recall, we normalize the words in the next step: year and percent specifications as well as numbers are substituted by placeholders.

For each context, the algorithm first tries to induce the most general pattern consisting of as less tokens as possible, i.e. <word1><study name><word2>. If it cannot find enough evidence for the validity of this patterns, it continues with more specialized patterns by expanding the range of surrounding tokens, e.g. <word1><word2><word3><study name><word4>. More specialized patterns have lower thresholds, but are only induced if the induction of more general patterns fails. For assessing pattern validity, we use a simple measure based on the relative frequency of matching contexts in the context set. We experimented with different thresholds and found the relative frequency of 0.25 (corresponding to 3 in 12 mentions) and a minimum number of occurrences of 2 to be optimal. Lower thresholds damage precision substantially. For more specialized patterns, the relative frequency threshold decreases by 0.025 for each additional word incorporated. We use a maximum number of 10 surrounding words.

Induction Strategies. In our experiments, we apply several strategies for pattern induction. At each iteration step, the algorithm can process each found seed and its contexts separately and afterwards merge the induced patterns (*separate*). Alternatively, the contexts of all found seeds can be merged first and the patterns induced based on this aggregated set (*merge*). These two strategies are depicted in Fig. 3. We will investigate the effects of the different strategies in Sect. 4.

Iterative Bootstrapping. In the next iteration step, the induced patterns are used to find new study names whose contexts are in turn retrieved from the corpus. At this step, ambiguous study names might be queried and wrong contexts might be added to the training set. If they do not resemble eacher other too much, the algorithm will not induce any patterns for those mentions.

The procedure is repeated until no new study names and patterns can be retrieved. Finally, a set of patterns, contexts and study names is returned.

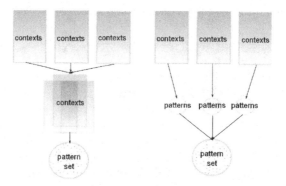

Fig. 3. Pattern induction strategies: merge (left) vs. separate (right)

4 Evaluation

4.1 Corpus and Preprocessing

For our experiments, we used publications contained in SSOAR. Due to the free availability, the transparency and replicability of our results is ensured. We selected all publications indexed with the keyword "empirisch-quantitativ" ("empirical-quantitative") which usually contain references to research data. To facilitate the manual verification of our results, we constrain our evaluation corpus to documents from the DGS corpus[5] which results in a total number of 259 documents. All considered documents are written in German. They are all available in PDF-format which requires several preprocessing steps. To extract the plain text of the PDF-files, we used the Python library PdfMiner[6]. Afterwards, we fixed the hyphenation, removed erroneous whitespaces and eliminated the bibliographies to avoid finding study names in citations. Even after these steps the documents still contain errors such as misclassified characters from OCR.

4.2 Experiments

We ran our algorithm with different seeds and pattern induction strategies to evaluate their influences. First, we compared the different induction strategies *separate* and *merge*. We hypothesized that strategy *separate* leads to higher recall but lower precision because the context sets are small and each context has a high influence on the resulting patterns. Therefore, even infrequent patterns might be found. At the same time, incorrectly classified seeds could cause the induction of wrong patterns. The more conservative strategy *merge* should prevent the induction of low-precision patterns because more evidence is needed for each pattern. In order to assess the influence of different seeds, we tested the seeds "ALLBUS," "SOEP," and "Eurobarometer." All of them are well known studies that are each referenced more than 20 times in our corpus.

---

[5] http://www.soziologie.de/

[6] http://www.unixuser.org/~euske/python/pdfminer/

4.3 Precision

High precision is an important criterion for the applicability of our approach: if a document containing an ambiguous study name which does not refer to the study in the particular context is linked to the dataset of the same name, this would hinder the information retrieval task for the user.

4.4 Recall

For lack of a gold standard, we generally apply the following approach to assess the recall of our method:

1. choose a study as reference study and use its contexts as reference set
2. run algorithm: if the reference study is used as seed, remove patterns induced from its contexts after first iteration
3. compare the algorithm's output to the reference set:
   how many of the mentions in the reference set were found by the algorithm?
4. do so for multiple reference studies to estimate the average recall

More precisely, we assess recall regarding three different factors:

Disambiguated Recall. Unambiguous titles can be queried and all contexts assumed to be referring to the study. For the recognition of ambiguous study names, only those contexts matching one of the induced patterns can safely be assumed to be a dataset reference. For the measurement of *disambiguated recall*, we therefore measure how many mentions of the reference study name the algorithm is able to find after having trained on several contexts. Note that the algorithm may find the reference study in one of its iterations and use its contexts for training.

Singleton Recall. Finding mentions for studies that have already been identified and whose contexts have been used for training is easier than identifying new studies in the first place, especially if they appear in dissimilar contexts. To get a meaningful assessment on the ratio of distinct study names that can be detected, we prevent the algorithm from processing the contexts of the reference study by instantly removing it from the seed set if it is found in any iteration. This measure corresponds to the recall anticipated for studies occurring only once in the corpus.

Identification of Alternative Names. The importance of finding abbreviations and synonyms was illuminated in the preinvestigations. Our algorithm should find alternative names as different study titles and thereby improve recall both for ambiguous and unambiguous studies. We examine whether the algorithm indeed manages to find abbreviations and synonyms in the final part of our evaluation.

Table 1. Comparison of different seeds and strategies

Seed	Strategy	Precision	Recall (mentions)	Recall (docs)	Mentions (total)
"ALLBUS"	merge	1.00	0.24	0.50	134
"ALLBUS"	separate	0.97	0.29	0.60	303
"SOEP"	separate	0.97	0.29	0.60	299
"Eurobarometer"	separate	0.97	0.29	0.60	306

In addition to these measures, we supply the total number of mentions for all studies that have been found. Together, this information permits a good approximation of the coverage of our method. To our knowledge, there has been no other work on this task so far. Therefore, we cannot compare our results to other approaches.

4.5   Results

Table 1 shows the results for different seeds and strategies. For assessment of recall, we used "ALLBUS" as reference study and *disambiguated recall* as measure. In total, the reference study has 34 mentions in 10 documents.

In sum, all strategies and seeds produce very high values for precision. The established links are therefore reliable enough to allow their integration into an information system. It is important to note that about 14% of all mentions (for the first configuration) in fact are references to literature. This is due to the fact that authors sometimes cite the primary publication instead of a study. In some cases, they also refer to information from secondary publications. Since these indirect references are nevertheless crucial to find, we count these citations as correct study references. For these mentions, a list of datasets and their primary publications is required to make the connection.

Recall for mentions does not exceed 0.29. Since we are interested in linking publications and datasets, it is only important to retrieve at least one reference to a dataset per document. The algorithm succeeds to do so in 60% of the cases when processing each context separately and in 50% of the cases when processing merged context sets. Consistent with our hypothesis, applying the more conservative strategy increases precision but hurts recall. As the more liberal strategy already performs well with regard to precision, this strategy appears to be favorable for the identification of study references.

For the total number of mentions retrieved for all studies, we can only provide absolute numbers as the total number of different studies contained in each document is unknown.

The choice of a seed has only minimal influence on the results. For each of them, the algorithm terminates after four iterations with very similar outcomes. There is a small number of diverging patterns which leads to slightly different precision values. The recall values do not differ which indicates that the diverging patterns are not relevant for finding the particular reference study.

Table 2. Different reference studies and measures for the measurement of recall

Study	Recall singleton (mentions)	Recall singleton (docs)	Recall disamb. (mentions)	Recall disamb. (docs)
ALLBUS	0.14	0.30	0.29	0.60
SOEP	0.01	0.04	0.42	0.65
Sozio-oekonomisches Panel	0.78	0.50	0.78	0.50
Wohlfahrtssurvey	0.42	0.57	0.42	0.57
Westdeutsche Lebensverlaufsstudie	0.30	0.50	0.78	1.00
EVS	0.22	0.25	0.22	0.25
ESS	0.23	0.50	0.31	0.50

For the previous experiments, recall has been measured for one particular study. However, we have no information about the relative difficulty to retrieve this study compared to others. To get more reliable information about the recall of our method, we therefore measured recall for a variety of different studies in our next experiments. Here, we used "ALLBUS" as seed and applied the more liberal induction strategy. Additionally, we compared the recall measures *singleton* and *disambiguated*. The results are listed in Table 2. Disregarding the outliers, *disambiguated* recall lies mostly between 0.3 and 0.4 for mentions and between 0.5 and 0.6 for documents. Since the recall values for various studies are similar, these numbers should constitute a good assessment of the algorithm's total recall. *Singleton* recall differs considerably from *disambiguated* recall for some reference studies ("ALLBUS," "SOEP") and remain unaltered for others (e.g. "Sozio-oekonomisches Panel" and "Wohlfahrtssurvey"). This result is caused by the fact that some studies like "ALLBUS" are frequently referenced by mentioning the full title followed by the abbreviation in brackets. In the *disambiguated* condition, the algorithm learns to use the full titles as a cue to identify the mention in brackets as a study mention. In the *singleton* condition, this information is missing and these mentions cannot be found. In addition, the exclusion of studies may also prevent the induction of more general patterns and thus lower recall for studies in general. This is the case for "SOEP" which is the most frequently referenced study in our test corpus and thus an important source for new patterns. The solution for this problem would be the usage of a larger corpus: in our test corpus, "SOEP" appears 81 times in only 23 different documents. We anticipate higher recall values for all conditions and recall values when applying the algorithm on larger corpora.

The absence of year or version specifications in some references helps the algorithm to expand its patterns and include these specifications, if present, into the study titles (e.g. it learns to identify "ALLBUS 2000" as a study title, not only "ALLBUS"). This sometimes produces duplicates which, however, can easily be removed automatically. Duplicates were not counted for precision and recall measurement.

As discussed before, recall for every single study is also determined by the number of different titles found. The algorithm succeeds in expanding its search to abbreviations, alternative titles and different versions of studies. For example, it learned that "SOEP," "Sozio-oekonomisches Panel," "SOEP-Ost," "SOEP Sondererhebungen," "SOEP-Zuwandererstichprobe," and "SOEP Pretest" constitute references to datasets – they all denote the same study series. In the preliminary examination, we highlighted the importance of knowing alternative study names.

Altogether, our algorithm receives highly precise results with acceptable recall. Note that references to unambiguous studies can be retrieved by searching the study name in the corpus instead of only using the unambiguously found mentions. The alternative titles and abbreviations can then be used to expand the search and improve recall. For these cases, the supplied recall values serve only as a lower bound.

5   Conclusions and Future Work

Connecting publications and datasets is a non-trivial task due to the absence of standards for dataset references. Although a wealth of information is available in bibliographic metadata and abstracts, these are generally not sufficient to establish connections between the data types. We therefore introduced an iterative pattern induction method to recognize dataset references in full texts. We showed that our approach achieves very useful results while requiring only minimal supervision (the manual selection of a non-ambiguous seed) and shallow features (e.g. no layout information is needed).

As part of our future work, we plan to investigate the use of pattern and instance ranking (see for example [11]) to enhance our frequency-based measure. Incorporating this additional evidence might allow the frequency thresholds to be decreased without damaging precision. Additionally, we are going to apply our algorithm on English documents and publications from other scientific fields to verify the language- and domain-independency of our approach. Furthermore, it would be interesting to investigate the different patterns emerging in different languages and domains and to thereby investigate how research data is referenced in different languages and scientific communities.

As a next step, we will match the acquired study references with metadata available in the DBK[7], a research data catalogue maintained by GESIS. We will use this information to integrate the links between research data and publications into the scientific information retrieval systems Sowiport[8] by GESIS and Primo[9] by Mannheim University Library. We will then be able to implement a first prototype of an integrated information system for both data types.

---

[7] www.gesis.org/en/services/research/data-catalogue
[8] http://www.gesis.org/sowiport
[9] http://www.bib.uni-mannheim.de/133.html

Acknowledgements. This work is funded by the DFG as part of the InFoLiS project (SU 647/2-1). We would like to thank Benjamin Zapilko and Christian Meilicke for their great support.

References

1. Afzal, M.T., Maurer, H., Balke, W.T., Kulathuramaiyer, N.: Rule based autonomous citation mining with tierl. Journal of Digital Information Management 8(3), 196–204 (2010)
2. Councill, I.G., Giles, C.L., Kan, M.Y.: Parscit: An open-source crf reference string parsing package. In: Proceedings of the Language Resources and Evaluation Conference, European Language Resources Association (2008)
3. Gipp, B., Beel, J.: Citation Proximity Analysis (CPA) - A new approach for identifying related work based on Co-Citation Analysis. In: Proceedings of the 12th International Conference on Scientometrics and Informetrics, vol. 2, pp. 571–575 (2009)
4. Kubala, F., Schwartz, R., Stone, R., Weischedel, R.: Named entity extraction from speech. In: DARPA Workshop on Broadcast News Understanding Systems, pp. 287–292 (1998)
5. Etzioni, O., Cafarella, M., Downey, D., Popescu, A.M., Shaked, T., Soderland, S., Weld, D.S., Yates, A.: Unsupervised named-entity extraction from the web: An experimental study. Artificial Intelligence 165, 91–134 (2005)
6. Hearst, M.: Automatic acquisition of hyponyms from large text corpora. In: Proceedings of the 14th Conference on Computational Linguistics, pp. 539–545. Association for Computational Linguistics, Stroudsburg (1992)
7. Berland, M., Charniak, E.: Finding parts in very large corpora. In: Proceedings of the 37th Annual Meeting of the Association for Computational Linguistics on Computational Linguistics, pp. 57–64. Association for Computational Linguistics, Stroudsburg (1999)
8. Pennacchiotti, M., Pantel, P.: A bootstrapping algorithm for automatically harvesting semantic relations. In: Proceedings of the Inference in Computational Semantics, pp. 87–96 (2006)
9. Meusel, R., Niepert, M., Eckert, K., Stuckenschmidt, H.: Thesaurus Extension Using Web Search Engines. In: Chowdhury, G., Koo, C., Hunter, J. (eds.) ICADL 2010. LNCS, vol. 6102, pp. 198–207. Springer, Heidelberg (2010)
10. Thelen, M., Riloff, E.: A bootstrapping method for learning semantic lexicons using extraction pattern contexts. In: Proceedings of the 2002 Conference on Empirical Methods in NLP, pp. 214–221. Association for Computational Linguistics, Stroudsburg (2002)
11. Xu, R., Morgan, A., Das, A.K., Garber, A.: Investigation of unsupervised pattern learning techniques for bootstrap construction of a medical treatment lexicon. In: Proceedings of the Workshop on Current Trends in Biomedical Natural Language Processing, BioNLP 2009, pp. 63–70. Association for Computational Linguistics, Stroudsburg (2009)
12. Altman, M., King, G.: A proposed standard for the scholarly citation of quantitative data. D-Lib Magazine 13(3) (2007)
13. Green, T.: We need publishing standards for datasets and data tables. OECD Publishing White Paper (2009)

Collaborative Tagging of Art Digital Libraries:
Who Should Be Tagging?
A Case Study

M. Mahoui[1], C. Boston-Clay[2], R. Stein[3], and N. Tirupattur[4]

[1] Polis Center, [2] School of Informatics, [4] CIS, IUPUI, Indianapolis, USA
{mmahoui,bostonc,ntirupat}@iupui.edu
[3] Indianapolis Museum of Art, Indianapolis, USA
rstein@imamuseum.org

Abstract. Collaborative tagging is attracting a growing community in the arts museums as manifested by several initiatives such as the *Steve* Museum project and the *Posse* initiative at the Brooklyn Museum. The driving force for these projects is the quest for increased and improved access to artifact collections such as art collections. Previous results of studying the nature of tags provided by users reveal that these tags have little overlap with museum documentation; but on the other hand, there is good overlapping with terms from vocabulary sources such as the Art and Architecture Thesaurus (AAT). This paper reports a case study that we performed where the aim was to include tags provided by "average" users from the broader community, not necessarily closely related to the art field as it was the focus of the previous studies. The study we performed comparing tags generated by average users, expert users and metadata seems to indicate the unique role that tags provided by average users would play in facilitating the interaction with art digital libraries.

Keywords: User tags, annotation, art digital libraries, metadata, comparative study.

1 Introduction

One of the most significant consequences of Web 2.0 technologies is the use of collaborative annotation and tagging to increase the utility and comprehension of the vast wealth of information available via the Web. In aggregate, these annotations and tags are referred to as folksonomies, a bottom-up approach to building taxonomies by folks: everyday users with varying degrees of expertise. Popular social tagging Web sites include del.icio.us (http://delicious.com/) and Flickr (http://www.flickr.com/). A tag may consist of a word, abbreviation, acronym, or any combination thereof. Tags are assigned to resources by a user without any formalized ontology or vocabulary. Therefore, there is no steep learning curve for a user, which is one of the main reasons folksonomies are popular. However, this lack of constraints also introduces tags with

P. Zaphiris et al. (Eds.): TPDL 2012, LNCS 7489, pp. 162–172, 2012.

typos, special characters such as hyphens and underscores, and synonymous tags. As a result, one of the main challenges when dealing with tags is to clarify their meaning [1, 2]. Unlike expert generated taxonomies, folksonomies usually lack a structure, such as a hierarchy structure, to facilitate their utilization. Rather, the generated tags have been presented to users, to support information seeking, either as clouds of tags/phrases or accompanied with the most popular tags they co-occur with. Another main difference between folksonomies and taxonomies is that the former ones are collections of tags (e.g. "contemporary" tag and "painting" tag) while the latter are organized collections of concepts (e.g. "contemporary painting").

Despite these limitations, tagging continues to grow in popularity especially in targeting non-text material such as images (e.g. Flickr project) that have limited indexing, and where users' tags can therefore provide additional indexing entries for the resources. Towards this objective, several initiatives have been launched in the area of art digital libraries to integrate tagging both as part of the representation of the art work and also as part of the user access behavior to the online art museums. Examples include the Steve Museum project (http://www.steve.museum) and the Posse initiative at the Brooklyn Museum (http://www.brooklynmuseum.org/community/posse/).

These two examples highlight two directions that researchers have been exploring. The Posse project focus is to explore strategies that keep users engaged in the process of resource tagging while the initial focus of the Steve project was to explore the relationships between users' tags and art documentation as well as the impact of tags in improving user access to artwork. In particular, the findings in the Steve project [3] revealed that these tags have little overlap with museum documentation but on the other hand, there is good overlapping with terms from vocabulary sources such as the Art and Architecture Thesaurus (AAT). This paper reports on a qualitative and quantitative analysis that we performed that aimed at expanding some of the research questions addressed in the Steve project to include a broader dataset of user annotations. For this purpose this study targeted not only tags that are provided by art savvy users but also by "average" users from the broader community, not necessarily closely related to the art field. The contribution of the study includes a comparison study performed between two types of user tags: tags provided by savvy (i.e. expert) users knowledgeable about art, and tags provided by "average" users; and a second comparative study that compared metadata (e.g. title, artwork description) supporting the arts collection in one hand and user tags (expert and/or average) on the other hand. This study, unlike the previous studies seeks to gain more knowledge on how characteristics of each type of annotations can be combined to provide useful description of artwork to its audience.

The remaining of this paper is organized as follows: after the related work section, section 3 describes the methodology we adopted for this study. Section 4 presents the results that are analyzed in more details in section 5. In section 6, we showcase the potential impact of combining user's tags annotations and metadata in recommendation systems. Summary and future work are presented in section 7.

2   Related Work

Comparing metadata associated to a digital item such as title and author keywords to user provided tags has been investigated in studies related to medical and health domains mainly to assess the potential of user tags to better support the search of biomedical and health data. In particular, the study described in [4] worked with two social bookmarking systems Connotea (connotea.org) and CiteUlike (citeulike.org) where the tagging population tends to be from Academia. The study focused on a subset of Pubmed biomedical documents available from NCBI Entrez portal (www.ncbi.nlm.nih.gov/pubmed/) that were tagged from these social sites. The study used metrics such as coverage of the document space, the number of metadata terms associated with each document, rates of inter-annotator agreement, and rates of agreement with MeSH terms (medical taxonomy). Some of the results reported indicate that users tend to provide few tags when tagging documents, and that there is a low agreement between users tagging the same resources (inter-annotator agreement), as well as between user tags and mesh terms.

Similar results are reported in the studies conducted in [5, 6] using general health documents. In [5], the authors focused on studying the overlapping between the content of an expert indexed health digital library in France (CISMeF) and delicious social network; and used metrics such as the overlapping documents between the two collections and the overlapping of user tags with expert generated descriptors. First, the results show a very low overlapping (113 documents) of CISMeF and Delicious. To perform the manual qualitative study of the documents common to both datasets, they collected for each resource the descriptor and its thesauri relation (synonyms, related terms, and broader terms) as well as related terms that are not official set as descriptor but still associated to the resource. The results of the qualitative study show that more than half of user tags are present in the thesauri category with more than 85% of them being with exact terms or broader terms. Among the 34% of terms that are not related to the expert indexed terms, the manual analysis shows that 68% are terms that are related to the content of the document and therefore can be used to index the documents. In summary, this study while conducted on a small collection of documents reveals that user tags can compete with thesauri metadata.

Finally, when compared to the research work presented in [7], the study described in this paper includes tags of expert savvy users; therefore allowing the research community to better understand how expert users compare to average users, and thus how their provided tags can be combined in describing artwork.

3   Materials and Methods

The study was performed on a subset of art objects from the Indianapolis Museum of Art (IMA) digital collection (www.imamuseum.org). The sub-collection was selected mainly because it had two types of user tags associated with it. User tags available from the previous Steve project initiative [3] and a recent tagging initiative conducted by the IMA digital library in late 2009. Two main features differentiate the two data sets: the data set from Steve project is usually provided by users that are close to the domain of art "art savvy" and no limit on the number of annotations was set for each annotator. An "art savvy" user is likely to interact with artwork more often than casual average user either because of his/her work or for fun. The second type of annotations targeted average users in contrast and only one "average" user annotated each object. Hereafter we call these two types of annotations, expert tags and average tags respectively. Each object in the target collection has corresponding metadata such as title, type of object (e.g. painting, sculptures, and photographs), description (e.g. hanging scroll) and creator. The metadata associated to images are often used to query the art collections in addition to the standard search-based feature, as is the case for IMA digital library.

The "description" attribute as its name indicates is perhaps the most representative of the content of the object. However only 15% of the objects had a description associated to them. Moreover as the "type" attribute is a very broad type of attribute, only the title feature was selected as the main metadata to perform the comparative study against the user annotations.

Each user annotation of an object is composed of a combination of tags. For the case of expert tags and unlike many other Web sites that support collaborative tagging, tags entered using the Steve tagger (http://tagger.steve.museum/) used for the Steve project allow participants to enter more than one word (e.g. glazed stoneware) as a tag. However, this feature is not supported through "tagcow" (www.tagcow.com/) services that facilitated collecting the second set of average users' tags. As a consequence, the tags included in the average users' tags data set present more noise issues as users would try to encompass a multi-word keyword in one word ending up with expressions such as "nonrepresentative" and "avant-garde".

The comparative analysis mainly consists of comparing four data sets of annotations: the expert tags data set - expert annotations (*EA*), the average tags data set - average annotations (*AA*), the combined average and expert tags data set - combined annotations (*CA*), and the metadata data set - metadata annotation (*MA*).

As with any knowledge discovery process, an important preliminary step to consider is the preprocessing step. This step is particularly important in our case as user annotations are notoriously recognized by their high ratio of noise.
Standard cleaning steps in information retrieval were performed on each annotation data set and summarized as follows:

— Split tags into words (i.e. terms) using standard boundary words (e.g. space, dash). This step does not handle compound words (i.e., nonrepresentative[1])
— Remove stop words (e.g. the, and), punctuations and special characters
— Convert words into canonical form using Porter stemmer

In addition to the comparative analysis, we gathered descriptive statistics about the data including average and standard deviation of tags per object per collection, and minimum/maximum number of tags associated to an object per collection.

---

[1] We attempted to split compound words. However, this step was abandoned as the analysis of sample split data revealed that the overall quality of the data degraded.

The comparative analysis, aimed at determining how much agreement exist between different types of annotations, was based on the study of overlapping terms between pairs of annotation data sets. To study the overlapping between types of annotations we used the Positive Specific Agreement (PSA) [8] measure. PSA is a measure of the degree of overlap between two sets. The PSA value ranges from 0 to 1, where the value "1" is an indication of maximum overlapping between two sets.

More specifically, given an object from the target collection, and two sets of annotations S1 and S2 associated with the object:

$$PSA(S1, S2) = \frac{2a}{2a+b+c}$$

Where "a" is the number of common annotation terms to S1 and S2 and "b" and "c" correspond to the number of distinct terms to S1 and S2 respectively. Given two types of annotations, the agreement between them is determined by the average agreement computed over all objects in the study. The pairs of objects we considered for the comparison study are the following:

— Expert annotations (EA) versus average annotations (AA)
— Expert annotations (EA) versus metadata annotations (MA)
— Average annotations (AA) versus metadata annotations (MA)
— Combined expert and average annotations (CA) versus metadata annotations (MA)

4   Results

4.1   Quantitative Analysis

There were a total of 20,142 objects that had both types of user annotation EA and AA. Table 1 summarizes the number of unique tags per data set: AA, EA, CA and MA.

Table 1. Count of unique tags in annotations EA, AA, CA and MA

AA	EA	CA	MA
10,161	25,380	28,365	12,680

The summary of the descriptive statistics for each of the annotations AA, EA, CA and MA is depicted in Table 2.

Table 2. Descriptive statistics in annotations EA, AA, CA and MA

	AA	EA	CA	MA
Maximum Number of terms Per Artifact Object	24	111	113	27
Average Number terms Per Artifact Object	11.9	12.1	16	2.8
Standard Deviation	2.3	4.5	5.6	1.7

As expected, tagging provides digital libraries with a larger pool of terms to use to annotate resources such as artifact objects. Also, depending on which objects are annotated by users, the difference in the number of annotations between objects can be large when compared to using metadata.

Table 3 depicts the top 15 frequent terms and their frequency in each of the annotations AA, EA, CA and MA.

Table 3. Top 15 frequent tags in annotations EA, AA, CA and MA

AA		EA		CA		MA	
term	freq	term	freq	term	freq	term	freq
black	4992	black	7695	black	8206	led	776
white	4072	white	5598	white	6181	woman	535
tree	3491	brown	3562	tree	3800	dress	395
blue	2523	tree	3533	brown	3744	panel	394
man	2325	tan	3503	tan	3513	portrait	386
red	2212	red	2900	red	3115	de	373
brown	2058	blue	2642	blue	2952	shirt	350
green	1904	sienna	2612	sienna	2615	mola	342
art	1740	beig	2487	beig	2496	man	338
water	1644	orang	2184	green	2338	plate	295
design	1634	green	2135	orang	2222	head	245
sky	1586	yellow	2020	man	2108	even	227
color	1558	gold	1686	yellow	2105	tree	193
dress	1549	water	1434	gold	1735	dai	192
mon	1504	sky	1422	sky	1716	landscape	191

An overall analysis of the tags shows that the color of the artifact is a popular and therefore important feature to highlight by annotators (AA and EA), while object type is more distinctive of title annotation (MA).

4.2 Qualitative Analysis

Recall that we used the PSA measure to assess the overlapping between two sets of annotations, where the PSA value ranges in '0..1' interval; the value '1' indicating a maximum overlapping between the sets of annotations. Table 4 summarizes the PSA values (converted into percentages) for the four comparisons we performed.

The results show that while the overlapping between expert user annotations and average user annotations is more than two third of the total users' provided annotations, the overlapping between tittle metadata and users tags is very small (see next section for more details).

Table 4. Overlapping percentages between pairs of annotation data sets from EA, AA, CA and MA

	EA vs. AA	EA vs. MA	AA vs. MA	CA vs. MA
PSA (%)	66.95	3.55	4.39	3.53

Table 5 displays the top 15 frequently overlapping (stemmed) terms between the annotation dataset pairs (EA versus AA, EA versus MA, AA versus MA, and CA versus MA).

Table 5. Top 15 frequent overlapping terms between pairs of annotation data sets from EA, AA, CA and MA

EA vs. AA		EA vs. MA		AA vs. MA		CA vs. MA	
term	freq	term	freq	term	freq	term	freq
black	4351	dress	277	dress	310	dress	309
white	3367	tree	212	tree	167	tree	169
tree	3012	portrait	118	woman	132	woman	133
blue	2111	woman	110	man	129	man	128
red	1932	vase	91	portrait	114	portrait	122
brown	1832	bridg	89	bridg	101	bridg	101
green	1666	bowl	76	bowl	78	bowl	80
yellow	1385	chair	69	chair	74	chair	75
water	1329	red	59	hat	64	vase	69
sky	1270	hou	58	hou	64	hat	65
cloud	1264	flower	58	castl	57	hou	65
abstract	1179	river	56	rug	56	rug	58
design	1140	rug	56	vase	55	castl	58
line	1073	castl	54	river	52	red	57
dress	1072	mask	54	mask	51	mask	57

Finally, table 6 highlights the top 15 frequent (stemmed) terms that are *distinct* (i.e. exclusive) to AA and EA, when these two data sets are compared; and the top 15 frequent *distinct* (stemmed) terms in MA and CA when these two data sets are compared.

The results of tables 5 and 6 are in line with the results summarized in table 3, emphasizing the differences that exist between user annotations and title annotations in terms of size of the two types of annotations and in terms of the annotation categories.

5   Discussion

Within its scope as a case study, the analysis of the results provides several insights on the relationships between user tags and metadata in one hand, and between tags provided by expert users and average users on the other hand:

Table 6. Top 15 frequent distinct terms when comparing EA vs. AA and CA vs. MA

EA vs. AA				CA vs. MA			
EA term	freq	AA term	freq	CA term	freq	MA term	freq
black	3366	man	2113	black	8143	led	776
tan	3212	art	1594	white	6056	woman	403
sienna	2609	mon	1447	brown	3713	panel	399
white	2233	old	1255	tan	3510	de	352
beig	2112	bng	1223	tree	3170	shirt	349
brown	1734	peopl	1142	red	2982	null	345
orang	1463	men	1066	blue	2763	mola	342
mountain	1330	draw	1008	sienna	2609	figur	330
lightgrai	1229	paint	845	beig	2496	landscap	301
gold	1202	beal	744	green	2250	plate	280
royalblu	1111	face	733	yellow	2067	portrait	265
maroon	1075	long	711	orang	1971	head	234
slategrai	1066	hair	623	man	1908	even	218
build	1062	work	614	water	1586	man	215
red	972	white	584	sky	1575	dai	202

— Annotations expressiveness: the user tags data set is a very rich collection of terms when compared with the metadata dataset. This is supported by the number of distinct terms, which is at least twice in the user annotation dataset as compared to the metadata dataset. Moreover, when we observe the length of the annotation using the maximum and the average number of terms per object, the user annotation includes more terms to describe the art object. For example, the average number of user tags per object (artifact) is at least more than 5.7 times the average number of metadata terms per object.

— Term frequency: the term "black" appeared most frequently among three of the four annotations (AA, EA, and CA). This suggests that more users would possibly search for artifacts which contain the color black, thus searching for what they see in the artifacts. This word would be least likely retrieved when searching the metadata of various artifacts in the art collection.

— Overlapping terms between EA and AA: After computing the PSA values for the overlapping between pairs of annotations, it was evident that there were substantially more similar terms between the EA and AA pairs; thus indicating that expert knowledge is not exclusively required to annotate artifact objects. We suspect that this overlapping is more substantial as we noticed that in several situations an AA contains combined words tags (e.g. nonrepresentative) for which the split words exist as separate entries in EA dataset. Moreover, we also suspect that if more than one average user were allowed to annotate the same artifact, the overlapping between the two types of user tags will be more substantial. Overall, this result seems to indicate that the tags produced by average users are nearly as good as the tags produced by expert users. The overlapping could also be attributed to the fact that EA or AA is a user view even through one group of users is more knowledgeable about the subject; it is about how the user (consumer) sees the artwork piece versus how the author or the museum staff (producer) describes the artwork.

— Overlapping terms between user tags and metadata: The results of the PSA computation also confirm the results of previous studies: these two data sets describe different aspects of the documents (artifacts) they annotate. In particular, our study confirms that this difference is independent of the type of users submitting the tags; whether they are savvy users in their field or average users.

— The study of the overlapping terms in the comparison studies we performed show that EA and AA most frequent overlapping terms are related to color and outdoor landscape; while the most frequent overlapping terms between user tags (EA, AA, or CA) and metadata is more related with indoor landscaping.

— When we analyze the distinct terms in the metadata data set when compared to the user tags data set, we find that user tags (CA) dataset is more about color and outside landscape. No clear observation can be made on the most frequent distinct (i.e. exclusive) terms in the metadata dataset except that they tend to describe the type (e.g. chair, dress, hair) of the artifact object.

— We also observe that the most frequent terms in each data set overlap considerably with their corresponding frequent distinct terms when compared to other types of annotations. In particular this observation suggests that user annotations and metadata are both important to describe documents (artifacts).

6   Potential Applications of Users Tags with Metadata in Web Searching: Recommendation Systems

One potential application of users' tags and metadata annotations is their deployment in recommendation systems that utilize content-based filtering. Content-based filtering recommendation systems utilize the similarity of a given item (i.e. object) accessed by a user to other items in the system, in order to recommend new items that may be of interest for the user [9]. In the case of artifact items, the reduced amount of associated metadata available limits the use of content-based filtering. Enriching the description of items with more features provide similarity–based algorithms used for content-based filtering a rich set of features to compute similarity between items. This similarity would be more refined if users' tags are combined with metadata to form the set of features describing the artifact items - as the two set of annotations seem to cover different aspects describing the data (see previous section).

To assess the potential of leveraging users' tags and metadata in content-based filtering, we graphically represented the union of two datasets. The first dataset includes artifact objects and relationships between objects linked only by terms from metadata annotations (MA). The second data set includes artifact objects and relationships between objects linked only by terms from combined user tags annotations (CA). The results depicted in figure 1 show the highly connected network created between artifact objects that are generated by both CAs and MAs. The vertices (i.e. nodes) in the graph represent artifact objects. The dominant connections are established by CAs. However, the lines representing MAs and cutting through the "Web" generated by CAs are an indication of the complementary role that MAs play linking between objects. The richness of the resulting network as opposed to a network reduced to only MAs is a clear indication of the impact that the combined network will have in the quality of the similarity measure deployed by content-based filtering. Future studies building recommendation systems combining CA and MA will allow the evaluation of this impact on the quality of the recommendations.

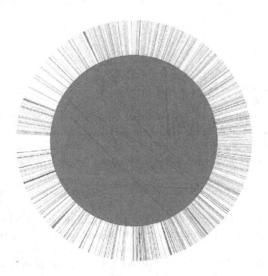

Fig. 1. Associations between artifact object linked using EA (pink color) and MA (dark green color) using Cytoscape software (*www.cytoscape.org/*)

7   Conclusion

In this paper we report on a case study we performed analyzing three types of annotations: metadata in the form of title information, user tags provided by average users and tags provided by expert users. Two main results are derived from this study. First, average users are almost as good in annotating artifact objects as expert users. The second result confirms other studies previously performed, showing that user annotation and metadata have little overlapping between them. As a consequence, they can be considered as complementary sources of information annotation.

This study while very descriptive of the relationships that exists between these different types of annotations, needs to be further refined. The results of this study may not be directly extended to make predictions about other museums that employ tagging because the case study only focus on a limited amount of annotations from one institution's art collection. A larger study, which involves annotations of art collections from various institutions, would improve external validity.

Future work includes deploying refined heuristics to split combined words. We also need more cleaning of stop words that are non-English words (e.g. le, de). Finally, a usability study testing the role of overlapping and distinct terms would provide better guidelines on how to leverage these different types of annotations for better search and retrieval of non-textual data.

Acknowledgements. This study is supported by the Office of the Vice President for Research at Indiana University with the grant title: New Frontiers in the Arts & Humanities.

References

1. Golder, S.A., Huberman, B.A.: Usage patterns of collaborative tagging systems. J. Inf. Sci. 32(2), 198–208 (2006)
2. Paolillo, J.C., Penumarthy, S.: The Social Structure of Tagging Internet Video on del.icio.us. In: Proceedings of the 40th Annual Hawaii International Conference on System Sciences. IEEE Computer Society (2007)
3. Trant, J.: Tagging, Folksonomy and Art Museums: Results of steve.museum's research (2009),
   http://conference.archimuse.com/files/trantSteveResearchRepor t2008.pdf
4. Good, B., Tennis, J., Wilkinson, M.: Social tagging in the life sciences: characterizing a new metadata resource for bioinformatics. BMC Bioinformatics 10(1), 313–313 (2009)
5. Durieux, V., Kerdelhué, G.: Looking for Health Information on the Internet: Can Social Bookmarking Systems Replace Expert Gateways. In: The European Association for Health Information and Libraries, EAHIL Workshop 2009, Dublin Castle, Ireland (2009)
6. Mahoui, M., et al.: Can Collaborative Tagging Improve the Searching of Health Web Sites? In: Proceedings of the AMIA 2010 Annual Symposium (2010)
7. Marshall, C.C.: No bull, no spin: a comparison of tags with other forms of user metadata. In: JCDL 2009 Proceedings of the 9th ACM/IEEE-CS Joint Conference on Digital Libraries 2009, New York, NY, USA (2009)
8. Hripcsak, G., Rothschild, A.S.: Agreement, the F-Measure, and Reliability in Information Retrieval. Journal of the American Medical Informatics Association 12(3), 296–298 (2005)
9. Bamshad, M.: The adaptive Web. Springer, Heidelberg (2007)

A System for Exposing Linguistic Linked Open Data

Emanuele Di Buccio, Giorgio Maria Di Nunzio, and Gianmaria Silvello

Department of Information Engineering, University of Padua, Italy
{dibuccio,dinunzio,silvello}@dei.unipd.it

Abstract. In this paper we introduce the Atlante Sintattico d'Italia, Syntactic Atlas of Italy (ASIt) enterprise which is a linguistic project aiming to account for minimally different variants within a sample of closely related languages. One of the main goals of ASIt is to share and make linguistic data re-usable. In order to create a universally available resource and be compliant with other relevant linguistic projects, we define a Resource Description Framework (RDF) model for the ASIt linguistic data thus providing an instrument to expose these data as Linked Open Data (LOD). By exploiting RDF native capabilities we overcome the ASIt methodological and technical peculiarities and enable different linguistic projects to read, manipulate and re-use linguistic data.

1 Introduction

Language defines a culture, through the people who speak it and what it allows speakers to say. Language is not just the medium of culture, it is also a part of it. Studying languages increases our understanding of how humans communicate and store knowledge. Several communities are now offering the online record of their language to be shared by any interested person around the world. A recent example is the National Geographic's Enduring Voices Project,[1] the goal of which is to preserve endangered languages by identifying language hotspots and documenting the languages and cultures within them.

For over a century, linguists have produced atlases showing the geographical distribution of linguistic features in the dialects of a language [12]. In the last two decades, several large-scale databases of linguistic material of various types have been developed worldwide. The World Atlas of Languages Structures (WALS) [8] is one of the largest projects with 160 maps showing the geographical distribution of structural linguistic features;[2] it is the first linguistic feature atlas on a world-wide scale. The study of dialectal heritage is the goal of many research groups in Europe as well. The Edisyn Project [11] is a project on dialect syntax which goal is to establish a European network of researchers in the area of language syntax that use similar standards with respect to methodology of data collection, data storage and annotation, data retrieval and cartography.

---

[1] http://travel.nationalgeographic.com/travel/enduring-voices/
[2] http://www.wals.info/

P. Zaphiris et al. (Eds.): TPDL 2012, LNCS 7489, pp. 173–178, 2012.

A full integration of all the studies carried out by each research team is hampered by the different choices made in each project. For example, the Edisyn search engine, the aim of which was to make differently dialectal databases comparable "in practice has proven to be unfeaseable".[3] In order to settle a common ground where linguistic material can be shared and re-used, the methodological and technological boundaries existing in each research linguistic project needs to be overcome. The research direction we pursue in this work is to move the focus from the systems handling the linguistic data to the data themselves. To this purpose the Linked Open Data (LOD) paradigm [9] is very promising, because it eases interoperability between different systems by allowing the definition of data-driven models and applications. LOD allows for interoperability because it is based on a unifying data model (i.e. Resource Description Framework (RDF)), a standardized data access mechanism, hyperlink-based data discovery, and self-descriptive data. In this context, a relevant initiative is ISOcat, the goal of which is to create a universally available resource for language-related metadata and provide uniform naming and semantic principles to facilitate the interoperability of language resources across applications and approaches [10].

In this paper, we introduce the ASIt enterprise [1] which has the following objectives: (i) it is a scientific project aiming to account for minimally different variants within a sample of closely related languages, (ii) it aims to share and make linguistic data re-usable. In order to create a universally available resource and be compliant with the ISOcat framework, we define a RDF model for the ASIt linguistic data thus providing an instrument to expose these data as LOD. By exploiting RDF native capabilities we overcome the ASIt methodological and technical peculiarities enabling different linguistic projects to read, manipulate and re-use linguistic data.

2   The ASIt Enterprise

The ASIt Enterprise builds on a long standing tradition of collecting and analysing linguistic corpora, which has originated different efforts and projects over the years [3,1,2]. ASIt accounts for minimally different variants within a sample of closely related languages, thus it does need a thorough part of speech (POS) disambiguation, since the 'trivial' identification of basic POS (e.g. Nouns vs Verbs) is not enough to capture cross-linguistic differences between closely related languages. To explain why the needs for ASIt are so special we have to take into consideration two different aspects: the nature of Italian dialects, and the kind of linguistic theory ASIt aims to interact with. The Italian dialectal area presents a kind of variation that involves parametric choices affecting many general aspects of syntax, morphology, and phonology. The kind of information we want to gather involves not only the presence of a certain element, but also the absence of an element; an element can be omitted only in some constructions and in conjunction with specific characteristics of the language. For this reason, ASIt proposed the creation of a specific set of tags starting from a universal core

---

[3] http://www.dialectsyntax.org/

shared by all languages (on the basis of the work done by DynaSAND [5, Ch. 4, pp.54–90]), and subsequently developing a language-specific periphery which is compatible with other projects.

ASIt Dialectal Data were gathered during a twenty-year-long survey investigating the distribution of several grammatical phenomena across the dialects of Italy [6]. These data and information were collected by means of questionnaires formed by sets of Italian sentences: dialectal speakers were asked to translate them into their dialects and write their translations in the questionnaire; therefore, each questionnaire is associated with many parallel dialectal translations. At present, there are eight different questionnaires written in Italian and almost 500 questionnaires, corresponding to the eight Italian questionnaires, written in more than 240 different dialects, for a total of more than 54,000 sentences and more than 40,000 tags.

In order to efficiently store and manage the amount of data recorded in the questionnaires, the interviews and the tagged sentences, ASIt has been realized as a digital library system relying on a relational database for which a specific relational schema has been designed. In the conceptual schema underlying the ASIt database [3] three areas considered of central interest for this work are:

- The *geographical area*, which is the place where a given dialect is spoken and where a speaker is born;
- the *derivation area*, which focuses on the background of the speaker: the level of knowledge of the dialect, the particular variety of the dialect, the birthplace, the ancestors, the document that she/he translated;
- the *tagging area*, which is how the document is structured and how it has been tagged (at a sentence level and at a word level).

In this work ASIt is modeled by means of an RDF/S which has been mapped from the ASIt conceptual model (Entity–Relationship (ER) schema) by employing the rules presented in [4,13]. In Figure 1 we report the `rdfs:Class` and `rdf:Property` realizing the RDF/S defined for the ASIt enterprise.

As far as the vocabulary adopted in this specification is concerned, we use the namespaces and prefixes reported in Table 1; `asit` is the only vocabulary which is not inherited from other domains. Instead, `dcterms` is used within the `Document` class, `foaf` within the `Actor` class, and `geo`, which is a vocabulary for representing latitude, longitude and altitude information, is used within the classes of the geographical area. We exploited this RDF/S to expose the linguistic data in ASIt as a Linked Dataset.[4]

3   Exposing Linguistic Linked Open Data

The existing ASIt infrastructure [2] has been extended in order to handle and publish linguistic data as LOD. Figure 2 highlights its two constituting layers.

The *linguistic layer* can be framed in three different levels: *datastore, application service,* and *Web service*. The *datastore* is responsible for the persistence of

---

[4] For the complete specifications of the dataset see `http://purl.org/asit/alld`

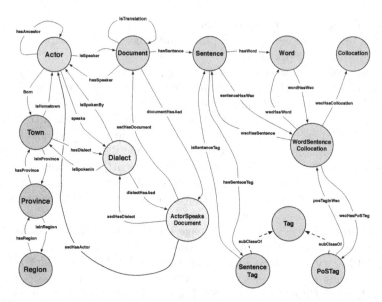

Fig. 1. Diagram representing the RDF/S defined for the ASIt enterprise

Table 1. Namespaces and Prefixes adopted by the ASIt RDF Specification

Prefix	Namespace	Description
asit	http://purl.org/asit/alld	ASIt vocabulary terms
dcterms	http://purl.org/dc/terms/	Dublin Core terms
foaf	http://xmlns.com/foaf/0.1/	Friend of a friend
geo	http://www.w3.org/2003/01/geo/wgs84_pos/	WGS84 Geo Positioning
owl	http://www.w3.org/2002/07/owl/	OWL vocabulary terms
rdf	http://www.w3.org/1999/02/22-rdf-syntax-ns/	RDF vocabulary terms
rdfs	http://www.w3.org/2000/01/rdf-schema/	RDF Schema

the linguistic resources and provides an interface to store and access linguistic data in the relational database. The *application service* is responsible for the interaction with the linguistic resources; it provides an Application Program Interface (API) to perform operations on the resources – e.g. list sentences in a document or words in a sentence, and add tags to sentences and words. When linguistic resources are created or modified, the application service exploits the datastore API for the persistence of data. The *Web service* provides functionalities to create, modify and delete resources, and gather their descriptions through appropriate HTTP requests based on a RESTful Web service [7].

The *RDF layer* is responsible for persistence and access of RDF triples of linguistic data instantiated on the basis of the defined RDF/S. The RDF layer has been developed by exploiting the functionalities of the open source library Apache Jena.[5] A *mapping service* has been developed to instantiate the

---

[5] http://incubator.apache.org/jena

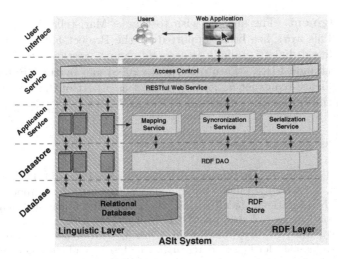

Fig. 2. Diverse layers and levels constituting the ASIt infrastructure

RDF/S starting from the data stored in the ASIt relational database. A request for creation, deletion or modification of an ASIt resource is processed by the linguistic layer that, through the proper module of the *application service*, allows the interaction with the resource and stores its new state. In parallel, by means of the *syncronization service*, the RDF layer processes the request and updates the RDF triples instantiating the ASIt RDF/S. This service allows for the interaction with the *RDF datastore* which is responsible for the persistence of the RDF/S instantiation in the *RDF Store*. When a request for resource access is submitted to the system, the RDF *serialization service* retrieves information on the requested resource from the RDF store; afterwards, the serialization service returns the retrieved RDF triples in the requested RDF output format.

4   Final Remarks

In this paper we described the variegated panorama of linguistic projects presenting the interoperability issues arising when we need to share and re-use the linguistic data they handle. To this purpose we proposed an approach aimed to enable interoperability at a data-level by overcoming the single project boundaries depending by different methodological and technological choices. This approach starts from a conceptual model and defines a mapping to a compatible Resource Description Framework Schema (RDF/S). This RDF schema allows us to map linguistic data into RDF triples which can be exposed as LOD. The possibility of serializing the data into different RDF output formats allows for the use of linguistic data regardless of the system exposing them. In the future, the flexibility of the proposed approach will allows us to extend the ASIt data model in order to be compliant with new outcomes of the ISOcat standard.

Acknowledgment. The authors wish to thanks Maristella Agosti for useful discussion. This work has been supported by the Project FIRB "Un'inchiesta grammaticale sui dialetti italiani: ricerca sul campo, gestione dei dati, analisi linguistica" (Bando FIRB - Futuro in ricerca 2008) and by the PROMISE network of excellence (contract n. 258191) project as a part of the 7th Framework Program of the European commission (FP7/2007-2013).

References

1. Agosti, M., Alber, B., Di Nunzio, G.M., Dussin, M., Pescarini, D., Rabanus, S., Tomaselli, A.: A Digital Library of Grammatical Resources for European Dialects. In: Orio, N., Agosti, M. (eds.) IRCDL 2011. Communications in Computer and Information Science, vol. 249, pp. 61–74. Springer, Heidelberg (2011)
2. Agosti, M., Alber, B., Di Nunzio, G.M., Dussin, M., Rabanus, S., Tomaselli, A.: A curated database for linguistic research: The test case of cimbrian varieties. In: Proc. of the Eight Int. Conference on Language Resources and Evaluation (LREC 2012). European Language Resources Association (ELRA), Istanbul, Turkey (2012)
3. Agosti, M., Benincà, P., Di Nunzio, G.M., Miotto, R., Pescarini, D.: A Digital Library Effort to Support the Building of Grammatical Resources for Italian Dialects. In: Agosti, M., Esposito, F., Thanos, C. (eds.) IRCDL 2010. CCIS, vol. 91, pp. 89–100. Springer, Heidelberg (2010)
4. Auer, S., Dietzold, S., Lehmann, J., Hellmann, S., Aumueller, D.: Triplify: Light-Weight Linked Data Publication from Relational Databases. In: Proc. of the 18th Int. Conf. on World Wide Web, pp. 621–630. ACM Press (2009)
5. Bael, J.C., Corrigan, K.P., Moisl, H.L.: Creating and Digitizing Language Corpora, Synchronic Databases, vol. 1. Palgrave Macmillan (2006)
6. Benincà, P., Poletto, C.: The ASIS Enterprise: A View on the Construction of a Syntactic Atlas for the Northern Italian Dialects. Nordlyd 34, 35–52 (2007), Monographic issue on Scandinavian Dialects Syntax (2007)
7. Fielding, R.T., Taylor, R.N.: Principled design of the modern web architecture. ACM TOIT 2, 115–150 (2002)
8. Haspelmath, M., Dryer, M.S., Gil, D., Comrie, B.: The World Atlas of Language Structures. Oxford University Press, United Kingdom (2005)
9. Heath, T., Bizer, C.: Linked Data: Evolving the Web into a Global Data Space. In: Synthesis Lectures on the Semantic Web. Morgan & Claypool Publishers (2011)
10. Kemps-Snijders, M., Windhouwer, M.A., Wittenburg, P., Wright, S.E.: ISOcat: Remodelling Metadata for Language Resources. IJMSO 4(4), 261–276 (2009)
11. Kunst, J.P., Wesseling, F.: The Edisyn Search Engine. Language Variation Infrastructure 3(2), 63–74 (2011)
12. Lameli, A., Kehrein, R., Rabanus, S.: Language and Space: Language Mapping: An International Handbook of Linguistic Variation. Walter de Gruyter (November 2010)
13. Myroshnichenko, I., Murphy, M.C.: Mapping ER Schemas to OWL Ontologies. In: Proceedings of the 2009 IEEE International Conference on Semantic Computing, pp. 324–329. IEEE Computer Society (2009)

Linking the Parliamentary Record: A New Approach to Metadata for Legislative Proceedings

Richard Gartner

Centre for e-Research, King's College London, London, United Kingdom
richard.gartner@kcl.ac.uk

Abstract. This paper discusses an on-going project which aims to develop an XML architecture for linking Parliamentary and other legislative proceedings. The project has developed a schema which allows key components of the record to be linked semantically and a set of controlled vocabularies to support these linkages. The project will convert two collections of proceedings to the schema and develop a prototype web-based union catalogue for them.

Keywords: metadata, Parliamentary proceedings, XML, controlled vocabularies.

1   Introduction

In the United Kingdom, records of Parliamentary proceedings go back as far as the fourteenth century. The vast majority of these have been digitised over the course of the last fifteen years and the current Hansard record is updated on a daily basis as debates take place in the United Kingdom's four Parliaments or Assemblies. This record, comprehensive though it is, remains a cumbersome resource to navigate and search owing to the diversity of delivery systems used by each hosting institution and, crucially, the diversity of approaches to metadata taken by each.

This inconsistency of metadata strategies results in it proving difficult to trace the contributions of given Parliamentarians across resources, follow the passage of bills through their stages of progress towards their entry into law, or even identify unambiguously the votes cast on each occasion. This failure to provide an integrated approach to metadata therefore renders this exceptional resource less useful to the scholarly community than it could be.

A number of projects have previously attempted to address the metadata requirements of Parliamentary or other legislative information, including the AKOMA NTOSO framework for African Parliamentary metadata[1], Dublin Core element sets devised for the records of Portugal [2] and Chile [3] and controlled vocabularies and subject taxonomies for electronic publications of the UK Parliament [4]. None of these, however, have yet produced a comprehensive architecture for linking diversely located components of the record, nor established a consistent methodology for implementing it.

P. Zaphiris et al. (Eds.): TPDL 2012, LNCS 7489, pp. 179–184, 2012.

The LIPARM (Linking Parliamentary Records through Metadata) project [5] aims for the first time to provide an integrated metadata platform for the complete Parliamentary record, historical and contemporary; by doing so, it aims to allow scholars access to its component elements in a more intelligent way than has previously been possible. The project aims to define a new XML schema for core components of the record, produce authority lists for these components and implement a proof-of-concept union catalogue for these resources which will allow sophisticated searching and browsing at diverse levels of granularity. This work in progress paper details the current state of the project at the time of writing and outlines future developments which should be in place by the end of 2012.

2   The Parliamentary Metadata Language (PML) Schema

The centre of the LIPARM project is an XML schema which aims to define a generic structure within which key elements of the Parliamentary record can be identified and tracked across collections. The project has chosen XML as its primary syntax rather than RDF which is commonly used to provide linkages within the context of the semantic web; the choice was made to employ this architecture because of the relative ease with which XML records can be administered in working environments. If desired, RDF can readily be generated from the XML schemas employed in the LIPARM project and SPARQL-based queries used to interrogate the corpus of data.

The LIPARM schema, named the Parliamentary Metadata Language (PML), does not attempt to duplicate the functionality of already existing standards, particularly those that encode bibliographic metadata. It therefore includes no components that can be offered by well-established schemas such as MODS or Dublin Core: it has, for instance, no core bibliographic components (such as title, author or subject information). Instead, the PML schema is intended to be used in conjunction with, or preferably embedded in, these pre-existing schemas.

PML has been defined specifically to act as an extension to MODS (Metadata Object Description Schema), and it is recommended that it is embedded in MODS files as an extension: in this way, key bibliographic elements (including subject terms) can be encoded using MODS elements and only those components specific to Parliamentary metadata will be covered by the PML extension. The schema may also be used with Dublin Core (DC) (preferably referenced by the DC Relation element), within a TEI (Text Encoding Initiative) header or embedded in a content packaging schema such as METS (Metadata Encoding and Transmission Standard).

The PML schema identifies seven core components of the legislative record. The first top-level element is that of **Persons** who participant in the legislative process. A further element, **Units**, delineates the major administrative components of the legislative authority and their hierarchical relationships, officially designated or otherwise: this includes political parties and constituencies. **Functions** records the positions of authority and other roles to which members may be assigned or which they may achieve: in the UK Parliamentary system, for instance, this may include the 'Prime Minister'.

Calendar objects are those temporal components into which the activities of the legislature are divided; these are nestable, so that in the UK Parliament, for example, individual house sittings are nested within sessions which are themselves nested within Parliaments. **Proceedings objects** are likely to be the most extensive components of the record: these are the components of the proceedings themselves, including any activity that occurs within the legislative or executive process. In the UK Parliament, for instance, these may include debates, motions, Early Day Motions and Prime Ministers Questions. **Vote events,** a special example of proceedings objects with their own metadata requirements, are individual votes of members at which motions are decided or decisions taken. **Proceedings Groups** are grouping elements for proceedings objects, allowing all of the proceedings which result in a given output to be linked together: an group may be, for instance, an Act or a Bill, all stages of which can be traced using this facility.

Each of these generic components is rendered more specific for the legislative context in which they are applied by a consistent set of attributes: for a unit, for example, these may take the form:-

```
<unit ID="hh-1980015-units-001"
    regURI="http://liparm.ac.uk/units/houseOfCommons"
    type="chamber"
    typeURI="http://liparm.ac.uk/unittypes/chambers">
    <label>IHouse of Commons</label>
</unit>
```

Here, a regURI attribute contains a reference to the URI for regularised form for the name of the unit, as defined in name authority files or taxonomies, while the label element provides a human-readable version of the unit which can be used for building browsable indexes. The type attribute provides a human-readable indication of the type of unit being described (in this case, the chamber of a Parliament) and a typeURI attribute provides a URI for a controlled regularlised form of the type.

To encode linkages between its constituent components, the schema makes extensive use of internal XML identifiers; every component requires an XML ID and almost every component allows pointing by IDREF to any other in the schema. The link between a Member and their voting record, for instance, is done by recording the ID of each member from their <person> element in the voting event element:-

```
<option type="yes"
    regURI="http://liparm.ac.uk/votingOptions/yes">
    <label>Ayes</label>
    <vote voterID="hh-1980015-person0001">Thatcher, Margaret</vote>
    <vote voterID="hh-1980015-person0002">Alexander, Richard</vote>
</option>
```

Two generic elements appear frequently in the schema to record contributions by members (other than voting) and references to the source materials from which the PML record is generated. A speech in a debate, for instance, may be encoded as follows:-

```
<contribution ID="hh-1980015 proceedings0002.contrib.00002"
    type="speech" typeURI="http://liparm.ac.uk/contributions/speech"
    contributorID="hh-1980015-person0002">
    <description>Speech on European Communities Bill</description>
    <sources>
        <source
            sourceRef="http://liparm.ac.uk/commons/column_1436"
            sourceType="URL"/>
        <label>"HC Deb 15 January 1980 vol 976 cc1436-41</label>
    </sources>
</contribution>
```

Contributions use standard *type* and *typeURI* and label attributes to indicate the type of contribution recorded, and the *contributorID* attribute to reference the contributor. The source element may use any form of referencing to the original: in this case, as simple URL reference is used to point to the section in an XML file where the speech referenced is recorded. Other possibilities for referencing include URIs, XPointer and XLink.

3   Controlled Vocabularies

The consistent use of URIs in *regURI* attributes within the PML schema is designed to allow unambiguous linkages between components in disparate PML records: using them will, for instance, allow the contributions of a given member recorded anywhere in the Parliamentary record to be searched, browsed or analysed together. Although any URI scheme will serve this purpose, few exist at present for Parliamentary metadata and so a key adjunct to the schema itself will be production of controlled vocabularies for each of these components.

These components are identified at as low a level of granularity as the time resources of the project allow. It is intended, for instance, that every proceedings item, or every vote will receive a unique URI of this kind. Each vocabulary will be made available as MADS (Metadata Authority Description Schema) XML files, and also as a MADS-RDF ontology generated directly from these files. A number of these, including vocabularies for constituencies, members of the Westminster and Stormont Parliaments, calendar object types and proceedings outcome types, have already been compiled at the time of writing [6] .

4   The Current State of the Project and Future Plans

At the time of writing, the definitive version of the PML schema (Version 1.0) has recently been published. This version revises an initial draft version drawn up in January-February 2012 and subsequently shown to three focus groups of historians, archivists, librarians, and developers from a diverse range of sectors including academia, the Westminster Parliament and Welsh Assembly and publishing. A number of changes were made in the light of suggestions from these groups, including multilingual support.

Now that the PML schema has been defined in a stable form, the following stages of the project will test the feasibility of its approach in the context of two substantial test collections. The collections used will be the Stormont Parliamentary Papers, a digitized edition of the proceedings of the lower house of the devolved Northern Ireland Parliament from 1921 to 1972 and proceedings from the Westminster House of Commons from the same period. A team at Queen's University, Belfast, will generate PML records for each volume of these collections and link each component of these to the LIPARM controlled vocabularies.

As these collections are already encoded in XML (although conforming to different schemas), much of the PML records can be generated using XSLT transformations: chronological and person information should be readily produced by this method and data on many of the proceedings objects should also be extractable in at least a basic form. Much of the data, including that generated by the XSLT, will, however require some manual editing: part of the project will aim to develop methods for streamlining these workflows, and guidance to future projects will be compiled into a report.

The final component of the project will be the production of a prototype union catalogue which will allow public access to the XML records encoded in the schema. This web-based resource will allow the ingest, searching, browsing and display of records in the Parliamentary metadata schema, and linkages back from entries to the original digitized versions of the papers from which they are compiled. This work will be carried out by the web team at the National Library of Wales who have already produced substantial XML-based resources of this type.

The most challenging component of the design of this catalogue will be the interface for the presentation of linkages between components. The complexity of this network of links will be addressed in the initial design stages, and will involve gathering advice from a range of potential users including historians and librarians. Further input will be solicited later in the project in a focus group which will examine the initial version of the interface and its functionality: this will be fed back into the final design to be made public towards the end of 2012.

5   Future Directions

Although the project is at such a relatively nascent stage, it is clear from the stakeholder engagement that has already taken place that, if successful, the PML schema should perform a role for which there is a common expressed need in several communities. An important follow-up to the project will involve extensive dissemination amongst potential user bases, from whom substantial expressions of interest have already been received. The final success of the project will depend to a major extent of the extent of this take-up, but there are initial grounds for optimism that this will be significant.

The generic design of the schema is intended to make it useful for legislative bodies outside the UK Parliamentary system; it is hoped, therefore, that it could serve as the basis for metadata architectures outside the United Kingdom. The substantial literature on Parliamentary metadata architectures indicates that this is an area with great potential research value, and it is hoped that future projects will extend the work of LIPARM into new applications and generate new research questions.

References

[1] Vitali, F., Zeni, F.: Towards a Country-independent Data Format: the Akoma Ntoso Experience. In: Proceedings of the V Legislative XML Workshop, pp. 67–86. European Press Academic Publishing, Firenze (2007)

[2] Pinto, J.S., Martins, J.A., Almeida, P., Fernandes, M., Zagalo, H.: Portuguese Parliamentary Records: a Multimedia Digital Library Distributed Architecture, Based on Web Services. In: International Conference on Next Generation Web Services Practices, pp. 57–62. IEEE, Los Alamitos (2005)

[3] Fuentes Martínez, M.A.: Metadata at the Library of the National Congress of Chile: a Multidisciplinary Experience. In: DC-2005: Proceedings of the International Conference on Dublin Core and Metadata Applications, pp. 163–165. Dublin Core Metadata Initiative, Dublin (2005)

[4] Marley, E.: Metadata at the UK Parliament: Use of Controlled Vocabularies and Indexing. Legal Information Management 10, 3–6 (2010)

[5] Linking Parliamentary Records through Metadata (WWW Document), http://liparm.cerch.kcl.ac.uk/

[6] Controlled Vocabularies | Linking Parliamentary Records through Metadata (WWW Document), http://liparm.cerch.kcl.ac.uk/?page_id=61

A Ground Truth Bleed-Through Document Image Database

Róisín Rowley-Brooke, François Pitié, and Anil Kokaram

Department of Electronic and Electrical Engineering,
Trinity College Dublin, Ireland
{rowleybr,fpitie,anil.kokaram}@tcd.ie

Abstract. This paper introduces a new database of 25 recto/verso image pairs from documents suffering from bleed-through degradation, together with manually created foreground text masks. The structure and creation of the database is described, and three bleed-through restoration methods are compared in two ways; visually, and quantitatively using the ground truth masks.

Keywords: Document database, bleed-through, document restoration.

1 Introduction

Bleed-through degradation poses one of the most difficult problems in document restoration. It occurs where ink has seeped through from one side of the page and interferes with text on the other side. There have been many proposed solutions to the bleed-through problem, and it is clear that researchers working in the area of bleed-through restoration are faced with two main challenges. Firstly, it can be difficult to obtain access to high resolution degraded images unless connected with a specific library or digitisation project. Secondly, for all document restoration techniques, problems arise when trying to analyse results quantitatively, as there is no actual ground truth available. This problem may be overcome either by creating synthetic degraded images with known ground truth, [5],[16], or by creating synthetic ground truth data for given real degraded images, [2]. Alternatively, performance may be evaluated without any ground truth by quantifying how the restoration affects a secondary step, such as the performance of an Optical Character Recognition (OCR) system on the document image, [17],[16]. A further issue with quantitative evaluations for performance comparison is that results of different methods are often in different formats, such as binary images [1], pseudo-binary images where the background is uniform with varying foreground intensities [10],[8], or a textured background medium with varying foreground and background intensities [17],[11]. We propose that a fair quantitative comparison between methods can only be achieved if they are converted to the same format then compared to a ground truth that is also of the same format, and the simplest way of achieving this is to binarise all the results and compare them to a binary ground truth. To our knowledge there

P. Zaphiris et al. (Eds.): TPDL 2012, LNCS 7489, pp. 185–196, 2012.

are no bleed-through datasets with ground truth freely available for researchers at this time. Since converting scanned manuscript images to a suitable format for use in bleed-through restoration algorithms can be a time consuming process, we hope that the database introduced here, where all necessary processing has been done already, will prove to be a very convenient tool.

The contributions of this work are: (i) A Document Bleed-Through Database, containing 25 recto/verso image pairs, and manually created foreground text ground truth masks. (ii) A quantitative comparison method for results of different manuscript restoration algorithms, where the results are converted to a comparable format and ranked based on probability error metrics.

Section 2 contains the details of the database. Section 3 describes the bleed-through problem, and the chosen restoration techniques. In Section 4 the implementation is described and results are presented, and then discussed in Section 5. Finally the conclusions are presented in Section 6.

2   The Database

A new bleed-through document image database has been compiled for this work, consisting of 25 registered recto/verso image pairs, taken as crops from larger manuscript images with varied degrees of bleed-through degradation. The average crop size is 573x2385 pixels. All images contained in the database are taken from the collections of the Irish Script On Screen Project (ISOS).[1] ISOS is a project of the School of Celtic Studies, Dublin Institute for Advanced Studies, Dublin, Ireland, and is funded by the Dublin Institute for Advanced Studies.[2] The object of ISOS is to create digital images of manuscripts written in Irish, and to make these images accessible as an electronic resource for researchers.

Image Capture. Each manuscript image was scanned at 600dpi, and also photographed. The images used for the database were the photographs, taken using a 5x4 format viewing camera with a Phase One P45 digital back. Both camera and manuscript were positioned on a specially adapted book-cradle. Each image was processed in Photoshop to crop to an optimum canvas size and superimposes a text header and footer to distinguish each page. A ruler was also placed alongside each image to indicate scale. Digital enhancement was not performed at this stage.

Crop Details. As mentioned in Section 1 some pre-processing is necessary to crop out binding, ruler markers, and digital labels that could influence the performance of intensity based algorithms. Also, as high resolution manuscript images are often very large in size it is not practical to use them for testing - smaller sections are preferable. For the database crops were taken from the larger images such that they would contain a sentence or phrase of text on both the recto and

---

[1] http://www.isos.dias.ie
[2] http://www.dias.ie

verso sides. The reason for this was to allow for the possibility of restoration evaluation using legibility improvement as a metric. All the images were converted to grayscale and saved in tif format. File names in the database follow the format "lib"."MS"."fol".tif. "lib" represents the library from which the manuscript contained in the image originates and can be one of eight labels: (i) AC - The Allan and Maria Myers Academic Centre, University of Melbourne, Australia. (ii) FH - The Benjamin Iveagh Library, Farmleigh House, Ireland. (iii) NLI - The National Library of Ireland. (iv) NUIG - The James Hardiman Library, National University of Ireland, Galway. (v) NUIM - The Russell Library, National University of Ireland, Maynooth. (vi) RIA - The Royal Irish Academy Library. (vii) TCD - Trinity College Dublin Library. (viii) UCD - University College Dublin Library. "MS" refers to the manuscript number (eg. "MS1333"), and "fol" refers to either the page number, or the folio number followed by "r" or "v" to denote the recto or verso side. The ground truth images are labelled as for the degraded images, but with appended "gt.tif" to differentiate between the two.

Registration. To perform non-blind bleed-through restoration (see Section 3), the recto and verso sides must first be registered so that the bleed-through interference on both sides is aligned with its originating text from the opposite side. The registration method used involved three stages. Firstly a set of corresponding control points on both recto and verso images were manually selected. These points indicated locations of the same textual features on each side. A global affine warp model was then derived from a least squares fit to the displacements between these locations. Secondly, this affine model was used as an initialisation to the affine warp optimisation method of Dubois et al. [4]. Finally, local adjustments were made to the registration manually, using the 'gridwarp' function in NUKE,[3] that defines a grid over the source image, and allows the user to reposition the corners of squares in the grid, warping the local image region correspondingly. Some difficulties were encountered in registration of images where the crops contained text close to the manuscript binding. In these regions the page deformation is nonlinear and an affine model is unsuitable. This problem was overcome by using manual registration only, with a very fine grid over the whole image.

Ground Truth Creation. The ground truth foreground images were created manually, by drawing around the outline of foreground text on both recto and verso sides. These outline layers were then extracted from the images and filled in to create binary foreground text images, with black representing text, and white representing background. In handwritten documents, the edges of characters can often be blurred or gradually fade into the background due to ink absorption by the medium, or due to the angle and pressure of the writing instrument. This makes marking the precise location of the boundary between text and background a very subjective decision. For all the images it was decided that the

---

[3] The Foundry's node-based compositor, http://www.thefoundry.co.uk/products/nuke

edge of characters would be defined where the last traces of ink were visible when viewed in close detail, as it was considered preferable to preserve as much of the foreground text shape as possible.

Those wishing to access the database should contact Róisín Rowley-Brooke at the email address listed above, alternatively information on accessing the database may be found at:
http://www.isos.dias.ie/libraries/Sigmedia/english/index.html

3   Bleed-Through Removal

Approaches to bleed-through restoration can be separated into two categories according to whether both sides of the document are used - non-blind methods - or one side only - blind methods. In the case of non-blind methods, there is clearly more information available to work with, however, registration is an essential pre-processing step that can pose many challenges, (see Section 2). Furthermore, there may only be an image of one side of the page available. Conversely, for blind methods registration is not necessary, but there is less image data available to work with.

Blind Methods. Blind methods are mostly based on the assumption that there are three distinct intensity groups in the degraded images - the darkest region corresponding to foreground, the brightest to background, and bleed-through somewhere in between. There are many approaches available for segmentation based on intensity, for example, hysteresis thresholding [5], iterative K-means clustering and principle component analysis (PCA) [3], independent component analysis (ICA), either on the colour channels [15], or different colour space images [14]. However, the main issue with using intensity information only is that this will not be sufficient in severe cases where the bleed-through is equivalent in intensity to the foreground text. In these cases some spatial information is needed. Wolf in [17] uses intensity based clustering initially, then includes spatial information in the form of smoothness priors for the estimated recto and verso hidden label fields, modelling the problem via a dual-layer Markov Random Field (MRF).

Non-Blind Methods. Many non-blind methods use comparative intensity information from both sides to improve the performance of thresholding and segmentation algorithms. For example Sauvola's adaptive thresholding algorithm [12] followed by fuzzy classification is used in [2], the Kullback-Leibler (KL) thresholding algorithm and the binarisation algorithm of Gatos et al. [6] are extended by adding in second threshold levels for the bleed-through interference in [1], and ICA is extended to double-sided documents in [16], using the recto and the flipped verso images as the sources. A model based approach is used in [9], where a function of the *difference* in intensities between the two sides is used to locate bleed-through regions. Physical diffusion-based models are defined for the foreground text, bleed-through interference, and the background medium, and then a reverse diffusion model is applied on bleed-through regions to remove interference.

Selected Methods. Three recent non-blind techniques that use both spatial and intensity information were chosen to test on the database. Firstly the method for bleed-through reduction proposed by Huang et al. in [7] and [8],(referred to as *Huang*), aims to classify pixels as foreground, background, or bleed-through based on the *ratio* of intensities between the recto and verso sides, and spatial smoothness is enforced in a dual-layer MRF framework. The data cost energy is defined from a small set of user input training data, in the form of coloured strokes drawn by the user in the regions of each pixel class on both sides. These data are then used to define the energy via K-Nearest Neighbour (KNN) and Support Vector Machine (SVM) classification of the intensity ratios. An intra-field prior energy is used to ensure spatial smoothness in the classification of each layer, but also an inter-field prior ensures that certain label combinations between layers cannot occur, such as bleed-through in the same location on both sides. The energy is minimised using graph cuts, and areas classified as background or bleed-through are replaced with the mean background intensity value.

Secondly Moghaddam and Cheriet incorporated the diffusion model idea in [9] to a unified framework in [10], (referred to as *Mogh*), using variational models for non-blind and blind bleed-through removal. Their double-sided wavelet method uses, again, a function of the *difference* in intensity between the degraded recto and verso sides as an indicator of bleed-through and foreground text regions, and spatial smoothness is enforced in the wavelet domain. The variational model for each side consists of three terms: a 'fidelity' term ensures the restored image is close to the original in foreground regions, and a 'reverse diffusion' term ensures the restored image is close to a uniform target background in background and bleed-through regions. These two terms are weighted by the function of the intensity difference. The third component of the model is the smoothness term, defined on the wavelet coefficients of the restored image, with a smoothing parameter λ chosen based on the estimated background of the degraded image. The smoothing term ensures that the restored image does not contain harsh cut-off at character edges and that fine details are preserved. The solution of the model is obtained using hard wavelet shrinkage.

Finally, in our previous work, (referred to as *RB*) [11], we proposed a method that is based on a linear additive mixing model for the degraded images. However, binary foreground masks are included in the model explicitly to limit the presence of bleed-through to certain regions only. The intensity information on each side is used to define the masks initially, via K-means clustering (the darkest of three clusters representing an estimate of foreground text). Spatial smoothness is enforced in a dual-layer framework, similar to [17] and [7], but instead of solving for a binary or ternary labelling, the clean images intensities themselves are estimated using the degradation model. Intra-layer smoothness priors are used for the masks and mixing parameters on each side, and an inter-layer prior is also used for the mixing parameters to ensure that bleed-through cannot occur in the same location on both sides. Estimates for the model parameters are obtained via a variation of Iterated Conditional Modes (ICM). A secondary linear model without limiting masks is substituted every 10 iterations for the clean image estimates to ensure that the resulting clean images appear smooth with minimal visible bleed-through artefacts remaining.

4   Results

The implementation details and results of three chosen non-blind methods are presented in what follows. For the *Huang* implementation user markup consisted of $9 - 12$ strokes drawn on both recto and verso sides. For some image pairs the recto or verso side were classed entirely as background, or background and bleed-through with no foreground. The result was therefore an image of constant mean background intensity. In these cases the markup was repeated with a greater amount of strokes highlighting the foreground regions until a visually good classification was achieved. In the case of *Mogh* also the restoration results were not optimal on some images; the smoothing parameter, λ, estimate in these instances was too low. However, unlike in the *Huang* results, this did not affect detrimentally the legibility of the resulting image. For these images therefore two versions of the *Mogh* results were examined, the results with the automatically selected λ, and a result where the value of λ was chosen manually to produce the best result visually. For *RB* implementation, the restoration was performed over 35 iterations for each image pair, with the alternative linear model for the clean images substituted four times - at iterations 10, 20, 30, and 35.

Visual Comparison. Some examples of results of the three methods on documents with different degrees of bleed-through degradation are shown in what follows. Fig. 1 shows results on three recto sides of documents with varied bleed-through degradations. The first group shows a document with light bleed-through and clear distinction in intensity between foreground and bleed-through texts. The extract is taken from a 16th Century Irish Primer in The Benjamin Iveagh Library, Farmleigh House, by kind permission of the Governors and Guardians of Marshs Library, Ireland. The top image shows the original degraded recto and then subsequent images show results using *Huang*, *Mogh*, and *RB* respectively. In the right hand column example, the degradation is more pronounced, and there is less distinction between foreground and bleed-through texts. The extract is from a 16th Century collection of Ossianic tales and poems in Irish, courtesy of the National Library of Ireland. The order of results is the same as for the light bleed-through. Finally, the bottom left shows results on a very severe example, where the foreground and bleed-through text intensities are indistinguishable. This extract is taken from a 17th Century Foras Feasa ar Éirinn (History of Ireland), from the University College Dublin (UCD) Franciscan 'A' Manuscripts, reproduced by kind permission of UCD-OFM partnership. The order of results is again the same.

Numerical Comparison. To compare results objectively, it is necessary to convert them to a similar format. The *Huang* and *Mogh* methods produce images with varied foreground text intensities on a uniform background so could easily be compared to the ground truth images . However, the *RB* method produces images with smooth transitions between foreground text and a textured background; these were less comparable with the ground truth images. Therefore the solution proposed was to binarise all the results are using the adaptive document binarisation method of Gatos et al. [6]. These binary images were then compared to the manually created ground truth images. Three probability error metrics for each method were created over the full database: **FgError**, the probability that a pixel in the foreground text was classified as background, **BgError**, the probability that a background or bleed-through pixel was classified as foreground, and **TotError** the probability that any pixel in the image was misclassified. These were calculated as follows:

$$\text{FgError} \qquad = \frac{1}{N} \sum_{GT(Fg)} |GT - B_Y|$$

$$\text{BgError} \qquad = \frac{1}{N} \sum_{GT(Bg)} |GT - B_Y| \qquad (1)$$

$$\text{TotError} \qquad = \frac{1}{N} \sum_{GT} |GT - B_Y|$$

Where GT is the ground truth, B_Y is the binarised restoration result, $GT(Fg)$ is the foreground region only of the ground truth image, similarly $GT(Bg)$ corresponds to the background region only, and N is the number of pixels in the image. The results obtained from applying the binarisation method (referred to as *Gatos*) to the degraded images, and then comparing these to the ground truth masks are also included in what follows. Figs. 2 and 3 show the BgError plotted against the FgError for each method over all 50 images, with overlaid ellipses defined by the mean error values and covariances between the two metrics. Comparisons of the BgError and FgError values between images can be misleading however, as these values depend on the relative size of the background and foreground regions in each image with respect to the text character size. For example an image that is mostly background, with small text, is likely to have a much smaller BgError value than an image with large text characters covering most of the image and proportionally less background region. It is more useful therefore to rank the performance of each method on each image, via the three error metrics (least probability of error to greatest), and then use these values to obtain an overall performance rank. Comparative error rankings between methods are shown in Table 1, where Opt refers to the optimised *Mogh* results, and each entry represents the percentage of times that the method listed vertically was ranked higher than the method listed horizontally. For instance *Mogh* (Opt) has a lower FgError than *Gatos* for 60% of the images. A further comparison of the ranks is shown in Fig. 4, where the percentage of images for which each rank was obtained is shown for each metric. Ranked Pairs Voting (RP) [13] was then used to obtain an overall rank for each of the methods. The mean error probabilities, and RP rank for all three methods are shown in Table 2.

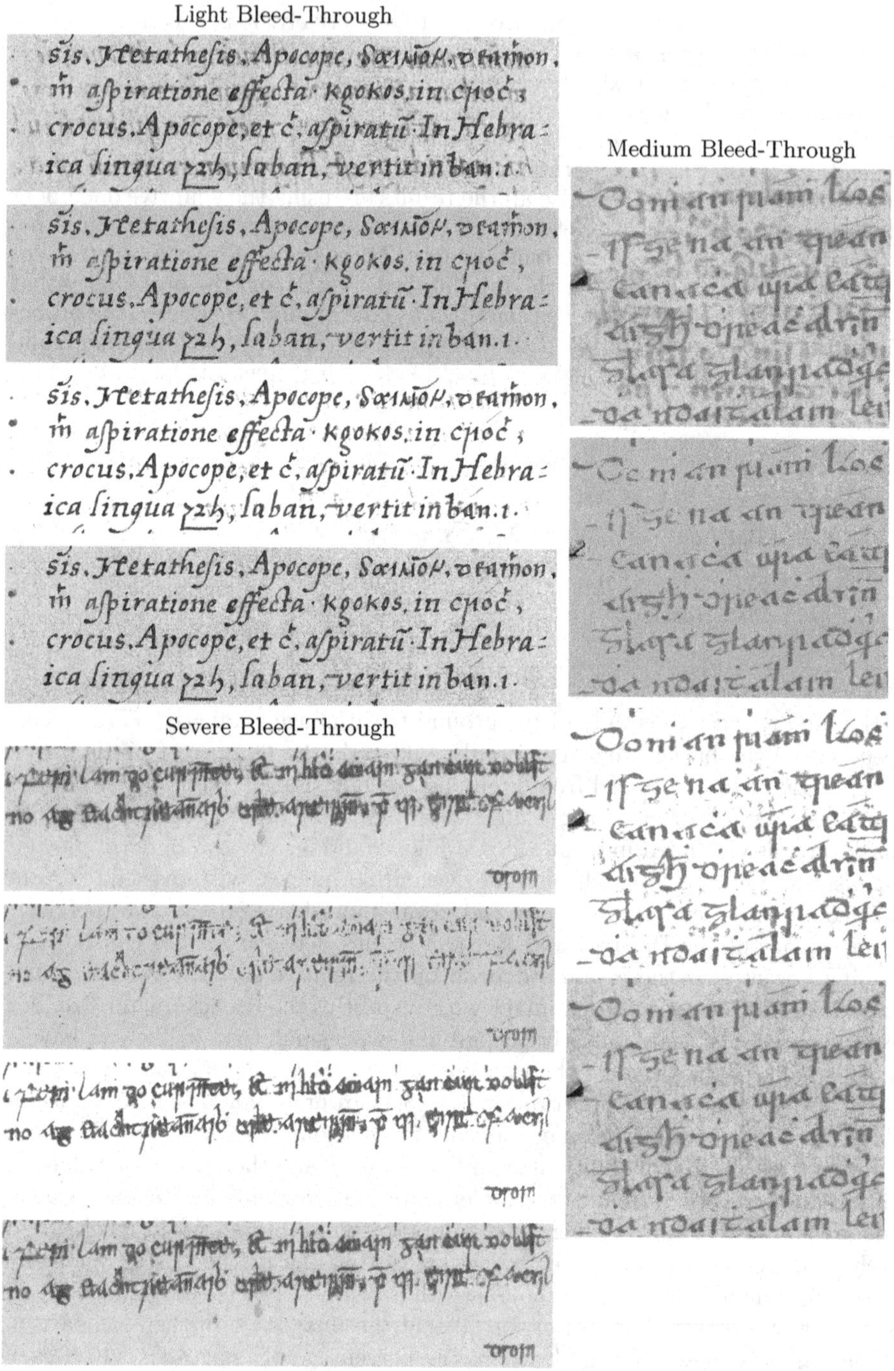

Light Bleed-Through

Medium Bleed-Through

Severe Bleed-Through

Fig. 1. Results on documents with varied bleed-through degradations (recto sides only shown). Order of images in each example from top to bottom: original recto crop, and results using *Huang* [8], *Mogh* [10] and *RB* [11].

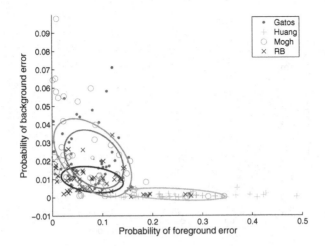

Fig. 2. BgError vs FgError for all dataset

Table 1. Pairwise Method Rank Comparison(%)

FgError	Gatos	Huang	Mogh/Opt	RB
Gatos	0	100	42/40	66
Huang	0	0	6/2	0
Mogh/Opt	58/60	94/98	0/0	58/50
RB	34	100	42/50	0
BgError	Gatos	Huang	Mogh/Opt	RB
Gatos	0	0	34/32	6
Huang	100	0	92/96	98
Mogh/Opt	66/68	8/4	0/0	28/24
RB	94	2	72/76	0
TotError	Gatos	Huang	Mogh/Opt	RB
Gatos	0	74	38/24	12
Huang	26	0	30/14	8
Mogh/Opt	62/76	70/86	0/0	26/26
RB	88	92	74/74	0

5   Discussion

The quantitative evaluations used were more objective than the visual comparisons, however the use of a binarisation technique on the clean image results was likely to favour images that have uniform background values, that is those from *Huang* and *Mogh*. *RB* however, maintains the image texture, and so was more likely to have misclassifications especially in images where the text was

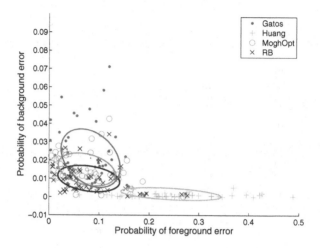

Fig. 3. BgError vs FgError for all dataset, with optimised *Mogh* results

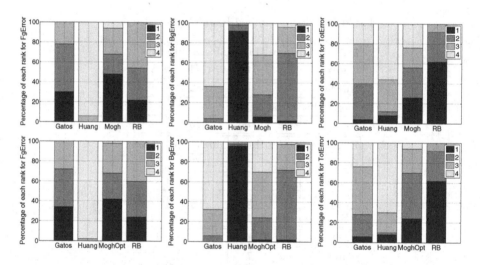

Fig. 4. Left to right: Percentage ranks for FgError, BgError, and TotError, with optimised *Mogh* results on the second row

close in intensity to the background region, or where the background medium was highly textured. The choice of binarisation method helped to mitigate this; *Gatos* performed well itself on the degraded document images as can be seen in the rankings (Table 2), therefore it was capable of obtaining good estimates of foreground text from a noisy background. The comparison of the probability error metrics for each method (Figs. 2 & 3, and Table 2) highlights the visual observations already made. *Huang* performed the best in terms of BgError and hence in terms of full bleed-through removal, however as a lot of foreground text

Table 2. Mean Error Probabilities and RP Ranks

	Gatos	Huang	Mogh/Opt	RB
Mean FgError	0.0828	0.2440	0.0804/0.0753	0.0793
RP FgRank	2	4	1	3
Mean BgError	0.0210	0.0016	0.0219/0.0149	0.0099
RP BgRank	4	1	3	2
Mean TotError	0.0304	0.0424	0.0302/0.0244	0.0216
RP TotRank	3	4	2	1

was removed the mean FgError was the highest. The *Mogh* mean error values improved significantly when the manually optimised smoothing parameters were used, this is also clearly highlighted in the graphs (Figs. 2 and 3) as the points are much more closely grouped in the optimised graph. The pairwise rank comparisons were only slightly altered with the optimised smoothing parameter, and the overall rankings were unaffected. The fact that the *RB* performed best in terms of the TotErr, but was ranked third for the FgError and second for BgError emphasises the differences in results obtained from the other methods. *Huang* uses a relatively harsh classification which allows for no ambiguity in the results, and while *Mogh* creates a smooth result, since the final results are on a plain background, this makes any remaining artefacts much more noticeable. These differences however result from different aims when creating a bleed-through removal solution; the *Mogh* and *Huang* results would be suitable as inputs for optical character recognition systems to improve recognition compared to the degraded images, and also would be useful to improve compression for storage since background texture and noise has been removed. *RB* however, would not be suitable for these applications as the resulting images aim to leave as much of the original document intact as possible, and are targeted at researchers studying the content contained within them. Knowing these differences a priori was essential in obtaining an objective comparison between the methods.

6   Conclusion

A bleed-through degraded document image database has been presented and its usefulness demonstrated by comparing the results of three non-blind bleed-through removal techniques against manually created ground truth foreground masks. Numerical comparisons between different methods will always be difficult due to the variability in the format of results produced. Binary masks were therefore created as ground truth, and the results of each method were binarised to ensure that numerical comparisons were as objective as possible. The overall ranking of results showed that the while *Huang* performed best in terms of complete bleed-through removal, and *Mogh* was ranked first in terms of foreground text preservation, *RB* performed best in terms of the overall error.

Acknowledgments. The authors would like to thank Prof. Pádraig Ó Macháin from ISOS for his assistance. This research has been funded by the Irish Research Council for Science, Engineering, and Technology(IRCSET), Science Foundation Ireland PI Programme: SFI-PI 08/IN.1/I2112, and Google, Inc.

References

1. Burgoyne, J.A., Devaney, J., Pugin, L., Fujinaga, I.: Enhanced bleedthrough correction for early music documents with recto-verso registration. In: Int. Conf. Music Inform. Retrieval, Philadelphia, PA, pp. 407–412 (2008)
2. Castro, P., Almeida, R.J., Pinto, J.R.C.: Restoration of Double-Sided Ancient Music Documents with Bleed-Through. In: Rueda, L., Mery, D., Kittler, J. (eds.) CIARP 2007. LNCS, vol. 4756, pp. 940–949. Springer, Heidelberg (2007)
3. Fadoua, D., Le Bourgeois, F., Emptoz, H.: Restoring Ink Bleed-Through Degraded Document Images Using a Recursive Unsupervised Classification Technique. In: Bunke, H., Spitz, A.L. (eds.) DAS 2006. LNCS, vol. 3872, pp. 38–49. Springer, Heidelberg (2006)
4. Dubois, E., Pathak, A.: Reduction of bleed-through in scanned manuscript documents. In: IS&T Image Process., Image Quality, Image Capture Syst. Conf., Montreal, Canada, vol. 4, pp. 177–180 (2001)
5. Estrada, R., Tomasi, C.: Manuscript bleed-through removal via hysteresis thresholding. In: 10th Int.l Conf. Doc. Anal. and Recogn., Barcelona, Spain, pp. 753–757 (2009)
6. Gatos, B., Pratikakis, I., Perantonis, S.J.: Adaptive degraded document image binarization. J. Pattern Recogn. 39(3), 317–327 (2006)
7. Huang, Y., Brown, M.S., Xu, D.: A framework for reducing ink-bleed in old documents. In: IEEE Conf. Comput. Vis. Pattern Recogn., Anchorage, AK, pp. 1–7 (2008)
8. Huang, Y., Brown, M.S., Xu, D.: User-assisted ink-bleed reduction. IEEE Trans. Image Process. 19(10), 2646–2658 (2010)
9. Moghaddam, R.F., Cheriet, M.: Low quality document image modeling and enhancement. Int. J. Doc. Anal. Recogn. 11(4), 183–201 (2009)
10. Moghaddam, R.F., Cheriet, M.: A variational approach to degraded document enhancement. IEEE Trans. Pattern Anal. Mach. Intell. 32(8), 1347–1361 (2010)
11. Rowley-Brooke, R., Kokaram, A.: Bleed-through removal in degraded documents. In: SPIE: Doc. Recogn. Retrieval Conf., San Francisco, CA (2012)
12. Sauvola, J., Pietikäinen, M.: Adaptive document image binarization. J. Pattern Recogn. 33(2), 225–236 (2000)
13. Tideman, T.N.: Independence of clones as a criterion for voting rules. J. Soc. Choice Welf. 4(3), 185–206 (1987)
14. Tonazzini, A.: Color space transformations for analysis and enhancement of ancient degraded manuscripts. J. Pattern Recogn. Image Anal. 20(3), 404–417 (2010)
15. Tonazzini, A., Bedini, L., Salerno, E.: Independent component analysis for document restoration. Int. J. Doc. Anal. Recogn. 7(1), 17–27 (2004)
16. Tonazzini, A., Salerno, E., Bedini, L.: Fast correction of bleed-through distortion in grayscale documents by a blind source separation technique. Int. J. Doc. Anal. Recogn. 10(1), 17–25 (2007)
17. Wolf, C.: Document ink bleed-through removal with two hidden markov random fields and a single observation field. IEEE Trans. Pattern Anal. Mach. Intell. 32(3), 431–447 (2010)

Identifying "Soft 404" Error Pages:
Analyzing the Lexical Signatures of Documents
in Distributed Collections

Luis Meneses, Richard Furuta, and Frank Shipman

Center for the Study of Digital Libraries and Department of Computer Science and Engineering
Texas A&M University
College Station, TX 77843-3112, USA
{ldmm,furuta,shipman}@cse.tamu.edu

Abstract. Collections of Web-based resources are often decentralized; leaving the task of identifying and locating removed resources to collection managers who must rely on http response codes. When a resource is no longer available, the server is supposed to return a 404 error code. In practice and to be friendlier to human readers, many servers respond with a 200 OK code and indicate in the text of the response that the document is no longer available. In the reported study, 3.41% of servers respond in this manner. To help collection managers identify these "friendly" or "soft" 404s, we developed two methods that use a Naïve Bayes classifier based on known valid responses and known 404 responses. The classifier was able to predict soft 404 pages with a precision of 99% and a recall of 92%. We will also elaborate on the results obtained from our study and will detail the lessons learned.

Keywords: Soft 404, Web resource management, distributed collections.

1   Introduction

Vannevar Bush in his pioneering 1945 essay "As We May Think" [1] envisions a time in which the world's knowledge is accessible by machine and in which the connections that describe the higher-level relationships among sources are themselves objects of scholarship that can be shared with colleagues. We can see this today on the Web, with the utility of resource lists such as Yahoo and the investigation of mechanisms such as our own Walden's Paths [2, 3]. Such interconnections of documents is a natural side effect of collaboration and cooperation, so as the problems to be solved grow beyond the technical abilities of an individual scholar and as social media becomes more embedded into our work practices, the presence of resources that situate knowledge into the broader environment will become ever more prevalent.

A factor not considered by Bush but critical in today's networked world is that of administrative ownership of data. Information today is not contained in neatly-defined book-like units that can be replicated and stored locally in libraries. Instead the administrative control of information related to a topic may be spread across

P. Zaphiris et al. (Eds.): TPDL 2012, LNCS 7489, pp. 197–208, 2012.

digital collections maintained by multiple scholars in multiple institutions. Administrative decentralization often is a critical factor in engaging a scholar to put in the work needed to create a valuable resource—the sense of ownership and control is motivating and often a necessary condition both for scholar and also for institution. Some of this need also centers on the desire to have a canonical copy of the resource—multiple copies in multiple locations can, and often do, diverge over time.

Administrative decentralization, though, leads to changes that are unexpected by the maintainer of a "meta-resource"—a resource created by tying together the existing resources. Individual collections can change in many ways, both intentional and unintentional. Change may be because of deliberate actions on part of the collector—for example, reorganization of the structure of the collection, switching to a different content management system, or changing jobs and institutions. They may be due to unexpected events—earthquakes, power outages, disk failures, and the like. They may be due to uncontrollable factors, as Cathy Marshall points out [4]—for example, death, seizure of computers by law enforcement, or termination of an ISPs services.

For some period of time, our group has been investigating the characterization of change in collections of administratively-decentralized Web-based resources [5-7]. These questions, of especial relevance to maintainers of these collections, foreshadow those that will be faced by anyone with similar collections. In this paper, we discuss some of our recent work on addressing issues raised when a Web server reports change in a human-readable, but not machine-readable format. This can occur intentionally (for example, attempting to provide a "friendly" or "soft" way to allow the reader to continue by redirecting failed pages to the site's index page), without knowledge of the original site (for example, the landing page provided by some ISPs when an address is given that refers to a host that is no longer available), or deceptively (for example, the registration of recently registered domain names and the "camping" of material there by entrepreneurs seeking to sell the domain name back to the original holder). We collectively call these "friendly" or "soft" 404s as they mask the return code of 404 normally returned when there is a failure to access a Web resource. Soft 404s also hinder the task of monitoring the changes in the structure and content of a Web document. Figure 1 shows an example of a soft 404 taken from http://badoo.com.

Previous work on 404 errors has focused on finding missing content [8-11], evaluating page changes [12] and reusing documents [13]. We are augmenting this work in order to provide a reliable mechanism to identify soft 404 error pages by analyzing the content of Web documents. Our approach uses the lexical signatures of Web pages [14], which are defined as a set of key identifying terms. In this remainder of this paper, we describe a study that focuses on the lexical signatures of soft 404 pages. We will also elaborate on the results obtained from the use of a text classifier to recognize soft 404 errors, which was built using the lexical signatures we synthesized from their study. We will also detail the lessons we learned.

2   Previous Work

Previous work on soft 404s is based around the premise that documents and information are not lost but simply misplaced [15] as a consequence of the lack of integrity in

Fig. 1. Soft 404 error page from http://badoo.com. Accessed on 4/3/2012.

the Web [16-19]. Other studies focused have also focused on finding the longevity of documents in the Web [20] and in distributed collections [21, 22].

On the other hand, Klein and Nelson [9] state that little research has been done to analyze the lexical signatures of Web resources. Phelps and Wilensky pioneered the use of lexical signatures to locate missing content in the Web [23]. Phelps and Wilensky claimed that if a Web request returned a 404 error, querying a search engine with a five-term lexical signature could retrieve the missing content. Park et al. used Phelps and Wilensky's previous research to perform an evaluation of nine lexical signature generators that incorporate term frequency measures [24]. Of the nine generators, eight were static and one was dynamic. As a result, Park et al. found that the dynamic generator generally outperformed the static versions. However, the lexical signature generators were affected with performance losses if the document corpus was altered or edited.

The problem of identifying soft 404 pages has been attempted before. Bar-Yossef et al. created an algorithm that analyzed the behavior of web servers to predict the occurrence of soft 404 error pages [25]. Their approach was based on generating a request for a document that does not exist on a server, thus triggering a 404 error. As webservers usually share similar rules for each hosted site, soft 404s can be predicted by extending the response patterns to a larger set of documents within the same host.

Our approach differs from previous efforts in its scope and application: we use the lexical signatures of Web documents to identify soft 404 errors. Additionally, previous research has relied heavily on HTTP status codes (mainly 404s) and server responses to identify error pages. However, these status codes do not always represent the nature and content of the retrieved document. Our work presents an approach to identifying errors not reported by status codes.

3   Methodology and Results

3.1   Initial Assessment

We started our study by analyzing the commonness of soft 404 errors in two collections of documents: resources that return an OK http response (code 200) and those returning soft 404 pages. For this purpose, we obtained a dataset for pages with OK responses by analyzing a random subset of 166,103 websites taken from the Open Directory Project database. Out of this sample, 147685 URLs returned OK responses when queried, which accounts for 88.91% of the sample. Only 206 returned a 404 error, which accounts for 0.12% of the sample. However, 404 errors in the web are very common. For instance, 475 out of 1000 sites in the sample were already "dead" (returning a soft or hard 404) when Bar-Yossef et al. ran their first experiment to measure the decay in the Web [25]. Thus, we interpreted the low number of 404 errors in the sample as an indicator of the prevalence of soft 404 errors in the Web.

We then proceeded to test the validity and correctness of the sample for the OK responses, which consisted of 147685 URLs. For this purpose, we forced a 404 error by appending a random sequence of 25 characters to each URL (i.e.: http://www.youtube.com/8q07a24ildpy3sbjmwnkgl0vs) and analyzed the response codes from the servers. 5,017 requests still returned an OK response, which accounts for the 3.41% of the subsample. 131,529 URLs returned a 404 error code, which accounts for 89.35% of the subsample. After conducting this cross-reference for validation purposes, we approximated that soft 404 errors could be found when accessing 3.41% of the documents in the web. Figure 2 describe the results we obtained from the sample where we forced a 404 error.

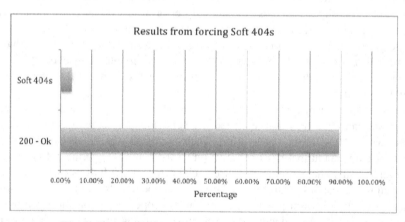

Fig. 2. Results obtained from the forced soft 404 subsample

3.2   Experiment Parameters

Following this initial assessment, we decided to use a different sample for the next phase in our experiment, which included the development and testing of a classification system. We used a snapshot of Alexa's daily 1,000,000 Top Sites taken on 3/22/2012. This dataset is freely available from Alexa.com and it is provided in CSV format. We chose to use the sites from Alexa's classification because of the increased certainty that the list would contain valid and up to date Web addresses.

Coincidentally, our study has some degree of similarity in the parameters used by Bar-Yossef et al. We summarize the parameters we used in four points. First, we attempt to fetch the "absolute URL" for each given website. It is a common practice nowadays for websites to use redirects depending on specific content, location, language or other factors. For example: nokia.com actually redirects to http://www.nokia.com/us-en/. Our methods allow 10 redirects per URL entry: if it exceeds 10 redirects the URL entry was discarded and not used in the study.

Second, to produce a soft 404 response from an actual working URL, we forced each Web server to return an error page by concatenating a random combination of 25 letters (a to z) and numbers (0 to 9) to the requested entry. The probability of a page actually existing in the server to fulfill the request is trivial: $N/(36^R)$ where N is the number of documents in the actual file path and R=25. To ensure that the server would return a soft 404, the random sequence of characters was appended to an absolute URL and file path. This ensured us that the retrieved document would be a soft 404 instead of a "hard" 404 error page.

Third, the content of the Web pages was retrieved using Python's HTTP protocol client with a timeout value of 10 seconds. The HTML cleanup, key term extraction and analysis was carried out with Python's Natural Language Toolkit (NLTK) and Beautiful Soup libraries.

Finally, using the analysis tools from the NLTK package, we found that the average soft 404-error page in the resulting collection contains 173 words; where 24 are stop words. As expected, the lexical signatures of the soft 404s provided indications of error responses: page, not, found, 404, requested, sorry, error. In the case of soft 404 error pages, these error terms synthesize the subject and purpose of the documents. With these characteristics, we proceeded to create two methods to identify soft 404 pages in collections of documents from the Web.

3.3   Running the Classifier

For our experiment, we applied the characteristics extracted from soft 404 pages towards building a text classifier that will analyze the contents of Web documents. We used a Naïve Bayes classifier that looked for specific lexical signatures in the content of the pages. We ran our text classifier against a dataset of 1000 random URLs from Alexa's daily 1,000,000 Top Sites snapshot for 10 iterations. Additionally, we used the classifier against two corpuses of data: complete HTML documents and metadata extracted from the title. Each run was completely independent from each other and the data was extracted at run time using a 54Mbps connection.

In the case of the HTML documents, we took into account the commonness of finding the lexical signatures and the overall term length of the documents. In the case of the title metadata, we only used the lexical signatures because the number of terms was trivial. On the other hand, we chose to concentrate on the titles embedded in the contents of Web documents based on the premise that titles synthesize the contents of the documents [26].

To evaluate our approach, the classifier used different sets of Web documents for training and testing. The training set was constituted with 100 documents: 50 soft 404 pages and 50 normal (not soft 404) responses. The documents in the training set were selected randomly from the top-sites corpus we obtained from Alexa.com. We initially thought of using a document proportion that was closer to a real world scenario, which is closer to 96% normal pages and 4% soft URLs. Given that our research deals with identifying and predicting soft 404s, we chose to use a ratio of 50% normal pages and 50% soft 404s.

Additionally, we implemented features for the classifier to minimize false positives: pages with OK responses that were predicted as 404 errors. For these specific cases, we chose to run the Naïve-Bayes classifier for a second time using a different Web address constructed randomly with the same URL to compare the document titles and their corresponding predictions. Consequently, this second pass of the classifier allowed us to increase the precision of the predictions of the classifier by approximately 10%.

For the first document corpus, dealing with full HTML documents, our text classifier was able to predict on average soft 404s with a precision of 95% and a recall of 87%. For the second corpus, which focused on the title metadata, our classifier was able to predict on average soft 404 pages with a precision of 99% and a recall of 92%. We found that using the titles as surrogates for the documents allowed the text classifier to make more reliable and accurate predictions. Figures 3 and 4 illustrate the individual precision and recall values we obtained identifying soft 404s on each of the ten iterations. Figure 5 shows the average values for soft 404 identification using complete HTML documents and the title metadata.

4   Discussion

Identifying soft 404 responses is not trivial. Our initial methods, which precede the results detailed in this paper, were based on an approach that allowed the text classifier to extract the most important features. This approach dealt with generating a large term concordance where weights and probabilities were assigned to each occurrence. This approach returned varied and inconsistent measures of precision and recall. There are two reasons for these results. First, the variability in the sample: Web pages vary greatly in content and layout across the Web. Second, the combined probabilities of occurrence from the terms in a Web page introduced noise into the classifier. However, we believe that a similar approach would work better in a specialized domain where the Web documents' concordance has a smaller number of key terms.

A central obstacle for this research was obtaining valid and reliable datasets. Nowadays, Web pages change constantly and their lifespan is limited. As expected, text classifiers are only as good as the training data. If the data fed into the training set of the classifier contains errors, false positives or false negatives, the classifier will not

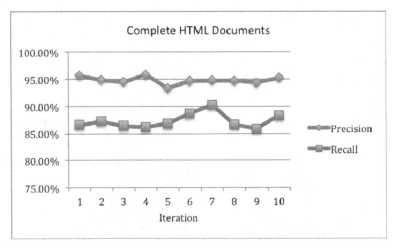

Fig. 3. Precision and recall measures per iteration for Soft 404 identification using complete HTML documents

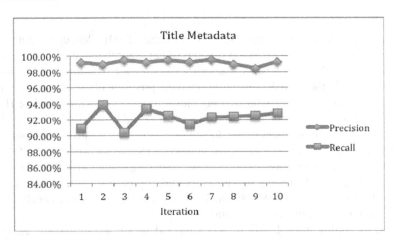

Fig. 4. Precision and recall measures per iteration for Soft 404 identification using the extracted title metadata

be accurate in its predictions. The continuous change in content and availability of documents in the Web created difficulties when we attempted to evaluate alternate approaches for the extraction of patterns and features. We believe that expanding the corpus of soft 404s allows identifying additional unique features of soft 404 pages and will improve the overall accuracy of the classifier.

During the early stages of our research, we perceived a boost in the performance of our algorithm by removing HTML markup and blocks of JavaScript code. We believed that these markup elements introduced noise into the classifier making it prone to erroneous results. However, removing HTML elements did not have a substantial effect on the performance of the classifier when we started detecting the lexical

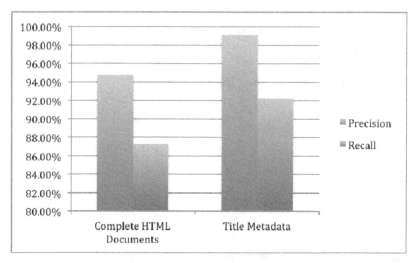

Fig. 5. Average precision and recall values for Soft 404 identification using complete HTML documents and title metadata

signatures in the document corpus. On the other hand, HTML cleanup became a fundamental piece of our approach in the versions of our algorithm that took into account the length of the Web documents. Factoring the normalized value that counts the number of terms in the Web documents boosted the precision of the classifier by 12%.

Our classifier was able to achieve a high level of precision with complete HTML documents because of the combination of two features: the analysis of lexical signatures and the overall term count of the documents. When we isolated each feature, our classifier achieved lower levels of precision when detecting soft 404 pages. However, we found the lexical signatures to be a better classifying feature than the overall term count. Factoring only the lexical signatures returned a precision value of 82% with the corpus of HTML documents. We obtained higher precision values when analyzing the lexical signatures from the title metadata.

Despite obtaining good levels of precision and recall, we found that analyzing large numbers of terms in the complete HTML document corpus introduced noise into the classifier. Consequently, these fluctuations had the potential to make the classifier produce erroneous results. For our second approach, which synthesized the titles of documents, the text classifier was less prone to noise by analyzing only document surrogates. We believe that the boost in performance was a consequence of the synthesized lexical signatures in the extracted data.

However, we found that the classifying procedure by itself favors 404 predictions. This perceived bias towards the predictions increased the number of false positives. In these specific cases, we addressed the problem by running the classifier again which ultimately increased the precision measure. Alternatively, we could have implemented additional classification features into the algorithm, which in turn could have further increased the complexity of our experiments and analysis. Additionally, by performing a second pass on individual documents with the text classifier, we addressed the problematic case where two different documents from the same site could share the same title.

We also found that page redirects in sites caused inaccurate predictions as false negatives. This issue was prevalent with sites whose content was different for different countries or audiences. For example, a request to skype.com originated within the United States will redirect the user to http://www.skype.com/intl/en-us/home. We ultimately addressed this issue by fetching and "absolute" URL and path. However, the number of redirects caused inaccurate predictions and ultimately affected the overall recall value of the algorithm in the early stages of our study.

Although somewhat similar, the two procedures we have described can have different application areas. Corpora of Web documents can benefit from using lexical signatures and overall term count, which yields a more thorough analysis but has a slightly lower precision and recall. Scenarios where only the title metadata is available can expect higher precision and recall values under the assumption that the title metadata contextualizes the contents from Web documents. Specific cases where higher precision and recall values are crucial should use the analysis of lexical signatures synthesized from the title metadata.

The advantages of our work are obvious: identifying error pages in a corpus can assist curators with the difficult task of maintaining a digital collection. However, we believe that the implementation of our approach can still be improved. In the future, we will aim to improve the ingestion mechanism that feeds the documents into classifier. Because of the propensity to changes and high volatility of Web objects, currently the documents are fetched and retrieved from the Web at run time. We could have avoided this issue by storing the documents in a centralized data structure. However, keeping the documents up to date would have required a reliable mechanism to detect and measure the degree of change in diverse object from the Web objects. Although related to our research, this issue falls beyond the scope of this study.

Given that identifying soft 404s was previously attempted in 2004 by Bar-Yossef et al [25], two questions still need to be answered. First, How do the methodologies compare? And second, which one is better?

How do the methods compare? The two methodologies have great differences. The method presented by Bar-Yossef et al. relied on server response codes to "learn" how websites react to different requests. On the other hand, our approach analyzes the contents of Web documents and relies on lexical signatures. We purposely used similar parameters and procedures in the experiments to facilitate comparisons and analysis. The similarities include: number or iterations (10), sample size per iteration (1000 URLs), retrieval timeout (10 seconds). Also, the algorithms for fetching the relevant data from each URL share some principles. For example: how to determine the absolute URL, how to circumvent redirects, and how to force a soft 404 error.

Ultimately, we believe that the two methods (ours and Bar-Yossef's) have different applications. Decentralized collections where documents are stored and cached can benefit from our approach using lexical signatures, while environments with network bandwidth limitations can utilize the approach from Bar-Yossef et al. that relies on response server codes.

Answering which methodology is better is complicated because of two reasons. First, Bar-Yossef et al. do not provide precision and recall measures; and second, there are differences in the samples that were used in the experiments. Additionally, we need to take into account their method for identifying soft 404s was not intended as an end result as they were ultimately attempting to measure the decay in the Web. However, we believe our approach presents some advantages given that it analyzes the lexical signatures of documents retrieved from the Web.

5   Future Work

We plan to continue exploring the lexical characteristics of the text contained in Web documents to improve the performance of our classifier. We plan to focus on four points. First, we will explore the relationships between the occurrence of collocations and bigrams in the lexical signatures and the probability of determining a soft 404 error page. Second, we plan on developing a standalone API that can be used by external Web services to determine the validity and correctness of documents in a decentralized collection. Third, we will analyze the prevalence of embedded objects in in common error documents such as images, scripts, external pages, and animations. Finally, we also plan on applying the lessons we learned during our research to analyze and classify Web documents that fall outside the scope of error responses. More specifically, possible applications include determining the characteristics of malicious pages and phishing sites. In conclusion, our future work will focus on discovering and identifying features of error documents that will improve the performance of our soft 404 error classifier and expand its possible applications domains.

6   Conclusions

We have described our approach towards identifying "soft" 404 errors in decentralized Web-based collections. The Web has brought new possibilities for access and interaction with digitized and born-digital documents. However, these new possibilities of interaction have also brought drawbacks. Soft 404s were originally conceived to make the Web more user-friendly and tractable, but they can introduce uncertainty when viewed as a parts of a document collection. Our work on identifying these responses helps to reduce this uncertainty by locating resources likely to be problematic and requiring the attention of collection managers.

Acknowledgements. This work was supported in part by National Science Foundation grants DUE-0840715 and DUE-1044212.

References

[1] Bush, V.: As we may think. The Atlantic (1945),
http://www.theatlantic.com/magazine/archive/
1945/07/as-we-may-think/3881/
[2] Logasa Bogen, P., Pogue, D., Poursardar, F., Li, Y., Furuta, R., Shipman, F.: WPv4: a reimagined Walden's paths to support diverse user communities. Presented at the Proceeding of the 11th Annual International ACM/IEEE Joint Conference on Digital Libraries, Ottawa, Ontario, Canada (2011)

[3] Shipman, F., Hsieh, H., Maloor, P., Moore, J.M.: The visual knowledge builder: a second generation spatial hypertext. Proceedings of the twelfth ACM conference on Hypertext and Hypermedia - HYPERTEXT 2001, Arhus, Denmark, pp. 113–122 (2001)

[4] McCown, F., Marshall, C.C., Nelson, M.L.: Why web sites are lost (and how they're sometimes found). Commun. ACM 52, 141–145 (2009)

[5] Francisco-Revilla, L., Shipman, F., Furuta, R., Karadkar, U., Arora, A.: Managing change on the web. Presented at the Proceedings of the 1st ACM/IEEE-CS Joint Conference on Digital Libraries, Roanoke, Virginia, United States (2001)

[6] Francisco-Revilla, L., Shipman, F., Furuta, R., Karadkar, U., Arora, A.: Perception of content, structure, and presentation changes in Web-based hypertext. Presented at the Proceedings of the 12th ACM Conference on Hypertext and Hypermedia, Arhus, Denmark (2001)

[7] Logasa Bogen, P., Francisco-Revilla, L., Furuta, R., Hubbard, T., Karadkar, U.P., Shipman, F.: Longitudinal study of changes in blogs. Presented at the Proceedings of the 7th ACM/IEEE-CS Joint Conference on Digital Libraries, Vancouver, BC, Canada (2007)

[8] Klein, M., Shipman, J., Nelson, M.L.: Is this a good title? Presented at the Proceedings of the 21st ACM Conference on Hypertext and Hypermedia, Toronto, Ontario, Canada (2010)

[9] Klein, M., Nelson, M.L.: Evaluating methods to rediscover missing web pages from the web infrastructure. Presented at the Proceedings of the 10th Annual Joint Conference on Digital Libraries, Gold Coast, Queensland, Australia (2010)

[10] Harrison, T.L., Nelson, M.L.: Just-in-time recovery of missing web pages. Presented at the Proceedings of the Seventeenth Conference on Hypertext and Hypermedia, Odense, Denmark (2006)

[11] Klein, M., Ware, J., Nelson, M.L.: Rediscovering missing web pages using link neighborhood lexical signatures. Presented at the Proceedings of the 11th Annual International ACM/IEEE Joint Conference on Digital Libraries, Ottawa, Ontario, Canada (2011)

[12] Dalal, Z., Dash, S., Dave, P., Francisco-Revilla, L., Furuta, R., Karadkar, U., Shipman, F.: Managing distributed collections: evaluating web page changes, movement, and replacement. Presented at the Proceedings of the 4th ACM/IEEE-CS Joint Conference on Digital Libraries, Tuscon, AZ, USA (2004)

[13] Johnson, D.B.: Enabling the reuse of World Wide Web documents in tutorials. University of Washington (1997)

[14] Park, S.-T., Pennock, D.M., Giles, C.L., Krovetz, R.: Analysis of lexical signatures for finding lost or related documents. Presented at the Proceedings of the 25th Annual International ACM SIGIR Conference on Research and Development in Information Retrieval, Tampere, Finland (2002)

[15] Baeza-Yates, R., Pereira, I., Ziviani, N.: Genealogical trees on the web: a search engine user perspective. Presented at the Proceedings of the 17th International Conference on World Wide Web, Beijing, China (2008)

[16] Ashman, H.: Electronic document addressing: dealing with change. ACM Comput. Surv. 32, 201–212 (2000)

[17] Ashman, H., Davis, H., Whitehead, J., Caughey, S.: Missing the 404: link integrity on the World Wide Web. Presented at the Proceedings of the Seventh International Conference on World Wide Web 7, Brisbane, Australia (1998)

[18] Davis, H.C.: Hypertext link integrity. ACM Comput. Surv. 31, 28 (1999)

[19] Davis, H.C.: Referential integrity of links in open hypermedia systems. Presented at the Proceedings of the Ninth ACM Conference on Hypertext and Hypermedia: Links, Objects, Pittsburgh, Pennsylvania, United States (1998)

[20] Kahle, B.: Preserving the Internet. Scientific American 276, 82–83 (1997)

[21] Koehler, W.: Web page change and persistence—a four-year longitudinal study. J. Am. Soc. Inf. Sci. Technol. 53, 162–171 (2002)

[22] Spinellis, D.: The decay and failures of web references. Commun. ACM 46, 71–77 (2003)

[23] Phelps, T.A., Wilensky, R.: Robust Hyperlinks Cost Just Five Words Each. University of California at Berkeley (2000)

[24] Park, S.-T., Pennock, D.M., Giles, C.L., Krovetz, R.: Analysis of lexical signatures for improving information persistence on the World Wide Web. ACM Trans. Inf. Syst. 22, 540–572 (2004)

[25] Bar-Yossef, Z., Broder, A.Z., Kumar, R., Tomkins, A.: Sic transit gloria telae: towards an understanding of the web's decay. Presented at the Proceedings of the 13th International Conference on World Wide Web, New York, NY, USA (2004)

[26] Jatowt, A.: Web page summarization using dynamic content. Presented at the Proceedings of the 13th International World Wide Web Conference on Alternate Track Papers and Posters, New York, NY, USA (2004)

User-Defined Semantic Enrichment of Full-Text Documents: Experiences and Lessons Learned

Annika Hinze[1], Ralf Heese[2], Alexa Schlegel[2], and Markus Luczak-Rösch[2]

[1] University of Waikato, Department of Computer Science
hinze@cs.waikato.ac.nz
[2] Freie Universität Berlin, Department of Computer Science
{heese,alexa.schlegel,markus.luczak-roesch}@inf.fu-berlin.de

Abstract. Semantic annotation of digital documents is typically done at meta-data level. However, for fine-grained access semantic enrichment of text elements or passages is needed. Automatic annotation is not of sufficient quality to enable focused search and retrieval: either too many or too few terms are semantically annotated. User-defined semantic enrichment allows for a more targeted approach. We developed a tool for semantic annotation of digital documents and conducted a number of studies to evaluate its acceptance by and usability for non-expert users. This paper discusses the lessons learned about both the semantic enrichment process and our methodology of exposing non-experts to semantic enrichment.

1 Introduction

Semantic technologies are of increasing importance in digital library (DL) research and practice [10]. Semantic enhancements have been used both at data level and at service level. At data level, FRBR provides an ontological scheme for bibliographic records [4] that increases the expressiveness of retrieval in library catalogues by incorporating information about user tasks. At service level, semantic enrichment has been used to give access to heterogeneous digital libraries and to support collaboration between location-based services and digital libraries. Supporting semantically-enriched services requires DL systems to handle different semantic models [8]. Semantic models and annotations at data and service level are typically defined by domain experts (*i.e.*, librarians or DL designers). Both types of models refer to the *conceptual aspect* of the data, *i.e.*, they annotate the meta-data of documents or document classes. So far, very little support is given for annotating the full-text body of documents.

Even though DL systems support full-text search, semantic enrichment is typically restricted to bibliographic data. Few approaches aim to enrich the full-text of DL documents; to the best of our knowledge, those approaches all use automatic text annotation methods [1,16,17]. Automatic annotations can deal with the large text corpora of Digital Libraries. However, selective annotations for domain-specific context and disambiguation of homonyms are challenging and require complex sentence analysis. Automatic tools provide excellent recall but poor precision. Furthermore, even though automatic tools are well developed for English language texts [16,17,5], other languages are poorly supported.

Our research therefore focuses on an alternative approach: we aim to support readers in manually enriching full-texts. We developed *loomp* – a tool to create user-defined

P. Zaphiris et al. (Eds.): TPDL 2012, LNCS 7489, pp. 209–214, 2012.

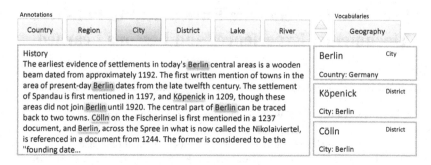

Fig. 1. *loomp* interface (stylized screen-shot for clarity)

semantic annotation of full-texts [12]. Although other tools exist for semantically annotating texts manually [2,15,9], those are typically extensions of wiki environments and
require considerable technical knowledge. Moreover, the processes of manually enriching texts have not been evaluated to date. In this paper we report on our experiences and
the lessons learned from observing how readers (*i.e.*, non-experts) create those semantic
annotations. Annotation tools for non-experts are essential for creating a large body of
high-quality annotations (*e.g.*, via crowd-sourcing) as required for the Semantic Web.

The paper is structured as follows: Sect. 2 briefly introduces *loomp*. Sects. 3 and 4
describe the setup and execution of our two user studies for annotating full-texts. Sect. 5
presents our lessons learned, and Sect. 6 the implications for digital libraries.

2   *loomp* Annotation Tool

loomp is an authoring platform for creating, managing, and accessing semantically enriched content.[1] Similar to content management systems that allow non-experts (*i.e.*,
people unfamiliar with HTML) to create websites, *loomp* supports non-experts (*i.e.*,
people unfamiliar with semantic technologies) in creating semantic annotations. It can
be used as a stand-alone tool or as a manual correction of annotations created by automatic tools which recognize named entities in analysed texts and add semantic annotations. Automatic annotation alone is not sufficient for scenarios requiring concise
annotations of high quality (*e.g.*, where precision is more important than recall).

To support non-experts, *loomp* was designed to resemble current word processors.
In a process similar to assigning formatting (*e.g.*, heading 1) to text passages, *loomp*
users can select vocabularies and assign annotations. Figure 1 shows the *loomp* UI with
its key elements of text pane, annotation toolbar, and annotation sidebar. The text pane
contains (part of) the full-text with highlighted annotations. *loomp*'s interface offers
references to concepts in different ontologies (shown as annotations and vocabularies in the annotation toolbar), highlights annotated text passages and shows existing
annotations (*e.g.*, 'Berlin' list in sidebar in Fig. 1). We explored alternatives to highlighting annotated text passages in a simplified user interface (discussed in Sect. 3)
and observed the readers' understanding of the annotation process using the full *loomp*
interface (Sect. 4).

---

[1] interactive *loomp* software online at demo.loomp.org

3   Annotation Process

As the tool was developed for use by non-exerts (*i.e.*, readers of a DL), effective feed-back about the process of creating semantic annotations is particularly important. Users need to be able to easily recognise which terms and phrases have already been anno-tated. Identification of different annotations (*i.e.*, ontological concepts) and clear dis-tinction of elements in overlapping annotations are important. Analysing systems for (non-semantic) annotation support,[2] we identified four characteristics for visual feed-back: (1) highlighting atoms and annotations, (2) position of annotations, (3) handling of overlapping atoms, and (4) connecting atoms and annotations.

For *loomp*, we explored the two alterna-tives of bar layout and border layout, imple-menting all four characteristics. In the bar layout, each atom within the text is indicated by a vertical bar in the left margin (Fig. 2, left). The colour of the bar reflects the an-notation concept. The bars are ordered by length and order in the text. Atoms in the text are highlighted by a mouse-over of the corresponding bar. The border layout high-lights annotations by enclosing an atom in a coloured frame (Fig. 2, right). Both lay-outs allow for many-to-many relationships between atoms and annotations. We observed 12 non-expert participants interacting with both interfaces (some starting with the bar and others with the border layout). During a learning phase, participants familiarized themselves with *loomp* using a short practice text. During the application phase, they had to execute a number of annotation tasks on a longer text.

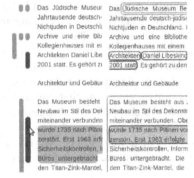

Fig. 2. Bar layout and border layout

4   Annotation Concept

We executed a second user study to evaluate *loomp*'s suitability for non-experts, with particular attention on how these users experience and apply the concept of seman-tic annotation. Here we focussed on user interaction and understanding (not interface design issues). Even though *loomp* is fully operational, we used a paper prototype to allow for greater flexibility in reacting to user activities and to elicit richer feedback. Its design allowed us to react easily to unexpected user behaviour and to make small changes to the user interface on the fly. It was prepared by printing the framework of the *loomp* UI and outlines of interaction elements; alternatives and pull-down menus were simulated by folding the paper into concertinas. Labels on interaction elements were handwritten so they could be changed dynamically. In the paper version, all UI components of *loomp* are present: the text pane, the annotation toolbar (consisting of annotation concepts and vocabularies), the annotation sidebar, and a resource selector.

---

[2] Amongst others, www.veeeb.com, atlas.ti, diego.com, http://itunes.
apple.com/us/app/bible+/id332615624

The resource selector, a separate pop-up window (see Fig. 3, not shown in Fig. 1), supports the selection of semantic identities via resource labels. For example, the atom 'Frankfurt' is annotated with the concept 'city' from vocabulary 'geography' and linked to the resource 'Frankfurt(Oder)' (see Fig. 3), which is internally referencing the resource-id 'http://dbpedia.org/resource/Frankfurt_(Oder)'.

Fig. 3. *loomp* paper prototype

The participants used a marker pen to simulate a computer mouse (used for highlighting text in the text pane and selecting UI elements by clicking with closed pen). This simulated mouse was readily accepted by the users; some additionally invented right clicks and alternate keys. The fast changing highlighting of UI elements (indicated by a pressed button and colour change in the *loomp* software) were indicated by pen caps being placed onto the elements. The study was performed by two researchers: the first one interacted with the participants while the second one acted as system simulator. The learning phase continued until each participant felt they understood the system and then the application phase continued until they had created sufficient annotations.. We observed 12 non-expert participants interacting with the *loomp* prototype; none of the participants had taken part in the first study.

5   Discussion

Both studies observed readers (thus, non-expert users) creating semantic annotations on full-texts. While the study on highlighting text passages (*i.e.*, focussing on the annotation process) required only computer literacy, the second study required a much greater conceptual understanding of semantic enrichment. In this section, we discuss our insights into both the semantic annotation process and our methodology of exposing non-experts to semantic enrichment. We found that for both studies, participants had few problems interacting with the annotation system. They openly embraced the concept of semantic annotations and aimed to create complex, partially overlapping annotations. Participants felt that indicating annotations using the bar layout was better suited to longer annotations whereas the border layout was more appropriate for shorter annotations. A combination of both forms needs to be explored.

As expected, interactions with the more complex user interface of the complete *loomp* system (annotation concept study) were more challenging. We observed that participants had difficulty recognising the implications of some of the more complex features of the user interface. In the simplified interface (study on annotation process), semantic annotation only required highlighting and selection of an annotation, whereas in the full interface (study on annotation concept), annotations required highlighting, annotation selection and assignment of resource identifier.

In the simpler study, all 123 (bar layout) and 116 (border layout) annotations were correctly formed and semantically meaningful. In the more complex second study (using a shorter text), 54 annotations were correct and meaningful, 14 were incorrectly formed but semantically meaningful, and 16 were both incorrectly formed and semantically incorrect. Two participants created several semantically meaningless annotations

(P10 and P11) and two others (P2 and P5) failed completely to create meaningful annotations. From our observations of the participants' interactions with *loomp*, we conclude that not every user group can be educated to be good annotators.

We observed that several participants of the second study had difficulty keeping in mind which passages they had selected and what their intention was (*e.g.*, people often wanted to refer back to the text, the flow was interrupted). Five participants forgot the task they were given and changed their perspective from being an information provider to an information consumer (*i.e.*, wanting to query the system). Five treated the system as a knowledge base (such as wikipedia) and wanted to insert additional information (*e.g.*, create cross-references, extend the vocabulary by synonyms, and insert unit conversions of kmph to mph). Three of the 12 participants tried to create summaries of the text by selecting whole sentences. We see one reason for this observed behaviour in the novelty of the annotation task. The wikipedia model is already well established but creation of semantic annotations (or even indexing) is not a typical task for a reader. This observation holds even for people familiar with the creation of keywords (such as librarians) and tags (such as web 2.0 users). Moreover, semantic search is not widely used and readers therefore do not have examples of well-established use-cases readily available.

Another problem is the concept of semantic identity, which is difficult for non-experts to grasp. Annotation of text has been used in digital libraries [7], for text mining [5] and for shared reading of texts [14,11]. All of these provide closed worlds of annotations (no linkage to vocabulary). Simple semantic markup of text has long been used in libraries (text keywords) as well as in web 2.0 markup of string literals [3]. Full semantic annotation requires the assignment of a semantic identity (*e.g.*, *loomp* uses DBpedia [12], LDP annotates biological texts with references to the Gene Ontology [6]). In *loomp*, this identity assignment is done implicitly by the resource selector (see Fig. 3). Other manual semantic annotation tools [2,9,15] require their users to assign this identity explicitly, thus making them unsuitable for non-expert users. SWickyNotes [13] provides a complex graphical interface that targets advanced non-expert users. The FRBR equivalent of semantic identity is the concept of *work*. The concept of semantic identity was problematic in our studies as some users did not understand the implications of freely assigning new identities to similar atoms.

Readers also do not necessarily feel bound to a particular vocabulary or do not share its understanding. For example, one participant wanted to mark some parts of the text as "political event" because she did not agree with its social implications. Thus the task of semantic annotation may be strongly bound to one's value system.

6   Conclusions

Semantic enrichment of DL full-texts (beyond FRBR markup) provides opportunities for rich and complex retrieval. However, semantic search is currently poorly supported and semantic enrichment almost completely absent. As a consequence, readers have not yet been able to form stable mental models of the markup and retrieval processes.

Using non-expert readers (*e.g.*, by crowd-sourcing) for the enrichment process is challenging. The resulting mark-up may be coloured by personal opinion and offers the opportunity to reflect diverse understandings of a text. However, because semantic enrichments are complex with potentially far-reaching consequences, testing the semantic annotation process requires clearly defined use-cases and better integration into the reader's context. The open definition of shared semantic annotations by non-expert readers may not be viable if a certain quality of annotations is required.

Semi-automatic creation of semantic enrichments is a variation of manual annotation: automatic tools create an initial markup, which is then confirmed, deleted or extended by manual annotations. This process may be best done by a single user – support for collaborative aspects needs further exploration.

In the context of smaller and well defined use-cases (*e.g.*, location markup), tools such as *loomp* are an attractive alternative to excessive markup through automatic tools. We are planning to study the use of *loomp* for creating location markup of full-texts in a mobile digital library setting with location-based access. We are currently developing educational tutorials for non-experts with the goal of raising the quality of user-defined semantic markup.

References

1. Auer, S., Bizer, C., Kobilarov, G., Lehmann, J., Cyganiak, R., Ives, Z.G.: DBpedia: A Nucleus for a Web of Open Data. In: Aberer, K., Choi, K.-S., Noy, N., Allemang, D., Lee, K.-I., Nixon, L.J.B., Golbeck, J., Mika, P., Maynard, D., Mizoguchi, R., Schreiber, G., Cudré-Mauroux, P. (eds.) ASWC 2007 and ISWC 2007. LNCS, vol. 4825, pp. 722–735. Springer, Heidelberg (2007)
2. Auer, S., Dietzold, S., Riechert, T.: OntoWiki – A Tool for Social, Semantic Collaboration. In: Cruz, I., Decker, S., Allemang, D., Preist, C., Schwabe, D., Mika, P., Uschold, M., Aroyo, L.M. (eds.) ISWC 2006. LNCS, vol. 4273, pp. 736–749. Springer, Heidelberg (2006)
3. Brickley, D., Miller, L.: FOAF vocabulary specification 0.91 (May 2007), http://xmlns.com/foaf/spec/20070524.html
4. Buchanan, G.: FRBR: enriching and integrating digital libraries. In: 6th ACM/IEEE-CS Joint Conference on Digital Libraries, pp. 260–269. ACM Press, New York (2006)
5. Cunningham, H.: Gate: general architecture for text engineering, http://gate.ac.uk
6. García-Castro, L.J., Giraldo, O.L., Castro, A.G.: Using the annotation ontology in semantic digital libraries. In: Polleres, A., Chen, H. (eds.) ISWC Posters&Demos, CEUR Workshop Proceedings, vol. 658 (2010), CEUR-WS.org
7. Gazan, R.: Social annotations in digital library collections. D-Lib. Magazine 14(11/12) (November/December 2008)
8. Hinze, A., Buchanan, G., Bainbridge, D., Witten, I.H.: Semantics in greenstone. In: Kruk, S., McDaniel, B. (eds.) Semantic Digital Libraries, pp. 163–176. Spinger (2009)
9. Krötzsch, M., Vrandečić, D., Völkel, M.: Semantic MediaWiki. In: Cruz, I., Decker, S., Allemang, D., Preist, C., Schwabe, D., Mika, P., Uschold, M., Aroyo, L.M. (eds.) ISWC 2006. LNCS, vol. 4273, pp. 935–942. Springer, Heidelberg (2006)
10. Kruk, S.R., McDaniel, B. (eds.): Semantic Digital Libraries. Springer (2009)
11. Kruk, S.R., Woroniecki, T., Gzella, A., Dabrowski, M.: JeromeDL - a semantic digital library. In: Golbeck, J., Mika, P. (eds.) Semantic Web Challenge, CEUR Workshop Proceedings, vol. 295 (2007), CEUR-WS.org
12. Luczak-Rösch, M., Heese, R.: Linked data authoring for non-experts. In: Proceedings of the Linked Data on the Web Workshop (co-located to WWW 2009). LNCS (March 2009)
13. Morbidoni, C.: SWickyNotes Starting Guide. Net7 and Universita Politecnica delle Marche (April 2012), http://www.swickynotes.org/docs/SWickyNotesStartingGuide.pdf
14. Pearson, J., Buchanan, G.: CloudBooks: An Infrastructure for Reading on Multiple Devices. In: Gradmann, S., Borri, F., Meghini, C., Schuldt, H. (eds.) TPDL 2011. LNCS, vol. 6966, pp. 488–492. Springer, Heidelberg (2011)
15. Schaffert, S.: Ikewiki: A semantic wiki for collaborative knowledge management. In: WS on Semantic Technologies in Collaborative Applications (STICA 2006), Manchester, UK (June 2006)
16. Thomson Reuters Inc. Open Calais website, http://www.opencalais.com/
17. Zemanta Ltd. Zemanta website, http://www.zemanta.com/

Semantic Document Selection

Historical Research on Collections That Span Multiple Centuries

Daan Odijk[1], Ork de Rooij[1], Maria-Hendrike Peetz[1],
Toine Pieters[2], Maarten de Rijke[1], and Stephen Snelders[2]

[1] ISLA, University of Amsterdam, The Netherlands
{d.odijk,o.derooij,m.h.peetz,derijke}@uva.nl
[2] Descartes Center for the History and Philosophy of the
Sciences and the Humanities, Utrecht University, The Netherlands
t.pieters@uu.nl, s.a.m.snelders@umcutrecht.nl

Abstract. The availability of digitized collections of historical data, such as newspapers, increases every day. With that, so does the wish for historians to explore these collections. Methods that are traditionally used to examine a collection do not scale up to today's collection sizes. We propose a method that combines text mining with exploratory search to provide historians with a means of interactively selecting and inspecting relevant documents from very large collections. We assess our proposal with a case study on a prototype system.

1 Introduction

The availability of vast amounts of publicly accessible digital data motivates the development of techniques for large-scale data analysis and brings about methodological changes in disciplines that are shifting from data-poor to data-intensive. So far, the humanities have profited only marginally from large-scale digital collections that possibly span decades or even centuries. This may be a result of the distinct nature of data in the humanities, which are often historically specific, geographically diverse, and culturally ambiguous [6]. Hence, a qualitative analysis of documents is a first choice. In historical research one closely studies a manually determined sample of the material [5, 9, 10]. These traditional historical research methods can be used for studying large-scale digital libraries, but they impose substantial limitations. Adapting these research methods and combining them with computationally-based methods—for document selection and the analysis of the selected documents—may yield new research questions for historians.

In this paper we describe the intertwined development of a method to select documents and a system to access a digital archive spanning several centuries, designed to provide an historian with valuable insight. Specifically, we consider the digital newspaper archive of the Koninklijke Bibliotheek (KB).[1] Today, the archive comprises over three million pages from hundreds of titles from 1618–1995, consisting of more than 36 million articles, about a third of its planned final size of 100 million articles.

The document selection method we propose is based on a combination of text mining and exploratory search methods. Text mining is an umbrella term for a wide range of techniques for obtaining useful information from large textual data sources [2, 8].

---

[1] The Dutch Royal Library, National library of the Netherlands, http://kranten.kb.nl/

P. Zaphiris et al. (Eds.): TPDL 2012, LNCS 7489, pp. 215–221, 2012.

Exploratory search is a form of information retrieval where users do not precisely know what they want beforehand, or where they can find it, but prefer to explore a collection to uncover new associations between documents [7]. Here, users explore the collection, and iteratively fine-tune their queries until they find what they are looking for. Typically, exploratory search interfaces have methods that quickly provide an overview of large parts of collections and tools to quickly zoom into details [3]. Comparing selections from a collection was shown to benefit television history researchers [4]. Exposing and presenting temporal information, in particular using timelines, are considered to be open challenges in exploratory search systems [1]. Our combination of interactive exploratory search and text mining supports historians to set up systematic search trails: the tooling helps them interpret and contrast the result sets returned. By exploring word associations for a result set, inspecting the temporal distribution of documents, and by comparing selections historians can make a more principled document selection.

In Section 2 we describe traditional sampling-based methods for document selection. In Section 3 we describe semantic document selection and Section 4 puts this into practice. In Section 5 we discuss implications and conclude.

2   Traditional Historical Research Methodology: Manual Sampling

Corpora available for historical research are often too large to be examined entirely. Researchers select subsets and closely examine only the selection. This leads to the fundamental choice: what to select? In this section we describe the most common approach to do this in historical research: through sampling. We start with three examples.

Van Vree [9] studied Dutch public opinion regarding Germany in the period 1930–1939. This was one of the first studies that explored the possibility to use media history as a form of mentality history. Van Vree [9] assumed newspapers to be the most important mass media at that time and selected four newspapers that represented major population groups (such as Catholics and Protestants). All issues of these newspapers were browsed manually, yielding a selection almost 4000 articles expressing an opinion on the subject. Neutral press, with a marketshare of about 45%, were not considered. Witte [10] followed a similar approach to study the image of the nation in the Belgian Revolution. Six newspapers from different cities and political signatures were manually selected and browsed, yielding 350 articles that expressed an opinion, possibly omitting many more. Condit [5] studied public expositions on heredity and genetics; 650 articles were selected from a period of 95 years based on indexes provided by publishers. Considering the dynamic and often inaccurate nature of indexes, one can question the representativeness of this sample. Moreover, only public expositions were studied, implicit assumptions regarding heredity were not studied.

These studies provide important insights on public opinion, but there are important practical and theoretical disadvantages to the methodological approach they employ. One can argue that not all relevant articles were selected, yet checking this would mean redoing the entire laborious selection process. Insight gained from inspecting the collection cannot be used to obtain a more representative sample. Even though the fields of digital libraries and information retrieval have come a very long way, this arguably subjective and rigid method of manual sampling is common practice—the selection of documents from a collection spanning over four centuries calls for an alternative.

3 Revisiting the Document Selection Process

Without dismissing the benefits of traditional document sampling methods, we advance document selection by including all relevant documents in a qualitative analysis.

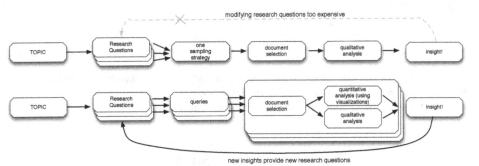

Fig. 1. Two implementations of the document selection process. We compare a manual sampling method (top) with semantic document selection method (bottom).

We have modeled manual sampling in the top part of Fig. 1. Given a research topic, a historian poses research questions that are then used to form a single sampling strategy. This strategy determines which documents to include in the selection. The researcher then develops an insight into the topic based on a qualitative analysis of the selection. This is a one way process: exploring new research questions means redoing the entire laborious process of sampling.

In semantic document selection (bottom part of Fig. 1) research questions are associated with queries against a collection. This takes away the limitation of having a single sampling strategy. A query can consist of keywords, a specific time period, a particular document source, or any combination. Each query yields a document selection, with no laborious sampling needed. Through text mining and visualizations, new insights can be gained from an initial selection. This can lead to an improved query and, therefore, a more representative document selection. This can be done by exploring word associations and metadata and through a visualization of the number of documents over time. This quantitative analysis leverages the knowledge of the historian. A clear benefit is that the historian can use the gained insight to investigate new research questions. Moreover, comparing document selections using quantitative analysis helps to validate these selections, making them less biased and more representative. With manual sampling validating the document selection is impractical or even impossible as replicating the manual selection process is too time intensive.

To support the document selection process, we developed ShoShin, an exploratory search interface that guides the user to interesting bits of information, leveraging the fact that the users are experts on the topic of interest. Due to the scale of the collection ShoShin does not store all articles locally, but collects articles on-demand and processes these on-the-fly. Fig. 2 details the architecture of ShoShin. Fig. 3 shows both an abstract overview and a screenshot of the interface. ShoShin provides the user not just with a list of relevant documents, but also with visualizations that allow inspection of, and

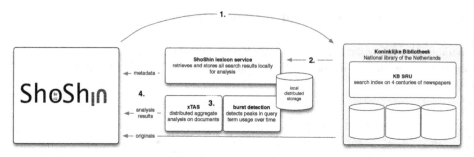

Fig. 2. (1) When a user enters a query, the query is sent to the KB. (2) The resulting document set is sent to the ShoShin lexicon service. (3) Here, the documents are analyzed by xTAS. (4) The documents and analysis results are visualized in the user interface.

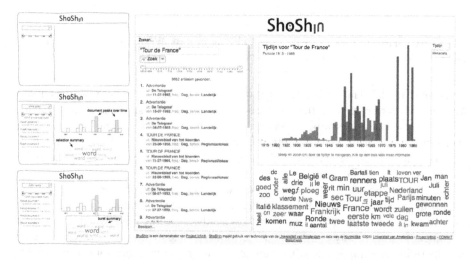

Fig. 3. Interface sketches of ShoShin. A user enters a query (top left), resulting in a list of relevant documents, a term cloud and temporal distribution visualization (center left). Users can click on bursts to see word associations of that burst (bottom left). The screenshot on the right is the actual ShoShin user interface.

navigation through, the document selection. Next, we describe the visualizations of word associations and temporal distribution of documents in more detail.

First, the word association visualization allows historians to glance over the content of the document selection. This visualization is a term cloud based on the relative frequencies of the words occurring in documents within a selection. Clicking on words in the cloud modifies the selection. The historian can thus inspect individual documents that contain both the clicked word and the original query.

Second, the temporal distribution visualization allows historians to discover patterns in documents's publication dates. This visualization is a histogram of publication dates that can be explored interactively; it provides more fine-grained data when zooming in on specific parts of the histogram. To enable quick recognition of atypical patterns, bursts within the histogram—time periods where significantly more documents were published compared to neighboring periods—are highlighted. Clicking on a burst yields a visualization of word associations of that burst alone and a list of documents contained within that burst. This allows the historian to get an in-depth understanding of what each burst is about. Together, these interactions facilitate exploration of the document selection in order to detect patterns, improving the representativeness of the selection.

4 A Worked Example

A full-blown assessment of the ShoShin demonstrator is work-in-progress. Below, we report on one of several case studies with individual historians. Our subject, a senior historian, wants to analyze the public opinion on drugs, drug trafficking, and drug users, as represented in newspapers, in the early twentieth century (1900–1940). He wants to know whether the view on drugs is predominantly based on medical aspects (addictions, health benefits) or on social aspects (crime). How does our subject use ShoShin? First, he needs to create a lexicon of terms related to drugs. The term *narcotica* is a Dutch umbrella term for several narcotics. The actual terms describing narcotics may have changed over time and not all may be known to the historian. For a high recall of documents, a lexicon is required that captures all possible relevant terms. When a researcher uses his *domain knowledge* to create a list of words, ShoShin supports the researcher to find terms that are not readily available to him by showing a *term cloud* based on all retrieved documents (see Fig. 4c for term clouds for drug related queries). The historian can expand the original query with terms he recognizes as drugs.

To inspect the representativeness of the document selection, the historian looks at the temporal distribution of documents. He sets the time period to 1900–1940 (see Fig. 4a) and queries for several names of drugs and compares the resulting temporal distributions. With his domain knowledge he marks key events, like the Opium Treaty from Shanghai (1912), the introduction of Dutch Opium laws (1920) and the tightening thereof (1928). He then concludes that before the Dutch Opium laws came into effect the term *chloroform* was dominantly used; afterwards, the terms *opium, heroïne*, and *cocaïne* are more prominent (see Fig. 4b).

To gain a better understanding of the aspects associated with drugs, the historian looks at what terms were associated with the drugs over time, by examining the associated term cloud. He compares term clouds of several time periods at several scales in time. These associated term clouds (Fig. 4c) show a shift from health issues (*geneesmiddelen, vergiften, wetenschap, apotheken*[2]) to crime related issues (*politie, smokkelhandel, gearresteerd*[3]). By inspecting the actual word counts, the historian can find quantitative evidence for an increased use of the terms associated with *narcotica* after the Dutch Opium laws came into effect and that they are decreasingly associated with health related terms while becoming increasingly associated with crime related terms.

---

[2] English: medications, poison, science, pharmacies.
[3] English: police, smuggling, arrest.

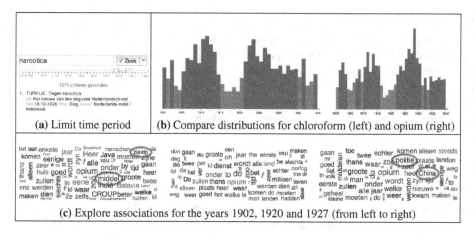

(a) Limit time period

(b) Compare distributions for chloroform (left) and opium (right)

(c) Explore associations for the years 1902, 1920 and 1927 (from left to right)

Fig. 4. Case study of a historian

5   Conclusion

We have described semantic document selections as a methodology for historical research on large repositories; it addresses three problems of traditional manual sampling: representativeness, reproducibility and rigidness. Word associations improve representativeness of the document selection as these associations are produced from the data, not from prior knowledge. Comparing selections and inspection of specific timespans in the data further support understanding the representativeness of a document selection.

The use of text mining and exploratory search allows document selections to be reproducible and remove the rigidness that stems from a single sampling strategy. Associations, longitudinal search and comparisons allow the researcher to return to document selection with new insights, at any time. In small-scale trials, we found that semantic document selection fits well in historical research methodology as an alternative to manual sampling, improving the representativeness and reproducibility of document selection and, thereby, the validity of the conclusions drawn.

Acknowledgments. This research was supported by the European Union's ICT Policy Support Programme as part of the Competitiveness and Innovation Framework Programme, CIP ICT-PSP under grant agreement nr 250430, the European Community's Seventh Framework Programme (FP7/2007-2013) under grant agreements nr 258191 (PROMISE) and 288024 (LiMoSINe), the Netherlands Organisation for Scientific Research (NWO) under project nrs 612.061.814, 612.061.815, 640.004.802, 380-70-011, 727.011.005, 612.001.116, the Center for Creation, Content and Technology (CCCT), the Hyperlocal Service Platform project funded by the Service Innovation & ICT program, the WAHSP and BILAND projects funded by the CLARIN-nl program, the Dutch national program COMMIT, and by the ESF Research Network Program ELIAS.

References

[1] Alonso, O., Strötgen, J., Baeza-Yates, R., Gertz, M.: Temporal information retrieval: Challenges and opportunities. In: TWAW Workshop, WWW 2011 (2011)

[2] Au Yeung, C., Jatowt, A.: Studying how the past is remembered: towards computational history through large scale text mining. In: CIKM 2011 (2011)

[3] Bron, M., van Gorp, J., Nack, F., de Rijke, M.: Exploratory search in an audio-visual archive. In: EuroHCIR 2011 (2011)

[4] Bron, M., van Gorp, J., Nack, F., de Rijke, M., Vishneuski, A., de Leeuw, S.: A subjunctive exploratory search interface to support media studies researchers. In: SIGIR 2012 (2012)

[5] Condit, C.: The meanings of the gene: Public debates about human heredity. University of Wisconsin Press (1999)

[6] Courant, P., Fraser, S., Goodchild, M., et al.: Our cultural commonwealth: The report of the american council of learned societies commission on cyberinfrastructure for humanities and social sciences (2006)

[7] Marchionini, G.: Exploratory search: from finding to understanding. Commun. ACM 49(4), 41–46 (2006)

[8] Michel, J., Shen, Y., Aiden, A., et al.: Quantitative analysis of culture using millions of digitized books. Science 331(6014), 176 (2011)

[9] van Vree, F.: De Nederlandse pers en Duitsland 1930-1939. Historische Uitgeverij (1989)

[10] Witte, E.: De constructie van België: 1828-1847. Lannoo Uitgeverij (2006)

Finding Quality Issues in SKOS Vocabularies

Christian Mader[1], Bernhard Haslhofer[2], and Antoine Isaac[3]

[1] University of Vienna, Faculty of Computer Science, Austria
christian.mader@univie.ac.at
[2] Cornell University, Department of Information Science, USA
bernhard.haslhofer@cornell.edu
[3] Europeana & Vrije Universiteit Amsterdam, The Netherlands
aisaac@few.vu.nl

Abstract. The Simple Knowledge Organization System (SKOS) is a standard model for controlled vocabularies on the Web. However, SKOS vocabularies often differ in terms of quality, which reduces their applicability across system boundaries. Here we investigate how we can support taxonomists in improving SKOS vocabularies by pointing out quality issues that go beyond the integrity constraints defined in the SKOS specification. We identified potential quantifiable quality issues and formalized them into computable quality checking functions that can find affected resources in a given SKOS vocabulary. We implemented these functions in the qSKOS quality assessment tool, analyzed 15 existing vocabularies, and found possible quality issues in all of them.

1 Introduction

The Simple Knowledge Organization System (SKOS) [13] is a standard model for sharing and linking controlled vocabularies (thesauri, classification systems, etc.) on the Web. Organizations like, e.g., the European Union[1], the United Nations[2], or the UK government[3] publish SKOS representations of their vocabularies on the Web so that they can easily be accessed by humans and machines.

However, quality issues can affect the applicability of SKOS vocabularies for tasks such as query expansion, faceted browsing, or auto-completion, as in the following examples:

- AGROVOC defines concepts in 25 different languages. While most concepts have English labels attached, only 38% have German labels. This can be a problem for multilingual applications that rely on label translations.
- An earlier version of the STW thesaurus (v8.06) contained 5 pairs of concepts with identical labels. As a result, the auto-complete function of the online search interface suggested identical entries without disambiguation information.

---

[1] EuroVoc, http://eurovoc.europa.eu/
[2] AGROVOC, http://aims.fao.org/standards/agrovoc/about
[3] Integrated Public Sector Vocabulary (IPSV), http://doc.esd.org.uk/IPSV

P. Zaphiris et al. (Eds.): TPDL 2012, LNCS 7489, pp. 222–233, 2012.
© Springer-Verlag Berlin Heidelberg 2012

– The non-public thesaurus of the Austrian Armed Forces (LVAk) contains 11 disconnected concept clusters. When confronted with these structures, the maintainers recognized them as data without practical significance.

The SKOS specification defines a set of integrity conditions that state whether given data patterns are consistent with the SKOS model. Yet the SKOS integrity conditions fail to capture quality aspects like the ones above. The main reason lies in SKOS' "minimal commitment" approach. A standard that aims at cross-domain interoperability should refrain from defining constraints that impose on one domain the requirements of another. SKOS is thus very liberal with respect to data quality. On the other hand, each vocabulary should fulfill domain- and application-specific quality aspects and taxonomists often follow standard guidelines specific to given types of vocabularies (cf., [1,15]) or apply their own hand-crafted checks [5]. Existing guidelines consider these aspects, but currently rely on human judgment, which is subjective and does not scale for larger vocabularies. The SKOS context, where vocabularies can be linked together on the Web, also brings issues hitherto unforeseen by traditional checking approaches.

We aim at contributing to the ongoing community efforts to bridge that gap between model-level integrity constraints and domain-specific quality aspects. Our goal is to help taxonomists in identifying possible quality issues in SKOS vocabularies and to give them a set of computable quality checking functions that, in combination with the taxonomists' experience and domain expertise, can serve as quality indicators for vocabularies. Finding such quality issues also gives important feedback on the overall vocabulary design process and should, in the end, lead to better vocabularies. With defining computable quality checking functions we tackle the problem of quality assessment from an objective perspective. Subjective perception of quality is not within the scope of this work. Our contribution can be summarized as follows:

– We identified 15 quality issues for SKOS vocabularies by examining existing guidelines and formalized them into computable quality checking functions that identify possibly affected resources in a vocabulary.
– With the qSKOS quality assessment tool we provide a reference implementation of these functions.
– We tested these functions by analyzing a representative set of 15 existing SKOS vocabularies to learn about possible quality issues.

In the following, we will first discuss what "quality" means in the context of SKOS vocabularies and how it is currently supported by the SKOS specification and existing tools. Then we introduce the quality issues we have identified and describe how we implemented them in the qSKOS quality assessment tool. Finally, we report on the results of an analysis we performed on 15 existing SKOS vocabularies and show that the quality issues we discussed are real and can lead to the improvement of existing vocabularies. All supplemental materials[4] are available online.

---

[4] qSKOS: `https://github.com/cmader/qSKOS/`, wiki: `https://github.com/cmader/qSKOS/wiki`, dataset: `https://github.com/cmader/qSKOS-data/`

2   Background and Related Work

The problem of "vocabulary quality" is closely related to the more general one of "data quality" and has been discussed in data and information systems research (cf. [4]). Pipino et al. [16] argue that dealing with data quality should involve both "subjective perceptions of the individuals" and "objective measurements based on the data". We see our work as a contribution to the latter.

The SKOS specification does not mention the notion of quality, but defines in total six integrity conditions [13], each of which is a statement that defines under which circumstances data are consistent with the SKOS data model. For example, "a resource has no more than one value of skos:prefLabel per language tag". Tools that can check whether these conditions are met are already available: two of the six conditions are defined formally in the OWL representation of SKOS and can therefore be validated by any OWL reasoner. For validating a SKOS vocabulary against the other integrity conditions, one can use tools such as the PoolParty Thesaurus Consistency Checker[5], or the Skosify[6] validator, which can also correct some detected quality problems.

Typical applications of controlled vocabularies are classification, indexing, auto-completion, query reformulation, or serving as a glossary. As we discussed in detail in earlier work [14], these areas impose specific requirements on vocabulary features, such as structure, availability, and documentation. Quality aspects of controlled vocabularies have already been discussed in standardized guidelines [1,15], manuals [2,6,8,21], tutorials [18], and scholarly articles [5,12]. These most often rely on manual, precise analysis of individual statements in the data, as in [19]. Our work builds on this literature, but focuses on the less intellectually loaded checks, which can be automatized to assist vocabulary users or publishers.

Data quality is also being discussed in Semantic Web and Linked Data research. Hogan et. al [9] identify four categories of common errors and shortcomings in RDF documents and Heath and Bizer [7] summarize best practices for publishing data on the Web. Ontology evaluation, i.e., measuring the quality of an ontology, has also been discussed extensively [23]. However, the authors focus on RDF datasets and ontologies in general. While we could use some criteria suggested here, such as consistent tagging of literals, these need to be completed by considering SKOS-specific properties.

One issue when assessing the quality of SKOS data is the so-called "Open World Assumption", which underlies the Web of Data itself. Established quality notions from closed-world systems, such as referential integrity or schema validation, do not hold anymore, because available information may be incomplete and non-explicitly stated facts cannot be determined as true or false. Work-arounds, often ad-hoc, are thus currently used to evaluate quality in Linked Data sets [7], as is done in the (rule-based) SKOS tools mentioned above, or the Pellet ICV[7], which re-interprets OWL axioms with integrity constraint semantics.

---

[5] http://demo.semantic-web.at:8080/SkosServices/check
[6] http://code.google.com/p/skosify/
[7] http://clarkparsia.com/pellet/icv/

3 Quality Issues in SKOS Vocabularies

We identified an initial set of possible quality issues in SKOS vocabularies by reviewing literature, manually examining existing vocabularies, and focusing on issues that can be measured automatically. Some measures, such as hierarchy depth or node centrality, have been omitted due to lack of evidence on their general influence on vocabulary quality. We published our findings in the qSKOS wiki and requested feedback from experts via public mailing lists and informal face to face discussions. Based on the received responses, we translated a subset of these issues into computable quality checking functions. Each function takes a given SKOS vocabulary and an optional vocabulary namespace as input and finds all resources that match the corresponding quality issue. For the purpose of this work, we define a SKOS vocabulary as follows:

Definition (SKOS Vocabulary). Let a SKOS vocabulary be a tuple of the form $V = \langle IR, C, AC, SR, LV, CS \rangle$, with $IR = I_{CEXT}(\textit{rdfs:Resource}^{\mathcal{I}})$ being the set of **resources**, $C \subseteq IR$ with $C = I_{CEXT}(\textit{skos:Concept}^{\mathcal{I}})$ being the set of **concepts**, $AC \subseteq C$ being the set of **authoritative concepts**, which are all concepts that are identified by URIs in the vocabulary namespace, $SR = I_{EXT}(\textit{skos:semanticRelation}^{\mathcal{I}})$ being the set of **semantic relations** associating concepts with one another, $LV \subseteq I_{CEXT}(\textit{rdfs:Literal}^{\mathcal{I}})$ being the set of untyped **plain literals**, and $CS = I_{CEXT}(\textit{skos:ConceptScheme}^{\mathcal{I}})$ being the set of **concept schemes**. Further, we let V be the fully entailed RDFS interpretation of the underlying RDF graph. We enrich V by entailment of $\textit{owl:inverseOf}$ properties as well as instances of $\textit{owl:TransitiveProperty}$ and $\textit{owl:SymmetricProperty}$ defined by the formal OWL semantics of SKOS [13].

In the following, we explain the origins and design rationale for each quality issue and explain how the corresponding quality checking function works. For better readability and due to lack of space we provide only semi-formal definitions and refer to the source code of the qSKOS tool for further details.

3.1 Labeling and Documentation Issues

Omitted or Invalid Language Tags. SKOS defines a set of properties that link resources with RDF Literals, which are plain text natural language strings with an optional language tag. This includes the labeling properties `rdfs:label`, `skos:prefLabel`, `skos:altLabel`, `skos:hiddenLabel` and also SKOS documentation properties, such as `skos:note` and subproperties thereof. Literals should be tagged consistently [23], because omitting language tags or using non-standardized, private language tags in a SKOS vocabulary could unintentionally limit the result set of language-dependent queries. A SKOS vocabulary can be checked for omitted and invalid language tags by iterating over all resources in IR and finding those that have labeling or documentation property relations to plain literals in LV with missing or invalid language tags, i.e., tags that are not defined in RFC3066[8].

---

[8] Tags for the Identification of Languages http://www.ietf.org/rfc/rfc3066.txt

Incomplete Language Coverage. The set of language tags used by the literal values linked with a concept should be the same for all concepts. If this is not the case, appropriate actions like, e.g., splitting concepts or introducing scope notes should be taken by the creators. This is particularly important for applications that rely on internationalization and translation use cases. Affected concepts can be identified by first extracting the global set of language tags used in a vocabulary from all literal values in LV, which are attached to a concept in C. In a second iteration over all concepts, those having a set of language tags that is not equal to the global language tag set are returned.

Undocumented Concepts. Svenonius [20] advocates the "inclusion of as much definition material as possible" and the SKOS Reference [13] defines a set of "documentation properties" intended to hold this kind of information. To identify all undocumented concepts, we iterate over all concepts in C and collect those that do not use any of these documentation properties.

Label Conflicts. The SKOS Primer [11] recommends that "no two concepts have the same preferred lexical label in a given language when they belong to the same concept scheme". This issue could affect application scenarios such as auto-completion, which proposes labels based on user input. Although these extra cases are acceptable for some thesauri, we generalize the above recommendation and search for all concept pairs with their respective skos:prefLabel, skos:altLabel or skos:hiddenLabel property values meeting a certain similarity threshold defined by a function $sim : LV \times LV \rightarrow [0, 1]$. The default, built-in similarity function checks for case-insensitive string equality with a threshold equal to 1. Label conflicts can be found by iterating over all (authoritative) concept pairs $AC \times AC$, applying sim to every possible label combination, and collecting those pairs with at least one label combination meeting or exceeding a specified similarity threshold. We handle this issue under the Closed World Assumption, because data on concept scheme membership may lack and concepts may be linked to concepts with similar labels in other vocabularies.

3.2   Structural Issues

Orphan Concepts are motivated by the notion of "orphan terms" in the literature [8], i.e., terms without any associative or hierarchical relationships. Checking for such terms is common in thesaurus development and also suggested by [15]. Since SKOS is concept-centric, we understand an orphan concept as being a concept that has no semantic relation $sr \in SR$ with any other concept. Although it might have attached lexical labels, it lacks valuable context information, which can be essential for retrieval tasks such as search query expansion. Orphan concepts in a SKOS vocabulary can be found by iterating over all elements in C and selecting those without any semantic relation to another concept in C.

Weakly Connected Components. A vocabulary can be split into separate "clusters" because of incomplete data acquisition, deprecated terms, accidental deletion of relations, etc. This can affect operations that rely on navigating a connected vocabulary structure, such as query expansion or suggestion of related terms. Weakly connected components are identified by first creating an undirected graph that includes all non-orphan concepts (as defined above) as nodes and all semantic relations SR as edges. "Tarjan's algorithm" [10] can then be applied to find all connected components, i.e., all sets of concepts that are connected together by (chains of) semantic relations.

Cyclic Hierarchical Relations is motivated by Soergel et al. [18] who suggest a "check for hierarchy cycles" since they "throw the program for a loop in the generation of a complete hierarchical structure". Also Hedden [8], Harpring [6] and Aitchison et al. [2] argue that there exist common forms like, e.g., "generic-specific", "instance-of" or "whole-part" where cycles would be considered a logical contradiction. Cyclic relations can be found by constructing a graph with the set of nodes being C and the set of edges being all skos:broader relations.

Valueless Associative Relations. The ISO/DIS 25964-1 standard [1] suggests that terms that share a common broader term should not be related associatively if this relation is only justified by the fact that they are siblings. This is advocated by Hedden [8] and Aitchison et al. [2] who point out "the risk that thesaurus compilers may overload the thesaurus with valueless relationships", having a negative effect on precision. This issue can be checked by identifying concept pairs $C \times C$ that share the same broader or narrower concept while also being associatively related by the property skos:related.

Solely Transitively Related Concepts. Two concepts that are explicitly related by skos:broaderTransitive and/or skos:narrowerTransitive can be regarded a quality issue because, according to [13], these properties are "not used to make assertions". Transitive hierarchical relations in SKOS are meant to be inferred by the vocabulary consumer, which is reflected in the SKOS ontology by, for instance, skos:broader being a subproperty of skos:broaderTransitive. This issue can be detected by finding all concept pairs $C \times C$ that are directly related by skos:broaderTransitive and/or skos:narrowerTransitive properties but not by (chains of) skos:broader and skos:narrower subproperties.

Omitted Top Concepts. The SKOS model provides concept schemes, which are a facility for grouping related concepts. This helps to provide "efficient access" [11] and simplifies orientation in the vocabulary. In order to provide entry points to such a group of concepts, one or more concepts can be marked as top concepts. Omitted top concepts can be detected by iterating over all concept schemes in CS and collecting those that do not occur in relations established by the properties skos:hasTopConcept or skos:topConceptOf.

Top Concept Having Broader Concepts. Allemang et al. [3] propose to "not indicate any concepts internal to the tree as top concepts", which means that top concepts should not have broader concepts. Affected resources are found by collecting all top concepts that are related to a resource via a skos:broader statement and not via `skos:broadMatch`—mappings are not part of a vocabulary's "intrinsic" definition and a top concept in one vocabulary may perfectly have a broader concept in another vocabulary.

3.3 Linked Data Specific Issues

Missing In-Links When vocabularies are published on the Web, SKOS concepts become linkable resources. Estimating the number of in-links and identifying the concepts without any in-links, can indicate the importance of a concept. We estimate the number of in-links by iterating over all elements in AC and querying the Sindice[9] SPARQL endpoint for triples containing the concept's URI in the object part. Empty query results are indicators for missing in-links.

Missing Out-Links. SKOS concepts should also be linked with other related concepts on the Web, "enabling seamless connections between data sets"[7]. Similar to *Missing In-Links*, this issue identifies the set of all authoritative concepts that have no out-links. It can be computed by iterating over all elements in AC and returning those that are not linked with any non-authoritative resource.

Broken Links. As we discussed in detail in our earlier work [17], this issue is caused by vocabulary resources that return HTTP error responses or no response when being dereferenced. An erroneous HTTP response in that case can be defined as a response code other than 200 after possible redirections. Just as in the "document" Web, these "broken links" hinder navigability also in the Linked Data Web and and should therefore be avoided. Broken links are detected by iterating over all resources in IR, dereferencing their HTTP URIs, following possible redirects, and including unavailable resources in the result set.

Undefined SKOS Resources. The SKOS model is defined within the namespace `http://www.w3.org/2004/02/skos/core#`. However, some vocabularies use resources from within this namespace, which are unresolvable for two main reasons: vocabulary creators "invented" new terms within the SKOS namespace instead of introducing them in a separate namespace, or they use "deprecated" SKOS elements like `skos:subject`. Undefined SKOS resources can be identified by iterating over all resources in IR and returning those (i) that are contained in the list of deprecated resources[10] or (ii) are identified by a URI in the SKOS namespace but are not defined in the current version of the SKOS ontology.

4 Analysis of Existing SKOS Vocabularies

We used the qSKOS quality assessment tool to find possible quality issues in existing SKOS vocabularies. From each quality checking function we obtained detailed reports listing possibly affected resources.

---

[9] `http://sindice.com/` indexes the Web of Data, which is composed of pages with semantic markup in RDF, RDFa, Microformats, or Microdata. Currently it covers approximately 230M documents with over 11 billion triples.

[10] See `http://www.w3.org/TR/skos-reference/#namespace`

4.1   Vocabulary Data Set

Table 1 summarizes some basic statistical properties of our vocabulary selection: the number of concepts and authoritative concepts, all skos:prefLabel, skos:altLabel, and skos:hiddenLabel relations involving concepts (Concept Labels), all asserted semantic relations, as well as the number of concept schemes. From these properties we can, for instance, already see that approximately 3,000 DBpedia Categories concepts do not have labels (e.g., Category:South_Korean_social_scientists), which is an indicator for missing natural language descriptions in some Wikipedia categories.

Table 1. Analyzed SKOS vocabularies

Vocabulary	Abbreviation	Version/ Last Modified	Concepts	Authoritative Concepts	Concept Labels	Semantic Relations	Concept Schemes
United Nations Agricultural Thesaurus	AGROVOC	1.3	32,035	32,035	620,629	65,934	1
DBpedia Categories	DBpedia	3.7	743,410	743,410	740,352	1,490,316	0
The EU's Multilingual Thesaurus	Eurovoc	5.0	6,797	6,797	457,788	18,491	128
Geonames Ontology	Geonames	2.2.1	671	671	671	0	9
Gemeenschappelijke Thesaurus Audiovisuele Archieven	GTAA	2010/08/25	171,991	171,991	178,776	50,892	9
Integrated Public Sector Vocabulary	IPSV	2.00	4,732	3,080	7,945	13,843	3
Library of Congress Subject Headings	LCSH	2012/03/29	443,164	408,009	750,219	598,134	1
Austrian Armed Forces Thesaurus	LVAk	0.9	13,411	13,411	17,250	16,346	0
Middle Kingdom Tombs of Ancient Egypt Thesaurus	Meketre	2011/07/07	422	422	569	1,698	2
Medical Subject Headings	MeSH	[22]	24,626	24,626	150,617	38,858	0
North American Industry Classification System	NAICS	2012	4,175	2,213	0	8,684	1
New York Times People	NYTP	2010/06/22	4,979	4,979	4,979	0	1
University of Southampton Pressinfo	Pressinfo	2011/02/24	1,125	1,125	0	0	0
Peroxisome Knowledge Base	PXV	1.6	2,112	1,686	3,628	2,695	1
Thesaurus for Economics	STW	8.10	25,107	6,789	58,441	91,816	3

4.2   Results and Discussion

The results of this analysis are summarized in Table 2, which shows the absolute number of possibly affected resources for each quality checking function and vocabulary. Numbers marked with an asterisk (*) were obtained by extrapolating from subsets containing 5% of the respective vocabulary resources.

We found labeling and documentation issues in all vocabularies. MeSH, PXV, Pressinfo, and LVAk omit language tags with their labeling properties, LCSH

Table 2. Results of the quality checking functions

Issue	AGROVOC	DBpedia	Eurovoc	Geonames	GTAA	IPSV	LCSH	LVAk	Meketre	MeSH	NAICS	NYTP	Pressinfo	PXV	STW
Omitted or Invalid Language Tags	0	0	219	0	0	0	100,316	13,411	0	23,950	0	0	1,224	1,578	2
Incomplete Language Coverage	32,035	0	6370	0	0	0	0	0	420	0	0	0	0	0	25,050
Undocumented Concepts	32,035	743,410	5,341	0	96,850	4,551	342,848	13,411	422	1,807	3,259	4,094	1,125	1,918	23,752
Label Conflicts	2,949	0	48	18	12,404	0	10,862	13	4	0	0	0	0	7	0
Orphan Concepts	0	77,062	7	671	162,000	0	173,149	21	0	0	0	4,979	1,125	2	70
Weakly Connected Components	4	1,506	4	0	621	1	22,343	11	5	4	1	0	0	10	141
Cyclic Hierarchical Relations	0	1,132	0	0	0	0	0	5	0	4	0	0	0	0	0
Valueless Associative Relations	282	8,839	6	0	9,448	253	0	5	0	550	0	0	0	0	5,004
Solely Transitively Related Concepts	0	0	2,652	0	0	0	0	0	36	0	2,189	0	0	0	0
Omitted Top Concepts	0	0	1	9	9	0	1	0	0	0	0	1	0	0	2
Top Concept Having Broader Concepts	0	0	0	0	0	0	0	0	0	0	0	0	0	1	0
Missing In-Links	32,035	733,800	6,796	19	171,980*	3,080	408,000*	13,411	422	24,625	2,213	20	1,125	1,686	6,781
Missing Out-Links	32,035	743,410	6,797	671	171,991	0	344,054	13,411	273	24,626	1	0	1,116	1,046	0
Broken Links	238	0*	0*	0	0	1	780	0	425	1	3,169	7	11	163	575
Undefined SKOS Resources	0	0	0	0	0	1	0	0	0	1	0	0	0	0	0

with the skos:note property. STW does not use language tags with 2 instances of
skos:definition. AGROVOC covers 25 languages but no single concept is labeled
in all languages, in Meketre all concepts have English but only some of them
French labels assigned. STW, which is expressed mainly in English and German,
has many concepts with incomplete language coverage because it (i) links to non-
authoritative concepts that are only labeled in German and (ii) uses the private,
but valid language tag x-other with some of its concept labels. Geonames, which
defines a concept scheme of "feature codes", is the only vocabulary in our dataset,
which has at least one documentation property assigned to all of its concepts.
All other vocabularies have a significant number of undocumented concepts. We
also detected possible label conflicts in half of the vocabularies. PXV, for instance,
uses the string "primary peroxisomal enzyme deficiency" with two concepts in
the same concept scheme, but once with a skos:prefLabel and another time
with a skos:altLabel property. In NAICS we could not detect any labeling issues
but found that it expresses statements with skosxl:prefLabel as predicate and
plain literals as object, which contradicts the SKOS-XL[11] specification.

When analyzing the vocabularies for structural issues, we found that certain
results can be seen as indicators for the types of vocabularies. In the Pressinfo,
Geonames, and NYTP vocabulary, all concepts are orphan concepts, which means
that these vocabularies are authority files rather than thesauri or taxonomies.
This also implies that these vocabularies have no weakly connected components.
GTAA is a mixture of name authority file (approx. 162K concepts) and thesaurus
(approx. 10K concepts). The 70 orphan concepts in STW are deprecated concepts.

---

[11] The SKOS eXtension for Labels (SKOS-XL) provides additional support for identifying, describing and linking lexical entities. [13]

Three vocabularies show no weakly connected components (WCCs) because all concepts are orphan concepts and thus no relations between them are established. Two vocabularies (IPSV, NAICS) consist of only one "giant component", which is often considered the ideal vocabulary structure. STW forms one giant component (containing 24,572 concepts), but has also 140 additional WCCs, which all contain linked authoritative and non-authoritative concepts. All other vocabularies split into several clusters of semantically related concepts, each of which represents a certain subtopic. Eurovoc, for instance, has 4 WCCs, containing 4, 5, 6 and 6775 concepts. In the large WCC it uses a custom ontology to organized numerous micro-thesauri and domains and cross-connects concepts by skos:related properties. However, this is not the case for the three small WCCs, indicating a quality flaw. WCCs divide the Meketre vocabulary into different topics, e.g., museums or concepts reserved for internal use. GTAA consists of 621 highly unbalanced WCCs. One component contains 8413 subjects from a thesaurus with carefully curated semantic relations. Most of the other components contain less than 10 entities from other categories, e.g., locations, person names, and genres, for which the "traditional" information management practices involve much less explicit linking. PXV splits into 10 topic-related WCCs, such as "deficiencies", "defects" or "signals". Some of the 11 concept clusters contained in the LVAk thesaurus are obviously forgotten test data.

Hierarchical cycles are not a common issue except in the collaboratively created DBpedia vocabulary, where many concepts have reflexive skos:broader relations. The cycles in MeSH and LVAk could, in our opinion, be resolved by replacing hierarchical with associative relations or synonym definitions. Valueless associative relations occur in 8 vocabularies, with their total number being relatively low compared to the total number of all semantic relations in the respective vocabularies. Solely transitive related concepts occur in 3 vocabularies, establishing relations using properties that, according to [13], should not be asserted directly. This indicates a possible misinterpretation of the SKOS specification and could result in a loss in recall on hierarchical queries. GTAA and Geonames omit top concepts in all concept schemes they define. Eurovoc uses 128 concept schemes but has one without top concept, which simply contains all concepts defined in the vocabulary. Such an "umbrella concept scheme" without top concept is also present in LCSH and NYTP. Only the PXV vocabulary is affected by top concepts having broader concepts in its current version. In earlier versions more of them could be found which were, according to the vocabulary creator, abandoned but still available in the triple store, probably caused by some bug in the vocabulary management software.

The difference between the number of concepts and the number of authoritative concepts in Table 1 already indicates which vocabularies are linked with other SKOS vocabularies. However, except NYTP and Geonames, no vocabulary has a high number of estimated in-links from other web resources. Also the number of out-links is rather low: NYTP, IPSV, and STW are the three exceptions, which are fully linked to other Web resources. The one concept with missing out-links in NAICS is the object of a skos:broaderTransitive relation. One reason for a high number of missing out-links is that the links were not available in the main thesaurus file, which is at least the case for AGROVOC. Even though we could not determine the exact number of broken links because of the large number of links to resolve (over 400K in Eurovoc, over 500K in LCSH), we found that broken links are a common issue in most vocabularies. Undefined SKOS resources seem to be a minor issue, because we could only find two of them in all vocabularies: MeSH introduces skos:annotation and IPSV uses the deprecated skos:prefSymbol property.

5   Conclusions and Future Work

We presented possible quality issues in SKOS vocabularies and described how we implemented them as quality checking functions in our qSKOS quality assessment tool. We analyzed a representative set of existing SKOS vocabularies and found issues in all of them. Labeling and documentation issues were omnipresent and also structural issues, which require further investigation by the vocabulary maintainers, were found in most vocabularies. Although SKOS is designed for Linked Data, many existing vocabularies still resemble their closed-system origin, which results in a relatively low number of in- and out-links. Broken links are a major issue and call for synchronization mechanisms in order to maintain navigability between concepts in different vocabularies.

We are aware that these issues are purely quantitative quality indicators. To learn more about the real-world impact of our work like, e.g., the relative importance of the identified quality issues, we will conduct a qualitative follow-up study, in which we discuss these results with more taxonomists. We will also set up a Web-based SKOS quality checking service to further collect community feedback and enhance the issues list. These enhancements may encompass a finer-grained evaluation on some issues like, e.g., documentation quality indicators which could also include average number or length of documentation statements and their standard deviation across concepts.

We already reported initial results from our analysis to some of the maintainers of the vocabularies we analyzed. At the time of this writing, we know that our findings led to improvements in at least two SKOS vocabularies.

Acknowledgements. We thank Andrew Gibson and Tom Dent for providing PXV and IPSV data and Joachim Neubert for his valuable feedback. The work is supported by the FWF P21571 Meketre project and EU Marie Curie Fellowship no. 252206.

References

1. ISO 25964-1: Information and documentation – Thesauri and interoperability with other vocabularies – Part 1: Thesauri for information retrieval. Norm, International Organization for Standardization (2011)

2. Aitchison, J., Gilchrist, A., Bawden, D.: Thesaurus construction and use: a practical manual. Aslib IMI (2000)

3. Allemang, D., Hendler, J.: Semantic Web for the Working Ontologist: Effective Modeling in RDFS and OWL. Morgan Kaufmann (2011)

4. Batini, C., Cappiello, C., Francalanci, C.,, M.: Methodologies for data quality assessment and improvement. ACM Computing Surveys 41(3), 16 (2009)

5. de Coronado, S., Wright, L.W., Fragoso, G., Haber, M.W., Hahn-Dantona, E.A., Hartel, F.W., Quan, S.L., Safran, T., Thomas, N., Whiteman, L.: The NCI Thesaurus quality assurance life cycle. J. Biomed. Inform. 42(3), 530–539 (2009)

6. Harpring, P.: Introduction to Controlled Vocabularies: Terminology for Art, Architecture, and Other Cultural Works. Getty Publications, Los Angeles (2010)

7. Heath, T., Bizer, C.: Linked Data: Evolving the Web into a Global Data Space. Morgan & Claypool (2011), http://linkeddatabook.com/

8. Hedden, H.: The accidental taxonomist. Information Today (2010)

9. Hogan, A., Harth, A., Passant, A., Decker, S., Polleres, A.: Weaving the pedantic web. In: Proc. WWW 2010 Workshop on Linked Data on the Web, LDOW (2010)

10. Hopcroft, J.E., Tarjan, R.E.: Algorithm 447: efficient algorithms for graph manipulation. Commun. ACM 16(6), 372–378 (1973)

11. Isaac, A., Summers, E.: SKOS Simple Knowledge Organization System Primer. Working Group Note, W3C (2009), http://www.w3.org/TR/skos-primer/

12. Kless, D., Milton, S.: Towards quality measures for evaluating thesauri. Metadata and Semantic Research, 312–319 (2010)

13. Miles, A., Bechhofer, S.: SKOS Simple Knowledge Organization System Reference, W3C Recommendation (2009), http://www.w3.org/TR/skos-reference/

14. Nagy, H., Pellegrini, T., Mader, C.: Exploring structural differences in thesauri for SKOS-based applications. In: I-Semantics 2011, pp. 187–190. ACM (2011)

15. NISO: ANSI/NISO Z39.19 - Guidelines for the Construction, Format, and Management of Monolingual Controlled Vocabularies (2005)

16. Pipino, L., Lee, Y., Wang, R.: Data quality assessment. Commun. ACM 45(4), 211–218 (2002)

17. Popitsch, N.P., Haslhofer, B.: DSNotify: handling broken links in the web of data. In: Proc. 19th Int. Conf. World Wide Web, WWW 2010, pp. 761–770 (2010)

18. Soergel, D.: Thesauri and ontologies in digital libraries: tutorial. In: Proc. 2nd Joint Conf. on Digital libraries, JCDL (2002)

19. Spero, S.: LCSH is to Thesaurus as Doorbell is to Mammal: Visualizing Structural Problems in the Library of Congress Subject Headings. In: Proc. Int. Conf. on Dublin Core and Metadata Applications, DC (2008)

20. Svenonius, E.: Definitional approaches in the design of classification and thesauri and their implications for retrieval and for automatic classification. In: Proc. Int. Study Conference on Classification Research, pp. 12–16 (1997)

21. Svenonius, E.: Design of controlled vocabularies. Encyclopedia of Library and Information Science 45, 822–838 (2003)

22. van Assem, M., Malaisé, V., Miles, A., Schreiber, G.: A Method to Convert Thesauri to SKOS. In: Sure, Y., Domingue, J. (eds.) ESWC 2006. LNCS, vol. 4011, pp. 95–109. Springer, Heidelberg (2006)

23. Vrandecic, D.: Ontology Evaluation. Ph.D. thesis, KIT, Fakultät für Wirtschaftswissenschaften, Karlsruhe (2010)

On MultiView-Based Meta-learning for Automatic Quality Assessment of Wiki Articles

Daniel H. Dalip⋆, Marcos André Gonçalves[1], Marco Cristo[2], and Pável Calado[3]

[1] UFMG, Dept of Computer Science, Belo Horizonte/MG, Brazil
{hasan,mgoncalv}@dcc.ufmg.br
[2] UFAM, Institute of Computing, Manaus/AM, Brazil
marco.cristo@ic.ufam.edu.br
[3] Instituto Superior Técnico/INESC-ID, Porto Salvo, Portugal
pavel.calado@tagus.ist.utl.pt

Abstract. The Internet has seen a surge of new types of repositories with free access and collaborative open edition. However, this large amount of information, made available democratically and virtually without any control, raises questions about its quality. In this work, we investigate the use of meta-learning techniques to combine sets of semantically related quality indicators (aka, *views*) in order to automatically assess the quality of wiki articles. The idea is inspired on the combination of multiple (quality) experts. We perform a thorough analysis of the proposed multiview-based meta-learning approach in 3 collections. In our experiments, meta-learning was able to improve the performance of a state-of-the-art method in all tested datasets, with gains of up to 27% in quality assessment.

1 Introduction

The Web 2.0 has brought deep changes to the Internet, as users are now able not only to consume, but also to produce content. This change gave rise to new ways for creating knowledge repositories, to which anyone can freely contribute. Some examples of these repositories include blogs, forums, or collaborative digital libraries, whose collections are maintained by the Web community itself. Some of these repositories became very popular, such as *Wikipedia*[1], containing more than 3 million articles in English, and *Wikia*[2], a repository of collaborative encyclopedias having thousands of collections.

These repositories, which can be edited without control by anyone, raise concerns about the quality of their content. For instance, in order to improve the quality of Wikipedia, members of its community created guidelines on how to properly write an article, and on what can be considered a good article [1]. Furthermore, using these guidelines, Wikipedia articles can be labeled with a score regarding the perceived quality of its content (e.g., "stub" or "featured article") [2]. However, manual assessment not

---

⋆ This research is partially funded by InWeb - The Brazilian National Institute of Science and Technology for the Web (MCT/CNPq/FAPEMIG grant number 573871/2008-6), and by the authors's individual research grants from CAPES, CNPq, and FAPEMIG.
[1] http://www.wikipedia.org
[2] http://www.wikia.org

P. Zaphiris et al. (Eds.): TPDL 2012, LNCS 7489, pp. 234–246, 2012.
© Springer-Verlag Berlin Heidelberg 2012

only does not scale to the current rate of growth and change of Wikipedia, but is also subject to human bias, which can be influenced by the varying background, expertise, and even a tendency for abuse [3].

To facilitate automatic quality assessment, some approaches have been proposed [4,5,6], specially for Wikipedia. These approaches define quality features extracted from the article, and propose ways to combine them, to produce a final score capturing its global quality. An article usually has multiple sources of evidence from which to extract such features. Examples are the history of reviews, the network of links among the articles, and the textual content and its structure. Particularly, in the approach proposed in [6,7], 68 quality indicators were extracted and combined using Support Vector Regression [8]. Experimental results showed significant gains over previous state of the art.

Common to all these studies is the fact that only a single training model, based on all available sources of information, is generated to predict quality. Differently, in this paper, we propose to group the indicators used in [6,7] in semantically meaningful *views of quality*, i.e. groups of attributes, each representing a different type of evidence (textual information, link information, etc.). The quality predictions produced with such views are then combined, by means of meta-learning techniques, into one single quality value[3]. Organizing features in such a way, allows us to better exploit their different properties, thus improving the final prediction.

This idea was motivated by work such as [9], which demonstrated that the combination of views may improve the performance of machine learning methods. Since views represent different perceptions of a same concept (in our case, the relative quality of an article), the combination of models created specifically for each view may improve results in a way similar to the combination of the opinions of different experts.

In sum, the main contributions of this paper are: (1) an approach to assess the quality of collaboratively created content by organizing quality indicators into semantically related views; (2) the combination of these views by means of meta-learning; and (3) an in-depth analysis of this method and of the impact of the views in quality assessment of articles. Our experimental results show that the proposed meta-learning method is able to improve quality assessment over a state-of-the-art method in all tested collections, with gains of up to 27%.

This article is organized as follows. Section 2 covers related work. Section 3 describes in details the proposed approach. Section 4 presents and discusses our experimental evaluation. Finally, Section 5 summarizes and concludes the paper.

2   Related Work

Issues about the quality of collaboratively created content on the Web have motivated several previous studies. Since Wikipedia is the most popular of these collaborative sites, and its repository is available for download, these studies frequently use it as case study. Such work usually explores several sources of evidence to analyze its quality, such as the edit history [3,10], the network created by the links of the articles [11], and the textual content of the article [4,5,12].

---

[3] We stress that our focus is on the quality prediction task, not on the particular type of meta-learning technique employed.

Based on these earlier studies, some authors have proposed to combine different sources of evidence into a unique value to represent quality. For instance, [3] proposes to measure the quality of an edit based on the quality of its reviewers. Recursively, the quality of the reviewers is based on the quality of the articles they reviewed. In addition, in [4] several pieces of evidence (e.g., from revision history, textual content, and hyperlink structure) are combined to build an article ranking that tries to capture certain aspects of quality, such as stability, editing quality, and importance. Differently from [4], which used simple linear combination methods, a few other efforts were proposed to combine the available evidence using machine learning. One example is [5], which proposed the use of natural language features such as the number of phrases, auxiliary verbs, and other readability features [13] along with a Maximum Entropy method in order to estimate the quality of the articles. Other work that uses machine learning to predict article quality is [12], in which lexical clue words and a decision tree are used.

Since a score can be viewed more naturally as a continuous scale, the work described in [6,7] proposed to treat quality estimation as a regression problem. To accomplish this, the authors used a Support Vector Regression method [8]. Their main contribution was a detailed study of the various sources of evidence and their impact on the prediction of the quality of a Wikipedia article. In [6], the problem was revisited and the same approach was applied to two other collaborative DLs: Star-Wars and Muppets[4]. Furthermore, the method proposed in [6] was shown to achieve overall better results than the best approaches previously proposed in literature. For that reason, we will use the method proposed in [6] as our baseline.

Previous work applies a single learned model exploiting all available features in isolation for quality prediction. In this work we propose to organize sets of semantically-related features into coherent logical *views*, learn a model for each view, and combine them by means of a meta-learning technique inspired on stacking [14]. Traditionally, stacking learns the relation between the output of distinct learning algorithms and the target class. In our case, instead of using models generated by distinct algorithms, we will use models generated from the distinct views. As we shall see, our approach obtains significant gains over strong baselines in all tested scenarios.

3   Assessing Article Quality

Suppose six experts assess the quality of a wiki article, each according to a different perspective (or *view*) of quality. In our case, each expert views the article according to one of six perspectives (1) article length, (2) writing style, (3) structure, (4) text readability, (5) review history, and (6) the citation graph. The final quality assessment will be a combination of these multiple opinions. Particularly, if each opinion is given as a degree of certainty, it is possible to learn the global quality, taking this certainty into account.

Thus, the problem of estimating quality can occur in two learning phases. In the first phase (*learning level 0*), each article is represented by six sets of features, one for each view. Using these features, a regression model is learned for each view, thus obtaining an assessment of quality for each aspect of the article. In the second phase

---

[4] http://starwars.wikia.com;http://muppet.wikia.com

(*learning level 1*), each article is represented by the quality prediction obtained from each regression model (i.e., for each view). A global model of quality is then learned and, as a result, we can return a final (combined) assessment of quality. This process is depicted in Figure 1. The next sections detail each of those representations.

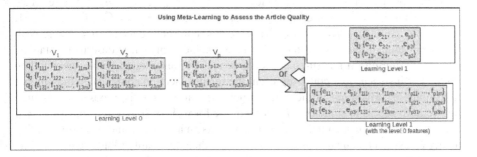

Fig. 1. The quality assessment process. V_i is view i; q_i is the target quality value, representing the score given by the user to each article; f_{vij} is feature j for article i in view v; and e_{vi} is the estimated quality in each view v, for article i, at learning level 0. This is then used as a feature for learning level 1. Alternatively, we can combine all level 0 features with those from level 1 (i.e., the view predictions) for learning the final decision.

3.1   Learning Level 0

In Wikipedia, the quality of an article is defined as a value on a discrete scale. Articles are classified, by the users, as [2]: (1) Featured Article (*FA*),the best Wikipedia articles according to the evaluators; (2) A-Class (*AC*), articles considered complete, but with a few pending issues that need to be solved in order to be promoted to FA; (3) Good Article, (*GA*), articles without problems of gaps or excessive content – these are good sources of information, although other encyclopedias could provide better content; (4) B-Class (*BC*), articles useful for most users, but researchers may have difficulties in obtaining more precise information; (5) Start-Class (*ST*), articles still incomplete, although containing references and pointers for more complete information; and (6) Stub-Class (*SB*), draft articles, with very few paragraphs and few or no citations[5].

We note, however, that in general quality can be seen as a value on a continuous scale. In fact, this is the most natural interpretation for the problem, if we consider that there are better or worst articles, even inside the same discrete category. For instance, in Wikipedia, class AC articles are defined as those that: (a) have recently been promoted and await expert evaluation; (b) have been evaluated by experts and await corrections; or (c) have been corrected and await promotion to featured article, which are clearly different quality levels. Also, in the case of other Wikis, a continuous scale is commonly used, where users score each article with a value from 1 through 5 and the final quality value is the average of all scores. For these reasons, in this work, we consider quality

---

[5] There is currently an intermediate class between ST and BC, the *C-Class*. We do not use this class because it did not exist at the time we performed our crawling.

as a continuous value. Consequently, the problem of learning to evaluate quality will be modeled as a numerical regression task. More specifically, we will apply a state-of-the-art method for regression learning—Support Vector Regression (SVR) [15].

Quality Assessment with SVR. To apply SVR to the quality estimation task, we represent the articles as follows. Given a view v, let $A_v = \{a_{v1}, a_{v2}, ..., a_{vn}\}$ be a set of articles. Each article a_{vi} is represented by a set of m features $F_v = \{F_{v1}, F_{v2}, ..., F_{vm}\}$, such that $a_{vi} = (f_{vi1}, f_{vi2}, ..., f_{vim})$ is a vector representing article a_{vi}, where each f_{vij} is the value of feature F_{vj} in a_{vi}. In this work, the term *feature* describes a statistic value that represents a measurement of some indicator associated with a view and an article. For instance, f_{vij} could represent the feature "section count" (i.e., the number of sections) of article a_{vi} in the view "article structure". In our proposal, we assume that we have access to some *training data* of the form $A_v \times \mathbb{R} = \{(a_{v1}, q_1), (a_{v2}, q_2), ..., (a_{vn}, q_n)\}$, where each pair (a_{vi}, q_i) represents an article a_{vi} and its corresponding quality assessment value q_i, such that if $q_1 > q_2$, then the quality of article a_{v1}, as perceived by the user, is higher than the quality of article a_{v2}. The solution we propose to this problem consists in: (1) determining the set of views v; (2) determining the set of features $\{F_{v1}, F_{v2}, ..., F_{vm}\}$ used to represent the articles in A_v; and (3) applying a regression method to find the best combination of the features, for each view v, to predict the quality value q_i for any given article a_{vi}.

The problem of regression is to find a function $\gamma : A \rightarrow \mathbb{R}$ that has at most ϵ deviation from the actually obtained targets q_i, for all the training data. Let our target function $\gamma(a) = \kappa(w, a) + b$, where κ represents an inner product function (or *kernel function*) in a given vector space, w represents a vector of m feature values, $b \in \mathbb{R}$ is a constant, and a represents an article whose quality will be estimated. The regression process consists in learning w and b from the training data. To reduce the global error, we minimize the norm of w ($||w||$). Furthermore, if we want to allow some deviation from a perfect fit (that is, we accept that some articles behave as outliers), we can represent such deviation by means of slack variables ξ_i and ξ_i^* (that is, outliers below or above the separating surface we are looking for). More formally, we can formulate this as the convex optimization problem of minimizing:

$$\frac{1}{2}\|\vec{w}\|^2 + C\sum_{i=1}^{\ell}(\xi_i + \xi_i^*) \tag{1}$$

subject to:

$$q_i - \kappa(\vec{w}, \vec{d}_i) - b \leq \epsilon + \xi_i$$
$$\kappa(\vec{w}, \vec{d}_i) + b - q_i \leq \epsilon + \xi_i^*, (\xi_i, \xi_i^* \geq 0)$$

where constant $C > 0$ determines the trade-off between the prediction error of γ and the amount up to which deviations larger than ϵ (outliers) are tolerated.

Here, we solve the quadratic optimization problem given by Eq. 1 using the LIBSVM package. In our experiments we have used a *radial basis function* (RBF) as κ. Other parameters were chosen using cross-validation within the training set, with the data scaling and parameter selection tool provided by LIBSVM.

Article Representation. Determining which features should be used to represent an article is a key decision in a regression-based quality assessment. Such features were chosen based on the criteria used by Wikipedia and Star-Wars guidelines to manually assess the quality of an article.

Length features are indicators of article size. The general intuition behind them is that a mature and good quality text is probably neither too short, indicating an incomplete topic coverage, nor excessively long, indicating verbose content. We have three length features: word, sentence and character count. *Style features* are intended to capture the way the authors write the articles through their word usage. The intuition is that good articles should present some distinguishable characteristics related to word usage, such as short sentences. To compute them (and also the Readability features), we use the *Style and Diction* software[6]. Some examples of style features is the ratio between short and large paragraphs, and the largest phrase size. *Structure features* are indicators of how well the article is organized. According to Wiki quality standards a good article must be organized such that it is clear, visually adequate, and provides the necessary references and pointers to additional material. For this, we have features such as the citation, image and section count, average number of citations per section, and so on. *Readability features*, first used in [5], are intended to estimate the age or US grade level necessary to comprehend a text. The intuition behind these features is that good articles should be well written, understandable, and free of unnecessary complexity. Some examples include the Automated Readability Index [16], and Coleman-Liau [13]. *Review features* are those extracted from the review history of each article. Some examples are the age, reviews per day, and ProbReview [3], which estimates the quality of an article based on the quality of its authors. Finally, *network features* are those extracted from the interconnections between articles. Examples of such features are the in and out–links, and PageRank [17]. In total, this yields 68 features, as used in [6,7], where a more detailed description of each individual feature can be found.

3.2   Learning Level 1

Once the quality of an article has been predicted for each view, we can describe it using a different representation. Thus, each article a_i is represented by a set of six level 1 features $\{e_{i1} \dots e_{i6}\}$. Each feature represents the article quality estimate given for each view. In other words, $e_{i1}, \dots e_{i6}$ represent the predictions according to the structure, length, style, readability, history of reviews, and network views, respectively. Given a training set $\{(a_1, q_1), \dots, (a_n, q_n)\}$, where (a_i, q_i) represents the article a_i and its quality q_i, the quality of the article can be learned by again applying SVR using the same kernel function and the previously described approach to compute the additional parameters, but using the level 1 representation. Note that, by doing so, we are in fact learning to combine the estimates obtained from each different view of quality. Alternatively, we can combine the level 0 features with those from level 1, as shown in Figure 1, in order to learn a model that exploits both sets of features. The idea is that it would be possible to learn interesting patterns or relationships between the view predictions and some of the level 0 features, potentially incurring in gains in quality prediction.

---

[6] http://www.gnu.org/software/diction/

4   Experimental Evaluation

4.1   Datasets

In our experiments, we used a sample of Wikipedia in English and samples of two other collaborative encyclopedias provided by the Wikia service. We chose the English Wikipedia because it is a large collaborative encyclopedia, with more than three million articles, where more than half have their quality evaluated by users[7]. From now on, we refer to this encyclopedia as WPEDIA. From the Wikia service, we selected the encyclopedias *Wookieepedia*, about the Star Wars universe, and *Muppet*, about the TV series "The Muppet Show". These are the two Wikia encyclopedias with the largest number of articles evaluated by users regarding their quality.

The Wookieepedia collection provides two distinct quality taxonomies. The first is a subset of the Wikipedia quality taxonomy. It comprises the classes *FA*, *GA*, and *SB*. The second is a star-based taxonomy where the worst articles receive one star and the best articles receive five stars. Unlike Wikipedia, the final rating of a Wookieepedia article is obtained as the average of the ratings provided by all the users that evaluated it. Since these taxonomies are not compatible with each other, we extracted two different samples of Wookieepedia. The first sample was built according to the Wikipedia-based taxonomy and, from now on, we refer to it as SW3. The second sample was derived according to the star-based taxonomy and we refer to it as SW5. Finally, the Muppet collection, which we refer to as MUPPET, provides only a star-based taxonomy. Note that all the feature sets used in this work are available for download[8].

Table 1. Sample size, for each dataset used in our experiments

Dataset	# Articles	# Reviews	# Edges	# Nodes	Version date
WPEDIA	3.294	1.992.463	86.077.675	3.185.457	Jan/2008
MUPPET	1.550	38.291	282.568	29.868	Sep/2009
SW3	1.446	127.551	1.017.241	106.434	Oct/2009
SW5	9.180	369.785	1.017.241	106.434	Oct/2009

Table 1 shows the size of each sample. To create our sample for each Wiki collection, we first extracted all articles from the smallest class and then randomly drew the same number of articles from the remaining ones. Note that, to accomplish this for the star-based taxonomies, we rounded the ratings[9]. We chose to use balanced samples since SVR can be biased towards the larger class, which could harm our analysis regarding the relative difficulty of classification in each class. Moreover, recent research has shown that training samples do not need to follow the original class distribution in order to obtain good results, and that a balanced distribution is a good alternative [18].

For all datasets, we also collected the links between the articles, in order to extract network attributes. Table 1 shows the total number of articles and revisions of each sample and information about the network graphs derived from the datasets.

---

[7] According to the quality taxonomy detailed in the Section 3.1.

[8] http://www.lbd.dcc.ufmg.br/lbd/collections/wiki-quality

[9] We only rounded the values to create a balanced sample. In the experiments, we used the original quality ratings.

4.2 Evaluation Methodology

Since we are dealing with regression methods, we evaluate their effectiveness by using the *mean squared error* measure (MSE) for all articles. We compute error e as the absolute difference between the quality value predicted and the true quality value, extracted from the database. In our experiments, we used quality values from 0 (Stub article) through 5 (Featured Article) for WPEDIA, 0 (Stub article) through 2 (Featured Article) for SW3, and 1 (one star) through 5 (five stars) for SW5 and MUPPET.

In order to evaluate the concordance among views, we adopt the percentage of agreement (P_{agree}) used to calculate Fleiss' kappa [19]. We chose this metric because it computes the agreement among multiple raters, in our case, the views. First, we compute the percentage of agreement P_i for each instance i, $P_i = \frac{1}{n(n-1)} \sum_{j=1}^{k} n_{ij}(n_{ij} - 1)$, where n is the number of quality classes and n_{ij} is the number of views that were assigned the same class j for instance i. Note that, to obtain the integer value j, we discretized the quality score predicted for each view, by rounding it. Then, the final value is the average of P_i, as given by $P_{agree} = \frac{1}{N} \sum_{i=1}^{N} P_i$, where N is the number of instances.

To perform our experiments, we used 10-fold cross validation, i.e., each dataset was randomly split into ten parts, such that, in each run, a different part was used as test set while the remaining ones were used for training. Note that, in order to make sure that the same training and test instances are used by the methods in each cross validation turn and that test information of level 0 is not used as training information of level 1, we carry out the experiments according to Algorithm 1. For the baseline, which uses just the level 0, we computed the results using PTs_{vp}, defined in Algorithm 1[10], considering that there is just one view V, with all the features, and for the meta-learning method we computed the results using P_p. For all comparisons reported in this work, we used the signed-rank test of Wilcoxon to determine if the differences in effectiveness were statistically significant. The symbol "*" is used to indicate comparisons with p-value less than 0.05 whereas "+" indicates p-value less than 0.1.

Algorithm 1. Training and testing procedures

Require: V is a set of views
Require: For each view $v \in V$, X_v is a dataset divided into 10 partitions, $X_v = \{x_{v0}...x_{v9}\}$
  for $p = 1$ to 9 **do**
    for all $v \in V$ **do**
      $train \leftarrow X_v - x_{vp}$
      $test \leftarrow x_{vp}$
      Do a cross validation in $train$ to obtain the predictions Ptr_{vp} and the training models $\{M_{vp0}...M_{vp8}\}$
      Apply the model M_{vp0} in $test$ to obtain the predictions PTs_{vp}
  for $p = 0$ to 9 **do**
    Create the training set T_p using the predictions Ptr_{vp} for each $v \in V$
    Create the test set S_p using the perditions Pts_{vp} for each $v \in V$
    Do a cross validation in T_p to obtain the training models $\{M_{p0}...M_{p8}\}$
    Apply the model M_{p0} in the test to obtain the predictions P_p

---

[10] Note that, to simplify our explanation, in Section 3, we represented the predictions in level 0 using the variable e_i, but here, to make it clearer, we used Pts_{vp} for the predictions in the test set and Ptr_{vp} for the predictions in training set.

Table 2. Mean Squared Error in each sample, for each method

Sample	Method	MSE	% Improvement
WPEDIA	SVR	0.837	-
	6_VIEWS	0.824	1.6%
MUPPET	SVR	1.691	-
	6_VIEWS	1.645	2.7%
SW5	SVR	1.692	-
	6_VIEWS	1.661*	1.8%
SW3	SVR	0.074	-
	6_VIEWS	0.054*	27.0%

4.3   Results

General Results.   We start by describing the impact of using meta-learning. Table 2 presents the results of the experiments for each collection. From now on we refer to our *baseline* method as SVR. We call 6_VIEWS the meta-learning method described in the Section 3 in which only the view predictions were used in level 1. From Table 2, we see that WPEDIA is the hardest collection to predict, whereas SW3 is the easiest. As we can observe, 6_VIEWS was able to significantly improve the results in both STARWARS collections, with large gains in SW3. However, no significant gains were observed in WPEDIA and MUPPET. To better understand these results, we analyze the degree of agreement among the views in Table 3 and the error per class in Table 7.

To analyze the agreement between views, we divided the samples into two sub-samples. The first consists of articles whose views were considered in agreement, while the second to those considered in disagreement, according to the P_{agree} metric (see Section 4.2). As mentioned, to calculate P_{agree} we first rounded the real rating predicted by each view for an article a to get an integer value that could be used as a vote. Then, we consider that the views are in agreement about a if $P_{agree} >= 0.46$. This means that one quality class has been voted by, at least, two more views than any other class.

Table 3. MSE obtained for the agreement and disagreement sub-samples. Column *size* indicates the proportion of articles in each sub-sample.

Sample	Method	View Agreement		View Disagreement	
		Result	Size(%)	Result	Size(%)
	SVR	0.776	42.56%	0.887	57.44%
WPEDIA	6_VIEWS	0.780		0.862	
	SVR	1.725	78.84%	1.605	21.16%
MUPPET	6_VIEWS	1.698		1.471*	
	SVR	1.686	79.38%	1.724	20.62%
SW5	6_VIEWS	1.658*		1.680*	
	SVR	0.066	92.39%	0.176	7.61%
SW3	6_VIEWS	0.045*		0.166	

From Table 3, we note that the view agreement was higher for collections with fewer classes. Also, meta-learning is progressively better as the number of instances in agreement is larger, but was able to improve the results in the disagreement subsets of collections MUPPET and SW5. This may be explained by the fact that meta-learning can take advantage of independent views to improve performance and that conflicting votes are indication of independent opinions. We further discuss these issues in the following, by performing a per class and per view analysis of the method.

Per View Analysis. We present here a detailed analysis of the impact of each view for the final performance of meta-learning. To accomplish this, we have experimented with all possible combinations of the six views, totalizing 63 different configurations. Due to space constraints, we highlight only the main results. In Table 4, we analyze the impact of each view by removing it from the dataset. Thus, the greater is the increase in MSE, the greater is the importance of the removed view.

As we can see, Structure and History views played the most important roles in most of the collections. The exception was SW3, where views Readability and Length had more impact. In particular, the good performance of Readability in SW3 was somewhat surprising, since in [6], Dalip et al. have found that the features used to learn the Readability view were not very useful for assessing quality. Note that, Readability is the only textual view that does not contain length related features. This makes it likely to be the most independent from the others, allowing the meta-learning model to get more useful information from it. Thus, Readability, in some cases, such as SW3, can be a good source of information about quality, when combined with the remaining available evidence. The Style view presented a mixed performance, slightly increasing MSE in WPEDIA and decreasing it in SW5. This difference may be due to the different quality criteria used in the evaluation of both collections.

Table 4. MSE (and % error increasing over baseline) when removing a particular view from the dataset. We use all the views (All) as our baseline.

| View | Sample | | | |
	WPEDIA	MUPPET	SW5	SW3
All	0.824 (–)	1.645 (–)	1.661 (–)	0.054 (–)
Structure	0.883* (7.1%)	1.675* (1.8%)	1.683* (1.3%)	0.058 (6.3%)
History	0.854* (3.7%)	1.719+ (4.4%)	1.687* (1.5%)	0.055 (1.5%)
Readability	0.829 (0.6%)	1.646 (0.0%)	1.654+ (-0.5%)	0.058* (7.0%)
Style	0.828+ (0.5%)	1.639 (-0.4%)	1.650* (-0.7%)	0.055 (1.1%)
Network	0.828 (0.5%)	1.648 (0.1%)	1.665 (0.2%)	0.054 (0.2%)
Length	0.826 (0.2%)	1.645 (-0.1%)	1.665 (0.2%)	0.057* (5.1%)

By analyzing all the possible combinations of views, we observe that in the samples SW5 and MUPPET, several subsets of views have performed better than using all the views. We show in Table 5 the results obtained with the best subsets of views. In case of the SVR baseline, results correspond to using all the features, since this was the best configuration for this method in all collections.

Table 5. MSE by method for each sample, using the best set of views. SLHN stands for structure, length, history, and network.

Sample	Method	MSE	% Gain
WPEDIA	SVR	0.837	-
	6_VIEWS (all views)	0.824	1.6%
MUPPET	SVR	1.691	-
	6_VIEWS (SLHN)	1.635*	3.3%
SW5	SVR	1.692	-
	6_VIEWS (SLHN)	1.651*	2.4%
SW3	SVR	0.074	-
	6_VIEWS (all views)	0.054*	27.0%

Improving Results with an Extended set of Features in Level 1. We now analyze the results using the extended set of features described in Section 3.2, where we combine the view predictions with the level 0 features. Table 6 shows the results (line $6\_VIEWS + F0$). We see that, with this configuration, we obtain significantly better results in all collections. We can also see that, with the exception of SW3, in which there was a slight decrease in effectiveness, there were always improvements over the original level 1 representation. This confirms our hypothesis that useful patterns can be learned about the relationships between the view predictions and some of the original features. We want to investigate these patterns deeper in the near future.

Per Class Analysis. Finally, we conducted a detailed per class analysis of our experimental results, in order to better understand them. To this effect, (i) in case of SVR, we use all available level 0 features; (2) for the original 6_VIEWS configuration, we use the best combination of views shown in Table 5, which we call $6\_VIEWS(Best)$; and (3) for the extended level 1 configuration, which we call $6\_VIEWS + F0$, we use the results reported in Table 6[11]. Table 7 presents the MSE per class, obtained by SVR, $6\_VIEWS(Best)$ and $6\_VIEWS + F0$.

As we can see, in WPEDIA both meta-learning strategies were able to improve the extreme classes SB, ST and FA, but performed worse in the intermediate classes. Due to the Wikipedia quality policies, FA articles are evaluated more rigorously, with the adherence to quality guidelines being enforced. Further, they normally have many citations, a large review history, and are more likely to have been evaluated by experts. All these facts contribute to reliable and independent views and, by extent, a better performance of meta-learning. As for the SB class, its articles are small and have very few history information, a pattern which meta-learning also seems to learn well. Notice also that $6\_VIEWS + F0$ was able to significantly reduce some losses in the intermediate classes when compared to $6\_VIEWS(Best)$, which explains its better performance.

Table 6. MSE in each sample, for each method with the extended set of features in Level 1

Sample	Method	MSE	% Gain
WPEDIA	SVR	0.837	-
	6_VIEWS	0.824	1.6%
	6_VIEWS+F0	0.794*	5.2%
MUPPET	SVR	1.691	-
	6_VIEWS	1.645	2.7%
	6_VIEWS+F0	1.632*	3.5%
SW5	SVR	1.692	-
	6_VIEWS	1.661*	1.8%
	6_VIEWS+F0	1.657*	2.1%
SW3	SVR	0.074	-
	6_VIEWS	0.054*	27.0%
	6_VIEWS+F0	0.058*	21.5%

We also note in Table 7 that the proposed methods produced better results for the extreme classes in collections which use a star taxonomy (SW5 and mainly MUPPET).

---

[11] We should note that we tried an additional combination of $6\_VIEWS(Best)$ and $6\_VIEWS + F0$ for those collections in which a subset of all views performed better, but we obtained mixed results which are not reported here for space reasons. We intend to analyze these results further in the near future.

These classes are hard to predict (see the larger MSE) due to the very subjective nature of these collections where most of the articles are about fictional characters. In these collections, votes to classes 1 and 5 are biased towards character popularity and personal preferences of the users regarding these characters. Finally, in SW3, the meta-learning methods were not able to significantly improve results only in the SB class, which already presented an extremely low error rate.

Table 7. MSE per class, for each sample using the best representation of each method

Sample	Class	SVR	6_VIEWS (Best)	% Gain	6_VIEWS+F0	% Gain
	SB	0.338	0.267*	20.81%	0.284*	15.8%
	ST	0.606	0.567	6.37%	0.526*	13.2%
WPEDIA	BC	0.854	0.999*	-17.01%	0.887*	-3.9%
	GA	0.716	0.773	-7.96%	0.729	-1.9%
	AC	1.121	1.130	-0.77%	1.105	1.4%
	FA	1.388	1.207*	13.04%	1.232*	11.3%
	Average	0.837	0.824*	1.57%	0.794*	5.1%
	1	3.079	2.974	3.41%	2.839*	7.8%
	2	0.719	0.757	-5.24%	0.692	3.8%
MUPPET	3	0.347	0.371	-7.00%	0.401*	-15.4%
	4	0.866	0.793	8.47%	0.883	-2.0%
	5	3.443	3.278*	4.52%	3.344*	2.9%
	Average	1.689	1.634*	3.22%	1.632*	3.3%
	1	2.709	2.596*	4.17%	2.635*	2.7%
	2	0.848	0.801*	5.56%	0.772*	8.9%
SW5	3	0.342	0.333	2.69%	0.302*	11.6%
	4	0.835	0.812	2.73%	0.833	0.2%
	5	3.728	3.715	0.35%	3.745	-0.5%
	Average	1.692	1.651*	2.42%	1.657*	2.4%
	SB	0.025	0.021	13.26%	0.020	18.6%
SW3	GA	0.112	0.096*	14.46%	0.095*	15.3%
	FA	0.084	0.045*	46.37%	0.059*	30.6%
	Average	0.074	0.054*	26.46%	0.058*	27.0%

5   Conclusions

In this work, we carried out a thorough analysis of the application of meta-learning over multiple views of quality derived from a large set of features, to automatically assess quality of Wiki articles. We found that a meta-learning strategy that combines the view predictions with the original quality features was able to improve the performance of a state-of-the-art method in all four test datasets. The highest gains (up to 27%) over the state-of-the-art method were obtained using only the combination of the views in a star-based Wiki collection. We further noted that meta-learning is better when dealing with instances for which different views reached higher agreement. According to our study, the best views are, in general, text structure and revision history. For two datasets, Muppets and Star-Wars, we found that it is better to use a subset of the views.

As future work, we intend to exploit instance selection methods with active learning. We also want to study the impact of quality in other information services, such as searching and recommendation, besides trying to learn how reliable is the quality labeling provided by human reviewers.

References

1. Wikipedia: Version 1.0 editorial team/release version criteria (2012),
   http://en.wikipedia.org/wiki/Wikipedia:1.0/Criteria
2. Wikipedia: Version 1.0 editorial team/assessment (2012),
   http://en.wikipedia.org/wiki/
   Wikipedia:Version_1.0_Editorial_Team/Assessment
3. Hu, M., Lim, E.P., Sun, A., Lauw, H.W., Vuong, B.Q.: Measuring article quality in wikipedia: models and evaluation. In: CIKM 2007, pp. 243–252 (2007)
4. Dondio, P., Barrett, S., Weber, S., Seigneur, J.-M.: Extracting Trust from Domain Analysis: A Case Study on the Wikipedia Project. In: Yang, L.T., Jin, H., Ma, J., Ungerer, T. (eds.) ATC 2006. LNCS, vol. 4158, pp. 362–373. Springer, Heidelberg (2006)
5. Rassbach, L., Pincock, T., Mingus, B.: Exploring the feasibility of automatically rating online article quality (2007),
   http://upload.wikimedia.org/wikipedia/wikimania2007/
   d/d3/RassbachPincockMingus07.pdf
6. Dalip, D.H., Gonçalves, M.A., Cristo, M., Calado, P.: Automatic assessment of document quality in web collaborative digital libraries. ACM JDIQ 2(13) (2011)
7. Dalip, D.H., Gonçalves, M.A., Cristo, M., Calado, P.: Automatic quality assessment of content created collaboratively by web communities: a case study of Wikipedia. In: JCDL 2009, pp. 295–304 (2009)
8. Vapnik, V.N.: The nature of statistical learning theory. Springer (1995)
9. Kakade, S.M., Foster, D.P.: Multi-view Regression Via Canonical Correlation Analysis. In: Bshouty, N.H., Gentile, C. (eds.) COLT 2007. LNCS (LNAI), vol. 4539, pp. 82–96. Springer, Heidelberg (2007)
10. Han, J., Wang, C., Jiang, D.: Probabilistic Quality Assessment Based on Article's Revision History. In: Hameurlain, A., Liddle, S.W., Schewe, K.-D., Zhou, X. (eds.) DEXA 2011, Part II. LNCS, vol. 6861, pp. 574–588. Springer, Heidelberg (2011)
11. Korfiatis, N., Poulos, M., Bokos, G.: Evaluating authoritative sources using social networks: An insight from wikipedia. Online Information Review 30(3), 252–262 (2006)
12. Xu, Y., Luo, T.: Measuring article quality in wikipedia: Lexical clue model. In: 2011 3rd Symposium on Web Society (SWS), pp. 141–146 (October 2011)
13. Coleman, M., Liau, T.L.: A computer readability formula designed for machine scoring. Journal of Applied Psychology 60(2), 283–284 (1975)
14. Wolpert, D.H.: Stacked generalization. Neural Networks 5, 241–259 (1992)
15. Drucker, H., Burges, C.J.C., Kaufman, L., Smola, A.J., Vapnik, V.: Support vector regression machines. In: Mozer, M., Jordan, M.I., Petsche, T. (eds.) NIPS, pp. 155–161. MIT Press (1996)
16. Smith, E.A., Senter, R.J.: Automated readability index. Aerospace Medical Division (1967)
17. Brin, S., Page, L.: The anatomy of a large-scale hypertextual web search engine. Computer Networks and ISDN Systems 30(1-7), 107–117 (1998)
18. Weiss, G.M., Provost, F.: Learning when training data are costly: The effect of class distribution on tree induction. Journal of Artificial Intelligence Research 19, 315–354 (2003)
19. Fleiss, J.L., Cohen, J.: The equivalence of weighted kappa and the intraclass correlation coefficient as measures of reliability. Educational and Psychological Measurement 33(3), 613–619 (1973)

A Methodology for Folksonomy Evaluation

Spyros Daglas[1], Constantia Kakali[2,3], Dionysis Kakavoulis[1],
Marina Koumaki[1], and Christos Papatheodorou[2]

[1] Department of Informatics, Athens University of Economics and Business, Athens, Greece
[2] Department of Archives and Library Science, Ionian University, Corfu, Greece
[3] Library & Information Service, Panteion University, Athens, Greece
{daglass,kakavoulisd,koumakhm}@aueb.gr, nkakal@panteion.gr,
papatheodor@ionio.gr

Abstract. In recent years, the folksonomies were created and maintained in libraries and other information organizations alongside the traditional subject indexing systems. Folksonomies, consisting of tags, often express the "wisdom of the crowd". Despite their weaknesses and unstructured form, they reveal the language of users or even the terminology of the experts. This knowledge could be exploited in order to update and enrich the indexing vocabularies, thus improving information services. This paper deals with the design of an evaluation model for social tags. It introduces four quality indicators for an information service which offers social tagging functionalities, a weighted metric for the tag assessment process and a set of evaluation criteria that support information professionals to select meaningful tags as new descriptors to a set of bibliographic records.

Keywords: social tagging, folksonomy, subject indexing, evaluation.

1 Introduction

Information organizations, in line with the dominant paradigm of web 2.0, have developed innovative services and they currently apply a set of features, which encourage user's interaction and participation. Among them, the social tagging systems stand out as important tools for metadata production by the users. Social tagging permits users to generate a new vocabulary as a supplement of the library's subject index, presented in a tag cloud, which is well known as folksonomy.

An OCLC research exploring users' expectations from libraries' catalogues revealed the need of "more subject information" [1]. Recently another OCLC survey referred that: "the use of tags is supported by 39% of the sites reviewed—but by 60% of the LAM (Libraries, Archives & Museums) sites" [2]. At the same time, next generation library catalogues have been empowered by the development and use of user-contributed content (tags, reviews, annotations) [3] and the recent literature refers to the widespread use of folksonomies.

However some scientists faced critically the use of folksonomies in information organizations; by comparing it with traditional indexing systems they explored their

P. Zaphiris et al. (Eds.): TPDL 2012, LNCS 7489, pp. 247–259, 2012.
© Springer-Verlag Berlin Heidelberg 2012

strengths and weaknesses [4, 5]. Some others analyzed and categorized taggers' behavior, and the impact of folksonomies on information retrieval [6, 7]. Moreover researchers are interested in studying the relationship between folksonomies and traditional vocabularies of subject indexes and the overlap between them. Though the significant research efforts, there has not been any proposal for a methodology of evaluating and exploitating a folksonomy by an information organization and hence folksonomies operate as independent systems to the existing information services.

The objective of this paper is to investigate whether the dynamically evolving language of the users expressed by the social tags of a folksonomy could be integrated to the indexing language of information organizations. The paper examines the hypothesis whether the social tags of a folksonomy could enrich the thematic description of the records of an On-Line Public Access Catalogue (OPAC). For this purpose a methodology is introduced to assess the impact of social tags to the thematic description of a document. The assessment process takes weights two facets of a tag in a particular document: (i) its popularity, as recorded in social bookmarking systems, such as LibraryThing[1] (LT) and (ii) its frequency, if it exists as term in the document; the document is retrieved in digital collections, such as Google Books[2] (GB), which offer, among others, automated indexing services. The high-valued tags could be added either as enhanced descriptors in a bibliographic record, or as new records in an authority file. In the next section of the paper its motivation is documented by surveying the current state of the art, while section 3 presents the stages of the proposed methodology. In section 4 the implementation of the methodology is presented and in section 5 the derived results are discussed, revealing several alternatives and issues for further research. Finally section 6 concludes the results of this work.

2   Research Background

This paper is motivated by the research trend of incorporating social tags into the cataloguing and subject indexing process. In general several researchers have tried to correlate the social tags with the subject description. For instance Smith [8] and Bartley [9] showed that the majority of tags are identical terms from the fields of MARC: 245 (Title) and 600 (Subject fields). Rolla [10] compared the subjects of 45 bibliographic records with the tags assigned by the LT users to the same documents and found that the tags reflect mainly topical information, though a large proportion of them are personal without value to information retrieval.

Moreover the scientists compare and correlate the tags with knowledge organization systems and especially the Library of Congress Subject Headings (LCSH) [11, 12, 13]. Thomas, Caudle and Schmitz [14] presented a comparison of social tags with thematic headings (LCSH), based on a sample of 10 books from different libraries and thematic areas with a high degree of subjectivity in their use, e.g. literature, religion. Lu, Park and Hu [15] compared a social tags subset from LT with subject terms

---

[1] LibraryThing is an electronic service that users create their personal electronic catalogues of books and other material (http://www.librarything.com/)
[2] http://books.google.com/

from LCSH assigned by experts and explored the differences between the two vocabularies. The results revealed that the social tags improve the accessibility of a library collection, but the existence of non-subject-related tags can also create serious obstacles. In the same vein Yi and Chan [16] reviewed more than 4,500 websites tagged by the Delicious bookmarking system. The comparison showed that the 61% of the tags with usage frequency of more than two matched the subject headings. The authors represented the LCSH along with their synonyms, narrower and broader terms as trees and investigated the similarity between the subject headings and tags. In a later study [17] Yi compared the LCSH with LT tags based on five similarity measures aiming to predict subject headings from the tags.

Given these results a question that arises is the selection of the tags that enhance the thematic description of a document. Dealing with this issue Peters and Stock proposed a method for cutting off the "long tail" of the tag distribution for a document and keeping the "power tags" that describe the document precisely [18]. Another approach [7] computes the similarity between the queries and the social tags attached to the documents that formulate the answer to the queries. Finally, a tag analysis methodology [19, 20], based on a small number of social tags and following manual processes, provides a set of stages for the development of a policy for the convergence of user-based and expert-based subject indexing.

This paper is based on the principle that the users' collaboration and their vocabulary provide useful feedback for the enhancement of the thematic description of the documents of a collection. Therefore it introduces and validates a methodology that supports decision making concerning the incorporation of a subset of social tags in the thematic description of a collection.

3   The Proposed Methodology

Within this context, the proposed methodology examines the added value of tags within the bibliographic records they are related to. The primary goal is to achieve the "golden ratio" and select tags that reveal the users' vocabulary and the free of any unnecessary information, preserving, thus, the OPAC functionality.

The main entities for the analysis are defined as follows:

— The folksonomy, the set of tags $F = \{t_1, t_2,...,t_n\}$ in the form of a tag cloud.
— The authority file defined as a set of subject headings $A = \{h_1, h_2,...,h_m\}$. The subject headings describe thematically the bibliographic records and consequently the documents of the information organization collection.
— The query log file $Q = \{q_1, q_2,...,q_k\}$, which depicts the vocabulary of the user queries.

The proposed methodology consists of four basic stages summarized as follows:

Data Quality Indicators Estimation. During this initial exploratory stage we seek for the overlap between the mentioned basic data entities (folksonomy, authority file and query log file). The outcome of this work brings out a conclusion regarding the margins for the authority file enhancement. Actually, it investigates the overlap of the users vocabulary as expressed by the social tags of the folksonomy and the user queries with the official vocabulary of the information organization, as expressed by the authority file's subject headings.

Worthy Tag Selection. This stage aims to exclude the tags that are highly correlated with the subject description or the access points of the corresponding bibliographic records. Our objective is to keep tags that offer added value to the items they refer to.

Tag Assessment. After having a list of "worthy" tags, we search for their importance in two external web resources with bibliographic data: (a) Google Books and (b) Library Thing. For this purpose we introduce a weighted metric.

Evaluation of the Tag Weighted Values. The last step aims to evaluate the derived results and therefore introduces three indicators assisting the information organization staff in selecting the tags that enhance the subject description of a subset of bibliographic records.

3.1   Data Quality Indicators

As a first step, we establish some basic data quality indicators that are based on the overlapping of the three entities, the authority file (A), the queries log file (Q) and the folksonomy (F). These indicators constitute a baseline for the investigation of the improvement carried out by the adoption of the proposed methodology.

Folksonomy-Authority Overlap: Through the examination of the overlap degree between the folksonomy and the authority file, we intend to highlight the percentage of social tags, which represent new terminology for the subject description of the information organization documents. This indicator is defined by the relation: $FA = |F \cap A| / |F|$.

Authority-Queries log Overlap: Two indicators describe the overlap between the authority file (A) and the query log file (Q) and provide a picture for the acceptance of the authority file by the users: (a) $QA = |Q \cap A| / |Q|$ that reveals the percentage of queries satisfied by the authority file and (b) $AQ = |Q \cap A| / |A|$ that represents the extent the authority file is used by the user queries.

Folksonomy – Queries log Overlap: The overlap between the folksonomy and query log file emerges the percentage of the queries that constitute a new terminology for the subject description. To analyze this overlap it is necessary to discriminate two query categories: The subset of the queries posed only to the folksonomy (Q_0) and the total set of queries posed to all other access points (title, author, subject, etc.). Hence we define 4 different indicators: (a) $FQ_0 = |Q_0 \cap F| / |F|$ which defines the percentage of the folksonomy covered by the queries posed to the social tags index, (b) $Q_0F = |Q_0 \cap F| / |Q_0|$ determining the percentage of the queries answered by the social tags index, (c) $FQ = |Q \cap F| / |F|$ indicating the total number of the social tags in the queries and (d) $QF = |Q \cap F| / |Q|$, representing the number of the queries answered by the folksonomy.

3.2   Worthy Tag Selection

This stage investigates the relation of a tag (t_i) to each bibliographic record (e_j) it refers to, aiming to keep only the "worthy" tags for the rest processing and to exclude tags that do not serve to the enrichment of the subject description of the records e_j. The pairs (t_i, e_j) are categorized to the Unworthy pairs class, when the t_i is member of the set of the terms appeared in the access points of the record e_j, i.e. if t_i is subset of the union of the words in the *Title, Author, Subject, Notes* fields of e_j. In this case it is highly probable that the tag is useless, since other words in the record prevail semantically over the tag.

Two different algorithms are developed for this stage and the main idea of each of them is sketched as follows:

WorthyTagSelection Algorithm: Firstly we semantically enrich the subject description of the record e_j by adding to the content of the *Subject* field, all the related subject headings (equivalent, broader, narrow and related subject headings) from the authority file, if any exist (in the case of MARC Authorities the related terms of a subject heading are tagged by the fields 400 and 500). Afterwards we form the union of all the terms in the record's access points, we tokenize and stem it. Also the tags associated with the record, are tokenized and stemmed. Tokenization includes the removal of stopwords and punctuation as well as the conversion of all capital letters to small ones. Then we check the existence of the tag t_i, included in the pair (t_i, e_j), in the union of the access points of the record. If this condition holds, then we consider the pair as "Unworthy", otherwise we considered it as "Worthy".

Lucene Algorithm: This algorithm considers every record e_j as a vector with two attributes. The first one includes the unique identifier of the record and the second consists of the union of all the terms of the access points of the record (the access point *Subject* has also been semantically expanded by all the related subject heading terms from the authority file, and all words were tokenized and stemmed). Then each tag t_i of the pairs (t_i, e_j) is considered as query term for the documents e_j; if the similarity between tag t_i and the record e_j, equals to zero, then the pair is classified to the "Worthy" tags category, otherwise if the similarity is different than zero, the pair is considered as "Unworthy".

3.3 Tag Assessment Process

At this stage of the methodology, we estimate the value of each "worthy pair" and therefore, we developed wrappers that search for the pair in two external sources, Google Books and Library Thing, and retrieve related information.

Google Books provides the number of pages p_{ij} that the tag t_i appears in the preview of a document e_j. The procedure, whereby we find the p_{ij} value, is as follows:

For every pair (t_i, e_j) we perform an isbn-based search to retrieve the document e_j. If the search returns no results or the isbn of the document e_j is not known, then we perform a title-based search. In this case, usually there are more than one returned results and the desired result is the one whose title and authors are exactly the same with these of record e_j. Having accessed the document e_j, we search for the tag t_i in its preview; the GB response provides the p_{ij} for the pair (t_i, e_j). Then we estimate the normalized p_{ij} as $np_{ij} = p_{ij} / p_j$, where p_j is the number of the pages appeared in the preview of the document e_j.

On the other hand, the value u_{ij} of a pair (t_i, e_j) in LT depicts the importance that the users attribute to the tag t_i for a specific document e_j and is defined by the number of users that assigned t_i to e_j. The algorithm whereby we find the value of u_{ij} is as follows:

For every pair (t_i, e_j) we perform a search based on the isbn of the document e_j, in order to retrieve it in LT. In addition, we compare the title of the returned item with the title of the document e_j, and only if the two titles match, we proceed to the next step of the process. After retrieving the appropriate document, we search the tag t_i in its tag cloud. If found, there is a number that follows the tag, which is the desired u_{ij}. Afterwards we estimate the normalized u_{ij} as $nu_{ij} = u_{ij} / \Sigma_j u_{ij}$, where $\Sigma_j u_{ij}$ is the sum of all the u_{ij} of the tags assigned to the document e_j.

In the last stage of this process, we define the tf-idf weights of each u_{ij} and p_{ij} as:

$$tf\text{-}idf_G = np_{ij} \times idf_G , \; tf\text{-}idf_L = nu_{ij} \times idf_L, \; idf_G = log(N_L/n_{iL}), \; idf_L = log(N_G/n_{iG})$$

where N_G and N_L the number of the documents retrieved by GB and LT respectively and n_{iG}, n_{iL} the number of the GB and LT documents in which the tag t_i exists respectively. The weighted mean of the tf-idf values, provides the average total value V_{ij} of the pair (t_i, e_j):

$$V_{ij} = \frac{a(np_{ij} \times idf_G) + (1 - a)(nu_{ij} \times idf_L)}{idf_G + idf_L}$$

where $\alpha \in [0, 1]$. If $\alpha > 0.5$, the authors' vocabulary is considered more important, otherwise if $\alpha < 0.5$ the users' vocabulary is taken into account as more significant.

3.4   Evaluation of the Weighted Values

Given the value V_{ij} of each pair (t_i, e_j), we still have three questions that need to be answered. Firstly, we need to know which factor α is the best to use in this metric. Secondly, we have to choose the most suitable threshold θ on the value V, so that the tags t_i of the pairs (t_i, e_j) with $V_{ij} > \theta$, would be candidates for the enhancement of the subject description of the records e_j. Finally, we demand to evaluate the effectiveness of our methodology. For this purpose, we defined three evaluation metrics:

Users' Satisfaction Improvement (USI): The purpose of this metric is to investigate whether the satisfaction of users' queries (Q) increased after the enrichment of the authority file (A) with the tags (Rt) that passed the threshold θ. In detail, this metric examines the value of the indicator $QA = |Q \cap A| / |Q|$, after including the (Rt) tags in the authority file (A) and removing the queries posed to the social tags index (Q_0) from the query log file (Q), so as to be examined only the queries that are involved in the new authority file AUR_t and not in taxonomy. The formula for the first metric is: $USI = |(Q\text{-}Q_0) \cap (AUR_t)| / |Q\text{-}Q_0|$.

Document Coverage (DC): The objective of this metric is to estimate the percentage of the documents affected by the enrichment. The formula for the second metric is: $DC = |Passed\ Documents| / |WorthyDocuments|$, where *Passed Documents* is the set of the documents e_j that participate in at least one pair (t_i, e_j), whose value V_{ij} is greater that the threshold θ and *WorthyDocumens* is the set of the documents e_j that participate in at least one "Worthy" pair.

Document Enrichment Percentage (DEP): The aim of this metric is to bring out the extent, whereby the documents of previous metric have been enriched. The formula for the third metric is: *DEP = |Passed Tags| / |Tags of Passed Documents|*, where the *Passed Tags* is the set of tags t_i that participate in a pair whose value V_{ij} is greater than the threshold θ and the *Tags of Passed Documents* is the set of tags that participate in at least one pair (t_i, e_j) where $e_j \in$ *Passed Documents*.

The desired value of the threshold θ lies between the points where the slope of the DEP curve starts increasing and the slope of the DC curve starts decreasing.

4   Experimental Study

The data for the experimental study that validates the proposed methodology were derived from the Panteion University Library, Athens, Greece, which manages a collection of 80,000 titles of books, serials, video, and grey literature specialized on social sciences and humanities. The Library has developed a social OPAC, called OPACIAL[3], which is enhanced with several Web 2.0 technologies [19] such as tagging functionalities, folksonomy-based navigation and searching in tag cloud environment. The total number of tags in the folksonomy file (F) is 17,451, from which almost 11,000 tags have been imported from LibraryThing. The size of the authority file (A) is 164,819 subject headings; most of them are translated into the Greek language from LCSH and the majority of them is interlinked with references of broader, narrower and non-preferred terms. We processed a query log file (Q) covering a period of one year, from 27/1/2010 to 26/1/2011, consisting of a number of 368,966 queries, while the number of queries that refer only to the field "Social Tag" (Q_0) is 23,076. The processing of the OPACIAL bibliographic records resulted to a number of 96,908 tag-document (t_i, e_j) pairs.

After stemming and tokenizing the terms included in the datasets, we estimated the data quality indicators presented in section 3.1. The first indicator FA provides the overlap between the folksonomy and the authority file. Its value equals to 48.65% meaning that almost the half of social tags are terms of the authority file. This is an expected result since the users are interested in social sciences and they share a common vocabulary. The difference between the terms of the two sets might denote that the folksonomy carries either new terminology or personal, not useful information for the information needs of the user community.

The next two indicators AQ and QA represent the overlap between the authority file and the queries log file. Their values are equal to 81.3% and 73.96% respectively and they bring out the extent of the authority file terms usage in the queries. We believe that the high values of these indicators emerge from the fact that the Panteion library has developed a rich authority file, and the users are trained to use the authority file in their queries.

---

[3] http://library.panteion.gr/opacial/index.php?language=en

The last four indicators deal with the overlapping of the folksonomy and the queries log file. The first indicator Q_0F is equal to 42.74% denoting that the vocabulary the users use to pose queries to the folksonomy differs from the folksonomy *per se*. This could be explained by the fact that the majority of social tags has been imported from the LibraryThing and they do not reflect the vocabulary of the particular community. On the other hand the FQ_0 indicator equals to 88.13%, which means that the folksonomy satisfies the queries posed to it. However, in general, the users do not prefer to use the social tags as query terms and this is evident by the other two indicators FQ and QF which are equal to 39.37% and 13.41% respectively. Hence it could be supposed that the folksonomy is mainly used for navigation and it covers different/additional user needs to the search service. Thus, the hypothesis that the social tags could enrich the subject description of the bibliographic records seems to be valid.

In the worthy tag selection stage we run the WorthyTagSelection and Lucene algorithms to determine the set of the worthy pairs (t_i, e_j) which will be assessed in the next stage. The former algorithm characterizes 21,402 pairs as "unworthy", out of the 96,908, in contrast to the 30,518 "unworthy" pairs of the latter. It should be clarified that the unworthy pairs classified by the Lucene algorithm denote a similarity value different to zero. Actually the similarity values for the pairs need further processing so as to be checked and interpreted. On the other side the WorthyTagSelection is considered configurable enough for the needs of the experiment and its results are easily verifiable. Totally 75,506 pairs were classified to the "Worthy" category, according to the results of the WorthyTagSelection algorithm.

The next stage concerns the assessment of the tags of the "Worthy" pairs by acquiring additional information about their frequency in the corresponding documents and their popularity in the tagging process. This information was drawn on from GB and LT respectively. Totally 64,113 pairs were retrieved from LT and 41,471 from GB that correspond to 6,365 and 4,839 bibliographic records respectively. All the pairs from GB are included in the LT pairs. Afterwards we applied the weighted metric, as described in section 3.3, to compute the total value V_{ij} for every pair (t_i, e_j). The co-efficient α offers the flexibility to boost either the users' language or the authors' language. Moreover the threshold θ help the subject cataloguers to determine the bibliographic records e_j which will be enriched thematically by the tags t_i. The selection of a low θ and α values implies a large number of selected tags, many of them might be terms with no impact on the improvement of the subject description and consequently might affect negatively the information retrieval, such as broader terms, personal tags, etc.

We estimated the three evaluation metrics for three α values (0.3, 0.5 and 0.7) and for a range of θ values. The values of the *USI* metric, presented in Figure 1, decrease as the θ value increases, and vary between 74.47% and 74.48%. Besides, the value of the QA indicator is 73.96% implying that the authority file augmentation improves the coverage of the queries at a percentage between 0.51% and 0.52%. In any case the positive difference between *USI* and QA denotes a potential improvement of the authority file and hence the increase of the number of answered queries. The observed limited increase verifies that the OPACIAL authority file is well formed and maintained and the users are trained to use it in their searches. It is worth observing in the

Figure 1 that the usage of a the authors' vocabulary (α=0.7), for θ less than the value 0.16, provides a greater number of tags to be added to the records' subject description, while for greater θ values the recommended tags are reduced.

Concerning the curve of *DC* metric, presented in Figure 2, the candidate records for subject enhancement decrease as the threshold θ increases. Moreover, it is remarkable that it changes its slope at a value of the threshold θ around 0.12 and 0.14. Furthermore the number of the recommended tags for subject enhancement increases when the users vocabulary is preferred, i.e. for values α less than the value 0.5. The DEP curve is increasing especially when α equals to 0.5 and 0.7 and its slope presents peaks for all the values of α.

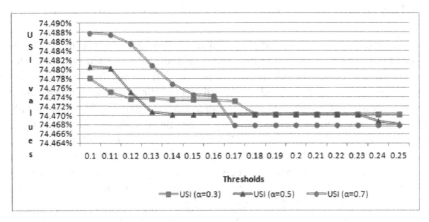

Fig. 1. USI metric

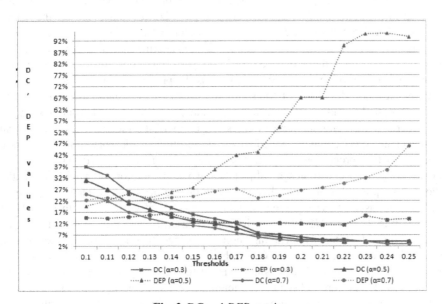

Fig. 2. DC and *DEP* metrics

Concluding an information organization could adjust its policy by choosing alternative values for the co-efficient α and the threshold θ and examining the values of the *USI*, *DC* and *DEP* metrics. Regarding the case of the Panteion University Library the evaluation metrics are optimized at θ values in the range between 0.13 and 0.16 and an α value equal to 0.5, indicating that the user and author vocabularies should be balanced. This policy seems to be recommended for an academic environment in which the users follow a standard vocabulary. In domains where the terminology is dynamically evolving the priority would be given to the users language, opting for lower α values.

5   Discussion

Given the results provided by the proposed methodology, information professionals still face the problem of assessing the quality of the recommended tags. Therefore a Tag Exploitation phase should further refine the derived results. In this phase the resulted tags should be compared to the subject headings of the authority file and could be categorized to the following groups: (a) tags that are identical to subject headings, (b) tags that are incomplete subject headings, such as "Jung", while there exists a subject heading "Jung, C. G., 1875-1961", (c) tags that match to subject heading subdivisions, e.g. "1920", (d) tags that do not exist in the authority file. The terms of the last category might introduce new terms or might be personal or subjective tags, e.g. 'toread", "fantastic", and therefore they should be retrieved from knowledge organization systems and further examined. According to the literature [13, 15, 18], the majority of the tags usually overlap with the authority file. As far as the Panteion University Library is concerned, the tags of only 8,665 pairs, out of 41,471, are not included in the authority file. Indicatively 610 pairs are recommended for the values $\theta \geq 0.13$ and $\alpha = 0.5$ for which 330 (54.09%) are subject headings, 146 (23.93%) are incomplete subject headings, while 77 (12.62%) are terms that should be further examined. Furthermore there exist 57 (9.34%) personal terms, including acronyms, without any value to the subject indexing. Thus the proposed methodology faces sufficiently the main drawback of folksonomies stated by Guy and Tonkin as "sloppy tags … "misspelt", badly encoded, such as unlikely compound word groupings" [5].

Concerning the manipulation of the rest tags, obviously, the tags that are identical to subject headings could be added to the description of the corresponding record, without further processing. For instance the tag of the pair (urban planning, doc_id_10064363) constitutes a subject heading and therefore it would enrich the record description. For the tags that are not complete headings, further processing is needed. For instance the tag Jung, should be checked if corresponds to the subject heading Jung, C. G., 1875-1961. An indicative case of new terms is the pair (ethnic violence, doc_id_10065351). The tag "ethnic violence" is related to the broader term "political violence" in Wikipedia. Given these issues the future work should focus on examining whether a tag that does not exist in the authority file constitutes new terminology and therefore substantial enrichment of the authority file. The available tools that promote collaboration between information organizations, such as the cooperative cataloguing networks, could facilitate this process. It should be mentioned that international cooperative cataloguing initiatives permit the generation of new multilingual subject headings.

Another issue concerns the alternatives the selected tags to be incorporated the bibliographic records thematic description. In the case of a MARC-based OPAC, the tag could be inserted directly using the field 610 field for UNIMARC records, or the field 653 for MARC21 records. The scope of these fields is defined as "this field is used to record subject terms that are not derived from controlled subject heading lists". MARC21 provides also the fields 690-699 which "are reserved for local subject use and local definition" [21].

The present methodology can also be implemented in an environment of a digital library with social tagging functionalities. In that case, the assessment of the tags would be based on a mechanism for automatic indexing of the digital objects, instead of an external content provider such Google Books. Additionally if there exists a rich folksonomy the users' vocabulary assessment could be replaced by the local folksonomy logs, instead of retrieving information from the LibraryThing. Finally it should be underlined that the proposed methodology could be adopted by information organizations that do not offer social tagging and folksonomy functionalities, but they recognize that the subject indexing operation needs to be adapted to the users' vocabulary. Thus the assessment of the tags of a well-known folksonomy and the incorporation of them in the subject description of the related documents would be for the benefit of the information seeking. However it is evident by the experimental results that the impact of the social tags should be moderated by the document terminology, i.e. the authors' vocabulary.

The proposed methodology will be further validated by experiments that determine the degree of the improvement of the information retrieval performance caused by the proposed enhancement of the subject indexing. In particular an issue for further research would be the refinement of the tag evaluation metrics under the prism of the information retrieval evaluation. However the usage of information retrieval performance evaluation metrics requires the development of an experimental setting able to classify thematically the user queries as well as the documents according to the query categories. Such experimental settings are difficult to be implemented in organizations whose collections and daily workload is huge and do not have available resources and infrastructures for laboratory work. Besides, these organizations need easily applicable decision support tools to improve the quality of the offered services.

6   Conclusions

Social tagging is a powerful tool enabling the information organizations to select collective knowledge appeared in the users' vocabulary and preferences. The proposed methodology intends to support the decision of selecting a number of tags that enhance the subject description of a set of bibliographic records. The methodology was validated by an organization with "heavy" bibliographic and indexing tools, such an academic library. The users' language, balanced by the authors' vocabulary, was compared to a well-organized authority file, which includes rich links - cross references - among subject descriptors. The results reveal that even in environments in which the users are trained and follow a strict vocabulary, the subject description of the bibliographic records could be enhanced by accurate terms derived by the folksonomy. Furthermore the main idea of this process is that the tag evaluation should take into account the authors vocabulary. The goal of the information organizations should be the cooperation between the folksonomies and the authority files. The indexing practices followed by the users and information professionals are "opposite" but they could operate complementary and reduce their weaknesses: the lack of authority control for the folksonomies and the delay of the new terminology incorporation and the authority file update, especially in academic environments.

References

1. Calhoun, K., Cantrell, J., Gallagher, P., Hawk, J.: Online catalogs: What users and librarians want. OCLC Online Computer Library Center, Inc., Dublin (2009)
2. Smith-Yoshimura, K., Cyndi, S.: Social Metadata for Libraries, Archives and Museums Part 1: Site Reviews. OCLC Online Computer Library Center, Inc., Dublin (2011)
3. Chua, A.Y.K., Goh, D.H.: A study of Web 2.0 applications in library websites. Library & Information Science Research 32(3), 203–211 (2010)
4. McCulloch, E., MacGregor, G.: Collaborative tagging as a knowledge organisation and resource discovery tool. Library Review 55(5), 291–300 (2006)
5. Guy, M., Tonkin, E.: Folksonomies: Tidying up tags. D-Lib Magazine 12(1) (2006), http://www.dlib.org/dlib/january06/guy/01guy.html
6. Peters, I.: Folksonomies: Indexing and Retrieval in Web 2.0. De Gruyter, Saur, Berlin (2009)
7. Pera, M.S., Lund, W., Ng, Y.-K.: A sophisticated library search strategy using folksonomies and similarity matching. Journal of the American Society for Information Science and Technology 60(7), 1392–1406 (2009)
8. Smith, T.: Cataloguing and you: Measuring the efficacy of a folksonomy for subject analysis. In: Lussky, J. (ed.) 18th Workshop of the American Society for Information Science and Technology Special Interest Group in Classification Research, Milwaukee, Wisconsin, USA (2007)
9. Bartley, P.: Book Tagging on LibraryThing: How, why, and what are in the tags. In: ASIS&T Annual Conference, Vancouver (2009)
10. Rolla, B.P.J.: User Tags versus Subject Headings Can User-Supplied Data Improve Subject Access to Library Collections. Library 53(3), 174–185 (2009)
11. Heymann, P., Ramage, D., Garcia-Molina, H.: Social tag prediction. In: Proceedings of the 31st Annual International ACM SIGIR Conference on Research and Development in Information Retrieval, pp. 531–538. ACM (2008)
12. Adler, M.: Transcending Library Catalogs: A Comparative Study of Controlled Terms in Library of Congress Subject Headings and User-Generated Tags in LibraryThing for Transgender Books. Journal of Web Librarianship 3(4), 309–331 (2009)
13. Iyer, H., Bungo, L.: An examination of semantic relationships between professionally assigned metadata and user-generated tags for popular literature in complementary and alternative medicine. Information Research 16(3) (2011)

14. Thomas, M., Caudle, D., Schmitz, C.: To tag or not to tag. Library Hi Tech. 27(3), 411–434 (2009)
15. Lu, C., Park, J.-r., Hu, X.: User tags versus expert-assigned subject terms: A comparison of LibraryThing tags and Library of Congress Subject Headings. Journal of Information Science 36(6), 763–779 (2010)
16. Yi, K., Chan, L.M.: Linking folksonomy to Library of Congress subject headings: an exploratory study. Journal of Documentation 65(6), 872–900 (2009)
17. Yi, K.: A Semantic Similarity Approach to Predicting Library of Congress Subject Headings for Social Tags. Journal of the American Society for Information Science 61(8), 1658–1672 (2010)
18. Peters, I., Stock, W.G.: Power tags in information retrieval. Library Hi-Tech. 28(1), 81–93 (2010)
19. Kakali, C., Papatheodorou, C.: Exploitation of folksonomies in subject analysis. Library & Information Science Research 32(3), 192–202 (2010)
20. Kakali, C., Papatheodorou, C.: Could social tags enrich the library subject index. In: LIDA 2010, Zadar (2010)
21. DeZalar-Tiedman, C.: Exploring user-contributed metadata's potential to enhance access to literary works: Social tagging in academic library catalogs. LRTS 55(4), 221–233 (2011)

Advanced Automatic Mapping from Flat or Hierarchical Metadata Schemas to a Semantic Web Ontology
Requirements, Languages, Tools

Justyna Walkowska and Marcin Werla

Poznań Supercomputing and Networking Center,
ul. Noskowskiego 12/14, 61-704 Poznań, Poland
{justyna.walkowska,mwerla}@man.poznan.pl

Abstract. This paper is dedicated to the issue of automatic mapping from flat or hierarchical metadata schemas to Semantic Web data formats. It proposes a checklist of requirements for such mappings and, based on this checklist, tries to compare functionalities of existing mapping tools. Finally, it introduces jMet2Ont, an open source mapping tool created by PSNC during the SYNAT research project.

Keywords: Metadata Mapping, Semantic Web, Dublin Core, MARC21, CIDOC CRM, FRBRoo, EDM, OWL, RDF.

1   Introduction

This paper is dedicated to the issue of automatic mapping of large amounts of records from flat (DC-like) or hierarchical metadata schemas to ontology-described formats. Semantic Web (RDF/OWL) and Linked Open Data formats are gaining popularity with a number of national (e.g. German WissKI, Finnish FinnONTO, Polish SYNAT) and international (e.g. Europeana and related projects) efforts to make large amounts of interconnected data accessible to the public in a highly usable form. However, many cultural heritage institutions, including those holding electronic resources, for years have been describing their collections with metadata schemas which do not necessarily translate well or easily to Semantic Web ontologies.

Consequently, a new group of metadata mapping tools has emerged recently. This paper tries to compare their functionalities and juxtapose them with a checklist of mapper requirements. The requirements list has been created based mostly on the analysis of the Dublin Core, PLMET[1] and MARC21 source formats, and the CIDOC CRM, FRBRoo and EDM target formats, analyzed and processed during the SYNAT[2]

---

[1] PLMET is a metadata schema designed for the purpose of aggregation of data from Polish digital libraries by the PIONIER Network Digital Libraries Federation (http://fbc.pionier.net.pl).

[2] SYNAT is a Polish national research project aimed at the creation of universal open repository platform for hosting and communication of networked resources of knowledge for science, education, and open society of knowledge. It is funded by the National Center for Research and Development (grant no SP/I/1/77065/10).

P. Zaphiris et al. (Eds.): TPDL 2012, LNCS 7489, pp. 260–272, 2012.

project. The project also led to the creation of jMet2Ont, a general mapper that externalizes mapping rules and allows for mapping from any XML-based (flat or hierarchical) schema to any Semantic Web ontology format. This open source tool is available at `http://fbc.pionier.net.pl/pro/jmet2ont`.

The paper is structured as follows: section 2 describes popular metadata schemas used by digital libraries. Section 3 explains the mapping requirements gathered while analyzing the source and target schemas. Section 4 presents and compares existing mapping tools. Section 5 covers jMet2Ont, a tool created by PSNC to address requirements that are not implemented by the existing tools. The paper closes with an evaluation and conclusions section.

2   Popular (Digital) Library Metadata Schemas

This section contains a quick overview of metadata formats used to describe library data, that were used as input in the requirements formulation process.

2.1   Dublin Core

Dublin Core Metadata Element Set[3] is a set of vocabulary terms which can be used to describe (electronic) resources for the purposes of discovery. It contains 15 elements, such as *title*, *creator*, *subject* or *date*. The elements have been refined at a later stage to form the DCMI Metadata Terms set. The goal of the DC schema is to facilitate resource description and exchange. It is widely used, also in many Polish digital libraries. Unfortunately, practice proved that the element set may be interpreted differently, leading to significant inconsistencies in metadata from different institutions [9].

When confronted with the need to represent data for which a schema element has not been defined, many institutions propose their own DC application profiles, adding proprietary metadata elements. A standardization effort undertaken by Polish Digital Libraries Federation resulted with the introduction of PLMET[4]. PLMET unifies application profiles developed independently by dozens of Polish digital libraries, with the aim of ensuring high level of metadata interoperability at least on the country level.

2.2   MARC 21

MARC (MAchine-Readable Cataloging) [8], [10] is used in many libraries and catalogues. It has different varieties: authority, bibliographic, classification, community information, and holdings. It is a very complete and structured format, but also complex and not very human-readable, as the fields have numeric names, and many types of information are encoded. MARC 21 is built upon MARC formats and practices used in the US and Canada. MARC XML is an XML format for representation of MARC 21 records. It lies quite far from the Semantic Web principles (e.g. often

---

[3] `http://dublincore.org/documents/dces/`
[4] `http://dl.psnc.pl/community/display/FBCMETGUIDE`

related data can be found in seemingly unrelated elements; one has to know the schema very well to discover those relations). Mapping from MARC 21 to other formats can be challenging, which is pointed out in section 3. An arguable advantage of MARC 21 is its unambiguity.

2.3   FRBR and FRBRoo

Functional Requirements for Bibliographic Records (FRBR) [5] is a conceptual entity-relationship model developed by the International Federation of Library Associations and Institutions (IFLA) to describe retrieval and access in digital libraries and catalogues from user perspective. One of FRBR's distinctive features is the representation of a single bibliographic object on four levels (so-called group A entities): work, expression, manifestation, and item.

FRBRoo is a formal ontology based on FRBR, created as a joint effort of the CIDOC CRM Special Interest Group and IFLA. FRBRoo extends the cultural heritage-oriented CIDOC CRM ontology [3] with FRBR concepts. You can find an RDFS implementation of FRBRoo on the CRM website, and an OWL implementation (of both CIDOC CRM and FRBRoo) is available as Erlangen CRM / OWL [6].

2.4   EDM

Europeana Data Model has been recently introduced by Europeana[5] (cultural heritage metadata aggregator) as the main resource description format. As opposed to the ESE format [2] originally used by Europeana, which is a Dublin Core application profile, EDM can be treated as a *top ontology* to describe data. It encourages the reuse of publicly available vocabularies. The model differentiates between: *the intellectual and technical creation that is submitted by a provider, the object this structure is about, and the digital representations of this object, which can be accessed over the web* [4].

3   Requirements

This section names three different groups of requirements for mapping tools: basic requirements, advanced requirements, and advanced Semantic Web requirements. The requirements in Table 1. Are grouped according to this distinction.

3.1   Basic and More Advanced Requirements

The most important, necessary condition to call a tool a mapper is quite simple. *The tool should be able to map the contents of* elem1 *in the source schema as contents of* elem2 *in the target schema.*

---

[5] http://europeana.eu/portal/

There is a group of slightly more advanced requirements that have already been recognized and implemented in some of the currently available mappers. Coverage of requirements by different mappers is presented in Table 1 later on in the article. Those advanced requirements are explained in detail below.

Attribute-Value-Based Mapping. Often it is not sufficient to base mapping rules solely on the name of the metadata element. The element's XML attributes and their values may also influence the mapping decision and the content semantics. An extreme example is the MARC/XML format in which there are only four elements (*leader, controlfield, datafield,* and *subfield*). The meaning of the *controlfield* and *datafield* elements depends on the *tag* attribute, while the meaning of *subfield* is determined by *code.*

XML Structure-Based Mapping Rules. With hierarchical schemas, an element of the same name may occur on many paths, i.e. may be nested in different elements. A universal mapper should be able to properly define mappings for such cases.

XML Structure in the Output. When mapping between two XML-based formats, not only the source, but also the target metadata schema may have a hierarchical structure. The tool's mapping rule language should make it possible to express such structural relations in the output. This requirement corresponds to the element level identifiers described in section 3.2.

Patterns. In many situations, the contents of one metadata element may be mapped differently based on the pattern that the contents match. For instance, the *dc:format* element may contain units such as centimeters or megabytes; based on the unit the final mapping decision may be different (physical or electronic resource).

Substring Mapping. Substring mapping, i.e. the ability to map only a part of the source element to a target element, turns out to be necessary usually when the target schema is richer from the source one. A good example it the *dc:creator* element which often apart from a personal name contains additional information, such as the dates of birth and death. In a richer metadata schema the creator string may need to be divided between elements, based on a provided pattern.

Concatenation. Substring mapping, as explained above, is useful when the target schema is richer from the source schema. Concatenation is a requirement for the opposite situation: when information from different places in the source record has to be concatenated in a target schema record. Another variant of the concatenation requirement is the possibility to add a fixed string to the element contents.

Value Maps. Many metadata schemas use internal codes that are consistent and concise, but make no sense outside of the schema. An example of such a codes list is *MARC Instruments and Voices Code List.*[6] It is not instantly obvious that *tb* represents a guitar. The value maps mechanism may also be used to translate controlled vocabulary lists to another language or to map between two vocabularies.

---

[6] http://www.loc.gov/standards/valuelist/marcmusperf.html

3.2   Semantic Web Requirements

Semantic Web and Linked Open Data oriented metadata formats, such as FRBRoo (CIDOC CRM extension for libraries, based on the FRBR specification, and implemented in OWL), LIDO or EDM are gaining popularity for a variety of reasons.

A number of quite advanced mappers already exist, but mapping from certain older schemas (e.g. MARC 21) is still very challenging, especially while mapping to Semantic Web formats. The most popular solutions is to either apply a very simplistic mapping, losing information that could in fact be represented in the target schema, or to perform the mapping programmatically, with dedicated hardcoded algorithms, which of course has to be done by programmers. This section lists those advanced Semantic Web requirements.

Ontology Paths. Semantic Web ontologies are built with RDF triples. It is crucial to be able to express the fact that, for instance, *dc:title* should be mapped to the *E73 Information Object - P102 has title - E35 Title – P3 has note* path in a Semantic Web implementation of CIDOC CRM (the elements starting with E are classes, and those starting with P are properties). As a result of processing this path the following triples should be created by the mapper:

```
( X1    rdf:type           E73_Information_Object )
( X2    rdf:type           E35_Title              )
( X1    P102_has_title     X2                     )
( X2    P3_has_note        "source title"         )
```

[Listing 1. Triples resulting from processing the *E73 Information Object - P102 has title - E35 Title – P3 has note* path]

Record Level Identifiers. Ontology paths are probably the most important requirement for a tool that allows mapping to an ontology format. What those paths represent are in fact objects and relations between objects. The same object may appear on different paths, and the *repeatable* object will not always be the main/root one (e.g. the publication to which the metadata record corresponds). One path may contain more than one object that makes sense in many different parts of the mapping description – for instance, information about the publishing event (date, place, publishing house) may be found in different places in the original record, and thus the same event may appear in many paths. It is necessary to be able to express in the mapping rules that the publishing events (or creator, and so on) in different paths have the same identity. You can see an example of a record level identifier in Fig. 2.

Element Level Identifiers. There are objects that make sense (i.e. can be referred to) anywhere in the record, and thus they have an identifier with a record scope, as described in the paragraph above. However, some are only valid within one XML element and its sub-elements. One may think that record level identifiers deal with this requirement as well (the identifier would only be used from within the element), but this solution is not sufficient. In many metadata schemas elements may be repeated and the mapper cannot know in advance how many instances of an element the record will contain. Let us consider the following part of a MARC/XML record:

```
<datafield tag="700" ind1="1" ind2=" ">
  <subfield code="a">DeCenzo, David A.</subfield>
  <subfield code="d">(1955- ).</subfield>
</datafield>
<datafield tag="700" ind1="1" ind2=" ">
  <subfield code="a">Ehrlich, Andrzej.</subfield>
  <subfield code="e">Tł.</subfield>
</datafield>
```

[Listing 2. A fragment of a MARC/XML record representing two contributors].

Occurrences of the *datafield* tag provide information about two different contributors. Their sub-elements contain details (name, date of birth, role in the contribution process) that have to be connected to the main person (*E21 Person* in CIDOC CRM) object, but the person is different within the scope of the two *datafield* tags. This is where element level identifiers are necessary (together with information about their scope, i.e. the name of the element in which they are valid).

Iterable Identifiers. There are cases in which two separate elements that share no common parent other than the document root are related to each other in such a way that the first occurrence of element 1 corresponds to the first occurrence of element 2, the same hold for second occurrences and so on. This relation may hold e.g. between MARC 21 *datafield 300 (physical description)*, *subfield e (accompanying material)* and contents of *datafield 538 (system details note)*.

Static Paths. The ontology paths presented above are all of even length, which means that to be completed they needed to obtain the value taken from the source metadata element. However, in some cases the source value does not need to be included by itself, but it should lead to the creation of a specific set of triples. For instance, the occurrence of *datafield 534 (original version note)* leads to the assumption that the described resource is a reproduction, so a triple representing its type may be added, irrespective of the element contents. A static path like this may refer to classes or instances: a requirement concerning this distinction is described in the following paragraph.

Classes vs. Instances on Path. If you refer to the first example of a mapping path with the resulting set of triples (Listing 1), you will see that the subjects and objects from the path were treated as class names, which resulted in adding the *rdf:type* information to the triple set. However, in some situations the subject or object is not represented by a class name, but by an instance URI. The mapper should make it possible to mark some subjects and objects in the paths as instances.

Repeatable URI's. One of the very important requirements is for the mapper to generate repeatable URI's, i.e. URI's that will be the same for the same object during repetition of the mapping process. Some URI's could be given in advance (e.g. as externally set record level identifiers), but as in RDF every resource needs a unique identifier, some will have to be generated by the mapper. An alternative is using blank nodes instead of identifiers for the less meaningful resources, but this decision results in consequences that need serious consideration.

Unique vs. Shared URI's. While generating repeatable URI's, it is important to make sure that the same URI is not generated twice, at least not by accident. The repeatable URI's may be based on the main record identifier, element contents, and maybe also a kind of hash code of the whole record. However, in some cases repetition is intended: e.g. if a newly created object represents a type (e.g. *sheet music*), and the types are not known beforehand, you may want the mapper to create the same URI every time it encounters the same value of the type. The mapper should therefore make it possible to mark a resource on the path as "possibly shared".

Existential Conditions. As stated in the paragraphs above, element contents could be mapped differently based on: the element name and attributes, element position in document hierarchy, and the contents pattern. In case of Semantic Web mappings, one more important condition exists: mapping paths can be different based on whether an ontology object has already been created. For instance, while mapping *dc:title*, one may want to treat the first title in a different manner then another, possibly alternative one. It would be very helpful if it was possible to base the choice of path on the existence of a resource with a given record level id.

4   Existing Tools

This section is an overview of some of the most interesting mapping tools found during the described research. Table 1 summarizes the comparison. Its scope does not include aspects like usability or performance. Such aspects are important, but the work presented here focused mostly on the advanced mapping possibilities.

The mappers presented here are not all of the same type. Some map to XML formats, and some to RDF triples. They are put together because all of them are supposed to be useful in Semantic Web mappings (partly as a consequence of the Europeana decision to switch to EDM). Table 1 summarizes their functionalities.

4.1   AnnoCultor Converter

AnnoCultor[7] is a project that includes a tagger and a converter, that *allow converting databases and XML files to RDF, and semantically tag them with links to vocabularies, to be published on Linked Data and the Semantic Web*. The tagger is outside the scope of this publication.

The conversion rules for different tags need to be formulated in XML, optionally containing snippets of Java code. A small number of ready converters are available on the project's website. The project site lacks thorough documentation, so the information in Table 1 has been deduced from the available examples and a brief analysis of the source code. The table lists functionalities that do not require writing Java code.

---

[7] http://annocultor.eu/

4.2   MINT

MINT services[8] compose a web based platform that was designed and developed to facilitate aggregation initiatives for cultural heritage content and metadata in Europe. It provides a visual mappings editor that translates the mappings to XSL transformations. It offers a set of string transformation functions that make substring mapping and concatenation possible. Also, there is a visual tool to build conditions on source elements. A valuable feature is that the tool takes into consideration the XSD of both the source and target schemas, enforcing validity and suggesting elements in the view.

4.3   WissKI

WissKI[9] (Wissenschaftliche KommunikationsInfrastruktur, i.e. Scientific Communication Infrastructure) is a platform for storing and managing digital objects, data and information, created in Erlangen and Nurnberg. It is a user-friendly, Semantic-Web-ready, highly configurable tool to describe and present resources. Among other features (e.g. annotation), it allows for the creation of *shortcuts* that translate a piece of data, e.g. the input of the creator field, to ontology paths.

Wiss-ki is not purely a mapping tool, but the import mechanism applies techniques that have been an inspiration for the representation of ontology paths in jMet2Ont described later in this paper.

4.4   Pure XSLT

XSLT (Extensible Stylesheet Language Transformations) is a declarative, XML-based language used for the transformation of XML documents. Using XSLT one can translate data from one XML schema to another, so it is tempting to use while translating from one XML-based metadata. Traditionally though, XSLT's main goal was to render presentation versions of XML data. XSLT can be used to *add or remove elements and attributes, rearrange and sort elements, perform tests and make decisions about which elements to hide and display* [1].

While working with XSLT, XPath (XML Path Language) is used to identify elements, attributes, text and other nodes within an XML document. XPath provides a few string processing functions that are also used by MINT (see 4.2).

4.5   Tools Summary

Table 1 compares the functionalities of different mappers described above. The plus sign means that a requirement is implemented by a tool, minus means the opposite thing, and the question marks means "unable to determine". The information about tool requirements has been obtained from their documentation, source code (in some cases) and usage experience.

---

[8] http://mint.image.ece.ntua.gr
[9] http://wiss-ki.eu/

Table 1. Semantic web mapping requirements and their coverage in available mapping tools

	Anno Cultor	Mint	Wiss-ki mapper	XSLT
El1 to el2	+	+	+	+
Attr.-value-based rules	+	+	-	+
XML-structure-based rules	+	+	+	+
XML structure in the output	-	+	-	+
Patterns	+	+	-	+
Substrings	+	+	-	+
Concatenation	-	+	-	+
Code maps	+	+	-	+
Ontology paths	+	-	+	-
Record level ids	-	-	-	-
Element level ids	-	-	-	-
Iterable ids	-	-	-	-
Static paths	+	-	-	+
Blank nodes	-	-	-	-
Classes/instances	-	-	-	-
Repeatable URI's	-	-	?	-
Unique/shared URI's	-	-	-	-
Existential conditions	-	-	-	-

The outcome of the comparison is that there are tools which satisfy all the basic and more advanced requirements, but the analyzed tools do not provide sufficient support for the identified Semantic Web requirements (at least not without extending the software's code). This was the main reason for developing a new tool which meets the identified requirements and is based fully on mapping rules expressed in XML (no programming required).

5   jMet2Ont and Mapping Rules Language

It was one of the most crucial requirements in the SYNAT project to be able to perform mappings from different metadata formats (flat or hierarchical; all expressible in XML) to OWL ontology-described knowledge base (CIDOC CRM / FRBRoo in the case of this project). In light of the absence of a sufficient mapping tool able to fulfil all requirements defined in section 3 (see the comparison in Table 1), a new solution had been designed. At the early stage of the project the mapping was hardcoded, but later, in order to achieve better flexibility and adaptability, a mapping rules language (XML-based) was proposed, and a tool to parse the rules implemented.

The basic assumption was that the most natural way of representing ontology-described data is by declaring property paths of the form:

$$Obj_1\text{-}prop_1\text{-}Obj_2\text{-}prop_2\text{-}...\text{-}prop_n\text{-}\texttt{"mapped value"}.$$

To give a quick overview of the tool, Fig. 2 shows an example of a mapping rule. The rule describes a simplistic mapping of MARC 21's field 111. Fig. 1 shows a sample contents of the 111 field. Fig. 3 shows triples resulting from applying the rule.

jMet2Ont is available at `http://fbc.pionier.net.pl/pro/jmet2ont` together with complete user documentation in English and sample mappings. At this point it does not have a visual mappings editor, although development of such is seen as one of the potential directions of future works.

The remaining part of this section briefly discusses the features of the mapper and the corresponding expressions in the rule language.

Each XML element from the input data (Fig. 1) has its own mapping defined (`elementMapping`, Fig. 2). Mappings of nested input elements (`subfield`, in the example) are defined within the mapping of the parent. Attributes of the input element are also taken into consideration, so the same element with different attribute values may have a different mapping definition. The attributes (in the example: `tag` attribute of the `datafield` element, `code` attribute of `subfield`) are matched against a regular expression pattern, so [acden] works for values "a", "c" and so on (a|c|d|e|n would work too). In the example all those subfields are mapped in exactly the same way. A more fine-grained rule could treat the date (d) differently.

```
<datafield tag="111" ind1="2" ind2=" ">
    <subfield code="a">International Congress on Hermeneutics</subfield>
    <subfield code="n">(1 ;</subfield>
    <subfield code="d">2002 ;</subfield>
    <subfield code="c">St. Bonaventure).</subfield>
</datafield>
```

Fig. 1. Contents of field 111 in a sample record from NUKAT (Polish Union Catalogue)

```
<!--111 - MAIN ENTRY - MEETING NAME (NR)-->
<elementMapping elementName="http://www.loc.gov/MARC21/slim/datafield">
    <attrValues>
        <attrValue>
            <attr>tag</attr>
            <value>111</value>
        </attrValue>
    </attrValues>
    <elementMapping elementName="http://www.loc.gov/MARC21/slim/subfield">
        <attrValues>
            <attrValue>
                <attr>code</attr>
                <value>[acden]</value>
            </attrValue>
        </attrValues>
        <patternToPaths>
            <pattern prior="1">(?s)(.*)</pattern>
            <path no="1" type="singleton" joiner=" ">
                <pathElement ord="1" recordLevelId="creation">&frbroo;F27_Work_Conception</pathElement>
                <pathElement ord="2">&cidoc;P20i_was_purpose_of</pathElement>
                <pathElement ord="3" recordLevelId="conference">&ext;E7h_Meeting</pathElement>
                <reUsableSubPath ord="4" name="hasPureAppellation"/>
            </path>
        </patternToPaths>
    </elementMapping>
</elementMapping>
```

Fig. 2. One of the MARC21 to FRBRoo mapping rules

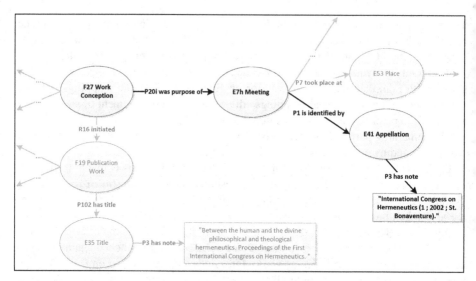

Fig. 3. Visual presentation of triples resulting from mapping the element from Fig. 1 according to rules from Fig. 2. The grey parts are sample triples added as a result of mapping other elements of the record; the black part is the path from Fig. 2 itself.

The contents of the element are also matched against a regex pattern. Patterns have priorities: if the contents do not match the first pattern (highest priority expressed by lowest number), the pattern will be checked. After finding a match, the corresponding paths are applied.

There is one mapping path in the example. One path corresponds to one capturing group in the regular expression (indicated with parentheses). In this case, there is only one group, matching the whole contents of the element, irrespective of what they are. The s flag allows for line breaks. In practice, different groups often represent different pieces of information, e.g. the name of a person and their dates of life.

The path consists of a list of classes and properties. The element contents will be added at the end of the path, as seen in Fig. 3. The last element in the example path is not a class or a property: it is a reference to a reusable *subpath* defined in an external file. The path is a *singleton path*, which means that there can only be one occurrence of it in the metadata record, so the values at the end need to be concatenated.

Among the most important features of the mapper are identifiers. There are two record level identifiers in the example: creation and conference. If the same record level identifier appears in any other path, the created object will be reused. In Fig. 3 (in grey) you can see other triples connected to objects with those identifiers as a result of mapping other metadata elements. The mapper also implements support for element level identifiers and iterable element level identifiers, defined in section 3.2.

Apart from the features described in relation to the mapping rule example, the mapper implements all requirements from section 3, excluding *XML Structure in the Output*, because the structure is instead preserved in RDF semantics.

6   Evaluation and Conclusions

A large number of institutions and international projects have decided to represent their (meta) data in a Semantic-Web-compliant format. In many cases the data needs to be transformed to an RDF or OWL format from an existing schema.

This paper has presented important requirements for such transformations and compared them with the offerings of modern mapping tools. It has also described a tool developed by PSNC during the course of the SYNAT project. This tool (jMet2Ont,) – a dedicated XML to OWL mapper – is now available as an open source software at `http://fbc.pionier.net.pl/pro/jmet2ont`. It does not require any programming experience, but some regular expressions knowledge is useful.

The mapper has been used to map full NUKAT[10] (Polish National Union Catalogue) and Polish Digital Libraries Federation[11] catalogue and library records (more than 3 million records) to the FRBRoo format. The mapping process took 3 hours on a standard desktop PC and created 254,064,115 explicit triples. The mapper has not yet been subject to a serious performance optimization, so this result should improve.

Additionally an extensive test suite has been created to verify that all requirements are met and the resulting triples are a correct interpretation of the source metadata according to the mapping rules.

jMet2Ont is a mapper, not an annotation or enrichment tool. In the PSNC's knowledge base building process, enrichment (i.e. named entity recognition and retrieval of corresponding resources from authority services) is performed at the next stage, on ontology data. Another possibility is to use a tool like Semium[12] and apply enrichment before mapping.

Possible directions of future works include: the creation of a graphical mapping rules definition interface, annotation and enrichment plug-in option, rule language syntax optimization, performance optimization (the mapper is fast, but most of the operations could be optimized by running them in parallel), and the implementation of additional requirements that have not been recognized based on the analyzed metadata formats.

References

1. Clark, J. (ed.): XSL Transformations (XSLT). Version 1.0. W3C Recommendation (November 16, 1999), `http://www.w3.org/TR/xslt`
2. Clayphan, R. (ed.): Europeana Semantic Elements Specification, Version 3.4, `http://pro.europeana.eu/technical-requirements`
3. Crofts, N., Doerr, M., Gill, T., Stead, S., Stiff, M.: Definition of the CIDOC Conceptual Reference Model, 5.0.2 edition (June 2005), `http://www.cidoc-crm.org/docs/cidoc_crm_version_5.0.2.pdf`

---

[10] `http://centrum.nukat.edu.pl`
[11] `http://fbc.pionier.net.pl`
[12] `http://semium.org/`

4. Definition of the Europeana Data Model elements, Version 5.2.3, (February 24, 2012), `http://pro.europeana.eu/edm-documentation`
5. Functional requirements for bibliographic records. Final report, `http://www.ifla.org/files/cataloguing/frbr/frbr_2008.pdf`
6. Görz, G., Oischinger, M., Schiemann, B.: An Implementation of the CIDOC Conceptual Reference Model (4.2.4) in OWL-DL. In: Proceedings of CIDOC 2008 — The Digital Curation of Cultural Heritage, ICOM CIDOC, Athens (2008)
7. Lewandowska, A., Mazurek, C., Werla, M.: Enrichment of European Digital Resources by Federating Regional Digital Libraries in Poland. In: Christensen-Dalsgaard, B., Castelli, D., Ammitzbøll Jurik, B., Lippincott, J. (eds.) ECDL 2008. LNCS, vol. 5173, pp. 256–259. Springer, Heidelberg (2008)
8. MARC 21 Format for Bibliographic. Data Field List 1999 Edition, `http://www.loc.gov/marc/bibliographic/ecbdlist.html`
9. Mazurek, C., Sielski, K., Stroiński, M., Walkowska, J., Werla, M., Węglarz, J.: Transforming a Flat Metadata Schema to a Semantic Web Ontology: The Polish Digital Libraries Federation and CIDOC CRM Case Study. In: Bembenik, R., Skonieczny, L., Rybiński, H., Niezgodka, M. (eds.) Intelligent Tools for Building a Scient. Info. Plat. SCI, vol. 390, pp. 153–177. Springer, Heidelberg (2012)
10. Paluszkiewicz, A.: Format MARC 21 rekordu kartoteki haseł wzorcowych: zastosowanie w Centralnej Kartotece Haseł Wzorcowych NUKAT. In: Nasiłowska, M., Śnieżko, L. (eds.) Wydawnictwo Stowarzyszenia Bibliotekarzy Polskich, Warszawa, Formaty, Kartoteki, vol. 17 (2009)

Ontological Formalization of Scientific Experiments Based on Core Scientific Metadata Model

Armand Brahaj[1,2], Matthias Razum[1], and Frank Schwichtenberg[1]

[1] FIZ Karlsruhe, Hermann-von-Helmholtz-Platz 1,
76344 Eggenstein-Leopoldshafen, Germany
{name,surname}@fiz-karlsruhe.de
[2] Humboldt-Universität zu Berlin, Unter den Linden 6
D-10099 Berlin, Germany

Abstract. This paper describes an ontology for the representation of contextual information for laboratory-centered scientific experiments based on Core of Scientific Metadata Model. This information describes entities such as instruments, investigations, studies, researchers, and institutions that play a key role in the generation of research data, thus forming an important source for understanding the provenance of the data. Formalization of this information in the form of an ontology and reusing existing and well-established vocabularies foster the publication of research data and accompanying provenance metadata as Linked Open Data. Core Scientific Model Ontology (CSMO) is part of a larger effort, which includes data acquisition in the laboratory and semi-automated metadata generation. It is intended to support cataloging, data curation and data reuse. A formal definition of the RDF classes and properties introduced for CSMO is provided. We demonstrate the efficacy of this ontology by applying it to two different research domains.

Keywords: Ontologies, research data management, data cataloging, contextual information, scientific experiments, science study, CSMD, CSMO.

1 Introduction

Research is increasingly data-driven. The ubiquity of computer technology and the ongoing digitization of research have an impact on how we do research. Computers allow for simulations and the analysis of large-scale datasets. Jim Gray coined the term 'fourth paradigm' for this type of data-intense scientific discovery. Instruments in laboratories produce vast amounts of born-digital data, even in the so-called 'small sciences'. The resulting "data deluge" necessitates new methods for the curation and organization of research data. With the increase volume of data and experiments, new challenges and opportunities arise in the process of collecting, organizing, interpreting and sharing the data. Archival and dissemination of research data relies heavily on the use of metadata. By use of good metadata models, scientists and researchers can improve the publishing and sharing outreach of their research information. Various institutions and projects have been exploring and implementing metadata models for

P. Zaphiris et al. (Eds.): TPDL 2012, LNCS 7489, pp. 273–279, 2012.

research data in recent years. Such models are often subject-specific and difficult to transfer to other disciplines. Proper organization of data should also encourage the reuse of data within and across scientific disciplines. With semantic web emerging, new methods of interchanging data and concepts are developed. In this paper we present a solution to formalize generic knowledge about scientific studies, investigations, results and other basic research components through use of an ontology model. The ontology aims to be a core ontology for studies with the flexibility to be extended and specialized by developers in particular domains. For our ontological formalization we have chosen to use Core of Scientific Metadata Model (CSMD) which is a well-established metadata model.

The ontology presented in this paper is named Core of Scientific Metadata Ontology (CSMO) and is defined as W3C OWL2 ontology. This initiative is connected to eSciDoc applications with a focus on building e-research repositories and virtual research environment that can assist researchers during their day-to-day work. The ontology should be best suitable in backing up applications that retrieve data automatically from experiments in eScience.

2   Modeling and Publishing Scientific Data

The formal description of experiments for efficient analysis, annotation and sharing of results is a fundamental part of the practice of science. Data generated by sensors or other instruments typically lacks a lot of important information for the correct understanding and interpretation by others. What instrument was used to capture the data? Was it calibrated? How was it configured? Who actually conducted the experiment? Experiments often require a combination of several instruments (rigs), which may create various artifacts. These artifacts are related, but these relations are most often implicit knowledge of the researcher. Making all these information (or contextual metadata) explicit helps addressing one of the most demanding challenges in data reuse: trust. Researchers must be able to understand the attributes, quality, and provenance of data in order to produce valid and reliable answers to scientific challenges.

Numerous institutions and projects have been exploring with metadata models that address contextual information. From a variety of research metadata models we have chosen the CSMD which proved to be flexible for reuse in different research domains through our experience in the BW-eLabs Project[1] (some other metadata models considered are documented in).

CSMD contains entities which describe actors, projects and resulting data. The model is organized around the *Study* unit, which can be used to represent a science research programme. Studies have *Investigators* that define who is undertaking the activity. Essential part of the model is the *Investigation* which can be an experiment, an observation, simulation or a measurement. In CSMD each investigation is associated with metadata describing the data holding associated with that investigation. A general view of the main entities of CSMD is presented in Figure 1.

---

[1] http://www.bw-elabs.org/

In this paper we present an ontology for expressing contextual metadata that are automatically acquired in laboratories during the progress of experiments. The ontology is modeled by strictly adhering to CSMD. Considering the *core* feature of the aforementioned, we modeled the CSMO with the same philosophy. Entities can be easily extended to include additional specific properties. We have shown a practical example in the BW-eLabs ontology [2] which extends some entities and additional

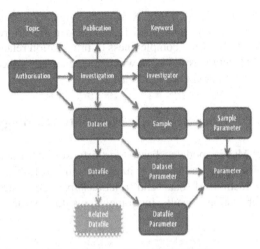

Fig. 1. Main entities of CSMD [4]

properties of CSMO. The BW-eLabs ontology is incorporated in the BW-eLabs extension[3] of the eSciDoc Browser (Figure 3).

3   Related Work

Formalization of knowledge acquisition has been a hot topic and a motivating force since ever in the semantic web movement. The KA² Ontology for example, was developed with a focus on modeling the knowledge acquisition of a community. It can be used to formalize research teams, projects, scientific documents and its primary focus was the annotation of WWW documents.

Other projects have dealt more specifically with research data. In the field of biology for example, several initiatives have produced quite a few ontology models. The Open Biological and Biomedical Ontologies [4] contain an extensive list of ontologies specific for experiments in the field of biomedical domain. Similar ontologies follow a vertical knowledge management schema and have been developed for a particular kind of situation. Some attempts have been made for a more generic ontological approach for experiments and research data. EXPO was modeled as ontology of science. It is an extension of the upper ontology SUMO and its potential lays in the fact that it can be further extended to domain specific ontologies of experiments.

As it can be seen significant amount of research has been targeting experiments as an entity. The data that should be made publicly available from research experiments is expected to be published as linked data and it should contain a full scope of information related to the experiment. Therefore information not only related to the experiment, but to parent entities such as Study or Programme, including

---

[2] https://www.escidoc.org/ontologies/bwelab/
[3] https://www.escidoc.org/wiki/ESciDoc_Browser#BW-eLabs_Extension
[4] http://www.obofoundry.org/

Organization, License Notes, Investigators etc. should be provided. These entities make the data published plenum and more informative. Our solution addresses this exact need for a completeness of information related to studies and experiments.

CSMO was modeled with Protégé and by adhering to the recommendations of the Green Linked Data guidelines.

4   Core Scientific Metadata Ontology

CSMO is modeled based on thirty key entities of scientific studies and more than seventy properties. An implementation of this ontology allows easy exploration, extraction of data and relationships about studies, investigators, investigations, publications, results, instruments, institutions etc.

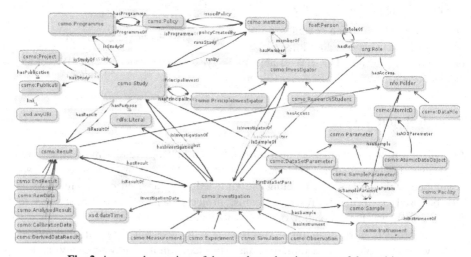

Fig. 2. A general overview of the ontology showing some of the entities

Our work with CSMO is an attempt to simplify integration of research data in the semantic web. Usage of semantic technologies allows improved access and better specification of data based on their meaning. In contrast to conventional separated instances of data sources, semantic web technologies can be used to implement a huge knowledge graph of research data which contains information from heterogeneous and distributed data from any public repository. CSMO benefits in supporting the organization of contextual research information beside primary experiment's data. Once these data are published, any information system can query through protocols like SPARQL and retrieve more information on each entity and go deeply in the graph to retrieve additional similar or related information.

The concepts modeled in the ontology are illustrated in Figure 2 and an exhaustive documentation can be found in CSMO documentation website[5]. An example of entity documentation can be found below:

---

[5] https://www.escidoc.org/ontologies/csmo/

Class: csmo:Study

Study – Studies investigate some aspect of science and have a Principal Investigator and/or institution, co-investigators and some specific purpose.

URI: https://www.escidoc.org/ontologies/csmo/#Study

Properties include: csmo:runBy, csmo:isStudyOf, csmo:hasResult, csmo:hasPublication, csmo:hasPrincipalInvestigator, csmo:hasInvestigation, csmo:hasPurpose

Used with: csmo:runsStudy, csmo:isSampleOf, csmo:isResultOf, csmo:isPrincipalInvestigatorOf, csmo:isInvestigationOf, csmo:hasStudy

Class: csmo:Investigation

Investigation – Investigation: forms the fundamental unit of the model, with a title, abstract, dates, and unique identifiers referencing the particular study. Also associated with the investigation are the facility and instrument used to collect data.

URI: https://www.escidoc.org/ontologies/csmo/#Investigation

Properties include: csmo:investigationAbstract, csmo:investigationDate, csmo:hasData, csmo:hasDataSetParameter, csmo:hasInstrument, csmo:hasInvestigator, csmo:hasResult, csmo:hasSample, csmo:isInvestigationOf

Used with:csmo:hasAccess, csmo:hasInvestigation, csmo:isDataOf, csmo:isInvestigatorOf, csmo:isResultOf

Subclasses: csmo:Experiment, csmo:Measurement, csmo:Observation, csmo:Simulation

As it can be easily noticed, both entities above are connected to each other and other entities through their properties. These connections provide a well-defined non constraining graph of information related to each entity. We have aimed to express basic concepts that are common across of different domains and leave the opportunity to extend the ontology wherever distinct concepts and vocabularies of individual domains are needed. Entities can be easily extended to include additional specific properties. We have shown a practical example in the BW-eLabs ontology which extends some classes of CSMO to include additional properties.

4.1   Implementation of CSMO

CSMD provides a good basis for describing the context of scientific investigations. Context description of experiments is heavily related to state-relations between hardware, locations, people, organizations and actual research data. In the context of VRE, such as the eSciDoc Infrastructure, these descriptions are not just enabling search, but also describing the structure of resources stored inside the repository.

To expose them as RDF graphs is clearly an advantage over document centric metadata. A digital repository with a resource oriented service interface is naturally predisposed to use linked-data beside records of metadata

We have successfully implemented and evaluated the usage of CSMO in the BW-eLabs project. In this project, we have coupled the ontology with eSciDoc Infrastructure, an e-research environment. Figure 3 shows a visual representation of a study whose information is stored as RDF/XML. Results and data generated during the runtime of an experiment are automatically added and visualized through the eSciDoc Browser[6]. The solution allows scientific data to be stored and published as linked data automatically. Data are stored as RDF/XML further implementations on the infrastructure will be improved to use an RDF Database as a storage endpoint. Such implementations will lead to a comprehensive semantic research environment platform.

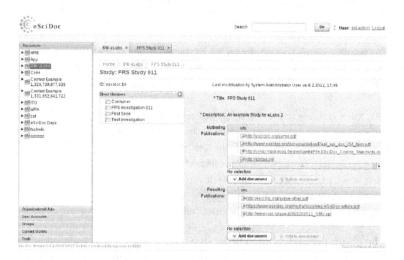

Fig. 3. eSciDoc Browser visualizing results of a Study from the BW-eLabs project

5   Conclusion

In this paper we present CSMO, an ontology for the representation of scientific metadata based on the Core Scientific Metadata Model. The ontology supports publishing of linked-data related to studies and investigations in research. It covers different entities within a study and should be considered as a core ontology that can be extended for use in specific domains.

The research is part of an effort to improve the data curation and data cataloging process in research. It should aid better organization of the metadata through the use of Semantic Technology; increase the access to contextual data in studies; increase the trust on the scientific published data, improve the access to contextual data on experiments. When used in automated systems that retrieve data automatically from experiments, it can also improve the process of publishing research data.

---

[6] The BW-eLabs defines few specific properties and couple of entities which are not part of CSMO. Therefore CSMO has been extended with a small *bw-elab* ontology. A full documentation of the BW-eLabs extension ontology can be found at https://www.escidoc.org/ontologies/bwelab/

References

[1] Hey, T., Tansley, S., Tolle, K. (eds.): The Fourth Paradigm: Data-Intensive Scientific Discovery, p. 284. Microsoft Research, Redmond, Washington (2009)

[2] Hey, T., Trefethen, A.: The Data Deluge: An e-Science Perspective. Grid Computing - Making the Global Infrastructure a Reality (January 2003)

[3] Matthews, B., Sufi, S., Flannery, D., Lerusse, L., Griffin, T., Gleaves, M., Kleese, K.: Using a Core Scientific Metadata Model in Large-Scale Facilities. The International Journal of Digital Curation 5(1), 106–118 (2010)

[4] Razum, M., Schwichtenberg, F., Wagner, S., Hoppe, M.: eSciDoc Infrastructure: A Fedora-Based e-Research Framework. In: Agosti, M., Borbinha, J., Kapidakis, S., Papatheodorou, C., Tsakonas, G. (eds.) ECDL 2009. LNCS, vol. 5714, pp. 227–238. Springer, Heidelberg (2009)

[5] Qin, J., D'ignazio, J.: The Central Role of Metadata in a Science Data Literacy Course. Journal of Library Metadata (10), 188–204 (October 2010)

[6] Razum, M., Schwichtenberg, F.: Metadatenkonzept für dynamische Date - BW-eLabs Report. FIZ Karlsruhe (2012)

[7] Benjamins, V.R., Fensel, D., Gomez-Perez, A., Decker, S., Erdmann, M., Motta, E., Musen, M.: The Knowledge Annotation Initiative of the Knowledge Acquisition Community (KA)2. In: Eleventh Workshop on Knowledge Acquisition, Modeling and Management, Voyager Inn, Alberta, Canada (1998)

[8] Soldatova, L.N., King, R.D.: An ontology of scientific experiments. Journal of the Royal Society Interface, 795–803 (2006)

[9] Hoxha, J., Rula, A., Ell, B.: Towards Green Linked Data. In: Proceedings of the Second International Workshop on Consuming Linked Data (CEUR 2011) CEUR Workshop Proceedings (Oktober 2011)

Domain Analysis for a Video Game Metadata Schema: Issues and Challenges

Jin Ha Lee[*], Joseph T. Tennis, and Rachel Ivy Clarke

Information School, University of Washington
Mary Gates Hall, Ste 370, Seattle, WA 98195
{jinhalee,jtennis,raclarke}@uw.edu

Abstract. As interest in video games increases, so does the need for intelligent access to them. However, traditional organization systems and standards fall short. Through domain analysis and cataloging real-world examples while attempting to develop a formal metadata schema for video games, we encountered challenges in description. Inconsistent, vague, and subjective sources of information for genre, release date, feature, region, language, developer and publisher information confirm the imporatnce of developing a standardized description model for video games.

1   Introduction

Recent years demonstrate an immense surge of interest in video games. 72% of American households play video games, and industry analysts expect the global gaming market to reach $91 billion by 2015 (GIA, 2009). Video games are also increasingly of interest in scholarly and educational communities. Studies across various scholarly disciplines aim to examine the roles of games in society and interactions around games and players (Winget, 2011). Games are also of interest to the education community for use as learning tools and technologies (Gee, 2003). Thus we can assert that video games are entrenched in our economic, cultural, and academic systems.

As games become embedded in our culture, providing intelligent access to them becomes increasingly important. Effectiveness of information access is a direct result of the design efforts put into the organization of that information (Svenonius, 2000). Consumers, manufacturers, scholars and educators all need meaningful ways of organizing video game collections for access. Current organizational systems for video games, however, are severely lacking. What organizational challenges emerge due to the unique nature of video games? How does the lack of standardization affect access to these games? A collaborative domain analysis for the development of a metadata schema specifically for video games reveals issues inherent in this domain.

2   Challenges and Critical Literature Analysis

Current models of video game organization come from two divergent sources: the field of knowledge organization which specializes in arranging, describing, and

---

[*] Corresponding author.

P. Zaphiris et al. (Eds.): TPDL 2012, LNCS 7489, pp. 280–285, 2012.
© Springer-Verlag Berlin Heidelberg 2012

presenting metadata for information objects and collections, and video game information from commercial systems on the internet.

Describing non-book artifacts—like video games—with knowledge organization standards has long been problematic. Hagler (1980) observed that imposing book-based characteristics on non-book materials creates inapplicable and unusable standards. Leigh (2002) notes this approach often leaves materials described by form rather than content. Even newer models like *Functional Requirements for Bibliographic Records* (FRBR) do not cover all types of materials and works: work, expression, manifestation, or item cannot be determined easily in a classic computer game (McDonough et. al., 2010a). Attributes derived from context, like mood or similarity to other objects—perhaps significant for video games—are not represented in the FRBR model (Lee, 2010). Other existing standards are similarly problematic. *Library of Congress Subject Headings* (LCSH) contains only 214 headings for describing video games by name (e.g., *Halo*, *Legend of Zelda*), with notable series missing (e.g., *Final Fantasy*, *God of War*). LCSH includes only 5 terms for computer game genre, limiting the ability to describe and therefore search or browse by genre.

Recent interests in video game preservation suggest metadata description as a preservation strategy (Winget 2008; McDonough et. al., 2010b). Besides emphasizing preservation rather than description, these projects focus on domain analysis from a data- or creator-centric point of view, rather than end users. Currently the only systematically designed game-specific descriptive framework comes from Huth (2004). However, this schema only addresses historical game systems, and like the previous examples, does not provide for the needs or behaviors of users. This limited understanding and focus of domain analysis of video games impedes development of useful information systems that meet the needs of real users.

Video game organization and description also comes from commercial systems on the internet. The web contains massive information about video games, scattered across many sites and sources. Websites such as Amazon, GameStop, GameFly, etc. are generally geared toward purchase decisions and mostly provide basic elements like title, genre, platform, release date, and publisher. Other sites provide abundant descriptive information, but it is often unstructured, cumbersome to navigate, and unverified. Users may have to visit multiple sites to find and cross-check information. All these challenges indicate the need for a more formal and standardized representation of video games based on a user-centered domain analysis approach.

3   Domain Analysis

At the University of Washington Information School, we are collaborating with the Seattle Interactive Media Museum (SIMM) to develop a metadata schema for describing all aspects of video games for improved organization, access, and preservation. The SIMM aims to contribute to the aggregation, research, preservation and exhibition of interactive media culture and the physical, digital, and abstract artifacts therein. In 2011, the authors, SIMM colleagues and selected students participated in a special topics course "Video Game Metadata" at the Information School. This course offered opportunities for students interested in organizing video games to collaborate with the authors and the SIMM founders to get hands-on experience creating a metadata schema that will be used in real life.

The bulk of the course focused on document- and user-based domain analysis activities to determine metadata elements crucial for describing video games. First, 5 different personas epitomizing the most common types of game players and consumers potentially interested in the SIMM were developed to represent the needs, behaviors, and goals of that particular user group (Cooper, 1999): Player, Parent, Collector, Academic, and Game Developer/Designer. Once these personas were described in detail, we recorded metadata elements essential to each persona and compiled them into one list. From this, the class distilled a set of 16 core elements perceived to be most useful to all 5 personas. The CORE included Title, Edition, Platform, Format, Developer, Retail Release Date, Number of Players, Online, Special Hardware, Genre, Series/Franchise, Region, Rating, Language, and UPC. We report on the schema in more detail elsewhere (Lee et al. under review). Here we highlight and discuss problems that arose during our domain analysis.

4   Discussion

After deciding upon the CORE elements in our metadata schema, the class spent several weeks cataloging video games to test the schema's usability and the domain analysis. As we worked, we identified several challenges for description, some unique to video games and others shared by other non-textual information objects.

4.1   Inconsistent, Vague, and Undefined Genre Labels

Genre is one of the few elements that describes content of a game rather than descriptive features (e.g., title, platform). Therefore it seems immensely useful for browsing a video game collection as well as finding new games to play. As we investigated hundreds of labels from different sources offering genre classification, it became evident that the genre metadata across these websites significantly vary with regards to the types and granularity of the terms. Most websites did not provide definitions for the genre labels, and those that did do not match across other sites. For instance, on Mobygames, both *Super Mario Bros.* and *Grand Theft Auto* are classified as "action" although most people would agree that they significantly differ. We found these current labels too broad and vague to be of use.

Establishing a controlled vocabulary for video game genres is an iterative process. We started with field-testing the cataloging process. We established a controlled list of genre and style labels taken from a number of websites related to games. We established instructions allowing multiple labels in an attempt to provide more specific information about game content. But this too was problematic: not only did it not solve the issue of label ambiguity, it introduced a new issue of how to order the multiple genre labels in a meaningful way. Due to this, we are pursuing further work in this area by working on a faceted scheme for video game genres.

4.2   Lack of Reliable Source for Retail Release Date Information

A game's release date was agreed to be important for all the user personas. However, as we cataloged the games, a lack of reliable source for this information became evident. The only date information obtainable from the game itself is the copyright date. Using copyright information for the release date is problematic, especially for games that belong to a series, because the copyright date typically indicates the date the series was first published and does not apply to the later games in the series.

We explored different ways to obtain this information. First, we reviewed the websites mentioned in 4.1. Using these multiple sources to find and cross-check the release date for the game worked for some cases, but we did occasionally find conflicting information. For instance, the release date for the North American version of *Shenmue* on Wikipedia is November 6, 2000, as opposed to November 7 on GameSpot, and November 8 on Allgame. While the difference in date might be insignificant for average users, it poses problems for identifying and preserving games from an organizational point of view like the SIMM's. The most reliable source of release date information came from game companies' websites, although many did not carry information about every game that they published. We contacted some companies (e.g., ATLUS) and were told that there is no single person managing that information. We suspect this may be a common issue because many game companies are short-lived or merge with other companies.

4.3   Inconsistent and Marketing-Oriented Description of Features

Game features was a highly debated metadata element and eventually excluded from the CORE elements. While we agreed that game features were a valuable addition, obtaining consistent information made cataloging difficult and time-consuming. Commercial websites do include feature descriptions, but the source of this information is often unknown. Some websites, like Allgame, have their own list of features whereas others do not list feature information at all.

During our cataloging exercise, most of the class used the features element for a variety of information that was potentially useful but unable to be represented in any other element. Thus the features element ended up more like a traditional "notes" field. Through faithfully transcribing features, we learned that many sources contain text geared toward marketing rather than objective description (e.g., "Unleash over 100 mind-blowing spells" from *Disgaea*). We concluded that either this element needs controlled vocabularies from which features can be chosen, or it should be uncontrolled to include any information that catalogers think would be useful for users.

4.4   Unclear Boundaries for Region and Language

Region information is necessary for players because most console games are locked to particular geographic areas. Some games, such as smartphone apps, are free of regional restrictions, but can still be targeted for particular language-speaking audiences. Thus it can be unclear how to describe the "region" of a game. In some cases, a game is released in a country without being localized, meaning a Japanese game can be released in Korea without being translated into Korean. If so, should the region information include Japan as well as Korea? There are also cases where a game is available in multiple languages although it is still locked to particular region: for instance, a game originally released in Japan and later published in North America may have an option for Japanese subtitles and/or voice track. In this case, should the main language be Japanese or English? These cases suggest a need for detailed rules to describe language and region information.

4.5   Difficulties in Distinguishing Developer vs. Publisher

Game containers usually have various logos representing companies involved in production. Unless other sources were consulted, we found it difficult to determine which companies represent publishers vs. developers. This is further complicated because some companies are publishers as well as developers. Sometimes this information can be found in the manual, but this was not consistently true for all cases. For older games, some companies have dissolved, making it difficult to find any information. In addition, there are multiple ways of describing a company (e.g. Nintendo, Nintendo Corp., Nintendo US), implying a need for better analysis of companies and a controlled vocabulary of organization names.

4.6   Other Issues

There were several other issues in describing games: mismatching titles and numbering of games released in multiple regions (e.g., *Final Fantasy VI* in Japan released in North America as *Final Fantasy II*); multiple titles and other names by which a game is known (e.g., *The Legend of Zelda* vs. *Zelda*); denoting actual differences among different versions/editions of games (e.g., Special, Classic, Limited, etc.); difficulty of determining series information unless the cataloger is familiar with the game, to name a few.

5   Conclusion and Future Work

The issues described in this paper emerged from our first step in creating a formal metadata schema for describing video games and interactive media. Through domain analysis and cataloging, we encountered several challenges, many unique to video games. These confirm the need for a standardized description for games, including metadata element definitions, instructions for description, and controlled vocabularies. We plan to further develop our schema by extending the CORE set of elements by defining a larger "recommended" set of potential use to gamers and developing controlled vocabularies for particular elements such as genre and publisher. Additionally, we plan to conduct systematic user studies to discover which information elements are perceived as useful and necessary for end-users such as gamers or parents of young gamers. We can also conduct quantitative analysis of metadata element frequencies to complement the user studies (cf. Tennis, 2003). We believe that our end results will be useful for any game related organizations: not only libraries, archives, and museums with video games in their collections, but also commercial enterprises like game developers, manufacturers, and distributors. Improving organization and access will enhance people's gaming experiences and also have substantial commercial and cultural consequences.

References

1. Cooper, A.: The Inmates are Runningthe Asylum. Sams, Indianapolis (1999)
2. Gee, J.P.: What Video Games Have to Teach Us about Learning and Literacy. Palgrave Macmillan, New York (2003)
3. Global Industry Analysts. Video Games—A Global Strategic Business Report (2009), http://www.strategyr.com/Video_Games_Market_Report.asp
4. Hagler, R.: Nonbook Materials: Chapters 7-11. In: Clack, D.H. (ed.) The Making of a Code: The Issues Underlying AACR2, pp. 7–11. ALA, Chicago (1980)
5. Huth, K.: Probleme and LösungsansätzezurArchivierung von Computer Programmen—am Beispiel der Software des ATARI VCS 2600 und des C64 (Unpublished master's thesis). Humboldt Universitat, Berlin (2004)
6. Lee, J.H.: Analysis of User Needs and Information Features in Natural Language Queries Seeking Music Information. JASIS & T 61(5), 1025–1045 (2010)
7. Lee, J.H., Tennis, J.T., Clarke, R.I., Carpenter, M.: Developing a Video Game Metadata Schema for the Seattle Interactive Media Museum. Submitted to International Journal on Digital Libraries (under review)
8. Leigh, A.: Lucy Is "Enceinte": The Power of an Action in Defining a Work. Cataloging & Classification Quarterly 33, 3–4 (2002)
9. McDonough, J., Krischenbaum, M., Reside, D., Fraistat, N., Jerz, D.: Twisty little passages almost all alike: Applying the FRBR model to a classic computer game. Digital Humanities Quarterly 4(2) (2010a)
10. McDonough, J., Olendorf, R., Kirschenbaum, M., Kraus, K., Reside, D., Donahue, R., Phelps, A., Egert, C., Lowood, H., Rojo, S.: Preserving Virtual Worlds Final Report (2010b), http://hdl.handle.net/2142/17097
11. Svenonius, E.: The Intellectual Foundation of Information Organization. MIT Press, Cambridge (2000)
12. Tennis, J.T.: Data Collection for Controlled Vocabulary Interoperability—Dublin Core Audience Element. Bulletin of the ASIS & T 29
13. Winget, M., Murray, C.: State of the Archive: A Review of Video Game Archives within the United States. In: Proceedings of the Annual Meeting of the ASIS & T, ASIS & T, Columbus (2008)
14. Winget, M.A.: Videogame Preservation and Massively Multiplayer Online Role-Playing Games: A Review of the Literature. Journal of the ASIS & T 62(10), 1869–1883 (2011)

A Benchmark for Content-Based Retrieval in Bivariate Data Collections

Maximilian Scherer[1], Tatiana von Landesberger[1], and Tobias Schreck[2]

[1] TU Darmstadt, 64283 Darmstadt, Germany
{maximilian.scherer,tatiana.von-landesberger}@gris.tu-darmstadt.de
[2] University of Konstanz, 78457 Konstanz, Germany
tobias.schreck@uni-konstanz.de

Abstract. Huge amounts of various research data are produced and made publicly available in digital libraries. An important category is bivariate data (measurements of one variable versus the other). Examples of bivariate data include observations of temperature and ozone levels (e.g., in environmental observation), domestic production and unemployment (e.g., in economics), or education and income level levels (in the social sciences). For accessing these data, content-based retrieval is an important query modality. It allows researchers to search for specific relationships among data variables (e.g., quadratic dependence of temperature on altitude). However, such retrieval is to date a challenge, as it is not clear which similarity measures to apply. Various approaches have been proposed, yet no benchmarks to compare their retrieval effectiveness have been defined.

In this paper, we construct a benchmark for retrieval of bivariate data. It is based on a large collection of bivariate research data. To define similarity classes, we use category information that was annotated by domain experts. The resulting similarity classes are used to compare several recently proposed content-based retrieval approaches for bivariate data, by means of precision and recall. This study is the first to present an encompassing benchmark data set and compare the performance of respective techniques. We also identify potential research directions based on the results obtained for bivariate data. The benchmark and implementations of similarity functions are made available, to foster research in this emerging area of content-based retrieval.

Keywords: bivariate data, benchmarking, content-based retrieval, feature extraction.

1 Introduction

Scientific disciplines that range from economics and sociology, to medical science, biology, physics, and others heavily rely on empirical research data, that are produced or collected in large amounts on a regular basis. Due to increased efforts in the digital library community, such research data are recently made available in public data repositories. This is important, as effective user access to research

P. Zaphiris et al. (Eds.): TPDL 2012, LNCS 7489, pp. 286–297, 2012.
© Springer-Verlag Berlin Heidelberg 2012

data repositories will eventually lead to a large increase in research productivity and efficiency [14]. In existing data repositories, user access methods are typically based on textual annotations. These are provided by experts who collected the data in the first place, or by data curators. Annotation-based retrieval is however limited to the availability and scope of textual annotations, which often are expensive and ambiguous to obtain. Moreover, annotations do not allow for retrieval of specific data based on its content (e.g., a specific relationship between two variables). Following the general ideas of content-based retrieval in multimedia data [24], first content-based retrieval methods are currently being developed and applied in research data repositories [25,3].

Four large and important categories of research data are univariate (time-series) data, biviarate data, multivariate data and multimedia data (e.g., 2D/3D image data from satellites, microscopes, MRT, etc...). Figure 1 shows an example for each of the first three categories. For collections of time-series data, the task of content-based retrieval has received a significant amount of research attention by the community in the last two decades. Subsequently, several efficient methods for indexing such databases and retrieving the data were established since [9]. For retrieval in collections of bivariate and multivariate data, research was carried out in a rather limited scope. In particular, no exhaustive evaluation of feature extraction techniques that support retrieval in bivariate data collections has been conducted so far. This can be mainly accounted to the fact, that no benchmark to support a quantitative comparison of different techniques has been proposed by now.

We attribute absence of respective data retrieval benchmarks to the difficulty of defining *similarity* for univariate, bivariate or multivariate data. For data like text, images, audio, 3D models or video, the notion of similarity usually follows a straight-forward concept. For example, similar text documents can be about the same subject or similar images show similar scenery or objects. For retrieval in sequential, bivariate or multivariate research data, such similarity concepts are not considered so far to judge relevance of retrieved data objects to a query. A meaningful quantitative evaluation for retrieval in research data collections requires meaningful annotations assigned by humans to the data objects, to allow

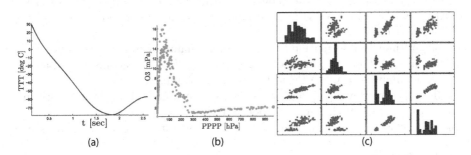

Fig. 1. Examples for time-series data (a), bivariate data (b) and multivariate data (c) (here the four-dimensional Fisher-Iris dataset, visualized as a scatterplot-matrix)

construction of similarity classes or relevance judgments [16]. So far, gathering such annotations was very expensive, as expert users are required to annotate large amounts of rather abstract objects. Due to recent efforts in the digital library community however, research data with accurate annotations by experts became publicly available on a large scale, e.g. in the PANGAEA project [8].

Thus, we use the wealth of publicly available, manually annotated research data to construct a benchmark for retrieval in bivariate data collections. This benchmark is composed of real-world scientific measurements in the domain of earth observation science, that were annotated by domain experts. Based on these metadata annotations – particularly type, location and time of measurement – we automatically group data objects to similarity classes (see Section 3). For example, all measurement data of type *altitude [m] vs. pressure [hPa]* measured at longitude 20.5 and latitude -30.7 in December can be expected to be similar [28] and are therefore assigned to the same similarity class. This results in a labeled collection of research-data, where data objects with similar content have the same label. We extensively evaluate performance of different feature extraction techniques for retrieval via precision and recall on this collection (see Section 4). Since this is the first benchmark of its kind, we also analyze and discuss the composition of similarity classes in the benchmark itself (see Section 5).

2   Related Work

For information retrieval and data mining tasks like regression, clustering, classification and retrieval benchmarking plays an important role. Ideally, such a benchmark consists of many test data sets, covering all aspects of a certain challenge, along with ground truth or a *gold standard* for each of these data sets.

For classical problems like clustering and classification, several established datasets have emerged, serving as benchmarks to compare effectiveness. For example, several datasets in the UCI Machine Learning Repository [12] are widely used to compare results of algorithms for classification and clustering. However, so far no datasets to compare techniques for retrieval are available there.

Several large-scale retrieval challenges exist for text and multimedia retrieval (TREC[1], CLEF [2] and NTCIR [3]), although a track for research-data has not yet been established.

For multimedia information retrieval, many manually annotated datasets exist and are used for benchmarking [19]. The MPEG-7 benchmark is used for 2D shape analysis [18]; the Princeton Shape Benchmark, among others, is used for 3D object retrieval [26]; and several large benchmarks for content-based image retrieval exist [6,7]. For these benchmarks, objects are usually assigned to similarity classes (either manually by humans, or automatically by using *social*

---

[1] http://trec.nist.gov
[2] http://www.clef-initiative.eu
[3] http://research.nii.ac.jp/ntcir/

tags, e.g., from Flickr), and precision-recall can be computed, to measure effectiveness of feature extraction algorithms for similarity assessment. However, their suitability is sometimes discussed [20], since such automatically designed benchmarks lack specified query sets and relevance judgments of retrieval results.

In 2003, Keogh and Kasetty [16] discussed the need for benchmarking retrieval in time-series data. They empirically show that given a sufficiently large number of datasets to choose from, the superiority of any technique can be shown when only considering numeric similarity of retrieval results. Thus, they argue for the need of similarity concepts to construct a meaningful benchmark. Only recent advances in the digital library community however, led to publicly available, manually annotated research data on a large scale [10,23,11], which is required for such a benchmark construction. Nevertheless, for retrieval in research data, and in particular for retrieval in bivariate data, no such benchmark has been proposed to date.

3   Benchmark Construction for Bivariate Data Retrieval

In this section, we present our approach to construct a benchmark for retrieval in bivariate data collections. Similar to benchmarking in multimedia retrieval, we assign data objects to similarity classes, based on metadata annotations by experts. In our case these data objects are bivariate measurement data in the area of earth observation. The annotations of the measurement data are done by scientists, that describe the type of measurement and the experimental conditions under which it was conducted. So far, such annotations were expensive to obtain, preventing construction of a benchmark large enough. Due to recent efforts in the digital library community however, repositories offering expert-annotated research data became available. We describe how to use this new data source to define similarity classes and subsequently construct a benchmark.

3.1   Data Source

We use earth observation data, which is publicly available from the PANGAEA Data Library [8,21]. PANGAEA archives, publishes, and distributes geo-referenced primary research data in the domain of earth observation (water, sediment, ice, atmosphere) from scientists all over the world. It is operated by the Alfred-Wegener-Institute for Polar and Marine Research in Bremerhaven, and the Center for Marine Environmental Sciences in Bremen, Germany. Most of the data available can be publicly accessed via http://www.pangaea.de and can be downloaded under the Creative Commons Attribution License 3.0. For content-based bivariate data retrieval, a subset of these measurement files was recently used as an application example for bivariate data retrieval [25]. Each file consists of a table of multivariate measurements, that include radiation levels, temperature progressions and ozone values, among many more. Each file available through PANGAEA is carefully annotated by the scientist who conducted the measurements. Quality control over this annotation process is taken care of by the

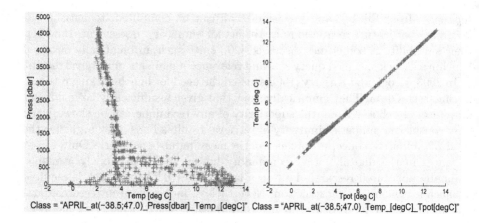

Class = "APRIL_at(-38.5;47.0)_Press[dbar]_Temp_[degC]"   Class = "APRIL_at(-38.5;47.0)_Temp_[degC]_Tpot[degC]"

Fig. 2. Exemplary data objects of the two largest similarity classes. For both classes (left and right), data points of all 58 objects are plotted into a single display in separate colors. The unique class labels consist of measurement type, location and time.

PANGAEA data curator. Most importantly for our purposes, these annotations include the type of measurement (standardized names along with base units (SI)[4] for each measurement variable in the data table), as well as the experimental conditions (time and location) under which the measurement was conducted.

To construct a test data set, we downloaded 490 publicly available measurement files from PANGAEA. Each file contains multivariate measurements with 10 to 100 columns each. By extracting every pair-wise variable combination from each of these measurement files, we obtained 24,700 bivariate data objects which form the test data set of our benchmark.

3.2 Definition of Similarity Classes

Based on the expert-annotations, we define similarity classes for the bivariate data objects. In particular, we assume data measuring the same relationship (e.g., *Temperature [deg C] vs Pressure [dbar]*) at the same time of year (e.g., December) at a close-by location (e.g., *longitude* ≈ 24, *latitude* ≈ 12) to be similar. To compute a unique class identifier, the pair of annotated variable names – already a categorical label – was used directly. The month part of the timestamp was extracted, and the geocode of the location was categorized in a 6x12 grid. By combining these three categorical labels, we assigned data objects to 1,608 different similarity classes. Such a spatio-temporal quantization is biased, as neghboring data points may be assigned to different similarity classes if they are close to the decision border. We discuss such implications on intra- and inter-class variability in Section 5.

This assumption for similarity class definition is based on Tobler's first law of geography: "'Everything is related to everything else, but near things are more

---

[4] As defined by the International System of Units.

Table 1. Statistics of proposed benchmark. Data point correlation is computed as Pearson's correlation coefficient for each bivariate data object and averaged over all objects of each class.

	Sum	Mean	Std	Median	Min	Max
objects: total	24,700	-	-	-	-	-
classes: total	1,608	-	-	-	-	-
objects: per class	-	15.36	10.9	11	5	58
data points: per class	-	657.98	1,043	319	51	9,770
data points correlation: avg per class	-	0.66	0.28	0.73	0.001	1.00

related than distant things"' [28]. The decision of the discretization parameters (temporal and spatial resolution) to construct the similarity classes influences similarity of data within a class and similarity of data among classes. We discuss these two benchmark statistics in detail in Section 5.

Table 1 gives a detailed overview of the most important benchmark statistics. Particularly interesting is the high avg data-point correlation per class, which indicates that objects exhibit some (linear) relationship, which can in principle be captured by feature extraction. Figure 2 shows data objects from the two largest similarity classes for illustration. We see that all objects within those two classes are numerically similar.

4   Evaluation

In this section, we evaluate the following nine feature extraction techniques on the benchmark described in the previous section. To measure performance of each technique for bivariate data retrieval, we compute precision-recall on the benchmark.

4.1   Feature Extraction for Retrieval in Bivariate Data

Feature extraction is the process of computing a descriptor, that mathematically represents one or several properties of an object under consideration. Such a descriptor allows to assess pairwise similarity between objects, by computing a distance measure between their respective descriptors. A prominent type of descriptors are feature vectors. As the name implies, we try to capture descriptive and discriminative object features as a vector of numerical values.

The following list provides an overview of the nine techniques, which we adapted to feature extraction of bivariate data and that we propose for evaluation. The techniques are based on time-series analysis, regression and image processing, respectively.

Euclidean Distance (SM). Baseline technique that resamples data by fitting a smoothing spline to the data [5] to allow for measuring Euclidean distance between data objects.

Correlation Coefficients (CORR). Another baseline technique that is composed of Person's sample correlation coefficient r, Kendall's tau rank correlation coefficient τ and Spearman's rank correlation coefficient ρ, to capture how strong a linear relationship exists in a bivariate data object.

Regressional Features (RF). Recently proposed [25] for indexing bivariate data. Based on the goodness-of-fit of data to several, predefined functional models.

Smoothing Splines (Spline). A feature extraction technique based on nonparametric fitting of a smoothing spline to the data [27,13]. The spline's coefficients in pp-form describe the data.

Discrete Fourier Transform (DFT).[5] A technique from signal processing to describe data by its first $k = 0, \ldots, M$ Fourier coefficients [1].

Piece-wise Aggregate Approximation (PAA)[6]. Describes sequential data by splitting a sequence of length n into m segments and compute the mean value of all data-points in each segment [29,15].

Symbolic Aggregate Approximation (SAX)[6]. A descriptor for time-series based on symbolic representation [17]. The time domain *and* the value domain of time-series data are discretized.

Kernel Density Estimation (KDE). Estimates the kernel function of the probability density of two-dimensional data as a Gaussian kernel [4]. This probability density function is used as a descriptor for the data.

Edge Histogram Descriptor (EHD). Prominent approach in image processing to describe the shapes seen in an image, by computing the distribution of the orientation of *edges* in that image [22]. Also included in the MPEG-7 standard.

4.2   Retrieval Results

For our quantitative evaluation of effectiveness, we compare retrieval performance for each of the nine considered feature extraction algorithms. In particular, we use a query-by-example, leave-one-out evaluation. This means that we use each object as a query and compute precision and recall for the ranking of all other objects in the data set. We compute r-precision (also known as firsttier precision or precision at r, see [2, section 3.2]), which is suitable since our similarity classes have a significantly different number of objects. To compute r-precision, we retrieve $k - 1$ objects from the data set for a given query, where k is the number of objects in the query's similarity class. Then the percentage of relevant objects within these $k - 1$ retrieved objects is the r-precision.

Figure 3 shows the boxplot of the r-precision for retrieval results obtained with each approach on the entire test data set. We see that average r-precision is between 1% and 7%, and that retrieval for several classes does not work at all (r-precision of zero). The difference in r-precision between the techniques is significant nonetheless, as an algorithm that randomly retrieves data objects

---

[5] Can be applied to bivariate data by sorting the data along either dimension to get a sequential representation.

Table 2. Overview of the considered feature extraction techniques for bivariate data. Average feature extraction time t_{sec} (lower is better) and average r-precision results μ_r (higher is better) indicate performance on the proposed benchmark. The low r-precision of randomized retrieval shows significance of changes in r-precision between the obtained results.

Descriptor	Abbr.	Dim	t_{sec}	μ_r (%)
Regressional Features	RF	43	2.43	2.5
Smoothing Splines	SPLINE	996	0.115	2.08
Discrete Fourier Transform	DFT	100	$0.1 \cdot 10^{-3}$	1.1
Piecewise Aggregate Approx.	PAA	100	$0.2 \cdot 10^{-3}$	2.3
Symbolic Aggregate Approx.	SAX	100	0.002	3.08
Kernel Density Estimate	KDE	1024	0.07	5.73
Edge Histogram Descriptor	EHD	80	0.26	**6.56**
Correlation Descriptor	CORR	3	0.028	1.7
L_2 of Resampled Data	L_2	100	0.009	1.35
Random Retrieval	-	-	-	0.059

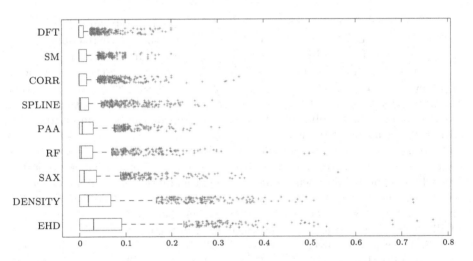

Fig. 3. Boxplot of average r-precision for each studied descriptor for the benchmark data set. The box visualizes the 95% confidence interval around the mean. The red, vertical bar indicates the median and the scattered plus-signs show outliers. Note that the difference in retrieval precision between the techniques is significant, as randomized retrieval only reaches 0.00059 r-precision.

only reaches an average r-precision of 0.059%. Particularly the image-based descriptors KDE (density) and EHD (shape) perform quite well. The only technique that performs below the two baseline techniques (CORR and SM) is the discrete Fourier transform based descriptor (DFT).

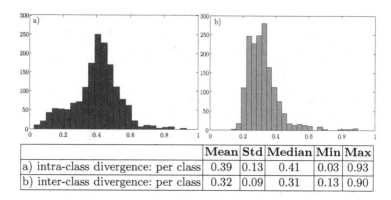

	Mean	Std	Median	Min	Max
a) intra-class divergence: per class	0.39	0.13	0.41	0.03	0.93
b) inter-class divergence: per class	0.32	0.09	0.31	0.13	0.90

Fig. 4. Intra-class (a) and inter-class (b) divergence based on individual data point distribution versus class data point distribution

5 Benchmark Discussion

In this work, we propose a new approach to define similarity among bivariate data objects, by using metadata annotations by experts in research data repositories. After assigning objects to similarity classes, an interesting questions is how numerically dissimilar objects within each class are (intra-class divergence) and how dissimilar objects among classes are (inter-class divergence). Given the motivation for the construction of this benchmark – evaluating feature extraction to retrieve similar objects – measuring these divergences computationally is difficult, as a similarity measure is required.

Looking at the 2D probability density of the bivariate data objects.w e compute intra-class divergence as the Euclidean distance between each data object's individual 2D probability density and the 2D probability density of all objects in the corresponding class. One can visualize this divergence as the difference of the scatter-plot density of a single data object versus the density of the scatter-plot of all bivariate data objects at once (visualized in Figure 2).

To judge inter-class divergence, we select all objects of two random classes and again compute the distance of the 2D probability density. This time, we compute the distance between the distribution of all the data points of these two random classes and the data point distribution of each individual object. By repeating this experiment until convergence for each class, we get an average distance of objects in different similarity classes – thus the inter-class divergence.

The results are presented in Figure 4. We see that on average intra-class divergence is similar to inter-class divergence, which explains the low average r-precision for the evaluated techniques.

We explore the r-precision of individual similarity classes and their respective inter- and intra-class divergence in detail. Figure 5 shows an overview of this relationship. As expected, we see that well performing classes (big, non-blue points) are primarily located in the upper left corner and thus exhibit a low intra-class divergence and a high inter-class divergence. However there are a few

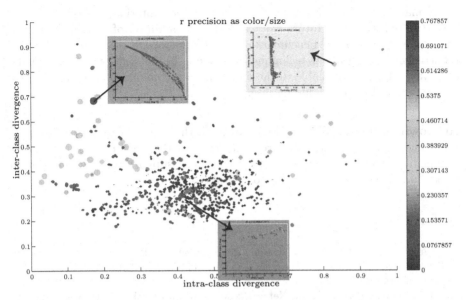

Fig. 5. Intra-class divergence plotted versus inter-class divergence for each similarity class in the benchmark. Color and size of data-points redundantly visualize mean average r-precision for each class. Data-point color in the small sub-images show class-membership.

well performing classes in the upper right corner. These classes exhibit a very high intra-class divergence (which makes retrieval difficult), but at the same a high (but not as high a) inter-class divergence. This indicates that for bivariate retrieval, numerical discrimination from other classes is more important for good retrieval results than numerical similarity within a class. nction, respectively. Retrieved results are shown for each technique detailed in the previous section in a side-by-side comparison.

6   Conclusion and Future Work

In this paper, we constructed a benchmark for retrieval in bivariate data collections, based on metadata annotations by domain experts. To the best of our knowledge, this is the first such benchmark for bivariate data retrieval. We used publicly available earth observation data for this purpose, by defining similarity classes based on available expert annotations. We verified that our definition of similarity classes – type, location and time of measurement – was meaningful, by computationally analyzing inter- and intra-class divergence. We exhaustively evaluated nine different feature extraction techniques for the task of retrieval in bivariate data collections on our benchmark, to give a tenable indication as to their respective retrieval performance. Results show that retrieval performance of all current techniques leaves lots of room for future improvements.

We make the benchmark and all reference implementations available[6] for future research. Our own future work includes the addition of new, more homogeneous datasets to the benchmark. Furthermore we are researching new and refined feature extraction techniques based on the obtained results, to make retrieval techniques available to the digital library community, which yield performance suitable for production-use.

Acknowledgments. We would like to thank the Alfred-Wegener-Institute for Polar and Marine Research in Bremerhaven, and the Center for Marine Environmental Sciences in Bremen for their continued support and collaboration in researching new access modalities to digital data libraries.

References

1. Agrawal, R., Faloutsos, C., Swami, A.: Efficient Similarity Search in Sequence Databases. In: Lomet, D.B. (ed.) FODO 1993. LNCS, vol. 730, pp. 69–84. Springer, Heidelberg (1993)
2. Baeza-Yates, R.A., Ribeiro-Neto, B.: Modern Information Retrieval. Addison-Wesley Longman Publishing Co., Inc., Boston (1999)
3. Bernard, J., Brase, J., Fellner, D.W., Koepler, O., Kohlhammer, J., Ruppert, T., Schreck, T., Sens, I.: A visual digital library approach for time-oriented scientific primary data. Int. J. on Digital Libraries 11(2), 111–123 (2010)
4. Botev, Z., Grotowski, J., Kroese, D.: Kernel density estimation via diffusion. Annals of Statistics 38(5), 2916–2957 (2010)
5. Cleveland, W.S.: The Elements of Graphing Data. Hobart Press (1985)
6. Datta, R., Joshi, D., Li, J., Wang, J.: Image retrieval: Ideas, influences, and trends of the new age. ACM Computing Surveys (CSUR) 40(2), 5 (2008)
7. Deselaers, T., Keysers, D., Ney, H.: Features for image retrieval: an experimental comparison. Information Retrieval 11(2), 77–107 (2008)
8. Diepenbroek, M., Grobe, H., Reinke, M., Schindler, U., Schlitzer, R., Sieger, R., Wefer, G.: Pangaea–an information system for environmental sciences. Computers & Geosciences 28(10), 1201–1210 (2002)
9. Ding, H., Trajcevski, G., Scheuermann, P., Wang, X., Keogh, E.: Querying and mining of time series data: experimental comparison of representations and distance measures. Proceedings of the VLDB Endowment 1(2), 1542–1552 (2008)
10. Dryad Digital Repository for Data Underlying Published Works, http://www.datadryad.org/
11. ELIXIR European Life Sciences Infrastructure for Biological Information, http://www.elixir-europe.org/
12. Frank, A., Asuncion, A.: UCI machine learning repository (2010), http://archive.ics.uci.edu/ml
13. Heckman, N., Ramsay, J.: Penalized regression with model-based penalties. Canadian Journal of Statistics 28(2), 241–258 (2000)
14. Hey, T., Tansley, S., Tolle, K. (eds.): The Fourth Paradigm: Data-Intensive Scientific Discovery. Microsoft Research, Redmond, Washington (2009), http://research.microsoft.com/en-us/collaboration/fourthparadigm/

---

[6] www.gris.informatik.tu-darmstadt.de/%7Emaschere/retrievalBenchmark/

15. Keogh, E., Chakrabarti, K., Pazzani, M., Mehrotra, S.: Dimensionality reduction for fast similarity search in large time series databases. Knowledge and Information Systems 3(3), 263–286 (2001)
16. Keogh, E., Kasetty, S.: On the need for time series data mining benchmarks: A survey and empirical demonstration. Data Mining and Knowledge Discovery 7(4), 349–371 (2003)
17. Keogh, E., Lin, J., Fu, A.: Hot sax: Efficiently finding the most unusual time series subsequence. In: IEEE International Conference on Data Mining, pp. 226–233 (2005)
18. Latecki, L.J., Lakämper, R., Eckhardt, U.: Shape descriptors for non-rigid shapes with a single closed contour. In: Proc. IEEE Conf. Computer Vision and Pattern Recognition, pp. 424–429 (2000)
19. Lew, M., Sebe, N., Djeraba, C., Jain, R.: Content-based multimedia information retrieval: State of the art and challenges. ACM Transactions on Multimedia Computing, Communications, and Applications (TOMCCAP) 2(1), 1–19 (2006)
20. Müller, H., March, S., Pun, T.: The truth about corel - evaluation in image retrieval. In: Proceedings of The Challenge of Image and Video Retrieval (CIVR), pp. 38–49 (2002)
21. PANGAEA Publishing Network for Geoscientific & Environmental Data, http://www.pangaea.de/
22. Park, D., Jeon, Y., Won, C.: Efficient use of local edge histogram descriptor. In: Proceedings of the 2000 ACM workshops on Multimedia, pp. 51–54. ACM (2000)
23. PsychData National Repository for Psychological Research Data, http://psychdata.zpid.de/
24. Rüger, S.M.: Multimedia Information Retrieval. Synthesis Lectures on Information Concepts, Retrieval, and Services. Morgan & Claypool Publishers (2009)
25. Scherer, M., Bernard, J., Schreck, T.: Retrieval and exploratory search in multivariate research data repositories using regressional features. In: Proceeding of the 11th Annual International ACM/IEEE Joint Conference on Digital Libraries, JCDL 2011, pp. 363–372. ACM, New York (2011)
26. Shilane, P., Min, P., Kazhdan, M., Funkhouser, T.: The princeton shape benchmark. In: Shape Modeling Applications, pp. 167–178. IEEE (2004)
27. Silverman, B.: Some aspects of the spline smoothing approach to non-parametric regression curve fitting. Journal of the Royal Statistical Society. Series B (Methodological) 47(1), 1–52 (1985)
28. Tobler, W.: A computer movie simulating urban growth in the detroit region. Economic Geography 46, 234–240 (1970)
29. Yi, B., Faloutsos, C.: Fast time sequence indexing for arbitrary Lp norms. In: Proceedings of the 26th International Conference on Very Large Data Bases, pp. 385–394 (2000)

Web Search Personalization Using Social Data

Dong Zhou[1,2], Séamus Lawless[2], and Vincent Wade[2]

[1] School of Computer Science and Engineering, Hunan University of Science and Technology,
Taoyuan Road, Xiangtan, Hunan, 411201, China
[2] Center for Next Generation Localisation, Knowledge and Date Engineering Group,
School of Computer Science and Statistics, Trinity College Dublin, Dublin 2, Ireland
dongzhou1979@hotmail.com,
{Seamus.Lawless,Vincent.Wade}@scss.tcd.ie

Abstract. Web search that utilizes social tagging data suffers from an extreme example of the vocabulary mismatch problem encountered in traditional Information Retrieval (IR). This is due to the personalized, unrestricted vocabulary that users choose to describe and tag each resource. Previous research has proposed the utilization of query expansion to deal with search in this rather complicated space. However, non-personalized approaches based on relevance feedback and personalized approaches based on co-occurrence statistics have only demonstrated limited improvements. This paper proposes an Iterative Personalized Query Expansion Algorithm for Web Search (*iPAW*), which is based on individual user profiles mined from the annotations and resources the user has marked. The method also incorporates a user model constructed from a co-occurrence matrix and from a Tag-Topic model where annotations and web documents are connected in a latent graph. The experimental results suggest that the proposed personalized query expansion method can produce better results than both the classical non-personalized search approach and other personalized query expansion methods. An "adaptivity factor" was further investigated to adjust the level of personalization.

Keywords: Personalized Web Search, Query Expansion, Social Data, Tag-Topic Model, Graph Algorithm.

1   Introduction

Over the past decade, the area of personalized web search has gained much attention in the literature [8, 12]. An important concern in personalized search systems is how to store and represent the gathered usage information. Some systems store this information in an individualized user model [18], while other systems maintain an aggregate view of usage information [1].

Personalization in search systems can be achieved by query adaptation, result adaptation, or both. Query adaptation attempts to expand (augment) the terms of the user's query with other terms, with the aim of retrieving more relevant results [8]. In terms of personalized query expansion, additional terms often come from individual user profiles to assist the user in formulating a better query.

P. Zaphiris et al. (Eds.): TPDL 2012, LNCS 7489, pp. 298–310, 2012.

Recent years have also witnessed the explosive growth of information on the World Wide Web (WWW). In social tagging systems such as del.icio.us[1], users are able to annotate each web resource with any number of free-form tags of their own choice. This type of system provides an ideal test bed for personalized search [17]. A user profile can be easily derived from their feedback, providing a good indication of the user's interests.

However, the uncontrolled manner of social tagging results in the use of an unrestricted vocabulary. This makes searching through the collection difficult and generally less accurate. In current social media systems, search algorithms also tend to be rather simplistic in nature, relying upon term matching methods, which often fail to deal with the vocabulary mismatch problem and result in poor ranking results.

Query expansion can partially solve the above mentioned problem. A classic technique is Pseudo relevance feedback (PRF) or local analysis [3]. This approach has been previously proven to work well. However, in the context of personalized search, the selected terms may be different from the users' true interests, so that the retrieved documents may not be relevant to a particular user. There have been few attempts at selecting the appropriate expansion terms from a user profile [4, 5, 6]. Past research appears to favor tag-tag relationships, by selecting the most related tags from a user's profile to enhance the source query. Given the fact that the tags might not be the precise descriptions of resources, the resulting retrieval performance has been markedly low [4, 5]. Borrowing from the traditional Information Retrieval (IR) field, local analysis and co-occurrence based user profile representation have been adopted to expand the query according to a user's interaction with the system [6, 8]. However, in this case the selection of expansion terms is solely based on lexical matching between the query and the terms which exist in the user profile. If the terms are not found in the user's profile, the query cannot be expanded at all.

In this work, an Iterative Personalized Query Expansion Algorithm for Web Search, called *iPAW,* is presented. This algorithm is based on individual user profiles. In the user profile, terms are modeled according to their relationships, which can be defined by co-occurrence statistics or defined by a tag-topic model, introduced below. Each term in the user profile will have an associated weighting score calculated based on its relationship with other terms in the profile and terms extracted from top-ranked documents. After calculation, the terms with the highest scores will be chosen to expand the original query. The intuition behind the model is the prior assumption of term consistency: *the most appropriate expansion terms for a query are likely to have the same weighting scores, be associated with, and influenced by terms extracted from the documents ranked highly for the initial query.* In other words, the selection of expansion terms for a given query is not solely based on lexical matching, but by iterative context enhancing and weight propagation. In addition, one important "adaptivity factor" was examined that could affect the expansion process.

---

[1] http://www.delicious.com

2   Related Work

Manual query expansion has been studied in early IR systems [11]. This approach demands user intervention and requires the user to be familiar with the search system, which is generally not true for the modern web. For these reasons, the overwhelming majority of search systems in existence today, function via automatic query expansion. One common technique employs a machine readable thesaurus to locate expansion terms in lists of synonyms [16]. Other approaches extract expansion terms from large collections of documents [13]. Local analysis involving relevance feedback is another popular category of approach. Explicit feedback is often difficult to obtain. An alternative method is implicit relevance feedback through pseudo relevance feedback (PRF) [3]. Web query logs are also used by researchers to bridge the gap between the user-centric query space and author-centric web page space [10]. However, in practice, acquiring web query logs is difficult for most researchers due to the various concerns of search companies.

Personalized web search has been extensively studied. There are approaches that utilize query log and click-through analysis [12]. There are systems that explore desktop data and external resources [8]. In personalized social media search, Personalization usually involves two general approaches. The first approach runs the unmodified original query for all users but re-ranks the returned results based on an individual user profile [7, 17]. Another group of work modifies or augments a user's original query, or query expansion. Researchers have frequently used co-occurring tags to enhance the source query [4, 5].

3   Personalized Query Expansion

3.1   Problem Definition

In social tagging systems such as del.icio.us, users can label interesting web pages with primarily short and unstructured *annotations* in natural language called *tags*. These web pages are denoted as a URL in the del.icio.us website. Textual content is crawled by following a URL that refers to a *document* or *web document*. Multimedia content is excluded in this research. In response to a *query*, an initial set of the most relevant documents is fetched. The top "c" ranked documents are assumed to be relevant, and therefore refer to the *top-ranked documents*. *Term* refers to a *word* in the vocabulary, these two terminologies are used interchangeably. Terms extracted from documents are specifically called *docTerm*, to be distinguished from general "terms" used in user profiles and from tags.

Formally, social tagging data can be represented by a tuple $\mathcal{P} := (\mathcal{U}, \mathcal{D}, \mathcal{T}, \mathcal{A})$, where $\mathcal{U}, \mathcal{D}, \mathcal{T}$ are finite sets of users, web documents and tags, and $\mathcal{A} \subseteq \mathcal{U} \times \mathcal{D} \times \mathcal{T}$ is a ternary relation, whose elements are called tag assignments or annotations. The set of annotations of a user is defined as: $\mathcal{A}_u := \{(t, d) | u, d, t \in \mathcal{A}\}$. The tag vocabulary of a user, is given as $\mathcal{T}_u := \{t | (t, d) \in \mathcal{A}_u\}$. The user's set of documents is $\mathcal{D}_u := \{d | (t, d) \in \mathcal{A}_u\}$. The docTerm vocabulary of a user is further defined to be

$docTerm_u := \{w|w \in \mathcal{D}_u\}$ where w denotes the words in the document corpus. $\mathcal{T}_u$ is the full list of tags that the user has used, and $docTerm_u$ is the vocabulary extracted from the documents that the user has tagged. So that terms in a user profile could be chosen from $\mathcal{T}_u$, $docTerm_u$, or $\mathcal{T}_u \cup docTerm_u$.

Given a source query q, a set of terms/words in the user profile $\{w_1, w_2 \dots w_n\}$, and a set of initial top-ranked documents $\mathcal{D}^{top} = \{d_1, d_2 \dots d_c\}$ the goal is to return a ranked list of profile terms to be added to the query, regularized by terms extracted from the top-ranked documents.

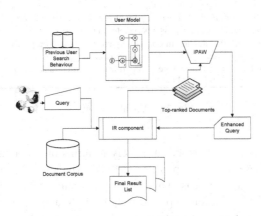

Fig. 1. Solution Architecture

3.2   Solution Overview

Figure 1 shows a sketch of the proposed solution architecture where personalized query expansion consists of two phases: (1) The construction of user models by using a co-occurrence matrix built for all the tags and documents in the training set or a tag-topic model where the tags and web documents are modeled simultaneously. This phase will be detailed in the next section. (2) The expansion of the given query by using the *iPAW* algorithm where the global term consistency is assumed over the word graph in addition to leveraging the top-ranked documents retrieved by the initial query. This process is described in the next section.

3.3   *iPAW* Algorithm

Let $G = (V, E)$ be a connected graph, where nodes V correspond to the n words in the user profile, and edges E correspond to the association strengths between words. An $n \times n$ symmetric weight matrix A on the edges of the graph is given, where a_{ij} denotes the weight between words w_i and w_j and M is a diagonal matrix with entries $M_{ii} = \sum_j a_{ij}$. A $n \times c$ matrix F is also defined with $F_{ij} = f(w, d)$ if a word w is presented in a document d and $F_{ij} = 0$ otherwise, where $f(w, d)$ denotes the weight of w in d.

Inspired by semi-supervised learning methods [19], here an iterative personalized query expansion algorithm for web search is developed. The algorithm is shown in Figure 2.

---

1.   Form the affinity matrix A, which is defined by co-occurrence statistics or relationships that are calculated by the tag-topic model proposed in the user model construction section.
2.   Construct the matrix $S = M^{-1/2}AM^{1/2}$.
3.   Iterate $F(t + 1) = \mu SF(t) + (1 - \mu)F^0$ until convergence, where F^0 is the initial weighting matrix obtained by terms extracted from the top-ranked documents (*tf-idf* weighting used in the current paper (Jones 1988)) and μ is a parameter in (0,1).
4.   Let F^* denote the limit of the sequence $\{F(t)\}$. Compute the final weighting scores for each word w as $w = \sum_{i=1}^{c} f(w, d_i)$, from which the top γ words can be acquired from the final ranked list of profile words to be added to the query.

Fig. 2. iPAW algorithm

The algorithm can be understood intuitively in terms of spreading activation networks [14]. A pairwise relationship is first defined, which can be viewed as the user model construction, and then it is symmetrically normalized, which is necessary for the convergence of the iteration step 3. During each iteration of step 3, terms in the user model will receive the information (weights) spreading from its neighbors, but also influenced by terms extracted from top-ranked documents (initial information). The parameter μ specifies the relative amount of the information from its neighbors and its initial weighting information.

Term consistency is assumed in the iteration phase. The first term of the iteration function in step 3 is the global consistency constraint, which means that a good weighting function should not change too much between nearby points. In this paper, nearby points are refined weighting scores with respect to initial relationships between words and context information (top-ranked documents) obtained by the initial query. They are likely to have the same effect over the graph. The second term is the fitting constraint, which means the weighting of words should fit the weighting scores of words extracted from the top-ranked documents retrieved by the given query.

After simplifying, a closed form solution can be derived as (see also [19, 20]):

$$F^* = (1 - \mu)(I - \mu S)^{-1}F^0$$

An important feature of such computation is that the weightings calculated here share similarities with the entries obtained in the F^* calculation in Zhou et al's paper [19], where they try to find the largest entry in order to get the corresponding label, while this approach attempts to find select terms with large added weights. The actual values of weights are not so critical as far as they have discriminative power to separate high potential words from low potential words.

4 User Model Construction

In order to capture accurate information for the construction of the user model, in this paper the tags and web documents are modeled simultaneously. Initially, a simple user model is defined using a co-occurrence matrix. A new model is then introduced where a latent graph is built among terms.

---

1. For each tag $t \in \mathcal{T}$, choose $\theta_t \sim Dirichlet(\alpha)$
   For each topic $o \in \mathcal{O}$, choose $\varphi_o \sim Dirichlet(\beta)$

2. For each document $d \in \mathcal{D}$
   Given the vector of tags t_d
   For each word w_i indexed by $i = 1, ..., N_d$
   Conditional on t_d choose an tag $x_i \sim Uniform(t_d)$
   Conditional on x_i choose a topic $z_i \sim Discrete(\theta_{x_i})$
   Conditional on z_i choose a word $w_i \sim Discrete(\varphi_{z_i})$

Fig. 3. Generative process of the Tag-Topic model

4.1 Co-occurrence Matrix

Firstly, a co-occurrence matrix was built according to [6, 8]. For all the tags and documents in the training set, important terms (or keywords) with high *tf-idf* scores (20% was used in the experiment described below) were selected. The cosine similarity between two words w_i and w_j was then calculated using:

$$cos(w_i, w_j) = DF_{w_i, w_j} / \sqrt{DF_{w_i} \cdot DF_{w_j}}$$

Where DF_{w_i} is the document frequency of word w_i. Note that here tags and doc-Terms are modeled into the matrix together. Graph G is built in which the nodes denote the terms and the edges E are weighted by their co-occurrence similariy.

4.2 A Tag-Topic Model

In the current paper, an Author-Topic model, introduced by Steyvers et al. [15], was adopted and a Tag-Topic model was proposed in order to learn topic-word and tag-topic distributions from the annotation data in an unsupervised manner. Then a latent graph is built based upon the features derived. This can be achieved through important docTerms, tags or a mixture of both.

To run the original Author-Topic model on the social tagging data at an individual user level, the tags can be viewed as authors in the new proposed model. When generating a document, a tag is chosen at random for each individual word in the document. This tag picks a topic from its multinomial distribution over topics, and then samples a word from the multinomial distribution over words associated with that topic. This process is repeated for all words in the document. This process is summarized in Figure 3.

In the figure, θ and φ denote topic-word distributions and tag-topic distributions respectively, while α and β denote Dirichlet priors. From the count matrices obtained during the modeling process [15], θ and φ can be easily estimated. The algorithm assigns words to random topics and tags (from the set of tags annotated to the document), and then repeats the Gibbs sampling process to update topic assignments for several iterations.

After the topic-word distributions and tag-topic distributions have been obtained, the adjacency graph of word associations and tag associations can be constructed. To illustrate how the model could be used in this respect, taking tags as an example, the distance between tags t_i and t_j was defined as the symmetric *KL divergence* between the topic distributions conditioned on each of the tags:

$$symKL(t_i, t_j) = \sum_{o=1}^{O} \left[\theta_{io} \log \frac{\theta_{io}}{\theta_{jo}} + \theta_{jo} \log \frac{\theta_{jo}}{\theta_{io}} \right]$$

Similarly we can compute associations between words.

So a latent graph G is defined using the latent feature obtained from the tag-topic model, where the nodes denote the terms and the edges E are weighted by $symKL$. After normalization, matrix S can be calculated. This process is executed offline, and then matrix S is saved for the query expansion model. Since terms extracted from the documents are not equally important, only the top δ terms from each topic are retained to form the graph. Tags are usually regarded as high quality descriptors of the web pages' topics and a good indicator of web users' interests, so they are all retained in the user profile. To compare the use of docTerms extracted from documents and tags assigned to the documents for query expansion, three sets of user profiles have been defined: selected docTerms, tags and a mixture of both.

5   Empirical Evaluation

5.1   Experimental Setup

In order to evaluate the methods on real-world data a crawl was conducted on the popular social tagging site delicious during 2010. A total of 5,943 users, 1,190,936 web pages and 283,339 tags were obtained. The average number of tags per user and pages per user in the corpus are 47.68 and 200.39, respectively. Users with less than 10 personal tags were removed from the sample.

Four groups of users were created according to the number of bookmarks (less than 50, 50-100, 100-500 and more than 500) associated with the users (see [17]). 50 randomly selected users from each group together with their tagging records were extracted to form a total collection of 200 test users. The English terms were processed in the usual way, i.e. down-casing the alphabetic characters, removing the stop words and stemming words using the Porter stemmer. All the pre-processed web pages were used in the experiments as the document corpus. For each user, 75% of his/her tags with annotated web pages were used to create the user profile and the other 25% were used as a test collection.

The evaluation method used by previous researchers in personalized social search [7, 17] is employed. The main assumption is as follows: Any documents tagged by u with t are considered relevant for the personalized query (u, t) (u submits the query t).

The following evaluation metrics were chosen to measure the effectiveness of the various approaches: the precision of the top 5 documents (P@5), mean reciprocal rank (MRR), mean average precision (MAP) and the recall of the top 5 documents (R@5). The first three measurements are commonly used to evaluate search algorithms while the last one is useful for evaluating query expansion systems as this method has been shown to improve both recall and precision in the past. Statistically-significant differences in performance were determined using a paired t-test at a confidence level of 95%.

Table 1. Overall Results

	MAP	MRR	P@5	R@5
BM25	0.0354	0.0411	0.0128	0.0483
BM25Prf	0.0278	0.034	0.0113	0.0398
LexMatch	0.0341	0.0432	0.0129	0.0471
iCo	0.042	0.0498	0.015	0.0533
iTerms	0.0413†*	0.0495†*	0.0155†*	0.0543†*
iTags	0.038*	0.0456†*	0.0138†*	0.0506†*
iMix	0.0448†*pl	0.053†* pl	0.0158†* pl	0.0574†* pl

5.2 Experimental Runs and Parameter Tuning

In order to usefully evaluate the performance of the personalized query expansion framework 2 different non-personalized baselines were selected: *BM25* – a popular and quite robust probabilistic retrieval method, and *BM25Prf* – a pseudo-relevance feedback oriented query expansion method based on the Divergence from Randomness theory [2]. In addition to non-personalized baselines, we also have several personalized baselines. Firstly, a method was adopted which processes the user model using lexical matching between query terms and terms that exist in the user model (constructed by using co-occurrence matrix) [8]. This method is denoted as *LexMatch*.

For our proposed approach, the method using *iPAW* and co-occurrence matrix as its user model representation is denoted as *iCo* (this can be viewed as an upper baseline). Finally, there are three variant approaches which use *iPAW* and the proposed tag-topic model. For those user models which only contain terms extracted from the documents, the algorithm is denoted as *iTerms*, for those user models which only contain tags, it is denoted as *iTags*. *iMix* is used to represent the method, which utilizes user models that contain a mixture of terms and tags.

For the tag-topic modeling, O and δ were set to 5 and 20 empirically. In the expansion framework, the optimal values were obtained when $\mu = 0.9$ and $\gamma = 10$ for terms (used in *LexMatch* as well) and $\gamma = 1$ for tags. The number of top documents $\mathcal{D}^{top}$ used in the query expansion framework was set to 10. The parameters for *BM25Prf* were also set to 10 documents and 10 terms.

5.3   Personalization vs Non-personalization

Firstly, we examine the experimental results that describe the performance of the three personalized query expansion runs (*iPAW* with the tag-topic model) proposed in this paper together with two non-personalized baselines on the overall test users, which are shown in Table 1. The statistically significant differences are marked as †️ w.r.t to the *BM25* baseline and * w.r.t to the *BM25Prf* baseline.

As illustrated by the results, the *BM25Prf* model was the lowest performer for all evaluation metrics. This result is not surprising because the evaluation described in this paper is based upon a personalized-approach rather than the non-personalized evaluation model normally employed in large evaluation campaigns. This further demonstrates that merely borrowing common techniques from traditional IR will not solve the personalized search problem. Pleasingly, the three personalized query expansion-based search models all outperform the simpler text retrieval model with the highest improvement of 28.95% (In terms of the *iMix* method with the MRR metric when compared to *BM25*), which is statistically significant. It should be noted that all three personalized query expansion methods provide an average improvement of 40.74% compared to the traditional query expansion method with the highest improvement at 61.15%.

There were noticeable improvements in retrieval effectiveness when using user profiles that consisted of terms and a mixture of terms and tags in query expansion, but a more modest increase for the user profiles consisting tags alone. This reinforces the earlier finding that using tags alone for expanding queries is not sufficient. Another exciting observation is that in many cases, the personalized query expansion methods can outperform the baselines for all the evaluation metrics, with statistically significant improvements in almost entire runs.

5.4   Personalization via Lexical Matching and Co-occurrence statistics

The goal of the second set of experiments was to evaluate the performance of personalized query expansion using lexical matching and co-occurrence statistics (*LexMatch*), in comparison with the *iCo*. The statistically significant differences in the table are marked as *l* w.r.t to the *LexMatch* baseline and *p* w.r.t to the *iCo* for the *iMix* method only.

As we can see from Table 1, query expansion solely based on co-occurrence statistics and lexical matching is unsatisfactory. Although the performance is better in terms of MRR and P@5 metrics when compared to the non-personalized baseline, however, in the MAP metric the performance is even lower. After examining the expanded terms in the *LexMatch* model, it was found that because of the nature of social tagging systems, many tags are freely chosen and different from the terms stored in the user models. This results in a large number of queries being left un-expanded. Furthermore, the expanded terms sometimes show noise, resulting in lower performance than the *BM25* baseline.

However, using the same co-occurrence matrix as in *LexMatch*, the *iCo* method works much better, with performance just slightly lower than the *iMix* method. In some metrics it appears to work better than *iTerms*. This shows the power of using pseudo-relevance feedback documents to enhance the word graphs. Also the effectiveness of using the Tag-Topic model is also empirically confirmed (in terms of *iMix* which works better than *iCo*). It should be noted that the improvements achieved are statistically significant.

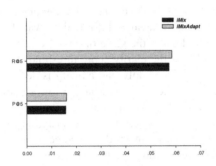

Fig. 4. Adaptivity Factor

5.5   Groups and Recall Results

A comparison of the results was also conducted across all the user groups. It is worth mentioning that there are improvements for users with low levels of activity. This demonstrates that the method is very effective even for users who do not have much available data in online social tagging systems.

In addition to the precision-based measurements, the personalized query expansion methods also showed significant improvements w.r.t the recall-based metric R@5. These improvements are on a similar scale when compared to the non-personalized and various personalized baselines. This reveals the benefits of adopting this method even in situations where recall is important.

6   Adaptivity Factor

In the previous section personalized query expansion was applied to expand all queries for each user. However, an optimal personalized query expansion algorithm should automatically adapt itself to various aspects of each query by leaving non-ambiguous queries un-expanded. This section examines one important factor that could affect the expansion process. Initial experimental results when using the factor, namely query clarity, are also presented.

It has been long known that the success of IR systems clearly varies across different topics. Therefore, a widely accepted method was adopted, called query clarity [9], to automatically tweak the amount of personalization fed into the algorithm. It measures the divergence between the language model associated with the user query and the language model associated with the collection. A simplified version was employed here [8] and it is defined as:

$$QC = \sum_{w \in q} P_{ml}(w|q) \cdot log \frac{P_{ml}(w|q)}{P_{coll}(w)}$$

where $P_{ml}(w|q)$ is the probability of the word w within the submitted query, and $P_{coll}(w)$ is the probability of w in the entire collection of documents.

In general, when there is no relevance information available for queries, Cronen-Townsend et al. [9] proposed using the scale of possible clarity scores for the collection at hand to heuristically set the clarity score threshold. In this research, the threshold is set heuristically to 90%. Simply put, a query is deemed "clear enough" if an estimated 90% or more of all queries would have a lower clarity score. If a query is defined to be clear, it is left unaltered in the retrieval process.

The same experimental setup was used as previously, however, only the *iMix* method is evaluated (where adaptation method is denoted as *iMixAdapt*), with P@5 and R@5. It can be seen from the results (Figure 4), there is more significant improvement in terms of recall than precision. In fact, the difference between the two methods in terms of R@5 is statistically significant. The major reason for this difference appears to be the arbitrary selection of queries to be expanded.

7   Conclusion

In this paper an Iterative Personalized Query Expansion Algorithm for Web Search, called *iPAW,* was described which is based on individual user profiles mined from the annotations and resources the users bookmarked. The intuition behind the model is the prior assumption of term consistency. A tag-topic model for the construction of user models was also introduced which simultaneously integrates the annotations and web documents through a statistical model in a latent space graph. The proposed personalized technique performed well on the social data crawled from the web, delivering statistically significant improvements over non-personalized and personalized representative baseline systems. An adaptivity factor was applied to the proposed algorithm and also demonstrated improvements in a separate set of experiments.

Acknowledgements. This research is supported by the Science Foundation Ireland (Grant 07/CE/I1142) as part of the Centre for Next Generation Localisation (www.cngl.ie) at Trinity College Dublin.

References

1. Agichtein, E., Brill, E., Dumais, S.: Improving web search ranking by incorporating user behavior information. In: Proceedings of the 29th Annual International ACM SIGIR Conference on Research and Development in Information Retrieval, pp. 19–26. ACM, Seattle (2006)
2. Amati, G., Rijsbergen, C.J.V.: Probabilistic models of information retrieval based on measuring the divergence from randomness. ACM Trans. Inf. Syst. 20(4), 357–389 (2002)
3. Baeza-Yates, R.A., Ribeiro-Neto, B.: Modern Information Retrieval. Addison-Wesley Longman Publishing Co., Inc. (1999)
4. Bender, M., Crecelius, T., Kacimi, M., Michel, S., Neumann, T., Parreira, J.X., Schenkel, R., Weikum, G.: Exploiting social relations for query expansion and result ranking. In: IEEE 24th International Conference on Data Engineering Workshop, ICDEW 2008, April 7-12, pp. 501–506 (2008)
5. Bertier, M., Guerraoui, R., Leroy, V., Kermarrec, A.-M.: Toward personalized query expansion. In: Proceedings of the Second ACM EuroSys Workshop on Social Network Systems, pp. 7–12. ACM, Nuremberg (2009)
6. Biancalana, C., Micarelli, A.: Social Tagging in Query Expansion: A New Way for Personalized Web Search. In: Proceedings of the 2009 International Conference on Computational Science and Engineering, vol. 04, pp. 1060–1065. IEEE Computer Society (2009)
7. Carmel, D., Zwerdling, N., Guy, I., Ofek-Koifman, S., Har'el, N., Ronen, I., Uziel, E., Yogev, S., Chernov, S.: Personalized social search based on the user's social network. In: Proceeding of the 18th ACM Conference on Information and Knowledge Management, pp. 1227–1236. ACM, Hong Kong (2009)
8. Chirita, P.-A., Firan, C.S., Nejdl, W.: Personalized query expansion for the web. In: Proceedings of the 30th Annual International ACM SIGIR Conference on Research and Development in Information Retrieval, pp. 7–14. ACM, Amsterdam (2007)
9. Cronen-Townsend, S., Zhou, Y., Croft, W.B.: Predicting query performance. In: Proceedings of the 25th Annual International ACM SIGIR Conference on Research and Development in Information Retrieval, pp. 299–306. ACM, Tampere (2002)
10. Cui, H., Wen, J.-R., Nie, J.-Y., Ma, W.-Y.: Query Expansion by Mining User Logs. IEEE Trans. on Knowl. and Data Eng. 15(4), 829–839 (2003)
11. Harter, S.P.: Online information retrieval: concepts, principles, and techniques. Academic Press Professional, Inc. (1986)
12. Joachims, T., Granka, L., Pan, B., Hembrooke, H., Gay, G.: Accurately interpreting clickthrough data as implicit feedback. In: Proceedings of the 28th Annual International ACM SIGIR Conference on Research and Development in Information Retrieval, pp. 154–161. ACM, Salvador (2005)
13. Qiu, Y., Frei, H.-P.: Concept based query expansion. In: Proceedings of the 16th Annual International ACM SIGIR Conference on Research and Development in Information Retrieval, pp. 160–169. ACM, Pittsburgh (1993)
14. Shrager, J., Hogg, T., Huberman, B.A.: Observation of phase transitions in spreading activation network. Science 236, 1092–1093 (1987)
15. Steyvers, M., Smyth, P., Rosen-Zvi, M., Griffiths, T.: Probabilistic author-topic models for information discovery. In: Proceedings of the Tenth ACM SIGKDD International Conference on Knowledge Discovery and Data Mining, pp. 306–315. ACM, Seattle (2004)
16. Voorhees, E.M.: Query expansion using lexical-semantic relations. In: Proceedings of the 17th Annual International ACM SIGIR Conference on Research and Development in Information Retrieval, pp. 61–69. Springer-Verlag New York, Inc., Dublin (1994)

17. Xu, S., Bao, S., Fei, B., Su, Z., Yu, Y.: Exploring folksonomy for personalized search. In: Proceedings of the 31st Annual International ACM SIGIR Conference on Research and Development in Information Retrieval, pp. 155–162. ACM, Singapore (2008)
18. Zhang, Y., Koren, J.: Efficient bayesian hierarchical user modeling for recommendation system. In: Proceedings of the 30th Annual International ACM SIGIR Conference on Research and Development in Information Retrieval, pp. 47–54. ACM, Amsterdam (2007)
19. Zhou, D., Bousquet, O., Lal, T.N., Weston, J., Scholkopf, B.: Learning with local and global consistency. In: Advances in Neural Information Processing Systems, vol. 16, pp. 321–328 (2004)
20. Zhu, X., Ghahramani, Z., Lafferty, J.: Semi-supervised learning using Gaussian fields and harmonic functions. In: The 20th International Conference on Machine Learning (ICML 2003), pp. 912–919 (2003)

A Model for Searching Musical Scores by Instrumentation

Michel Beigbeder

École Nationale Supérieure des Mines de Saint-Étienne
michel.beigbeder@emse.fr

Abstract. We propose here a preliminary study on the definition of
a search model that allows to look for musical scores that exactly or
approximately match a query where the query defines the exact instrumentation wanted by the user. We define two versions of approximate
matchings. In the first one, the ranking is done with a crisp matching
of the instruments. In the second one we relax this constraint and we
use a similarity between instruments. We present a first experiment and
envision future works.

1 Introduction

There are many library or bookshop sites on the Web that propose an access to
musical scores. Typically the access is provided either by browsing or searching
by the composer, the arranger, the publisher or the title of the piece or any
conjunctive combination of these fields. This search facility is efficient if the user
is aware of some precise musical piece and is able to provide enough information
to retrieve the corresponding score and parts.

Our goal is different and we want to give the possibility to search the pieces
that a given ensemble is able to play in terms of the instrumentalists that belong
to the ensemble. This would be particularly useful for nonprofessional ensembles
or for pupil groups in music schools where the composition of the ensemble is
not much under control. So the key point in our work is to match the description
of an ensemble against the descriptions of the pieces in terms of the number of
parts and their instruments.

2 Related Works

Music information retrieval is a large domain because music has many aspects
of interest for the searcher. Downie quotes the following facets: Pitch, Temporal,
Harmonic, Timbral, Editorial, Textual, Bibliographic[1]. He defends that the instrumentation[1] information is part of the Timbral facet but that it is sometimes
considered as part of the Bibliographic facet with an enumeration of the instruments used in a piece. We adopt here the latter point of view. The Downie's
survey paper does not quote any work on retrieval by instrumentation.

---

[1] The author uses the term *orchestration* in place of *instrumentation*.

P. Zaphiris et al. (Eds.): TPDL 2012, LNCS 7489, pp. 311–316, 2012.

The paper by Ferrara et al. is focused on genre classification of musical pieces and it defines four facets to describe them: Ensemble, Rhythm, Harmony, and Melody[2]. The Ensemble facet is the one that is related to the instrumentation of a piece, and it is described in terms of performers rather than in terms of parts in the score. In their semantic point of view, they want to transform a list of instruments in more general and compact information. The trivial example is to transform the list *2 Violins, 1 Viola, and 1 Cello* to *String Quartet*. In terms of retrieval they do not address the problem we tackle with, but they are interested in finding pieces that resemble to a given one and they argue that instrumentation is a useful hint for that purpose.

We do not know any publication that address our problem, so we looked at some Web sites to build a tentative state of the art on that matter. We present two commercial sites and a cooperative one that propose scores. Their functionnalities are representative of the best possibilities that we found on any other Web site.

2.1   Free Sites

Regarding instrumentation there is currently only a browsing facility in the site `http://imslp.org/`. This browsing facility is built over a taxonomy where the top level classes are "Keyboard", "Chamber-Instrumental", "Orchestral", "Vocal" and "Featured Instruments". The second level for "Chamber-Instrumental" is mainly composed of the number of instruments (with or without continuo). The second level of the "Orchestral" class distinguishes combinations of i) "orchestra" or "strings", ii) with (or without) soloist, iii) with (or without) continuo. There are also subclasses concerning "toy instruments". It is very difficult to search pieces for a given ensemble. For instance, there are 181 categories of "Chamber-Instrumental" "For 5 players". Moreover in this site it seems that there is not a clean description of the instrumentation, but some "categories" are assigned to each piece just like categories are assigned to articles in Wikipedia.

2.2   Commercial Sites

The site `http://www.di-arezzo.com/` is the site of a bookshop specialized in selling musical scores. Here one can specify that the scores have to contain parts for at least some instruments[2], and the count of these instruments can be specified. For instance, one can search a piece that needs two flutes, two violins, one viola and one violoncello. At the time of writing, this query returns four pieces. The first one needs two clarinets and two pianos too. The second one needs a keyboard in addition. The third one needs two oboes and two more violins and one more violoncello. The last one needs "13 players" whose list can be found on the page of this item. So we can suppose that none of them is a relevant answer for the given ensemble.

---

[2] More precisely, it is possible to specify at most four pairs of one instrument with a count. So it is not possible to add a trumpet, for instance, to the subsequent example.

If browsing is used, selecting one instrument lists the scores that needs at least one instance of this instrument. This it not very useful, for instance browsing for Flute returns 13212 items mixed with flute solos, many flute and piano pieces, concertos for flute and orchestra, etc.

The search possibility let us infer that the exact instrumentation is stored in the database underlying this site.

On the site `http://www.ewh.dk/` one can specity quite precisely an intrumentation which is developped as a list of instruments with a count. Moreover there are some shortcuts as for instance, "str4tet" (the short for String Quartet) which is developped as the list: 1 "Violin 1", 1 "Violin 2", 1 "Viola", 1 "Cello". Given this list four matchings are proposed:

- "exact match",
- "approximate match", which returns scores with at most one more instrument, or those where one instrument is not used, or those where one intrument is replaced by another one,
- "any combination", which matches all scores whose instrumentation is included in that of the query,
- "any size match", where any piece whose intrumentation is larger matches the query; this is the behavior of the `di-arezzo.com`'s search.

3   Modelization

In terms of mathematical modelling, the first and most simple approach is to model the instrumentation of a musical piece as a function from the set I of instruments to the set $\mathbb{N}$ of positive integers. We will call *multiplicity* this function. In fact the mathematical notion of *multiset* corresponds to that and we will use this vocabulary. So to describe a *string quartet* the multiplicity function maps *Violin* to 2, *Viola* to 1, *Cello* to 1, and any other instrument to 0. The same simple model is applicable to the ensembles: the composition of an ensemble is a multiset of instruments.

We distinguish two kinds of approximate matching. The first one concerns the number of parts and the number of instrumentists where each player has to play a part written for his instrument. The second one relax the latter constraint and will allow instrumentists to play parts written for another instrument. We will present a model for these two matchings in the following sections.

Cardinality Approximate Matching. The matchings used on the site `//www.ewh.dk/` are based on exact matching of instruments and on relationships between multisets that are analogous to those used in the boolean model of document retrieval. To refine the set theoretic point of view used on the site, we can adopt an usual method used in information retrieval to present the result with the most relevant ones in top of the list[3].

---

[3] *Relevance* as computed by the information retrieval system, which is not necessarily the human relevance.

To model the "any combination" match, the boolean answer is to return each piece p such that $p \subset e$. The larger is p (with $p \subset e$), the closer to the ensemble. So ranking the pieces with decreasing cardinality of p does make sense as the pieces will be ranked with those that employ the maximum of people of the ensemble ranked first. We can notice that the same ranking is obtained if the pieces are ranked with increasing cardinality of $e \setminus p$, and the latter multiset is equal to $e \Delta p$ as $p \subset e^4$.

Symmetrically, for the "any size match", pieces which verify $p \supset e$ should be ranked by increasing cardinality of p, which is also the increasing cardinality of $e \Delta p$ as $e \subset p$.

If both matchings are merged in a single one, ranking by the increasing cardinality of $e \Delta p$ will rank first 0) pieces which matches exactly the ensemble, then 1a) the pieces where one player don't play 1b) with those where one extra musician is needed, then pieces 2a) with two silent players, 2b) or two extra musicians, 2c) or one silent player and one extra musician, etc. Union of cases 0, 1a, 1b, 2c corresponds to the "approximate match" of site www.ewh.dk.

Instrument Approximate Matching. The second type of approximation is concerned when a part written for an instrument is played by another one. We introduce operators that can be used in a query, so a user might explicitly state her preferences.

We introduce now the OR operator as in the following example: flute OR oboe. There are at least three usages for this operator. The first one is for describing an ensemble where a player is able to play different instruments: for instance someone who can play the flute and the oboe. The second usage is a decision of the user for relaxing his query, probably because he did not find enough answers with a previous query. The last one is when the user knows that in a database two instruments are in fact the same, for instance flute and flauto. Such an operator would be trivially inserted in the previous modelling. But it also could be seen as a degenerated case of a similarity function in the instrument space, where the user explicitly says that the similarity between the oboe and the flute is one, and it is implied that all other similarities that are not explicitly mentioned in the query should be set to zero.

Finally, our most precise model considers as before a set of instruments I, and defines both ensembles and pieces as multisets of instruments. We now add a structure in the set of instrument with a symmetric similarity function $sim : I \times I \rightarrow [0,1]$ with the condition $(\forall x \in I)(\text{sim}(x,x) = 1)$. This similarity measures to what extent an instrument can replace another one. This function could be either user defined or system defined. We also introduce a pseudo-instrument ϵ which does not belong to I, a kind of *empty* instrument. We also extend the similarity function with $\text{sim}(i, \epsilon) = 0$ for each $i \in I$. This empty instrument will be useful in the mathematical formulation at the end of this section.

---

[4] The inclusion ($\subset$), the difference between sets ($\setminus$), the symmetric difference (Δ), and the cardinality should be understood in the multiset meaning.

Before giving the full formula for the similarity function between an ensemble e and a piece p, we will first present a formal example. Let us consider:

- an ensemble E with two instruments x and y, each with a multiplicity of 1;
- a similarity function with $\text{sim}(x,a) = x_a$ and $\text{sim}(y,a) = y_a$; $\text{sim}(x,b) = x_b$ and $\text{sim}(y,b) = y_b$;
- a piece P_1 with two instruments x and a;
- a piece P_2 with two instruments y and a;
- a piece P_3 with one instrument x;
- a piece P_4 with two instruments a and a.
- a piece P_5 with two instruments a and b.

The question is how to rank these five pieces for the ensemble E. With P_1, x matches exactly and y matches a with value y_a, so: $score(P_1, E) = 1 + y_a$. With P_1, y matches exactly and x matches a with value x_a, so: $score(P_2, E) = 1 + x_a$. With P_3, only x matches (and we can say that y matches ϵ with a null similarity), so: $score(P_3, E) = 1$. Finally, with P_4, x matches a with value x_a and y matches a with value y_a, so: $score(P_4, E) = x_a + y_a$. Thus in the final ranking, we will have P_1 and then P_2 or the reverse depending on the relative values of x_a and y_a. And then either P_3 or P_4 depending on the relative value of $x_a + y_a$ to 1. To compute the score of piece P_5 we can either affect part a to player x (and thus part b to player y) or the reverse *i.e.* part b to player x (and thus part a to player y). In the former case, the score would be $x_a + y_b$, and in the latter case the score would be $x_b + y_a$, so we can consider that the score of piece P_5 is the best score: $max(x_a + y_b, x_b + y_a)$.

We will now formalize for the general case. We have to consider every pair of one instrument in the ensemble with one instrument in the piece. To formalize that we artificially consider any numbering of the elements of the two multisets p and e. These numberings must have the same length. To do that the smallest multiset is filled with ϵ so as to have the same length as the largest one. After that their common length is $n = \max(|e|, |p|)$. For instance for $E = \{x, x, y\}$ and $P = \{a, b\}$, we have $n = 3$ and we can consider $e_1 = x, e_2 = x, e_3 = y$ and $p_1 = a, p_2 = b, p_3 = \epsilon$, but any other numbering of the elements would give the same result. Given these numberings, we define the score of the piece p for the ensemble e with:

$$score(p, e) = \max_{\sigma \in \mathcal{S}_n} \sum_{i=1}^{n} \text{sim}(e_i, p_{\sigma(i)})$$

where $\mathcal{S}_n$ is the set of all permutations of order n. With the same example we have to consider all the following affectations between parts and instrumentists: (x,a)(x,b)(y,ϵ), (x,a)(x,ϵ)(y,b), (x,b)(x,a)(y,ϵ), (x,b)(x,ϵ)(y,a), (x,ϵ)(x,a)(y,b), (x,ϵ)(x,b)(y,a).

4 Prototype

We have implemented a first version of the model with ranking on the cardinality of $p\Delta e$ as described in Sec. 3, *i.e.* without the generalization of approximation with a similarity function between instruments. For the collection of pieces we used the site `http://www.baroquemusic.it/`. This site is organized with a browsing by composer. We downloaded the pages of each composer and with simple regular expressions matching the HTML code we were able to extract the name of the pieces and their parts. The site provides 247 pieces. For instance, the very first piece lists the following five parts: "Flauto traverso", "Violino primo", "Violino secondo", "Viola", and "Cembalo". But sometimes the parts were not for a single instrument, for instance some parts (twenty exactly) are for "Violino I e II", and there are many other combination of instruments in the part names. So the main difficulty was to build the instrument sets for the pieces. We used ad'hoc rules to unify many variations of "Violino", "Violini", "Violino primo", "Violini ripieno", "Violons", etc. to a single instrument "Violino"; And, for instance, "Violino I e II" to a multiset with two "Violino".

The data extracted from the Web site were stored in a simple text file. A file containing the ad'hoc rules for converting part names to instrument multisets was manually built. When launching the query process, it first loads the rules and the list of pieces and builds the multisets, this is humanly instantaneous. Query processing considers each piece and compute the cardinality of the symmetric difference between the query and the pieces, and rank the pieces accordingly. With this number of pieces, this is also humanly instantaneous.

5   Future Work

The next step consists in the full implementation of the score formula between a piece and an ensemble for approximate matching. Particularly its complexity in terms of computation with a naive implementation would be in $O(n!)$, which is only usable for very small values of n. We will try a greedy algorithm which will first affect exact matches between an instrument and a part.

We also want to work on a larger database, and we think that `imslp.org` would be a good choice. Though it perhaps contains too many instrument/part names. An alternative choice would be a commercial site such as `di-arezzo.com` where the panel of scores is more limited and thus the instrumentations are controlled. We also think that the exact instrumentations are stored in their database and this would eliminate the problem of many names for the same instrument and also the splitting of part names to instrument names.

References

1. Downie, J.S.: Music information retrieval. Annual Review of Information Science and Technology 37, 295–340 (2003)
2. Ferrara, A., Ludovico, L.A., Montanelli, S., Castano, S., Haus, G.: A semantic web ontology for context-based classification and retrieval of music resources. ACM Trans. Multimedia Comput. Commun. Appl. 2, 177–198 (2006)

Extending Term Suggestion with Author Names

Philipp Schaer, Philipp Mayr, and Thomas Lüke

GESIS – Leibniz Institute for the Social Sciences,
Unter Sachsenhausen 6-8, 50667 Cologne, Germany
{philipp.schaer,philipp.mayr,thomas.lueke}@gesis.org

Abstract. Term suggestion or recommendation modules can help users to formulate their queries by mapping their personal vocabularies onto the specialized vocabulary of a digital library. While we examined actual user queries of the social sciences digital library Sowiport we could see that nearly one third of the users were explicitly looking for author names rather than terms. Common term recommenders neglect this fact. By picking up the idea of polyrepresentation we could show that in a standardized IR evaluation setting we can significantly increase the retrieval performances by adding topical-related author names to the query. This positive effect only appears when the query is additionally expanded with thesaurus terms. By just adding the author names to a query we often observe a query drift which results in worse results.

Keywords: Term Suggestion, Query Suggestion, Evaluation, Digital Libraries, Query Expansion, Polyrepresentation.

1 Introduction

When we look at specialized information systems like scientific digital libraries a long-known retrieval-immanent problem becomes clear: The same information need can be expressed in a variety of ways. This is especially true for scientific literature. Each scientific discipline has its own domain-specific language and vocabulary. For a long time indexers and digital library curators tried to encode this language into documentary tools like thesauri or classification systems, which are used to describe scientific documents. When we think of information retrieval as "fundamentally a linguistic process" like Blair [1] did, users have to be aware of these specialized documentation systems and their vocabularies. In order to get the most relevant results users should formulate their queries to best match these controlled thesauri terms. So-called search term recommenders (STR) try to map any search terms to controlled vocabulary terms [2] to support the user in the query formulation phase.

But terms from the title, abstract, full texts or controlled keywords are not the only entities that represent a digital library's document. Typical digital library systems incorporate many additional metadata representations of documents like author names, affiliation etc. These entities can be used during search but mostly in the form of free text search or during an "extended search". This pattern can be seen when studying the query formulations of actual users. We conducted a log file analysis with

P. Zaphiris et al. (Eds.): TPDL 2012, LNCS 7489, pp. 317–322, 2012.

Table 1. Query log analysis for the social science portal Sowiport. We analyzed the 1000 most popular user queries (n=129,251). Last line ("others") combines year, institution and location.

Search entities	Number of queries	% of total queries
Any Entity (all fields)	58,754	45,5%
Person(s)	40,979	31,7%
Keywords	26,959	20,9%
Title	1,108	0,9%
Source	581	0,4%
Others	354	0,3%

the social sciences digital library Sowiport[1]. Here we can see that in the 1000 most popular user queries nearly 1/3 of the users are explicitly searching for persons and only 1/5 are using controlled terms. Despite the fact that more users are looking for people rather than controlled terms, the included search term recommender in Sowiport (and the ones implemented in other DL systems) neglects this.

Therefore in this paper we argue for the explicit use of polyrepresentation of documents and different inter-document features to give a more complete term suggestion during query formulation. We present a general approach to obtain feasible author suggestions using simple co-occurrence analysis methods.

The paper is structured as follows: We give an overview on recent works in the field and the application of search term recommendation, expert finding and polyrepresentation in section 2. Section 3 presents an outline on how to make use of formerly unused metadata fields like author information in form of search term recommenders. This approach will be evaluated in section 4 using standard IR evaluation methods. We will conclude in section 5 with a general argumentation on why to develop the idea of polyrepresentation-aware recommenders further.

2   State of the Art in Digital Library Recommenders

In the following section we will give a very short overview on the field of term and author recommending systems in the domain of digital libraries and on the principle of polyrepresentation and its application.

2.1   Recommender in Digital Libraries

Recommending useful entities (e.g. terms) for search or browsing is a standard feature in typical search systems, and also in scientific DL. A couple of methodological different recommenders have been published in the DL domain. In the following we outline three typical examples extending the search process with additional document features or value-added services.

---

[1] http://www.gesis.org/sowiport/en/

Geyer-Schulz et al. [3] presented a recommender system based on log file analysis, which they implemented in a legacy library system. Their approach was based on the repeat-buying theory and could present related documents to a given document ("others also use..."), newer approaches extended their work [4]. Hienert et al. [5] presented another recommending approach, which focused on the mapping problem between specialized languages of discourse and documentary languages. General query terms and phrases were mapped onto a controlled vocabulary. Users could choose from these term suggestions during the query formulation phase to extend the initial user query. A third approach is called expert recommendation or expert search [6, 7]. The general idea is to support scientists in finding relevant literature from related peers. These recommended scholars don't necessarily have to be known to a searcher since it's a desired outcome to find new people's work.

Most of these systems allow the user to select some of their proposed entities (related documents, terms or author names) and most authors claim that their approaches help user in the search process, open alternative search paths and would optimize search systems performance. Nevertheless with few exceptions [8] these systems are not included in live retrieval systems (e.g. as query expansion mechanisms).

2.2   Polyrepresentation

The principle of polyrepresentation is described by Ingwersen and Järvelin [9]: "[...] the more interpretations of different cognitive and functional nature [...] that point to a set of objects in so-called cognitive overlaps, and the more intensely they do so, the higher the probability that such objects are relevant (pertinent, useful) to the information (need) situation at hand [...]".Cognitive differences in the representation derive from the interpretations from different actors, like keywords given by the author contrary to professional indexers keywords. A functional difference can – but does not have to – derive from the same actor and incorporate "contextual properties" like journal name, publication year, country, and others. Contrary to the principle of redundancy the principle of polyrepresentation focuses on the usage of cognitive and functional differences rather than to avoid them. It aims at a broad leverage in effectiveness by using and combining these differences compared to the single usage of each of the representations. Furthermore the idea is to surpass the drawbacks and weaknesses of some representations.

The general feasibility of this approach was shown in different studies like the one from Skov et. al [10]. They did query experiments using the Cystic Fibrosis data set extracted from Medline. In their work they transferred an initial query into different query representations/fields like title, abstract, MeSH major and minor controlled terms. While using a rather small document set with only 1,239 documents and 29 topics they could nevertheless show a significant improvement in retrieval effectiveness measured in precision, recall and cumulative gain. The approach from Skov is common in the domain of query term suggestion and query expansion: Query terms are expanded with semantically close terms from a thesaurus or other vocabulary. So a query on *What are the hepatic complications of cystic fibrosis?* is expanded with terms like *liver* or *hepatectomy*.

3   An Author-Aware Term Suggestion and Expansion Module

According to Larsen [11] the "polyrepresentation of the seeker's cognitive space is to be achieved by extracting a number of different representations from the seeker". He lists three representations or intentions of a seeker: a "what", a "why" and a work task or domain description. We argue that there are many other representations of a seeker's intention or information need. One of these might be a "who", meaning the potential desire to get results from a specific person or a person with reputation, knowledge or presence in the field. While a lot of work has been put into the exploitation of the functional polyrepresentation we argue that this cognitive site of a seeker's intention is ignored too often.

The general idea presented in this paper is to exploit the additional knowledge that is encoded into the documents' metadata and inter-document relationships, like author names belonging to a couple of relevant documents. In our approach we explicitly make use of the user's intention to get an answer for his "who" interest. We use co-occurrence analysis methods (Jaccard index) to compute a semantic distance between terms from title/abstract and the co-occurring authors and thesaurus terms in the whole document set. Using the Jaccard similarity measure we are able to compute the most similar authors and thesauri terms to a topic expressed through 1..n terms. In our social sciences setting, doing a query on *retirement* and *health* results in names like *Richard Hauser* or *Gerhard Bäcker*. Both of them are social scientists who did work related to social welfare and retirement topics. Additionally we get thesaurus terms like *social politics* or *elderly people*. We compute the top n=4 associated author names and thesaurus terms and expand the query with these recommendations. In addition we extend the search on different fields like title and abstract (TI, AB), controlled term (CT) and authors (AU).

The following example visualizes our approach (to keep the example readable we just add two names and terms respectively). An initial query on *Retirement and health issues*, would be transferred from a baseline query.

```
TI/AB = (retirement OR health)
```

to the extended query where italic terms are expanded thesaurus term and the last line originated from the list of co-occurring author names:

```
TI/AB = (retirement OR health OR "social politics" OR "elderly people")
OR CT = (retirement OR health OR "social politics" OR "elderly people")
OR AU = ("Richard Hauser" OR "Gerhard Bäcker")
```

4   Evaluation

In our evaluation setting we established four different queries, which then form TF*IDF ranked result lists using the Solr search engine. The simple result set B is

Table 2. Retrieval results of four different systems: baseline system B, baseline with term expansion (B+TE), baseline with author expansion (B+AE) and a combined system B+TA+TE. Statistical significant changes compared to B (measured using t-test with α = .05) are marked with *. Best values are marked with bold font.

run	MAP	rPrecision	p@10	p@20	p@100
B	0,139	0,182	0,442	0,353	0,172
B+TE	0,153	0,229 *	0,430	0,399	0,217 *
B+AE	0,122 *	0,184	0,400 *	0,350	0,175
B+TE+AE	**0,170** *	**0,239** *	**0,478**	**0,427** *	**0,218** *

formed by the unexpanded query generated from the topic description without stop words. This value is our baseline to which all other queries are compared. The second and third queries are the baseline combined with a thesaurus term expansion (B+TE) and author names respectively (B+AE). The fourth query is a combination of all three (B+TE+AE). In our experimental setup we used the GIRT4 corpus and the CLEF topics 76-125 to evaluate the performance of the proposed methods. We calculated MAP, rPrecision, p@10, p@20 and p@100 using the trec_eval toolkit.

The results of our evaluation are listed in table 2. The traditional approach B+TE, which adds controlled terms from a thesaurus to the query leads to a slight but not statistical significant increase in MAP (+9%) and a significant increase in rPrecision and p@100. B+AE performed significantly worse compared to B and B+TE, while rPrecision, p@20 and p@100 are comparable to the baseline. MAP (-12%) and p@10 are clearly below the values from B and B+TE. The combined method B+TE+AE performs best in all reported values. Generally speaking the B+AE approach in most cases was significantly worse than any other approach but combined with B+TE lead to significant gain in MAP, rPrecision, p@20 and p@100. When choosing B+TE as the common baseline the increase in B+TE+AE's MAP (+11%) is still significant.

5   Conclusion and Future Work

Looking back on the applications of the principle of polyrepresentation we see its use in the form of data fusion or ranking influential factors [10] rather than in the domain of query expansion and term suggestion. In this study we sketched out a term suggestion module that not only recommends controlled vocabulary terms but also the names of topical-related authors. We have shown that the explicit combined use of functional and cognitive polyrepresentational document features can increase the retrieval quality of a system. This is in line with traditional findings in the domain of query expansion but we could show that by using additional query entities that real users of a DL system were looking for, could bring a measurable benefit. This effect is only visible when the query is additionally expanded with thesaurus terms. A clear indicator for the strength of our polyrepresentational approach is the fact, that by only adding the author names we can observe a query drift which results in significantly worse results.

With our co-occurrence-based approach we can suggest other entities encoded in our digital library system. In this paper we focused on co-occurring and therefore potentially topic-related authors but in general other entities like journals, publisher, affiliation or other scientific community-related entities could be used to reformulate and enhance the original query. The open research questions would be: Are related journals, affiliations or any other DL's metadata capable of adding a positive retrieval effect or maybe a different view onto the data set? How can search effectiveness benefit from these co-occurrences? Does the polyrepresentational principle hold for these entities? While in this work we could only show that this principle holds for thesaurus terms (for a more detailed study on thesaurus terms see work by Lüke et al. in these proceedings) and authors names a more general evaluation and study on this has to remain as future work.

Acknowledgements. This work was partly funded by DFG, grant no. SU 647/5-2.

References

1. Blair, D.C.: Information retrieval and the philosophy of language. Annual Review of Information Science and Technology 37, 3–50 (2003)
2. Petras, V.: Translating Dialects in Search: Mapping between Specialized Languages of Discourse and Documentary Languages (2006),
   http://www.sims.berkeley.edu/~vivienp/diss/
3. Geyer-Schulz, A., Neumann, A., Thede, A.: Others Also Use: A Robust Recommender System for Scientific Libraries. In: Koch, T., Sølvberg, I.T. (eds.) ECDL 2003. LNCS, vol. 2769, pp. 113–125. Springer, Heidelberg (2003)
4. Nascimento, C., Laender, A.H.F., da Silva, A.S., Gonçalves, M.A.: A source independent framework for research paper recommendation. In: Proceedings of the 11th Annual International ACM/IEEE Joint Conference on Digital Libraries, pp. 297–306. ACM, New York (2011)
5. Hienert, D., Schaer, P., Schaible, J., Mayr, P.: A Novel Combined Term Suggestion Service for Domain-Specific Digital Libraries. In: Gradmann, S., Borri, F., Meghini, C., Schuldt, H. (eds.) TPDL 2011. LNCS, vol. 6966, pp. 192–203. Springer, Heidelberg (2011)
6. Gollapalli, S.D., Mitra, P., Giles, C.L.: Ranking authors in digital libraries. In: Proceedings of the 11th Annual International ACM/IEEE Joint Conference on Digital Libraries, pp. 251–254. ACM, New York (2011)
7. Heck, T., Hanraths, O., Stock, W.G.: Expert recommendation for knowledge management in academia. Proceedings of the American Society for Information Science and Technology 48, 1–4 (2011)
8. Mutschke, P., Mayr, P., Schaer, P., Sure, Y.: Science models as value-added services for scholarly information systems. Scientometrics 89, 349–364 (2011)
9. Ingwersen, P., Järvelin, K.: The turn: Integration of information seeking and retrieval in context. Springer, Dordrecht (2005)
10. Skov, M., Larsen, B., Ingwersen, P.: Inter and intra-document contexts applied in polyrepresentation for best match IR. Information Processing & Management 44, 1673–1683 (2008)
11. Larsen, B.: Practical Implications of Handling Multiple Contexts in the Principle of Polyrepresentation. In: Crestani, F., Ruthven, I. (eds.) CoLIS 2005. LNCS, vol. 3507, pp. 20–31. Springer, Heidelberg (2005)

Evaluating the Use of Clustering for Automatically Organising Digital Library Collections

Mark Hall[1,2], Paul Clough[2], and Mark Stevenson[1]

[1] Department for Computer Science, Sheffield University, Sheffield, UK
(m.mhall,r.m.stevenson)@sheffield.ac.uk
[2] Information School, Sheffield University, Sheffield, UK
p.d.clough@sheffield.ac.uk

Abstract. Large digital libraries have become available over the past years through digitisation and aggregation projects. These large collections present a challenge to the new user who wishes to discover what is available in the collections. Subject classification can help in this task, however in large collections it is frequently incomplete or inconsistent. Automatic clustering algorithms provide a solution to this, however the question remains whether they produce clusters that are sufficiently cohesive and distinct for them to be used in supporting discovery and exploration in digital libraries. In this paper we present a novel approach to investigating cluster cohesion that is based on identifying instruders in a cluster. The results from a human-subject experiment show that clustering algorithms produce clusters that are sufficiently cohesive to be used where no (consistent) manual classification exists.

1 Introduction

Large digital libraries have become available over the past years through digitisation and aggregation projects. These large collections present two challenges to the new user [22]. The first is resource discovery: finding the collection in the first place. The second is then discovering what items are present in the collection. In current systems, support for item discovery is mainly through the standard search paradigm [27], which is well suited for professional (or expert) users who are highly familiar with the collections, subject areas, and have specific search goals.

However, for the novice (or non-expert) user exploring, investigating, and learning [16,21] tend to be more useful search modalities. To support these modalities the items in the collection must be classified according to a relatively consistent schema, which is frequently not the case.

In many domains no standard classification system exists and, even if it does, collections are often classified inconsistently. Additionally, where collections have been formed through aggregation (e.g. in large-scale digital libraries) the items will frequently be classified using different and incompatible classification systems. Manual (re-)classification would be the ideal solution, however the time

P. Zaphiris et al. (Eds.): TPDL 2012, LNCS 7489, pp. 323–334, 2012.

and expense requirement when dealing with hundreds of thousands or millions of items means that it is not a viable approach.

Automatic clustering techniques provide a potential solution that can be applied to large-scale collections where manual classification is not feasible. The advantage of these techniques is that they automatically derive the cluster structure from the digital library's items. On the other hand the quality of the results can be variable and thus the choice of which clustering technique to employ is central to providing an improved exploration experience.

The research questions posed at the start of this work was: Do automatic clustering techniques produce clusters that are cohesive enough to be used to support the exploration of digital libraries? We define a cohesive cluster as one in which the items in the cluster are similar, while at the same time clearly distinguishable from items in other clusters. Our paper provides two major contributions in this area. Firstly, we propose a novel variant of the intruder detection task [6] that enables the measurement of the cohesion of automatically generated clusters. Secondly, we apply this task to evaluate the cluster model quality of a number of automatic clustering and topic modelling algorithms.

Our results show that the clusters are sufficiently good to be used in digital libraries where manually assigned classifications are not available or not consistent. The remainder of the paper is structured as follows: The next section provides background information on the use of clustering in digital libraries, their evaluation, and the clustering techniques evaluated in this paper. Section 3 describes the methodology used in the evaluation experiment and section 4 the experiment results. Section 5 concludes the paper.

2   Background

The issues large, aggregated digital libraries present to the user were first highlighted in [22] who suggested manual classification by the user and automated clustering as approaches for dealing with the large amounts of information provided by these digital libraries. Since then a number of digital library exploration interfaces based on clustering documents [26,9,8] and search results [11,28] have been proposed. Most of these approaches were evaluated in task-based scenarios and shown to improve task performance, however the cluster quality itself was not evaluated.

2.1   Cluster Evaluation Metrics

Cluster evaluation has traditionally focused on automatic evaluation metrics. They are frequently tested on synthetic or manually pre-classified data [17,1] or using statistical methods [29,13]. However, these do not necessarily capture whether the resulting clusters are cohesive from the user's perspective.

There have been attempts at using human judgments to quantify the cohesion of automatic clustering techniques. Mei et al. [18] evaluate the cohesion of Latent Dirichlet Allocation topics in the context of automatically labelling these topics.

The number of changes evaluators make to a clustering has also been used to judge cluster cohesion [24].

Chang et al. [6] devised the "intruder detection" task, where evaluators are shown the top five keywords for an LDA topic to which a keyword from a different topic is added. They then have to identify the added "intruder" keyword and the success at identifying the intruder is used as a proxy to evaluate the topic's cohesion. The more cohesive a topic, the more obvious it is which keyword is the intruder. The results of their work have been compared to a number of automatic similarity algorithms and Pointwise-Mutual-Information (PMI) was identified as a good predictor for the agreement between the evaluators [20].

2.2   Classification Models

This paper investigates three unsupervised clustering algorithms: Latent Dirichlet Allocation (LDA) [5], K-Means clustering [14], and OPTICS clustering [2]. Hierarchical and spectral clustering algorithms were also investigated, but not tested due to them either being too computationally complex for the data-set size or producing only degenerate clusterings.

Latent Dirichlet Allocation (LDA). is a state-of-the-art topic modelling algorithm, that creates a mapping between a set of topics T and a set of items I, where each item $i \in I$ is linked to one or more topics $t \in T$. Each item is input into LDA as a bag-of-words and then represented as a probabilistic mixture of topics. The LDA model consists of a multinomial distribution of items over topics where each topic is itself a multinomial distribution over words. The item-topic and topic-word distributions are learned simultaneously using collapsed Gibbs sampling based on the item - word distributions observed in the source collection [10]. LDA has been used to successfully improve result quality in Information Retrieval [3,30] tasks and is thus well suited to support exploration in digital libraries. Although LDA provides multiple topics per item, in this paper items will only be assigned to their highest-ranking topic and the topics will be referred to as clusters for consistency with the other algorithms.

K-Means Clustering. is a frequently used clustering method [31] that takes one input parameter k and assigns the n input items to k clusters and has been used in IR [12]. Items are assigned to the clusters in order to maximise the intra-cluster similarity while minimising the inter-cluster similarity. Cluster similarity is calculated relative to the cluster's mean value.

K-Means uses random initial cluster centres and then iteratively improves these by assigning items to the most similar cluster and moving the cluster centres to the mean of the items in the cluster.

OPTICS Clustering. is a density-based clustering algorithm that does not directly produce a cluster assignment, but instead provides an ordering for the items that can then be used to create clusters with arbitrary density thresholds. The algorithm defines a reachability value for each item which specifies the distance to the next item in the ordering. Large reachability values represent the boundaries between clusters and depending on what reachability threshold is chosen, a larger or smaller number of clusters is generated.

3 Methodology

In this paper we propose a novel version of the "intruder detection" task that evaluates the cohesion of the items in a cluster instead of just the cluster's keywords. To generate this "intruder detection" task, a cluster (the *main* cluster) is chosen at random from the clustering model and four items are chosen at random from the items allocated to that cluster. A second cluster, the *intruder* cluster, is also chosen at random and a random item chosen from that cluster as the *intruder* item. The five items, termed a *unit*, are then shown to the participants and they are asked to identify the *intruder* item. For each of the tested models the evaluation set consisted of 30 such *units*.

3.1   Source Data

The source data used in the experiments is a collection of 28,133 historical images with meta-data provided by the University of St Andrews Library [7]. The majority (around 85%) of images were taken before 1940 and span a range of 160 years (1839-1992). The images mainly cover the United Kingdom, however there are also images taken around the world. Most (89%) of photographs are black and white, although there are some colour photographs. Of the available meta-data fields only the the title, description, and manually annotated subject classification are used in the experiments. On average, each image is assigned to four categories (median=4; mean=4.17; $\sigma = 1.631$) and the items' title and description tend to be relatively short (word-count: median=23; mean=21.66; $\sigma = 9.5857$). Examples are shown in Figures 1 and 2.

The collection was chosen for a two main of reasons: first, the collection has a manually annotated subject classification that provides an evaluation baseline; second, the data provides a realistic test case, as it was taken from an existing library archive (enabling the generalisation of results to other digital libraries), is large enough to make manual classification time-consuming, and at the same time small enough that it can be processed in a reasonable time-frame.

3.2   Data Preparation

Each item's title and description were processed using the NLTK [15] to carry out sentence splitting and tokenization. The resulting bags-of-words are the input into the three clustering algorithms. All processing was performed on an Intel i7 @1.73 GHz with 8GB RAM. Processing times are shown in Tab. 1.

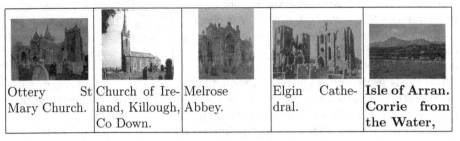

Ottery St Mary Church.	Church of Ireland, Killough, Co Down.	Melrose Abbey.	Elgin Cathedral.	**Isle of Arran. Corrie from the Water,**

Fig. 1. Example of a cohesive *unit* taken from the "*K*-Means TFIDF" model. The intruder is the last image, in **bold font**.

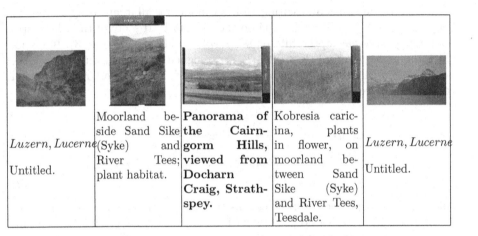

Luzern, Lucern Untitled.	Moorland beside Sand Sike (Syke) and River Tees; plant habitat.	**Panorama of the Cairngorm Hills, viewed from Docharn Craig, Strathspey.**	Kobresia caricina, plants in flower, on moorland between Sand Sike (Syke) and River Tees, Teesdale.	*Luzern, Lucern* Untitled.

Fig. 2. Example of a non-cohesive *unit* taken from the "*K*-Means TFIDF" model. The intruder is the middle image, in **bold font**.

Table 1. Processing time (wall-clock) for the tested clustering algorithms and initialisation parameters

Model	Wall-clock time
LDA 300 clusters	00:21:48
LDA 900 clusters	00:42:42
LDA + PMI 300 clusters	05:05:13
LDA + PMI 900 clusters	17:26:08
K-Means - TFIDF	09:37:40
K-Means - LDA	03:49:04
OPTICS - TFIDF	12:42:13
OPTICS - LDA	05:12:49

LDA. Two LDA-based clusterings were created using Gensim [23], one with 300 clusters ("LDA 300"), one with 900 clusters ("LDA 900"). The reason for testing two cluster numbers is that 300 clusters is in line with the number of clusters in other work using LDA [30]. At the same time our work on visualising clusters has hinted that clusters with around 30 items work best, which with 28000 items leads to 900 clusters. Although LDA provides a list of topics with probabilities for each item, the items are assigned only to their highest-ranking topic in order to maintain comparability with the clustering results.

Previous work [20] indicates that pointwise mutual information (PMI) acts as a good predictor of cluster cohesion. A modified cluster assignment model designed to increase the cohesiveness of the assigned clusters was developed. This approach is based on repeatedly creating LDA models and only selecting those clusters that have sufficiently high PMI scores.

The algorithm starts by creating an LDA model for all items using n clusters. The clusters are then filtered based on the median PMI score of their keywords $t_1 \ldots t_5$ (eq. 2), creating the filtered set T_g of "good" clusters (eq. 3). Items for which the highest ranked cluster $t \in T_g$ are assigned to that cluster. A new LDA model is then calculated using the items for which their highest ranked cluster $t \notin T_g$ using a reduced number of clusters $n - |T_g|$.

$$\text{pmi}(x, y) = \log \frac{p(x, y)}{p(x) p(y)} = \log \frac{p(x \mid y)}{p(x)} \tag{1}$$

$$\text{coh}(t) = \text{median}\{\text{pmi}(t_i, t_j) : i, j \in 1 \ldots 5 \land i \neq j\} \tag{2}$$

$$T_g = \{t \in T : coh(t) > 0\} \tag{3}$$

This process is repeated until either all items have been assigned to a cluster or the LDA model contains no clusters with a $coh(t) > 0$, in which case all items are assigned to their highest ranked cluster and the algorithm terminates. Two models using this algorithm with 300 and 900 clusters were created ("LDA + PMI 300" and "LDA + PMI 900").

K-Means. Two k-means classifications were produced, both with 900 clusters. The first used term-frequency / inverse-document-frequency (TFIDF) vectors, calculated from the items' bags-of-words, to define each item ("K-Means TFIDF"). As Tab. 1 shows the time required to create this model was very high, thus a faster k-means clustering was created using the item-topic probabilities from a 900-topic LDA model to define each item ("K-Means LDA").

OPTICS. The OPTICS clustering used the same input data as the k-means clustering, creating two models ("OPTICS TFIDF" and "OPTICS LDA"). The reachability threshold required to create a fixed set of clusters was automatically determined for both models using an unsupervised binary search algorithm.

Upper- and Lower-bound Data. An upper bound data-set was created based on the manually annotated subject classifications provided in the original meta-data. This classification has a total of 936 distinct clusters from which the 30 tested *units* were selected using the random algorithm described above.

To aid in the interpretation of the results a lower bound was determined statistically as the number of cohesive clusters where the binomial likelihood of seeing that number of correctly identified units out of 30 is less than 5%, resulting in a lower bound of 3 correctly identified *units*.

Control Data. To ensure that the participants took the task seriously and did not simply select an answer at random, a set of 10 control *units* were created. These were randomly selected from the manual subject classifications and then manually filtered to ensure that the intruder was as obvious as possible.

3.3 Experimental Set-Up

The experiment was constructed using an in-house crowdsourcing interface. In the experiment each *unit* was displayed as a list of five images with their captions, and five radioboxes that the participants used to choose the intruder. Participants were shown five *units* on one page, one of which was always a control *unit*. The four model *units* were randomly sampled from the full list of *units* (the model *units* and upper bound *units*). The sampling took into account the number of judgements already gathered for the *units* to ensure a relatively even distribution of judgments. The experiment was run using a population recruited from staff and students at our university.

A total of 821 people participated in the experiment, producing a total of 10,706 ratings. 121 participants answered less than half of the control questions they saw correctly and have thus been excluded from the analysis, reducing the number of ratings analysed to 8,840, with each *unit* rated between 21 and 30 times, with the median number of ratings at 27. The large variation is due to how the filtered participants' ratings were distributed, but has no impact on the results as the evaluation metric takes the number of samples into account.

3.4 Evaluating Cohesion

The human judgements were analysed to determine which *units* were judged to be cohesive. The metric used to determine cohesiveness is strict. A *unit* is judged to be *cohesive* if the correct intruder is chosen significantly more frequently than by chance and if the answer distribution is significantly different from a uniform distribution (Fig. 1). The first aspect is tested using a binomial distribution and testing whether the likelihood of seeing the observed number of correct intruder judgements relative to the total number of judgements for the *unit* by chance is less than 5%. This does not necessarily guarantee a cohesive cluster, as it does not take into account the distribution of the remaining answers. If these were evenly distributed, then even though the intruder was detected by a significant number of participants, the remaining participants were evenly split and thus the *unit* cannot be classified as cohesive. A χ^2-test was used to determine whether the answer distribution was significantly different ($p < 0.05$) from the uniform distribution. If both conditions hold then the *unit* and with it the *main* cluster the *unit* was derived from, are said to be cohesive.

In addition, a second metric was used to judge whether *units* were *borderline cohesive*. A *unit* (and its *main* cluster) are defined as *borderline cohesive* if the total number of judgements allocated to two of the five possible answers makes up more than 95% of all judgements for that *unit* and one of the two answers is the intruder. This covers the case where in addition to the intruder item there is a second item that could also be the intruder. The *main* cluster such a *unit* is derived from might not be ideal, but is probably acceptable to the user, especially as the evaluation will show that even manually created clusters can be non-cohesive. All remaining *units* are classified as "non-cohesive" (Fig. 2).

4   Results

Table 2 shows the number of "cohesive", "borderline" and "non-cohesive" clusters per model. The results clearly show that k-means clustering based on TFIDF produces the most cohesive clusters. LDA with a large number of clusters also works well. The impact of filtering by PMI seems to be negligible. OPTICS clustering does not work well on the tested data-set.

Table 2. Experiment results for the various clustering algorithms and initialisation parameters. "Cohesive" lists the number of clusters were the *intruder* was consistently identified, "borderline" the number of clusters with two potential intruders, and "non-cohesive" the number of clusters that are neither "cohesive" nor "borderline".

Model	Cohesive	Borderline	Non-Cohesive
Upper-bound	27	0	3
Lower-bound	3	0	27
LDA 300 clusters	15	6	9
LDA 900 clusters	20	4	6
LDA + PMI 300 clusters	16	4	10
LDA + PMI 900 clusters	21	2	7
K-Means - TFIDF	24	3	3
K-Means - LDA	20	0	10
OPTICS - TFIDF	14	2	14
OPTICS - LDA	16	0	14

Table 1 shows the time required to generate each of the clusterings. The pure LDA models are fastest, while LDA + PMI with 900 clusters is the slowest algorithm. OPTICS and k-means lie between these extremes. Using LDA item-topic distributions for item similarity is faster than using TFIDF vectors.

4.1   Discussion

All models show a clear improvement on the lower bound and can thus be said to provide at least some benefit if no manual classification is available. However, the OPTICS and "LDA 300" models achieve cohesion for only about 50% of the clusters. OPTICS is clearly not a good choice for the type of data tested, as it is either slower or less cohesive than the other techniques.

The upper bound achives a very high score (90% of clusters cohesive), however even here there are three *units* that were not cohesive and these were further investigated (Tab. 3). The non-cohesive *unit* #1 is mis-classified in the original data and the *intruder* item was from the same geographic area as the *main* cluster items, making it impossible to determine which is the *intruder*. That this was picked up by the experiment participants is a good indication that the "intruder detection" task can distinguish cohesive from non-cohesive clusters. Analysis of the other two non-cohesive *units* shows that for both neither the image nor the caption provide sufficient information to identify the *intruder*. However, when the classification label is known, then the *intruder* can be determined. What this implies is that as long as there is some kind of logical link between the items, a certain amount of variation between the items is acceptable to human classifiers.

Table 3. Subject labels attached to the non-cohesive upper-bound *units*

#	Main cluster	Intruder cluster
1	Renfrews all views	Isle of May all views
2	Garages - commercial	Colleges - technical
3	Soldiers	Belfries

"K-Means TFIDF" is clearly the best of the models, achieving cohesion for 80% of the *units* and including the 3 borderline cohesive *units* pushes the rate up to 90%, matching the manual classification. Post-hoc analysis of the three border line *units* shows that in all three cases the *main* cluster item also identified as an intruder is linked to the other *main* cluster items via the description text, which the participants did not see. This means that those three clusters should also be acceptable when the users have access to all of the items' meta-data. The drawback with k-means is the long processing time rquired to create the model. Using the LDA document-cluster distribution instead of TFIDF leads to a significant reduction in processing time, however the quality of the resulting model suffers and is lower than the pure LDA results. K-Means is thus only viable for smaller collections, although the exact limit depends on what optimisations [25] can be achieved through improved initialisation [4] or parallelisation [32].

Processing speed is the strength of the pure LDA models, with the best-performing "LDA 900" model faster than "K-Means TFIDF" by a factor of almost 14 (Tab. 1). It does not achieve quite as good classification results (66% of *units* cohesive), but for data-sets that are too large to be clustered using k-means it is a viable alternative. Of the four borderline *units* two were of similar type as the k-means borderline *units*, however in the other two the item from the *main* cluster identified as the intruder had no connection to the other *main* cluster items, leading to a total of 22 (73%) acceptable clusters.

The results of the LDA models also show the necessity for models where the individual clusters do not have too many items. The "LDA 300" and "LDA + PMI 300" models are significantly worse than their 900-cluster counterparts. This also validates the use of 900 clusters in the k-means and OPTICS models.

While [19,20] show that PMI is a good predictor for the inter-annotator agreement, using it to filter clusters shows only minimal improvement. For both the 300 and 900 cluster models a single additional cohesive *unit* is created. Additionally in the 900 cluster case, PMI reduces the number of borderline units. Considering the increase in processing time this is not a viable method.

5 Conclusion

Large digital libraries have become available over the past years through digitisation and aggregation projects. These large collections present a challenge to the new user who wishes to discover what is available in the collections. Supporting this discovery task would benefit from a consistent classification system, which is frequently not available. Manual re-classification is prohibitively expensive and time consuming. Automatic cluster models provide an alternative method for quickly generating classifications.

This paper investigated whether clustering algorithms can generate cohesive clusters, where a cohesive cluster is one in which the items in the cluster are similar, while at the same time being distinguishably different from items in other clusters. Latent Dirichlet Allocation (LDA), K-Means clustering, and OPTICS clustering were investigated. To enable the comparison we proposed a novel version of the "intruder detection" task, where the experiment participants have to identify an item taken from a cluster and inserted into a set of four items taken from a different cluster. The results show that this task provides a good measurement for the cohesion of cluster models and can successfully identify non-cohesive clusters and mis-classifications.

Using this evaluation metric we showed that k-means clustering on TFIDF vectors produces the highest number of cohesive clusters, but is computationally intensive and thus only viable for smaller collections. LDA-based models with large cluster numbers provide the best cohesion – processing time trade-off, allowing them to be applied to large digital libraries. We believe that both algorithms create models where a sufficiently large number of clusters are cohesive to allow them to be used where no (consistent) classification is available. We intend to investigate if post-processing the clusters can further improve cohesion.

Acknowledgements. The research leading to these results has received funding from the European Community's Seventh Framework Programme (FP7/2007-2013) under grant agreement n° 270082. We acknowledge the contribution of all project partners involved in PATHS (see: http://www.paths-project.eu).

References

1. Amigó, E., Gonzalo, J., Artiles, J., Verdejo, F.: A comparison of extrinsic clustering evaluation metrics based on formal constraints. Information Retrieval 12, 461–486 (2009), doi:10.1007/s10791-008-9066-8
2. Ankerst, M., Breunig, M.M., Kriegel, H.-P., Sander, J.: Optics: ordering points to identify the clustering structure. SIGMOD Rec. 28(2), 49–60 (1999)
3. Azzopardi, L., Girolami, M., van Rijsbergen, C.: Topic based language models for ad hoc information retrieval. In: Proceedings of the IEEE International Joint Conference on Neural Networks 2004, vol. 4, pp. 3281–3286 (July 2004)
4. Bahmani, B., Moseley, B., Vattani, A., Kumar, R., Vassilvitskii, S.: Scalable k-means++. In: VLDB 2012 (2012)
5. Blei, D.M., Griffiths, T., Jordan, M., Tenenbaum, J.: Hierarchical topic models and the nested chinese restaurant process. In: NIPS (2003)
6. Chang, J., Boyd-Graber, J., Wang, C., Gerrish, S., Blei, D.M.: Reading tea leaves: How humans interpret topic models. In: NIPS (2009)
7. Clough, P., Sanderson, M., Reid, N.: The eurovision st andrews collection of photographs. ACM SIGIR Forum 40(1), 21–30 (2006)
8. Eklund, P., Goodall, P., Wray, T.: Cluster-based navigation for a virtual museum. In: Adaptivity, Personalization and Fusion of Heterogeneous Information, RIAO 2010, Le Centre de Hautes Etudes Internationales d'Informatique Documentaire, Paris, France, France, pp. 211–212 (2010)
9. Granitzer, M., Kienreich, W., Sabol, V., Andrews, K., Klieber, W.: Evaluating a system for interactive exploration of large, hierarchically structured document repositories. In: IEEE Symposium on Information Visualization, INFOVIS 2004, pp. 127–134 (2004)
10. Griffiths, T., Steyvers, M.: Finding scientific topics. Proceedings of the National Academiy of Science 101, 5228–5235 (2004)
11. Handl, J., Meyer, B.: Improved Ant-Based Clustering and Sorting in a Document Retrieval Interface. In: Guervós, J.J.M., Adamidis, P.A., Beyer, H.-G., Fernández-Villacañas, J.-L., Schwefel, H.-P. (eds.) PPSN 2002. LNCS, vol. 2439, pp. 913–923. Springer, Heidelberg (2002)
12. Hassan-Montero, Y., Herrero-Solana, V.: Improving tag-clouds as visual information retrieval interfaces. In: Proceedings InfoSciT (2006)
13. He, J., Tan, A.-H., Tan, C.-L., Sun, S.-Y.: On quantitative evaluation of clustering systems. In: Information Retrieval and Clustering, pp. 105–133. Kluwer Academic Publishers (2003)
14. Lloyd, S.P.: Least square quantization in pcm. IEEE Transactions on Information Theory 28(2), 129–137 (1982)
15. Loper, E., Bird, S.: Nltk: the natural language toolkit. In: Proceedings of the ACL 2002 Workshop on Effective Tools and Methodologies for Teaching Natural Language Processing and Computational Linguistics, ETMTNLP 2002, vol. 1, pp. 63–70. Association for Computational Linguistics, Stroudsburg (2002)
16. Marchionini, G.: Exploratory search: From finding to understanding. Communications of the ACM 49(4), 41–46 (2006)
17. Maulik, U., Bandyopadhyay, S.: Performance evaluation of some clustering algorithms and validity indices. IEEE Transactions on Pattern Analysis and Machine Intelligence 24(12), 1650–1654 (2002)
18. Mei, X.S., Zhai, C.: Automatic labeling of multinomial topic models. In: Proceedings of KDD 2007, pp. 490–499 (2007)

19. Newman, D., Karimi, S., Cavedon, L.: External evaluation of topic models. In: Proceedings of teh 14th Australasian Document Computing Symposum, pp. 11–18 (2009)
20. Newman, D., Noh, Y., Talley, E., Karimi, S., Baldwin, T.: Evaluating topic models for digital libraries. In: JCDL 2010 (2010)
21. Pirolli, P.: Powers of 10: Modeling complex information-seeking systems at multiple scales. Computer 42(3), 33–40 (2009)
22. Rao, R., Pedersen, J.O., Hearst, M.A., Mackinlay, J.D., Card, S.K., Masinter, L., Halvorsen, P.-K., Robertson, G.C.: Rich interaction in the digital library. Commun. ACM 38(4), 29–39 (1995)
23. Řehůřek, R., Sojka, P.: Software Framework for Topic Modelling with Large Corpora. In: Proceedings of the LREC 2010 Workshop on New Challenges for NLP Frameworks, Valletta, Malta, pp. 45–50. ELRA (May 2010), http://is.muni.cz/publication/884893/en
24. Roussinov, D.G., Chen, H.: Document clustering for electronic meetings: an experimental comparison of two techniques. Decision Support Systems 27(1-2), 67–79 (1999)
25. Sculley, D.: Web-scale k-means clustering. In: WWW 2010 (2010)
26. Song, M.: Bibliomapper: a cluster-based information visualization technique. In: Proceedings of the Information Visualization, pp. 130–136 (1998)
27. Sutcliffe, A., Ennis, M.: Towards a cognitive theory of information retrieval. Interacting with Computers 10, 321–351 (1998)
28. van Ossenbruggen, J., Amin, A., Hardman, L., Hildebrand, M., van Assem, M., Omelayenko, B., Schreiber, G., Tordai, A., de Boer, V., Wielinga, B., Wielemaker, J., de Niet, M., Taekema, J., van Orsouw, M.-F., Teesing, A.: Searching and annotating virtual heritage collections with semantic-web technologies. In: Museums and the Web 2007 (2007)
29. Wallach, H.M., Murray, I., Salakhutdinov, R., Mimno, D.: Evaluation methods for topic models. In: Proceedings of the 26th International Conference on Machine Learning (2009)
30. Wei, X., Croft, W.B.: Lda-based document models for ad-hoc retrieval. In: Proceedings of the 29th Annual International ACM SIGIR Conference, SIGIR 2006, pp. 178–185. ACM, New York (2006)
31. Wu, X., Kumar, V., Ross Quinlan, J., Ghosh, J., Yang, Q., Motoda, H., McLachlan, G., Ng, A., Liu, B., Yu, P., Zhou, Z.-H., Steinbach, M., Hand, D., Steinberg, D.: Top 10 algorithms in data mining. Knowledge and Information Systems 14, 1–37 (2008), doi:10.1007/s10115-007-0114-2
32. Zhao, W., Ma, H., He, Q.: Parallel K-Means Clustering Based on MapReduce. In: Jaatun, M.G., Zhao, G., Rong, C. (eds.) CloudCom 2009. LNCS, vol. 5931, pp. 674–679. Springer, Heidelberg (2009)

A Unique Arrangement: Organizing Collections for Digital Libraries, Archives, and Repositories

Jeff Crow, Luis Francisco-Revilla, April Norris, Shilpa Shukla, and Ciaran B. Trace

School of Information, The University of Texas at Austin,
Austin TX, USA
{jcrow,revilla,anorris,shilpa,cbtrace}@ischool.utexas.edu

Abstract. Digital libraries increasingly host collections that are archival in nature, and contain digitized and born-digital materials. In order to preserve the evidentiary value of these materials, the collection organization must capture the general context and preserve the relationships among objects. *Archival processing* is a well-established method for organizing collections this way. However, the current archival workflow leads to artificial boundaries between materials and delays in getting digitized content online because physical and born-digital materials are processed independently, and digitized materials not at all. In response, this work explores the approach of processing materials in a digitized form using a large multi-touch table. This alternative workflow provides the first step towards integrating the archival processing of digital and physical materials, and can expedite the process of making the materials available online. However, this approach demands high quality digitization and requires that archivists perform additional tasks like matching multi-sided, multi-paged documents.

Keywords: Multi-touch, archival processing, digitized materials.

1   Introduction

The core mission of digital libraries is to facilitate the use of the information that they host. This mission is challenging because people create and use information in increasingly different ways. Furthermore, as technology evolves, digital objects increase in complexity and their file formats change. This creates dependencies on legacy hardware and software [1]. Consequently, researchers have been investigating how to provide long-term access [1], storage, and preservation [2, 3] of digital objects. Many of these solutions address the issue at object-level (e.g., creating smart digital objects that can automatically copy themselves [4]). However, focusing on the long-term use of these objects requires rethinking how collections are organized and managed as a whole.

Digital libraries and archives host innumerable digital objects of significant scientific, legal, economic, cultural, and historic value [1]. Currently, many digital libraries are built on the premise that these objects can be organized and made accessible as independent units. This makes sense when the items (e.g., books and journals) each

P. Zaphiris et al. (Eds.): TPDL 2012, LNCS 7489, pp. 335–344, 2012.

have a distinct internal cohesion and contain all the necessary information within them to be read, analyzed, and understood. However, digital libraries are increasingly incorporating digital objects (e.g., scientific data, and personal, organizational and government records) that are archival in nature. These archival records are distinguished by the fact that they are created as a by-product or instrument of some practical activity, and are set aside by their creator for future action or reference [5]. As such, these records constitute a "primary and privileged source of evidence about the activities and the actors involved in them" [6]. These archival records, though they can be read as individual units, lose much of their meaning as evidence when managed and accessed independently.

The unique nature of archival records has major implications for collection organization and system design. Records serve as evidence of the actions of the creating entity, and derive much of their meaning from the context in which they are created and filed. A key part of this context is the *archival bond* – the notion that a relationship exists between all records created as part of the same activity. Rather than treating records as standalone objects, archival thinking requires that the archival bond be maintained and preserved in order for records to retain their meaning and evidentiary nature [8]. While some digital library platforms support a certain level of grouping (e.g., volumes for journal issues), this is insufficient from an archival perspective.

Archival science addresses the requirement of preserving the evidential value of records through a well-established method for organizing and describing collections: *archival processing*. The activity of archival processing requires completing two steps: arrangement and description. *Archival arrangement* is the method for organizing the collection and involves establishing or re-establishing the original intellectual and physical order of records in a collection. In this iterative process the archivist is looking for clues of organization and order within the records and aggregations of records in order to restore or recreate the original filing system. *Archival description* is the "creation of an accurate representation of a unit of archival material by the process of capturing, collating, analyzing, and organizing information that serves to identify archival material and explain the context and records system(s) that produced it" [9]. Finding aids are similar to library catalogues in that they allow physical and intellectual management of the collection and facilitate user access to the collections. Supporting the activity of archival processing is crucial for digital libraries that aim to support activities such as scientific discovery and historical research.

In recent years, the need to reduce large backlogs of unprocessed collections has prompted a call for new ways of thinking about all aspects of this core activity [7]. Currently, digitized materials are not part of the processing workflow. Physical materials are processed before they are digitized. Born-digital materials are processed separately following a different methodology. In hybrid collections, this workflow can create an artificial boundary, potentially disrupting the archival bond.

This paper looks at the role of technology in supporting the activity of archival processing among practitioners as a key step in organizing groups of records before they become part of a digital library system. It focuses on recasting the workflow of archival practitioners by moving from a model where paper-based collections are processed first and digitized second, to one in which collections could be digitized first and then processed second such that all materials (physical and born-digital) can be considered together. Specifically, this paper introduces the Augmented Processing Table (APT) project. APT pioneers the use of surface computing devices for processing collections of digitized archival material. Just like a hybrid collection combines physical and digital materials, APT creates a space that allows for the processing of digitized materials in combination with born-digital material, integrating both modalities (paper and digital) in one workspace.

2 Interactive Surfaces

In addition to archival science the APT project is informed by previous work on interactive surfaces and tangible user interfaces (TUI). Interactive surfaces that support multi-touch interactions play an important role in a variety of settings where people are engaged in information intensive activities such as office work [10], disaster control management [11], and leisure activities [12, 13]. Recently, researchers have been investigating the use of interactive surfaces in complex information applications such as document review in legal cases [14] and collaborative search among co-located group members [15].

In the design of multi-touch interfaces, designers often draw from the physical world, whether it is through the use of metaphors to describe interaction or behavior, or using embodying aspects of the physical world that are thought to be relevant to human interaction [16]. Tangible User Interfaces (TUIs) go beyond the use of metaphors, utilizing physical objects to represent, display, and/or act as a physical control of the digital representations on the multi-touch surfaces. For example, PaperView uses pieces of plain paper that act as personal, location-aware, interactive screens [17]. Designing TUI interfaces can be difficult [13] because it is necessary to decide when to provide physical or digital interface elements, and to what degree digital elements should emulate real-world interactions. However, evaluations of TUI systems show that interfaces that rely on familiar objects (e.g., paper) provide predictable and straightforward interactions [17]. Furthermore, the approach of integrating digital material into established paper-centric processes such as literary criticism has been proven beneficial [18].

APT explores the use of interactive surfaces for archival processing and studies if archival processing is amenable to be conducted using digitized documents. Like Terrenghi et al. the APT project is interested in "understanding not only people's expectations and mental models about digital versus physical media, but also an understanding of the associated affordances for interaction in these different situations" [16].

3 Design

Previous research about the affordances of paper and digital media [16, 19, 20] indicate archival processing of digitized materials is viable, and interactive surfaces can ease this transition from paper to digital. Similarly to Family Archive [21], APT aims to let users interact with both born-digital and digitized material, and thus facilitates the study of how interactions with these objects differ. Although not the primary focus of this paper, this understanding will be crucial for developing a fully-hybrid environment for archival processing.

APT was built following a collaborative design process. The five-person research group consisted of three digital library/HCI researchers and two archival researchers (one of who has professional experience in archival processing). The team met weekly throughout the duration of the project, setting up tasks and tracking their progress. The archival researchers served as domain experts, providing crucial knowledge about the problem space. In processing collections, archivists need a quiet atmosphere and a large flat work surface (e.g., table). Typically, processing a collection requires several sessions to reconsider and fine tune the arrangement. In terms of time, the archivists expressed that for a disorganized collection of 40 items needing item level arrangement (this is a typical assignment in a graduate course on Archival Enterprise) it takes 2-3 hours, and is normally done in one or two sessions. In this scenario, archivists manipulate objects individually, and create and refine groups that reveal the relationships between objects (archival bond). This work is highly visual. Archivists often do not fully read the documents, but pay attention to the documentary form, general appearance, and certain internal metadata of the objects. While groups are expressed implicitly (e.g., piles) or explicitly (e.g., areas), archival arrangement has strict rules about group hierarchy, limiting the types of groups (*sub-group, series, files*) and the order in and among groups.

The archivists' domain knowledge facilitated identifying the following design implications:

- maximize the surface area on a dedicated workspace
- allow for the creation of ephemeral and permanent groups
- support user manipulations at the object and group level
- allow revisiting and/or reverting back to any previous states
- allow for note taking while processing
- allow metadata manipulation of objects and groups

In order to meet this requirements APT's core functionality was designed as a spatial hypermedia application [22]. Spatial hypermedia allows users to interact with objects and metadata, and can automatically infer groups (both implicit and explicit) based on visual structures such as piles and lists. Since a key goal behind the design of APT is creating a platform that allows for the study of archival processing, APT tracks the history of the workspace in a manner similar to spatial hypermedia systems such as VKB [22].

4   System

APT consists of a custom-made, large surface (5'x5' total, 47"x28" interactive), interactive tabletop computer that runs a specialized spatial hypermedia application for digital archival processing (see Figure 1).

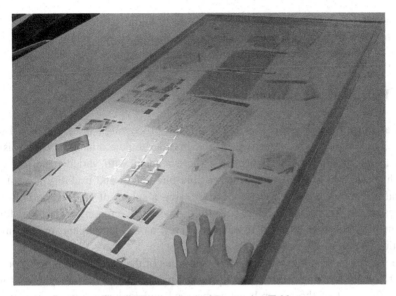

Fig. 1. The Augmented Processing Table

When document images are first imported into APT, they are tiled across the workspace to give an overall impression of the size of a collection. In the initial state all items are located at the root level of the workspace hierarchy. Users can move, scale and rotate items freely in 2D. Users can create groups, and add items to them, which in turn can be added to higher level groups, like series and subgroups. These groups correspond to specific levels of the archival hierarchy (see Figure 2).

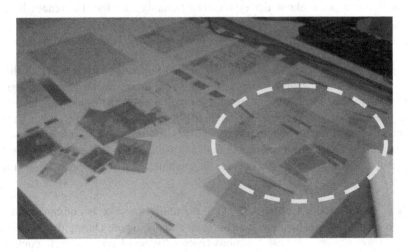

Fig. 2. Grouping and sub-groupings

Archivists may add metadata (e.g., title, description) to items and groups. APT saves the state of the workspace whenever any change occurs, requiring no input on the part of the user to save their work.

5   Pluralistic Walkthrough Evaluation

A key aspect of the project is to study if archival processing can be conducted using digitized materials on an interactive surface. In order to determine this, APT was evaluated using a pluralistic walkthrough (where a team with varied expertise walks through a scenario of use to uncover possible interaction and usability issues). One archival researcher created a test-collection using a subset of an existing collection. This researcher processed the test-collection physically, creating a baseline for the experiment. The other archival researcher served as the participant, processing the collection as part of the walkthrough.

The walkthrough studied the tasks and activities of the participant as she processed the test-collection using a think-aloud protocol to externalize her thoughts and motivations. The rest of the research team observed and took notes unobtrusively. The walkthrough was conducted in a single session that lasted four hours (including a break in the middle). After the task, an unstructured interview was conducted where the participant answered questions from the panel. Finally, after a week of individual reflection, the group met, compared notes and discussed the tasks and activities of processing using the interactive surface.

In general, the participant processed the digitized materials as if they were physical materials. The observation and analysis of the participant's activities and comments revealed some aspects about the system functionality and the differences between processing digitized collections and processing collections physically that are worth mentioning:

Quality of Digitization. Understanding the actual physical characteristics of the objects is extremely important for the task of archival arrangement. Improper digitization (e.g., deformed images) and lack of information about their physical characteristics (e.g., actual size of a document) can hinder the task. For example, when the participant was looking at an image, and particularly when she resized the image to read the content, the lack of information about the original dimensions of the document meant that she could not always decide, for example, if it was a postcard or a large painting.

Rotating and Resizing Controls. In APT the ability to scale is combined with the ability to rotate, and can be done at any corner of a document. During the experiment, the participant performed these functions repeatedly, and having a single control for both, resulted in unintentional rotation or resizing actions. The participant thought that the controls should be separated.

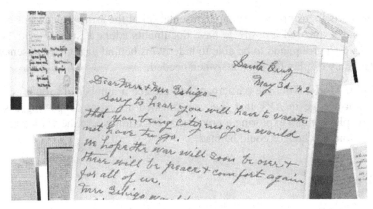

Fig. 3. A zoomed item

Lack of Digital Representation for Physical Characteristics. While the participant often resized documents in order to look at particular details, such resizing and zooming in groups made it hard to determine the size in relation to other documents in the workspace that potentially exist at many different scales (see Figure 3). This was aggravated in APT because the zoom factor was not shown and many of the digitized images lacked any information about scale. To address this issue, the next iteration of APT needs to show the zoom factor or include the physical size as metadata.

Matching Documents. Materials such as double-sided documents created a new *matching task* because each side of the document was captured in a different image. This led to issues as the participant had to determine if items were distinct documents or the front-and-back of a single document. The participant commented that a 'staple' function could solve the issue of multi-sided and multi-paged documents.

Creating and Utilizing Metadata. A common practice in archival processing is to take notes in order to facilitate the arrangement process. The participant specifically described the need for a digital equivalent, saying that she would use these temporary notes to capture titles, descriptions and group types (these were included in the system but not displayed outside of the editing menu), and key metadata from the documents themselves (such as enclosure annotations and reference initials). While note taking is common in traditional archival processing, APT can allow for notes to be directly appended to the images, something that is not done with physical items due to preservation concerns.

Searching and Fetching. Once the participant had done the initial sorting and added metadata to some items, her workspace had become "crowded" and she had difficulty finding specific items in the workspace. Even though APT can zoom in/out, she noted that it would be extremely beneficial to be able to find items using their metadata rather than having to move documents around trying to find a specific item.

Managing Groups. APT represents groups as distinct regions on the work space. Tasks like adding items to a group, manipulating the item order within a group, and creating a hierarchy of groups require additional support, especially when the participant starts focusing on creating a presentation of the arrangement. For example, the participant expressed a desire to have documents adopt a 'snap to grid' behavior once put inside a group, and to be able to hide items behind a single representation of a group and only display the contents when needed.

Presentation Mode. After the participant had created a number of groups, she asked how to acknowledge in APT that she was finished arranging, expecting a separate mode for displaying and exploring a processed collection where the workspace would be "locked" and no changes could be made.

6   Discussion

Overall, the walkthrough evaluation revealed that archival arrangement includes several stages:

1. Triage – quick sort of materials into temporary groups (e.g., piles).
2. Group refinement – revision of all groups (one by one), validating the inclusion of every item (or moving them if necessary). At this stage archivists also start working on presentation aspects.
3. Object matching – formalization of relationships between objects (e.g., matching front/back of postcards and "staple" them together). This stage requires a lot of searching and parsing to find documents.
4. Metadata and archival hierarchy – adjustment of groups according to the formal archival structure, and entering the permanent metadata.
5. Overall workspace organization – ordering 'messy' parts of the workspace and making it suitable for presentation.

For the archival researchers, this basic articulation of the stages of processing was significant, because this process has never been systematically studied. Instead the traditional focus in the archival literature is on articulating processing principles or on determining costs.

While there are changes that are needed in order to accommodate archival practices, the evaluation showed that archival processing can be conducted digitally using digitized materials. This supports the approach of digitizing first and processing second.

The evaluation highlighted the need to pay attention to the digitization process, as some information can be lost or distorted. However, some of these issues can be solved by providing additional functionality to APT. For example a *stapling* function could help users match double-sided documents, and combine multiple pages that make a single item (e.g., multiple pages in a letter) into a single representation.

In terms of functionality, the evaluation of the initial iteration of APT called for functions similar to those provided by spatial hypermedia systems [22] including searching and fetching, and visual presentation operations. The evaluation also revealed the need for functionality specific to archival processing, including visualizations for metadata, physical characteristics, and relative state of the items and the workspace.

7   Conclusions and Future Work

Digital libraries are increasingly hosting collections that contain digitized and born-digital archival materials. For collections of an archival nature it is critical to arrange the materials in a way that captures the context of the overall collection and the relationships between the individual objects in it. Archival science shows that archival processing produces a proper arrangement and collection description that protects the evidentiary nature of materials in the collections.

Archival processing has traditionally followed a workflow of "process first, digitize second". This workflow has some drawbacks that impact "principled practice" and work productivity. This workflow may also lead to delays in getting digitized content online, because digitization needs to wait for processing to take place. Further, in traditional archival processing, physical and born-digital materials are processed separately and differently. Arguably, this creates an artificial boundary between the objects.

The APT project shows that archival processing is amenable to be conducted digitally using interactive surfaces such as multi-touch tabletops. This is highly significant as it represents the first step to integrate the archival processing of digital and physical materials, and allow a workflow of "digitize first, process second". While processing is still a labor-intensive task, this approach has the advantage that it can potentially augment the availability of items in digital archives.

The "digitize first, process second" approach demands a high quality digitization phase, and requires that the processing archivist performs additional tasks such as matching multi-sided, multi-paged documents.

APT provides a platform that allows researchers to investigate future directions of digital archives. The APT project is interested in evaluating the effectiveness of APT as a tool for studying, documenting and teaching archival processing, as well as for exploring new ways to do archival processing including remotely and collaboratively.

The results of the pluralistic evaluation are guiding the design of a second APT prototype, which better represents the objects' physical characteristics, and provides advanced functionality for creating and presenting the archival arrangement. This second prototype will have a formal evaluation with a larger sample of archivists.

Acknowledgements. This work is partially funded by a University of Texas, School of Information Temple fellowship. An alphabetical author sequence is used to acknowledge the equal contribution made by each group member.

References

1. Woods, K.A.: Preserving Long-Term Access to United States Government Documents in Legacy Digital Formats. Ph.D. Dissertation, Indiana University (2010)
2. Galloway, P.: Preservation of Digital Objects. ARIST 38, 549–590 (2010)
3. Woods, K., Lee, C.A., Garfinkel, S.: Extending Digital Repository Architectures to Support Disk Image Preservation and Access. In: 11th Annual International ACM/IEEE Joint Conference on Digital Libraries (JCDL 2011), pp. 57–66. ACM, New York (2011)

4. Cartledge, C.L., Nelson, M.L.: Unsupervised Creation of Small World Networks for The Preservation of Digital Objects. In: 9th ACM/IEEE-CS Joint Conference on Digital Libraries (JCDL 2009), pp. 349–352. ACM, New York (2009)

5. Duranti, L.: The Long-Term Preservation of Authentic Electronic Records: Findings of the InterPARES Project (2001), http://www.interpares.org/book

6. Thibodeau, K.: Building the Archives of the Future. D-lib. Magazine 7, 2 (2001)

7. Greene, M.A., Meissner, D.: More Product, Less Process: Revamping Traditional Archival Processing. The American Archivist 68(2), 208–263 (2005)

8. Duranti, L.: The Archival Bond. Archives and Museum Informatics 11(3-4), 213–218

9. Society of American Archivists. Describing Archives: A Content Standard (DACS). Society of American Archivists, Chicago (2004)

10. Wigdor, D., Perm, G., Ryall, K., Esenther, A., Shen, C.: Living with a Tabletop: Analysis and Observations of Long Term Office Use of a Multi-Touch Table. In: Tabletop 2007, pp. 60–67 (2007)

11. Nebe, K., Klompmaker, F., Jung, H., Fischer, H.: Exploiting New Interaction Techniques for Disaster Control Management Using Multitouch-, Tangible- and Pen-Based-Interaction. In: Jacko, J.A. (ed.) HCII 2011, Part II. LNCS, vol. 6762, pp. 100–109. Springer, Heidelberg (2011)

12. Shen, C., Lesh, N., Vernier, F., Forlines, C., Frost, J.: Building and Sharing Digital Group Histories. In: Proceedings of CSCW 2002 Conference on Computer-Supported Cooperative Work, pp. 324–333. ACM, New York (2002)

13. Kirk, D., Sellen, A., Taylor, S., Villar, N., Izadi, S.: Putting the Physical into the Digital: Issues. In: Designing Hybrid Interactive Surfaces. People and Computers, pp. 35–44. British Computer Society (2009)

14. O'Neill, J., Privault, C., Renders, J.-M., Ciriza, V., Bauduin, G.: DISCO: Intelligent Help for Document Review. In: Global E-Discovery/E-Disclosure Workshop – A Pre-Conference Workshop at the 12th International Conference on Artificial Intelligence and Law, Barcelona (2009)

15. Morris, M.R., Wigdor, D., Lombardo, J.: WeSearch: Supporting Collaborative Search and Sensemaking on a Tabletop Display. In: 2010 ACM Conference on Computer Supported Cooperative Work (CSCW 2010), pp. 401–410. ACM, New York (2010)

16. Terrenghi, L., Kirk, D., Sellen, A., Izadi, S.: Affordances for Manipulation of Physical versus Digital Media on Interactive Surfaces. In: SIGCHI Conference on Human Factors Computing Systems (CHI 2007), pp. 1157–1166. ACM, New York (2007)

17. Grammenos, D., Michel, D., Zabulis, X., Argyros, A.A.: PaperView: Augmenting Physical Surfaces with Location-Aware Digital Information. In: 5th International Conference on Tangible, Embedded, and Embodied Interaction (TEI 2011), pp. 57–60. ACM, New York (2011)

18. Deininghaus, S., Möllers, M., Wittenhagen, M., Borchers, J.: Hybrid Documents Ease Text Corpus Analysis for Literary Scholars. In: ACM International Conference on Interactive Tabletops and Surfaces (ITS 2010), pp. 177–186. ACM, New York (2010)

19. Piper, A.M., Hollan, J.D.: Tabletop Displays for Small Group Study: Affordances of Paper and Digital Materials. In: CHI 2009, pp. 1227–1236. ACM, New York (2009)

20. Sellen, A., Harper, R.: The Myth of the Paperless Office. MIT Press, Cambridge (2002)

21. Kirk, D.S., Izadi, S., Sellen, A., Taylor, S., Banks, R., Hilliges, O.: Opening up the family archive. In: 2010 ACM Conference on Computer Supported Cooperative Work (CSCW 2010), pp. 261–270. ACM, New York (2010)

22. Shipman, F.M., Hsieh, H., Maloor, P., Moore, J.M.: The Visual Knowledge Builder: a Second Generation Spatial Hypertext. In: 12th ACM Conference on Hypertext and Hypermedia, pp. 113–122. ACM, New York (2001)

Mix-n-Match: Building Personal Libraries from Web Content

Matthias Geel, Timothy Church, and Moira C. Norrie

Institute for Information Systems, ETH Zurich
CH-8092 Zurich, Switzerland
{geel,norrie}@inf.ethz.ch, tim.church@gmail.com

Abstract. We present an approach to web content aggregation that allows information to be harvested from web pages, independent of specific markup languages. It builds on ideas from data warehousing and we present solutions to the well-known problems of data integration, namely detection of equivalences and data cleaning, adapted to this context. We describe how the content aggregation engine has been realised as an extensible framework in such a way that end-users as well as developers can use the associated tools to create personal libaries of content extracted from the web.

Keywords: content aggregation, data integration, data harvesting.

1 Introduction

The ability to harvest content published on the web and store it in personal libraries allows users to revisit and access that content as they please. Recent approaches to content extraction and aggregation tend to adopt a vision of the future web where content will be published together with some form of semantic markup to facilitate such processes. The problem, however, is that this has led to a situation where, not only is the content published in a wide variety of forms and content, but also the semantic markup. There are now several competing semantic markup languages, such as HTML 5 Microdata, RDFa and Microformats, in addition to the use of different, non-overlapping schema vocabularies. Further, platforms such as WordPress[1] are increasingly used to develop web sites resulting in millions of new web sites where content is published with little or no semantic markup but still exhibits a clear structure. Our goal was to develop a solution to content aggregation that would not assume a particular vision of the future, but rather be able to work with the heterogeneous nature of the web as it is today.

We propose an approach that borrows several concepts and techniques from traditional data warehousing and adapts them to the web context. Rather than building a corporate data warehouse from operational databases, we are interested in collecting and integrating *personal web content* from various online

---

[1] http://www.wordpress.com

P. Zaphiris et al. (Eds.): TPDL 2012, LNCS 7489, pp. 345–356, 2012.

resources. In contrast to data warehouses, the integration process should be dynamic and lightweight to allow new sources of content to be integrated as they are encountered on the web. Further, it should be possible for end-users as well as developers to specify what content should be integrated and how.

In this paper, we explain how data warehousing concepts have been adapted to meet these goals, with a focus on well-known data integration issues of identifying equivalences and performing data cleaning. We also describe how the approach has been implemented as an extensible framework so that the types of markup supported could easily be extended in the future. We start in Sect. 2 with a discussion of related work before presenting an overview of our approach in Sect. 3. The data similarity metrics used are presented in Sect. 4, followed by a description of the architecture and implementation in Sect. 5. Concluding remarks are give in Sect. 6.

2   Related Work

Bookmarks can be considered as one of the earliest ways of collecting web content. A study by Abrams and colleagues [1] found that bookmarks failed to effectively support the retrieval of information items. Furthermore, they only record the URL of a web page and do not allow data contained within the page to be manipulated or reused in another context. While some efforts were made to make it easier for users to interact with collections of web pages, including WebBook [2], Data Mountain [3], and TopicShop [4], studies showed that user were interested in creating their own collections of information extracted from web pages [5]. Approaches to extracting content from web pages range from manual selection of snippets within a web page to screen scraping tools that can automate the process based on extraction rules. Dontcheva et al. developed tools that would allow users to easily specify structural and content-based extraction rules as well as personalised presentation templates [6].

Another approach is taken by the Semantic Web community where they assume the presence of semantic markup. Thresher [7] is a tool integrated with the Haystack semantic browser that allows HTML documents to be annotated with semantic markup as users browse the web, based on user-defined examples. Piggy Bank [8] is a browser extension that allows users to collect and save content in semantic web format, relying on user-generated JavaScript screen scrapers to extract information from web sites that do not publish RDF.

Our goal was that developers should be able to harvest information from web content, but they should be able to do so with little programming effort, regardless of which, if any, semantic markup is present. We assume sources to be heterogeneous and focus on providing solutions for tackling heterogeneity rather than trying to avoid it. Our approach is therefore similar to data warehousing where data from many sources within an enterprise is integrated and aggregated in a single repository to support business analysis.

There have been other research projects that integrated web content with data warehouses. For example, in [9], data from the Semantic Web expressed in

RDF and OWL, is combined with OLAP techniques. Another example is Moya et al. [10] where they extract data from web feeds and integrated it with a data warehouse to include customer opinions in business intelligence models. It would be difficult to adapt these solutions for general purpose content aggregation as they tend to focus on a limited range of data formats and OLAP analysis.

A major challenge in data warehousing is data integration which includes the identification and merging of different data records representing the same real-world entity. Octopus [11] is an example of a tool for data integration for the web. It allows users to create data sets using data extracted from HTML tables and lists and attempts to infer missing values through best-effort operators. There are many techniques for detecting equivalences and data cleaning proposed in the research literature that can be applied in content aggregation tools. Many of these techniques rely on approximate string matching to detect the similarity or edit distance between strings. There have been many string similarity metrics and distance functions proposed over the years including, Levenshtein distance [12], Damerau-Levenshtein distance [13], Hamming distance [14], Jaro-Winkler distance, Jaccard coefficient, and Cosine similarity. But surprisingly, there is a lack of literature on similarity metrics for dates, even though these play an important role in matching content extracted from the web. We therefore had to develop our own metrics and detail these later in the paper.

3   Information Harvesting Framework

The design of our framework builds upon two basic assumptions. First, we assume that it is not feasible for most end-users or developers to create their own schema of the web data of interest. We have therefore decided to provide a built-in, though extensible, schema that relies on an ontology from the web community. Second, we assume that the web contains copious amounts of incomplete, inconsistent, invalid, and duplicate information, even if that data is published with semantic markup. By including support for data integration and cleaning as a primary concern, our solution systematically improves data quality levels and, therefore, results in more trustworthy output.

Central to the issue of data integration is the problem of mapping attributes extracted from web content to semi-structured data schemas. To define these mappings, we use an adapted interpretation of the extract, transform, and load (ETL) pattern from data warehousing. Our version allows users to extract information from the web in a number of diverse formats and then map that data to a single, shared schema. Figure 1 depicts an overview of the data integration pipeline that we are going to describe in this section. We will start by justifying our choice of a common data model and then continue by discussing each of the data integration steps in turn, highlighting how we adapted existing practices to deal with issues that arise in conjunction with web content and how expert users can leverage our data integration framework to fine-tune the content aggregation engine to adapt it to their needs.

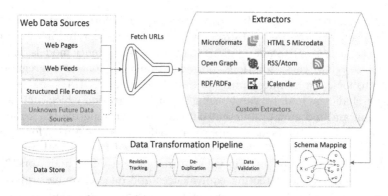

Fig. 1. Conceptual overview of the content aggregation engine

3.1 Data Model

We decided to use the Schema.org [2] vocabulary as a template for our common data model. Schema.org is a collaboration between several leading online search engine providers (Google, Microsoft Bing, and Yahoo!) to create a single unified vocabulary for structured data markup on web pages. Schema.org was selected because it is explicitly designed for web content, has been inspired by several previous semantic markup initiatives and is actively promoted by large search engine providers. We argue that having a common schema greatly facilitates not only the sharing of extraction and mapping but also the exchange of the extracted data itself. Another advantage is that it partly solves the cold-start problem and drastically reduces the time required for developers to make use of the framework, allowing them to concentrate on the mapping definitions instead. Since Schema.org offers a rather comprehensive collection of schemas, we have implemented only a small subset of it. Furthermore, our version of the schema uses a slightly simplified version of the underlying object model. There are no reference types and, thus, no composite objects are possible. Instead, we store a string representation in the correspoding reference fields. This was necessary because some types may refer to other types that are currently not in the implemented subset of Schema.org. Even with that limitation, we can still capture most entities as flat objects. This choice also makes the mapping from Schema.org-enabled websites to our internal schema trivial.

3.2 Extraction Process

A data source is identified by a URL which must to point to an actual resource. That resource has to be represented by a character-based document. In most cases, this will be HTML, but could also be an XML-based RSS feed or another supported format. The data extraction process consists of running a number of *extractor* components against a given source and then passing the combined

---

[2] http://schema.org/

output of these extractors to the next step in the ETL process. It begins by determining which extractors are required for the current source URL. The system then runs each of these extractors in succession. An extractor is a special component which takes a webpage as input, applies its unique extraction logic, and then outputs whatever metadata it was able to extract. If an extractor is not applicable to the current source, such as trying to extract RSS metadata from an iCalendar file, then the extraction will fail silently and no output will be produced for that extractor. Each extractor which successfully completes its extraction adds its output to a shared output hash table with its own unique key. The system includes a number of predefined extractors for several common web data formats and semantic metadata formats. It is also possible for developers to extend the system by creating their own custom extractor components.

3.3   Schema Mapping Specification

The main task that end-users and developers have to perform is to define custom mappings that map a subset of all information extracted by the extractor components to the common data schema/model. Figure 2 shows the formal definition of mapping rules. The type argument specifies the name of the Schema.org data type to which we want to map our data, while the property argument denotes the particular attribute of that type. The mapping expression starts by specifiying the "extractor key", a unique identifier of the extractor to be used. What follows is a sequence of strings that define the traversal path through the output space of the selected extractor. The evaluation of the path expression may resolve to a single value or an array of values. The latter might be the case if a particular semantic markup language allows the same property to be used repeatedly, e.g. the genre property of the Schema.org movie type. The specific order and possible values of these path expressions, as well as the name of the extractor, obviously depend on the concrete implementation of the corresponding extractor and should be documented properly by its developer. However, to facilitate the creation of new mappings, users may inspect the raw output of each extractor separately.

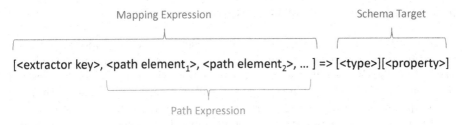

Fig. 2. Schema Mapping Specification

The set of effective mapping rules can be defined for each data source individually. Technically, all mapping rules of a given set are represented by a two-dimensional hash table where the keys define the target schema type and property, while the value denotes the path expression. Please note that this flexible approach allows users to define rulesets that complement partial information provided by one extractor with data from other extractors. For example, the title of a movie may be provided by HTML Microdata and the thumbnail URL by Open Graph markup.

3.4   Data Integration Pipeline

Data integrity is enforced by processing all incoming information items through a mandatory validation and transformation phase. This step cleans the data by performing normalisation, coercing information into the appropriate format, and ensuring certain required attributes are present. Because our content aggregation engine deals primarily with data extracted from web pages, most, if not all, incoming data is initially formatted as a `string`. However, in a personal library, we prefer to store this data in the most appropriate representation, such as a `datetime` object for dates and an `int` for integer numerical values. To ensure a consistent, universal encoding scheme for all strings stored in our database, they are automatically converted into Unicode encoding. Developers can influence this transformation step by customising the data model definitions which tell the system which data type to expect for each field. De-duplication is performed using a number of different techniques which are discussed further below. The final aspect of our data integration pipeline is the ability to track changes over time, a pattern known as change data capture. Any automatic data extraction process is prone to occasional errors, and there is a possibility that the data collection system could potentially override some existing data with invalid values. To counterbalance this, we log every update to every data record in order to provide the user with the ability to view the full revision history for each item. By automatically tracking all changes to every data object, our framework provides a built-in audit trail and enables the tracing of data lineage. This allows users to understand exactly where each piece of data originated, and also enables users to easily rollback any incorrect or erroneous updates.

3.5   Duplicate Detection Algorithm

Since the web contains several different datasets that often overlap, it is also important to be able to detect duplicate records. Information on various websites may differ slightly yet still refer to the same real world entity. When two textual representations do not match exactly but still denote the same entity, they are called approximate duplicates. There are many reasons for dirty data on the web including user input errors, typos, different abbreviation schemes, and dissimilar formatting. Consequently, approximate duplicates are often encountered when integrating data extracted from the web.

To support duplicate detection within our pipeline, we decided to combine several established fuzzy matching algorithms paired with fine-grained control over their weights for different entity types. Generally speeking, fuzzy matching, or probabilistic record linkage, determines a weighted probability that two records refer to the same entity based on the values of multiple characterising attributes of the records. Pairs with a probability above a certain threshold are deemed to be a match. We have defined a set of duplicate identification fields for each of the implemented data types. This set of fields acts as a composite primary key that can uniquely identify an entity occurrence. For example, an Event object can be uniquely defined by the combination of its name, start date, end date, and location.

The duplicate detection process consists of two stages. The first stage checks for the presence of exact duplicates. Note that a match will not be found if the candidate object contains values for any of the duplicate identification fields in which the existing object has no value. Therefore, the existing object must contain at least the same level of detail as the candidate object to be considered an exact match. The second stage checks for the presence of approximate duplicates. The approximate duplicate check calculates a weighted score of the similarity of each duplicate identification field. If the resulting score is above a configurable threshold, then the candidate object is considered to be a duplicate of the existing object.

Similarity comparisons for each field are type dependent and each return a value in the range [0,100]. The only field types that currently support fuzzy matching are strings and dates. All other field types return a score of 100 for a perfect match or 0 otherwise. Also, if either of the objects being compared is missing a value for a given field, the similarity score for that field is 0. For string matching, we have used a library that supports multiple approximate matching measures including edit distance, partial string similarity, sorted token ratio, and token set ratio. For each string similarity comparison performed, all four of these similarity measurements are calculated and the maximum value is returned. For date and datetime fields, our framework supports the similarity metrics proposed in Section 4. The most appropriate metric may depend on the specific context and can be configured per type.

4   Date Similarity Metrics

There are many reasons why two representations of the same date-in-time might contain differences, especially when dealing with user-generated content. Perhaps the most obvious cause is errors in user input such as inadvertent typos. It is possible that a user may accidentally hit a wrong key while manually entering the digits of a date. Some user interfaces, such as those on mobile devices, are especially prone to this type of mistake. The user may also enter the date information in a different order than expected by the system; formatting conventions of date values can frequently cause issues when inputting data or when converting data between different systems. For example, dates are typically entered in the MM-DD-YYYY format in some cultures, while people in other parts of the world commonly expect dates to be formatted as DD-MM-YYYY. Another possible source of approximate duplicates in dates is errors that result from OCR (Optical Character Recognition) scanning. These automated input systems are prone to mistakes and could cause problems with date values as well as strings. Finally, date discrepancies may be due to different levels of specificity. Some dates may include an exact time down to the microsecond while others may exclude the time component altogether. Also, the full specificity of the information might not be known when a date is first entered into a system, yet most systems commonly require values for all components of a date before that data can be saved. In this case, end-users often enter arbitrary data for the missing attributes as a placeholder in order to meet the requirements of the system.

Fuzzy date matching can be a useful technique in several different scenarios. For example, a recurring event may repeat multiple times with the exact same name and location. In this scenario, the only distinguishing features between event instances are the start and end dates. Current fuzzy matching techniques can help detect typos in string values such as the name attribute but would be unable to detect input errors in the date fields. For events or any other object types that rely on date attributes, the ability to detect the similarity of date values would help to improve duplicate entity resolution. Other potential use cases for approximate date matching include input validation of dates, input suggestions such as a spell check equivalent for dates, and data integration between different systems. In order to support fuzzy date matching, we propose three similarity metrics that are specific to date and date/time data types:

1. **Fractional time period**: This metric compares the distance between dates (in a certain unit of measure, e.g. hours, days, months) as a fraction of a fixed time interval. The time interval needs to use the same unit of measure. If the date difference is larger than the time period, a value of 1 is returned. Therefore, the output of this function is a continuous value in the range [0, 1]. This metric is useful for identifying dates that are close together in time, even if they occur in consecutive months or years. However, it is not suitable for correcting small typing errors or flipped numbers.

$$FractionTime(d_1, d_2, tp) = \left(\frac{|d_1 - d_2|}{tp} \right)$$

2. **Date component overlap**: This metric breaks down each date value into separate components and then compares the amount of overlap between these components. This includes day, month, year, and (if applicable) hour, minute, and second. Each component is compared against the same component of the other date, and the output is the fraction of components that match. For example, the date overlap betwen January 2nd, 2011 and February 1st, 2011 is 1/3 because the only date component that matches for both dates is the year. This metric is useful in detecting minor errors that only affect one component.

$$DateOverlap(d_1, d_2) = \left(\frac{(d_1[0] \leftrightarrow d_2[0]) + (d_1[1] \leftrightarrow d_1[0]) + ...}{\#\text{components}} \right)$$

3. **Jaccard similarity for dates**: This metric adapts the Jaccard coefficient for comparing dates. It breaks down each date value into sets of four components and then compares the similarity of these sets. The component set consists of the two-digit day, two-digit month, first two digits of the four-digit year, and the last two digits of the four-digit year:

$$Date(DDMMYYyy) \Rightarrow \{DD, MM, YY, yy\}$$

The Jaccard similarity between dates a and b is the ratio of the intersection of the component sets for those dates, A and B respectively, divided by the union of those same sets.

$$JaccardSim(a, b) = Jaccard(A, B) = \frac{|A \cap B|}{|A \cup B|}$$

The following example shows how the Jaccard similarity can be calculated for two date values, and it also illustrates this metric's usefulness for determining when parts of a date have been rearranged, which can easily occur due to formatting conflicts.

$$JaccardSim(01/02/2011, 02/01/2011) \Rightarrow$$

$$A = \{01, 02, 20, 11\}, B = \{02, 01, 20, 11\}$$

$$A \cap B = \{01, 02, 11, 20\} \Rightarrow |A \cap B| = 4$$

$$A \cup B = \{01, 02, 11, 20\} \Rightarrow |A \cup B| = 4$$

$$JaccardSim(a, b) = \frac{|A \cap B|}{|A \cup B|} = \frac{4}{4} = 1$$

Listing 1.1. Sample date similarity calculations

```
# Fraction of a year calculation
FractionTime(31/12/2011, 01/01/2012, '1 year') = 1/365
# Comparing exact duplicates in DateOverlap
DateOverlap(01/02/2011, 01/02/2011) = 3/3 = 1
# Severe penality for small permutations in DateOverlap
DateOverlap(01/02/2011, 02/01/2011) = 1/3
# Swapping days and months has no influence in Jaccard
JaccardSim(01/02/2011, 02/01/2011) = 4/4 = 1
# Low score for dates near each other in Jaccard
JaccardSim(31/12/2011, 01/01/2012) = 2/5
```

5 Architecture and Implementation

The framework for web content aggregation proposed in the previous sections has been implemented as a Python-based application which may be installed locally or on any dedicated web server. As illustrated in Figure 3, it exposes its functionality via a web API that follows the REST principle. The API offers several

HTTP GET and POST "'methods"' to access core operations of the content aggregation engine, such as adding web sources, installing new mappings or retrieving the extracted data. Data between user applications and the backend is exchanged via JSON [3], a lightweight alternative to XML, and converted internally to our model classes. The actual data, as well as the mappings and the URLs of the data sources, are stored in MongoDB [4], a schema-less, document-oriented database. In contrast to relational databases, document-oriented databases store information as grouped key-value pairs. MongoDB was chosen because its schema free nature is better suited to web content where information is expected to be incomplete, semi-structured and related to a quite diverse set of different entity types. Having the data schema defined exclusively in the application logic, our system is not only more flexible in terms of frequent schema modifications (mainly additions), but we also relieve developers from maintaining schema validation rules in two, possibly non-equivalent, schema definition languages.

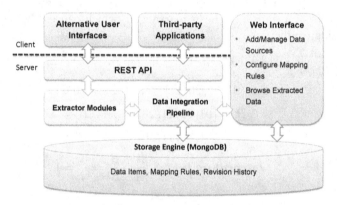

Fig. 3. System architecture overview

To define our document schema and validation constraints in code, we use the open-source MongoEngine [5] package. MongoEngine is an Object-Document Mapper for working with MongoDB from Python; this is conceptually similar to an Object-Relational Mapper (ORM) for traditional relational databases. MongoEngine defines field types for several data types including strings, URLs, email addresses, dates, numbers and booleans. Also, MongoEngine enables the developer to define additional validation checks such as setting a maximum length for strings or identifying mandatory fields. We have pre-defined document schemas for a small selection of data types from Schema.org, but developers are encouraged to add document schemas for additional Schema.org types.

We used the open-source FuzzyWuzzy[6] package for fuzzy string matching as part of the de-duplication step. We implemented revision history tracking

[3] http://tools.ietf.org/html/rfc4627

[4] http://www.mongodb.org/

[5] http://mongoengine.org/

[6] https://github.com/seatgeek/fuzzywuzzy

through the use of broadcast events and listeners. MongoEngine supports a limited set of broadcast 'signals' which are triggered before and after certain events including initialisation, save, and delete. We defined custom listeners for both the post_save and post_delete signals to update the revision history collection accordingly. All revision history is stored in a separate collection in the database rather than embedding history information within the object itself. Since revision history is not regularly queried except in special cases, keeping all history in a separate collection ensures that querying performance for the most common use cases is not affected as the number of revisions increases.

Extending the System. Developing a custom extractor is designed to be quick and painless for developers by minimising the amount of work necessary. Extractors are Python modules that extend one of the base classes provided by the framework. Each extractor is given a reference to either the raw data stream of the data source or, in the case of HTML, an already pre-parsed HTML document. Custom extractors are not limited to any specific techniques or methodologies but rather have the freedom to process data sources any way they choose. This means that, for example, one extractor can use XPath expressions, while another utilises semantic markup, and yet another relies on natural language processing. The results of each extractor are stored as key-value pairs in a global hash map which is eventually forwarded to the schema mapping step.

Web Interface. As an alternative to the web API, which is mainly intended for the development of third-party applications, we provide an integrated web interface that offers end-users some basic functionality to manage data sources, configure mappings and browse data items.

6 Conclusions and Future Work

We have presented a framework that allows users to create personal libraries from content harvested from the web in a lightweight manner. By adopting a data warehousing approach, it embraces hetereogeneity in markup rather than trying to enforce homogeneity, which we feel is unrealistic, at least in the near future. We therefore consider this to be a solution that offers a pragmatic, global approach to content aggregation. The toolkit was designed with end-users in mind. The framework has a number of built-in extractors for the majority of semantic markup languages and other common web data formats, so end-users can build their own personal libraries with little effort. Further, by building on a common core schema and providing a simple mapping language, users are not required to have data modelling experience. Basic data validation, history tracking and configurable duplicate detection are all supported.

In the future, we want to explore two main directions of research. One is to design tools on top of the framework to enable users to define mapping rules within the browser in an even easier manner. The other is to incorporate parts of the semantic web stack, most notably RDF, in order to be able to connect personal libraries to larger data repositories. The foundation for this approach has already been laid by choosing Schema.org as the common data model, because there already exists a canonical mapping from Schema.org terms to Web Data vocabularies.

References

1. Abrams, D., Baecker, R., Chignell, M.: Information archiving with bookmarks: personal web space construction and organization. In: Proceedings of the SIGCHI Conference on Human Factors in Computing Systems, CHI 1998, pp. 41–48. ACM Press/Addison-Wesley Publishing Co., New York (1998)
2. Card, S.K., Robertson, G.G., York, W.: The webbook and the web forager: an information workspace for the world-wide web. In: Proceedings of the SIGCHI Conference on Human Factors in Computing Systems: Common Ground, CHI 1996, p. 111. ACM, New York (1996)
3. Robertson, G., Czerwinski, M., Larson, K., Robbins, D.C., Thiel, D., van Dantzich, M.: Data mountain: using spatial memory for document management. In: Proceedings of the 11th Annual ACM Symposium on User Interface Software and Technology, UIST 1998, pp. 153–162. ACM, New York (1998)
4. Amento, B., Terveen, L., Hill, W., Hix, D.: Topicshop: enhanced support for evaluating and organizing collections of web sites. In: Proceedings of the 13th Annual ACM Symposium on User Interface Software and Technology, UIST 2000, pp. 201–209. ACM, New York (2000)
5. Schraefel, M.C., Zhu, Y., Modjeska, D., Wigdor, D., Zhao, S.: Hunter gatherer: Interaction support for the creation and management of within-web-page collections. In: Proc. 11th Intl. Conf. on World Wide Web, WWW 2002 (2002)
6. Dontcheva, M., Drucker, S.M., Salesin, D., Cohen, M.F.: Relations, cards, and search templates: User-guided web data integration and layout. In: Proc. of the 20th ACM Symposium on User Interface Software and Technology, UIST 2007 (2007)
7. Hogue, A., Karger, D.: Thresher: Automating the unwrapping of semantic content from the world wide web. In: Proc. 14th Intl. Conf. on World Wide Web, WWW 2005 (2005)
8. Huynh, D., Mazzocchi, S., Karger, D.: Piggy bank: Experience the semantic web inside your web browser. Web Semantics: Science, Services and Agents on the World Wide Web, 5(1) (2007)
9. Nebot, V., Berlanga, R.: Building data warehouses with semantic data. In: Proc. of the 2010 EDBT/ICDT Workshops, EDBT 2010 (2010)
10. Moya, L.G., Kudama, S., Cabo, M.J.A., Llavori, R.B.: Integrating web feed opinions into a corporate data warehouse. In: Proc. 2nd Intl. Workshop on Business intelligence and the WEB, BEWEB 2011 (2011)
11. Cafarella, M.J., Halevy, A., Khoussainova, N.: Data integration for the relational web. Proc. Endow. VLDB 2, 1090–1101 (2009)
12. Levenshtein, V.I.: Binary codes capable of correcting deletions, insertions, and reversals. Soviet Physics Doklady 10(8), 707–710 (1966)
13. Damerau, F.J.: A technique for computer detection and correction of spelling errors. Commun. ACM 7(3), 171–176 (1964)
14. Hamming, R.W.: Error detecting and error correcting codes. Bell System Technical Journal 29(2), 147–160 (1950)

Machine Learning in Building a Collection of Computer Science Course Syllabi

Nakul Rathod and Lillian N. Cassel

Department of Computing Sciences, Villanova University, Villanova PA USA 19085
{Nakul.Rathod,Lillian.Cassel}@villanova.edu

Abstract. Syllabi are rich educational resources. However, finding Computer Science syllabi on a generic search engine does not work well. Towards our goal of building a syllabus collection we have trained various Decision Tree, Naive-Bayes, Support Vector Machine and Feed-Forward Neural Network classifiers to recognize Computer Science syllabi from other web pages. We have also trained our classifiers to distinguish between Artificial Intelligence and Software Engineering syllabi. Our best classifiers are 95% accurate at both the tasks. We present an analysis of the various feature selection methods and classifiers we used hoping to help others developing their own collections.

Keywords: Syllabus, Feature Selection, Text Classification, Machine Learning.

1 Introduction

Syllabi have always played a central role in guiding the learning process in institutions of higher education. At their most basic level, they are a contract between the instructor and the learners about the learning objectives, topics and assessment policies. However, they also contain the most current, high quality learning resources in the form of readings, textbooks, presentations and assignments. Many educators make their syllabi freely available on the web in the hopes that others will find these materials useful. Interested learners can use syllabi to keep abreast of the new developments or learn about new topics. Professors planning their courses can get an idea of what others are teaching. Also, Computer Science is a very rapidly evolving field. That makes easy access to Computer Science syllabi even more important.

However, generic search results for Computer Science syllabi fail to meet the need. [1]. The first handful of links usually gives the most popular syllabi but the rest of the results are not very relevant. Also, it is very difficult to search for syllabi on a per-subject basis. With the emergence of efforts like Stanford led Coursera, MIT's MITx and Thrun's Udacity, which teach Computer Science courses for free over the internet, figuring out what exactly they teach is going to be even more important. Projects like the Ensemble [2] provide faculty with access to a wide variety of resources. We believe syllabus is a key resource in promoting learning and collaboration. We used machine learning here because a syllabus is not strictly defined. Two volunteers helped us hand categorize web pages as syllabi if they contained most of the typical

P. Zaphiris et al. (Eds.): TPDL 2012, LNCS 7489, pp. 357–362, 2012.

syllabus components like title, learning objectives, lecturer, schedule, resources (assignments or presentations), etc. Towards that end, the final goal of our research is to make a constantly updated syllabus repository available to the public by crawling the web and categorizing the syllabi by subject.

This paper describes our initial efforts at learning to recognize Computer Science syllabi from other web pages through machine learning as well as learning to distinguish between different types of syllabi, specifically, Artificial Intelligence (AI) and Software Engineering (SE) syllabi.

2   Data Collection

We built two corpora of web pages: one for identifying syllabi, and the other for distinguishing between AI and SE syllabi. To get our initial collection of training examples for the first corpus we searched on Google for "Computer Science departments" and noted the URL of the first 25 departments. We then searched for "syllabus <department name>" and retrieved the URLs of the first 100 syllabi that we saw. This allowed us to get a reasonably representative sample of syllabi. We manually classified a web page as a syllabus if it had typical components that one expects in a complete syllabus. We also got the URLs of 100 non-syllabi from the same page that the syllabus was linked to. Non-syllabi were web pages that were not syllabi and did not contain any of the syllabus components but were pages that could be mistaken for syllabi by a system doing simple text matching. These were professor's teaching pages, lists of workshops, research interests of the faculty, etc.

To build our second corpus we searched on Google for "Artificial Intelligence Syllabus" and "Software Engineering Syllabus" respectively and retrieved the first 35 complete syllabi we saw for each. We ended up with a total of 70 URLs.

3   Feature Selection

For our first corpus (Syllabus or not), we identified 15 typical components like overview, schedule, evaluation, etc. that occurred in a syllabus along with the various ways in which one could refer to them by brainstorming and using previous work done on the topic [3] [4]. So, for instance, the Evaluation component can be called evaluation, grading or assessment; the Professor component can be called professor, instructor or lecturer, etc. In the end, we identified 81 total synonyms that fell into 15 typical components. We then constructed three different feature sets:

1. topic_bool (fifteen 1s and 0s): whether a component occurred in a syllabus by way of presence of any of its identified synonyms.
2. topic_count (fifteen integers): the number of times each component occurred in a syllabus by way of presence of its identified synonyms.
3. word_bool (Eighty-one 1s and 0s): whether any of the pre-identified synonyms themselves appeared in a web page. This did not force a word to identify a component like in the previous two cases.

Finally, we tested how the classifier's performance varied when using just the top 1, 5, 10, 20, 40 or 81 (all) features. We used the Information Gain (IG) metric to identify the top components. IG measures the amount of information gained in predicting the class of an example by knowing a certain feature. Yang et al. [5] in their study of selection methods for text categorization found that IG along with Mutual Information performed the best for selecting features.

We wrote a Python script that would ingest a list of URLs and produce all three feature sets for each page. We could not extract the plain text from some of the web pages because they were either using iframes or had javascript controls. This reduced our training cases to 80 syllabi and 72 non-syllabi (152 total).

For our second corpus (AI or SE syllabi), we used a bag of words approach to build our feature set. We removed the stop words from each page and stemmed the words down to their root words using the more aggressive Lancaster stemming algorithm using Python's NLTK [6]. The unique list of roots for each page gave us a more manageable feature set. We used IG to rank the features and used the top (1, 5, 10, 25, 50 and 100) features as well as all the features (5848).

4   Building Classifiers

For identifying syllabi we used all the three feature sets to build Decision Trees (DT) [7], Naive-Bayes Classifiers (NB) [8], Support Vector Machines (SVM) [9] and Multi-Layer Feed-Forward Neural Networks (NN) [10]. For the DT, we split the nodes based on IG in line with previously mentioned research that has found it best for text categorization tasks [5]. We also stopped splitting into nodes when there were just two examples left. For the NB Classifiers we assumed prior class probabilities based on relative frequency which was calculated as the ratio of number of examples belonging to a certain class to the total number of examples in the training set. This ratio was pretty close to 1:1 for both our corpora. For our SVM, we always used a linear kernel to keep the classifier simple and consistent across different feature sets.

For the NNs we experimented with one hidden layer with 2 nodes and n-1 nodes, and two hidden layers each with n/2 nodes where n was the total number of features. This allowed us to increase the planes of classification and see whether that would have any effect on classification performance. We kept the learning rate at 0.2 but changing it to 0.02 to decrease the effect of back-propagation had no significant effect on accuracy.

5   Evaluation

We used repeated random sampling to test each of our classifiers. In every sampling, we would randomly select 30% of our examples and put it aside. We would then train our learners with the remaining 70% of the data. To test it we use the previously set aside 30% of examples as test cases. We did this 10 times and measured the average classification accuracy, Area under ROC curve (AUC) and F1 measures.

All the classifiers with any of the feature sets were more than 82% accurate at both the tasks. SVM's, NB's and DT's performance increased on average when going up from Top 1 to Top 20 features as scored by IG. After 20 features, SVM's performance continued to increase while NB and DT gradually started declining. Although DT had the weakest classification accuracy on average, its performance decline was not as steep as NB as number of features increased [Fig. 1].

The NN was extremely expensive to train as the number of features we used increased. With 81 features and 40 hidden nodes it took half an hour to do just fifty complete back-propagation cycles even with our small data set. The NNs were only practical for topic_bool and topic_count feature sets which had 15 features. In this case, they were 91% accurate. We decided that for the far larger feature sets we would be dealing with eventually in our syllabus repository a NN is not a practical solution and did not test it when distinguishing between different types of syllabi (5800+ features) or with the word_bool feature set (81 features).

For identifying syllabi, all the classifiers performed best (more than 90% accurate) when using the word_bool feature set. This shows we achieved better performance by using all the synonyms for syllabus components individually rather than bunching them up into 15 typical components. We achieved the best performance using a SVM with all the 81 features (Accuracy = 0.95, AUC = 0.99, F1 = 0.95). Also, just using one top feature as measured by IG, the word 'final', gave us 84% accuracy.

For distinguishing between AI and SE syllabi, SVM with 25 features (not all 5848 features) achieved the best classification accuracy of 97% [Table. 1]. Also, it was the most resistant to performance decline as number of features increased. When we tested it with all 5848 features rather than just 25 it dropped to 95% compared to DT which dropped to 84% (from 91% with 25 features) and worst of all NB which dropped to 6% (from 96% with 10 features). The top one word as ranked by IG was 'norvig' which all by itself identified 88% of the AI syllabi. This is after Peter Norvig's seminal book on AI.

Despite our small sample size, these results are very promising as they indicate that we may be able to identify most of the syllabi on the web.

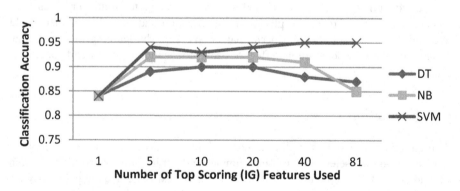

Fig. 1. Classification Performance of various classifiers at recognizing syllabi from other web pages over Number of Features used to train them when using the word_bool feature set

Table 1. Classifiers which performed best at distinguishing between AI and SE syllabi

Classifier	No. of Features	CA	AUC	F1
SVM	25, 50 or 100	0.97	0.99	0.97
NB	10	0.96	0.97	0.96
DT	25	0.91	0.92	0.91

6   Related Work

Educators have realized that there is valuable content locked away in syllabi and a few projects have tried to make this content more available. MIT's OpenCourseWare effort manually collects MIT syllabi in a central repository. However, this approach does not scale well and is very human-intensive. Syllabus Finder [11] tried to solve this problem by performing sophisticated regex searches on Google's cache. However, Syllabus Finder has been completely non-functional since 2002 when Google deprecated its SOAP Search API. Matsunga et al. [3] and Yu et al. [4] developed thesaurus based approaches which inspired ours to identify syllabi from other web pages. Yu et al. [4] also used contextual and content features in addition to the thesaurus based structural features and identified partial syllabi as well. None of the efforts, though, have gone so far as to distinguish between the subject areas like AI or SE. A few researchers have also directly attacked the problem of lack of standardization among syllabi that gives rise to this whole problem in the first place. Both the SylVia project [12] and Tungare et al. [13] have developed extensive ontologies to help instructors tag the various resources they use to make their syllabus. Both these projects have also developed syllabus annotation, creation and viewing tools to encourage adopting standard formats for syllabus representation. These tools have been around for some time now and face the problem of getting adopted by enough people to be useful. In light of this, genre-classification approach like ours and others [3] [4] will be more efficient at capturing resources locked away inside syllabi. Genre classification is being actively researched and has been used for all kinds for purposes including recognizing home pages from other web pages [14].

7   Conclusion and Future Work

The bigger feature sets combined with SVM worked best for both the tasks. However, even though SVMs were very resistant to performance decline as number of features increased, optimum performance was achieved using about 25 features. We also conclude that using the 15 typical syllabus components is not as useful as using the 81 individual synonyms that we divided into these 15 components.

Both our corpora had just 152 and 70 documents respectively. Therefore, the results we obtained could be heavily influenced by the training examples we collected. We realized this and used robust testing measures like repeated random sampling to ensure we were measuring the most stable performance. Still, we cannot make general observations from our study except that our results are very promising and require further investigation. We have started collecting more syllabi in different Computer Science subject areas to test our classifiers with a considerably bigger data set. This will allow us to see both, how the performance of our classifiers scales, and whether they can distinguish between more subject areas than just AI and SE syllabi reliably.

We are currently improving our scraper so that we can extract text from web pages that have iframes or javascript controls. We would also like to develop more sophisticated syllabus gathering techniques to cover syllabi that are distributed over a number of pages. Also, classification can be further improved by building an ensemble classifier which takes one vote from each of the classifiers. This classifier would classify an example accurately as long as the majority of the classifiers classified it correctly. Lastly, we would like to integrate our classifier with a web crawler so that we can start gathering Computer Science syllabi from all over the web and make them available as part of the Ensemble Computing Education portal [2], a collection of the National Science Digital Library (NDSL).

Acknowledgements. This material is based upon work supported by the National Science Foundation under Grant No. 0840713. We would like to thank Indu Bhargavi Guduri and Pranay Kumar Muthineni for helping us collect our training examples.

References

1. Baeza-Yates, R., Ribeiro-Neto, B.: Modern Information Retrieval. Addison Wesley (1999)
2. Ensemble Computing Portal, http://www.computingportal.org/
3. Matsunaga, Y., Yamada, S.: A web syllabus crawler and its efficiency evaluation. In: Proceedings of ISEE (2002)
4. Yu, X., Tungare, M., Fan, W., Yuan, Y., Pérez-Quiñones, M., Fox, E., Cameron, W., Cassel, L.: Automatic Syllabus Classification using Support Vector Machines. In: Handbook of Research on Text and Web Mining Technologies. Information Science Reference (2008)
5. Yang, Y., Pedersen, J.O.: A Comparative Study on Feature Selection in Text Categorization. In: Proceedings of the 14 th International Conference on Machine Learning, pp. 412–420 (1997)
6. Bird, S., Klein, E., Loper, E.: Natural Language Processing with Python. O'Reilly Media (2009)
7. Quinlan, J.R.: Induction of decision trees. Machine Learning 1(1), 81–106 (1986)
8. Kim, S., Han, K., Rim, H., Myaeng, S.: Some effective techniques for naive bayes text classification. IEEE Transactions on Knowledge and Data Engineering 18 (2006)
9. Joachims, T.: Text Categorization with Support Vector Machines: Learning with many relevant Features. In: Proceedings of the European Conference on Machine Learning (1998)
10. Anderson, C.: Learning and problem solving with multilayer connectionist systems. Technical Report, University of Massachusetts (1986)
11. Syllabus Finder, http://chnm.gmu.edu/syllabus-finder/syllabi/
12. SylViA: The Syllabus Viewer Application, http://groups.sims.berkeley.edu/sylvia/
13. Tungare, M., Yu, X., Cameron, W., Teng, G., Pérez-Quiñones, M., Fox, E., Fan, W., Cassel, L.: Towards a syllabus repository for computer science courses. In: Proceedings of the 38th Technical Symposium on Computer Science Education, SIGCSE (2007)
14. Kennedy, A., Shepherd, M.: Automatic Identification of Home Pages on the Web. In: Proceedings of the 38th Hawaii International Conference on System Sciences (2005)

PubLight: Managing Publications Using a Task-Oriented Approach

Matthias Geel, Michael Nebeling, and Moira C. Norrie

Institute for Information Systems, ETH Zurich
CH-8092 Zurich, Switzerland
{geel,nebeling,norrie}@inf.ethz.ch

Abstract. We report on the development of a powerful and task-oriented tool for the management of research publications. The work was motivated by a survey showing that researchers still rely heavily on basic tools such as text editors for managing bibliographic data. We present the approach as well as the resulting tool, PubLight, and compare the features of this tool with existing reference management systems.

Keywords: task-oriented information management, publications tool, reference management.

1   Introduction

Although advanced reference management tools exist, many computer science researchers still rely heavily on basic text editors for the management of bibliographic data. We believe that one reason for this is the fact that existing tools tend to take a *data-oriented* rather than *task-oriented* view and support only parts of the research workflow. Adobe Photoshop Lightroom[1] is a prime example of an advanced personal information management tool designed to offer a modular, task-based environment to support the workflow of professional photographers in a flexible way [1]. We therefore decided to investigate how such a task-oriented approach could be applied to the domain of managing publications.

We started with an extensive review of existing reference management systems to identify the particular research tasks that they support and how they are supported. We then carried out a survey among computer scientists to establish which tools they use and their experiences. The main contribution of this paper is to present the results of this analysis and show how adopting a task-oriented approach enabled us to rapidly develop a prototype, PubLight, that supports many of the more advanced features of existing tools, but with a more task-oriented view and greater direct manipulation. PubLight also adopts features promoted by the research community such as sophisticated tagging systems (e.g. [2,3]) and dynamic, faceted search [4,5]. We start in Sect. 2 with an overview of existing reference management tools and a report on the survey that we carried out. In Sect. 3 we then describe our prototype, PubLight. Concluding remarks are given in Sect. 4.

---

[1] http://www.adobe.com/products/photoshoplightroom

P. Zaphiris et al. (Eds.): TPDL 2012, LNCS 7489, pp. 363–369, 2012.

2   Reference Management Systems

Many reference managememt tools no longer simply manage references, but also manage copies of the publications and provide features to support tasks such as literature review. We carried out a survey to establish how researchers manage their references and which, if any, of these systems they use. We used an online questionnaire with a mix of multiple and single choice questions as well as matrix questions with a 5-point Likert-scale and some open questions. The participants were computer science researchers recruited through email invitations. We received 108 responses, of which 85 were complete. Only the completed surveys were analysed. From these 85 participants, 17% were female and 83% male. 49 participants (58%) were PhD students, 22 (26%) post-docs, 10 (12%) professors and 4 (5%) non-academic researchers (the sum exceeds 100% due to rounding). All participants had experience in authoring scientific papers or reports.

The results suggest that physical paper documents, hand-written notes and lists are still very widely used (64%) to capture related material. Interestingly, the manual editing of text and BibTeX files (57%) was slightly more common than the use of a dedicated reference management system (48%). In terms of references collected, only 18% typically collect individual quotes or paraphrases. Instead, references to complete articles were the most common (88%) together with the digital documents themselves (84%). Other types of information collected regularly were references to web pages (48%) and notes (39%).

Table 1. Top 7 tools

Tool	Used by
Text Editor	65%
Web Browser	34%
Papers	21%
JabRef	19%
BibDesk	14%
Mendeley	14%
Zotero	11%

Table 2. Top 7 Primary Tools

Tool	Used by
Text Editor	41%
JabRef	15%
Papers	11%
Zotero	9%
BibDesk	6%
Web Browser	5%
EndNote	4%

Participants were asked to specify which tools they use (possibly many) and which is their primary tool. The results are shown in Tables 1 and 2, respectively. A plain text editor was the most commonly used tool, with the BibTeX editor JabRef second even though it provides no support for collaboration or document management features. On average participants used at least two different tools in combination.

Although reference management is a highly specialised activity, most participants still rely to some extent on general-purpose tools such as web browsers or text editors. Interestingly, there are notable differences between the occasional and primary use of systems. While 8 out of 9 Zotero users use it as their primary tool, only 3 of the 12 Mendeley users reported it as their primary tool. It is worth noting that, although the web browser is used by many researchers (34%)

to manage references, only 4 use it as their primary tool which indicates that it is a popular tool to collect references but not necessarily to manage them.

We compiled a list of 11 common core functionalities of the different systems and asked participants to rate the importance of each regardless of whether their currently used tools support it or not. A standard, balanced five-point Likert-item was employed for each question with the answers ranging from *Very Important* to *Not Important*. A ranking of the functionalities is given in Table 3. The importance factor was calculated as the percentage of all users who rated a particular feature as either *Important* or *Very Important*. The results show that the ability to import references directly from web pages and search their content is crucial for any system. A quite unexpected result is the low importance of the ability to share references with other individuals and groups (8%). This indicates that reference management is still regarded as a rather individual activity than a collaborative effort.

Table 3. Feature Importance

Features	Importance
Full text search	76%
Import references directly from supported web pages	74%
Fetch PDFs for added references automatically	68%
Automatic metadata (i.e. title) extraction from PDFs	61%
Integrated online search in public repositories (i.e. PubMed, GoogleScholar)	56%
Ability to attach any document or file to references	54%
Online storage (cloud) and web access to library	54%
Synchronisation of reference libraries between different devices	53%
Annotations of PDFs	44%
Sharing of references with individuals and groups	39%
Support for other documents/media (videos, images, web pages etc.)	27%

Table 4. Feature Frequeny

Feature	Used	Never Used	Not Supported
Create static groups or collections of references	64%	20%	16%
Import references directly from supported web pages	60%	13%	27%
Adding arbitrary notes/comments	65%	24%	12%
Attach PDFs, documents or other files to references	48%	20%	32%
Tagging or labeling of references with arbitrary terms	52%	25%	24%
Integrated online search	31%	28%	41%
Marking references as read/not read	35%	25%	40%
Create smart or filter-based collections of references	26%	42%	32%
Marking references as favourites	24%	42%	34%
Rating of References	24%	38%	39%
Create Links between references	24%	41%	35%

Users were also asked how often they used certain features based on a three-point answer frequency scale of *Often, Sometimes* and *Never*, together with a *Not Supported* option. Table 4 summarises the results. From the classic management features of rating, tagging, favourites, only tagging is frequently used. Static collections are used more often than smart or filter-based ones, which suggests that most researchers prefer to manually organise their references. However, it was also the case that many users either were not aware of the feature or use tools that do not support it.

3   PubLight

PubLight is a modular system for browsing, managing and sharing publications with the goal of supporting the research workflow of individuals or groups. While the goal was to mitigate some of the shortcomings identified in existing reference management systems, from a methodological point of view, PubLight serves as proof-of-concept for applying the task-oriented approach to the design of new information management tools. We identified four key activities that together essentially represent the main tasks supported by modern reference management systems and organised them into the following modules in PubLight:

- **Library** – supports the browsing and management of publications
- **Author** – provides simple tools for maintaining different versions of research publications, e.g. drafts, submissions and camera-ready, along with a review and discussion interface to allow co-authors to share comments
- **Share** – provides means for sharing parts of the publication library in the form of references or reading lists with other users
- **Publish** – allows partial or complete export of the library to BibTeX or XML formats such as RSS, as well as integration with existing web sites

Figure 1 shows the library module. On the left, there is the author panel with author information, collection management and import tools. The search panel and library view is shown in the central pane. Details of selected publications are available in a context-sensitive panel on the right (not shown).

Similar to most existing reference management systems, publications can be imported from existing libraries. We currently support the EndNote XML format as well as an extended BibTeX format referencing the corresponding PDF files, so that the system has access to the publication document and metadata such as *title, author, booktitle, year* and *location*. The tool offers a faceted search interface to filter according to a combination of criteria and users can colour, rate or flag items in various ways to help organise and search publications. For example, a user might colour papers according to topic and use flags to indicate those read.

The most obvious difference between PubLight and existing reference management systems is the visual and direct manipulation interface. Studies on the organisation of paper documents have shown the importance of visual cues [6] and the colour labelling together with the summaries can help users recognise and categorise publications according to various criteria. While it would be possible to generate thumbnails for PDF documents, they would not be very informative or distinctive. We therefore chose to create summaries based on metadata such as title, authors, abstract and publication venue (Fig. 2a). The grid layout of

Fig. 1. PubLight prototype

summaries provides a very visual means of organising publications. Yet most existing reference management systems rely on textual lists with relatively poor visual clues to the various attributes used to manage them and limited support for direct manipulation. By contrast, in PubLight, reference thumbnails are enhanced with small, in-place edit controls which enable users to directly manipulate the user-defined attributes such as colours or ratings, as well as ordering publications within a collection. The latter could, for example, be used to guide the writing process by deciding the order in which to present related work.

PubLight was implemented as a web-based information system to support access from a range of devices, including mobile scenarios. This meant that one of the challenges was to handle direct manipulation actions and the resulting changes to the database while keeping the system responsive. The client-side builds on the rich user interface widget support of the jQuery framework[2]. This allowed us to implement facet calculation and filtering in the client based on AJAX and HTML DOM manipulation techniques to reduce the loading time of publication entries and dynamically show and hide them according to the search criteria. Moreover, a lightweight version of the system offering the faceted search browser can be integrated with existing web pages of researchers to allow visitors to quickly filter and browse their publication databases in a flexible way. Details of this functionality have been presented in previous work [7,8].

---

[2] http://jquery.com

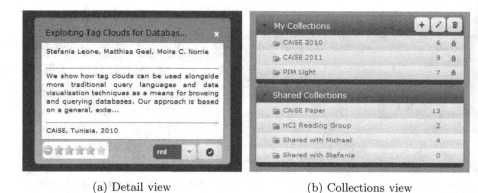

(a) Detail view   (b) Collections view

Fig. 2. Screenshots showing a selected paper and different collections

4   Conclusions

Although our investigations on task-oriented publication management using direct manipulation and visual cues is still in an early phase, the initial results in terms of speed of development and innovation show great promise. Our future work with respect to the PubLight system will follow two directions. First, we plan detailed evaluations of PubLight in terms of usability studies and also comparative analysis of the underlying models. Second, we are currently developing a tabletop version to investigate the potential for touch-based interaction as a means of direct manipulation and gesture-based control.

Acknowledgements. We thank Amir Tafreshi, Julien Ribon and Alessandro Zala for their implementation work. This project is supported by the SNF grant 200021_121847.

References

1. Kim, G.: Early Strategies in Context: Adobe Photoshop Lightroom. In: Proc. ACM Intl. Conf. on Human-Computer Interaction (CHI 2007 Extended Abstracts) (2007)
2. Cutrell, E., Robbins, D., Dumais, S., Sarin, R.: Fast, Flexible Filtering with Phlat. In: Proc. of Conf. on Human Factors in Computing Systems, CHI 2006 (2006)
3. Oleksik, G., Wilson, M., Tashman, C., Mendes Rodrigues, E., Kazai, G., Smyth, G., Milic-Frayling, N., Jones, R.: Lightweight Tagging Expands Information and Activity Management Practices. In: Proc. of Conf. on Human Factors in Computing Systems, CHI 2009 (2009)
4. Basu Roy, S., Wang, H., Das, G., Nambiar, U., Mohania, M.: Minimum-Effort Driven Dynamic Faceted Search in Structured Databases. In: Proc. of 17th ACM Conf. on Information and Knowledge Management, CIKM 2008 (2008)
5. Lee, B., Smith, G., Robertson, G., Czerwinski, M., Tan, D.: FacetLens: Exposing Trends and Relationships to Support Sensemaking within Faceted Datasets. In: Proc. of Conf. on Human Factors in Computing Systems, CHI 2009 (2009)

6. Sellen, A., Harper, R.: The Myth of the Paperless Office. MIT Press (2002)
7. Geel, M., Nebeling, M., Leone, S., Norrie, M.C.: Advanced Management of Research Publications based on the Lightroom Paradigm. In: Proc. CAiSE Forum, London, UK, pp. 97–104 (2011)
8. de Spindler, A., Leone, S., Nebeling, M., Geel, M., Norrie, M.C.: Using Synchronised Tag Clouds for Browsing Data Collections. In: Mouratidis, H., Rolland, C. (eds.) CAiSE 2011. LNCS, vol. 6741, pp. 214–228. Springer, Heidelberg (2011)

Improved Bibliographic Reference Parsing Based on Repeated Patterns

Guido Sautter and Klemens Böhm

Karlsruhe Institute of Technology, Computer Science Department,
Am Fasanengarten 5, 76131 Karlsruhe, Germany
{guido.sautter,klemens.boehm}@kit.edu

Abstract. Parsing details like author names and titles out of bibliographic references of scientific publications is an important issue. However, most existing techniques are tailored to the highly standardized reference styles used in the last two to three decades. Their performance tends to degrade when faced with the wider variety of reference styles used in older, historic publications. Thus, existing techniques are of limited use when creating comprehensive bibliographies covering both historic and contemporary scientific publications. This paper presents RefParse, a generic approach to bibliographic reference parsing that is independent of any specific reference style. Its core feature is an inference mechanism that exploits the regularities inherent in any list of references to deduce its format. Our evaluation shows that RefParse outperforms existing parsers both for contemporary and for historic reference lists.

Keywords: Parsing, Bibliography Data, Algorithms.

1   Introduction

Bibliographies that comprehensively cover both historic and contemporary publications are more and more important in many scientific domains. In biodiversity, for instance, two current projects [1, 17] aim at the creation of such bibliographies. While indexing services like Citeseer [3] and Google Scholar [7] are very helpful with contemporary documents, they are less useful with historical ones, which might not even be available in digital form. However, many historic publications have been cited many times, so their meta data is often available from the bibliographies of other publications. There is a good chance of finding a digital version of at least some of them.

Extracting bibliographic references in a more fine-grained form than plain strings requires parsing, i.e., identifying the attributes of the referenced work such as the title and the author names. Only this level of detail enables advanced processing, e.g., reconciliation of different reference strings that point to the same work.

There are several essential terms we use throughout this paper: A *(bibliographic) reference string* is the unparsed reference as a whole, as found in the References Section of this paper. The *data elements* are the individual attributes of the referenced work, e.g., author names or year of publication. The *reference style* is the arrangement of the data elements in a reference string, i.e., their order and their separating

P. Zaphiris et al. (Eds.): TPDL 2012, LNCS 7489, pp. 370–382, 2012.
© Springer-Verlag Berlin Heidelberg 2012

punctuation. A *reference list* is the content of the references section of a publication; it comprises bibliographic references formatted with the same style.

Several reference parsing algorithms have been proposed, falling into two main categories: (1) Pattern based parsers like ParaCite [12] use regular expression patterns to match the elements of references; these patterns are manually arranged in meta patterns that reflect different reference styles. (2) Training based techniques like [2, 4, 5, 6, 8, 9, 10, 16] use supervised machine learning to generate models of reference strings and the details they consist of from pre-parsed training examples.

However, all these parsing algorithms mostly aim at the bibliographies of contemporary publications, which are highly standardized and follow one of only a few styles. Historically, reference styles are more manifold, so there is a lot more variation. This strongly increases the amount of training data required by training based techniques to achieve good results. Furthermore, older references often make excessive use of abbreviations, which pose additional challenges to pattern matching and hamper knowledge base lookups. Ensuring data quality with the existing algorithms requires manual corrections of significant extent.

To improve parsing accuracy for the bibliographies of historic publications, this paper presents a new parsing algorithm named RefParse. It integrates and generalizes approaches taken in previous work: (1) RefParse uses regular expression patterns to extract morphologically distinctive elements of references, optionally adding fuzzy lexicon lookups to recognize known instances of data elements and abbreviated forms thereof; and (2) it represents arrangements of data elements in meta patterns. RefParse has two distinctive novel features: (1) It handles entire lists of references together. This facilitates exploiting the redundancy contained in such lists, on several levels. (2) It is truly independent of any particular reference style because it infers the style at hand at runtime.

The core idea of RefParse is the following: Data elements with a distinctive morphology, e.g., numeric elements, are easy to identify in the general case and thus helpful for delimiting other elements. However, such delimitation tends to fail if the data elements relied upon are themselves ambiguous, e.g., in references whose title includes numbers. To resolve ambiguous cases, RefParse compares possible arrangements of data elements across reference lists, exploiting that all references very likely follow the same style. By doing so, it can tell, for instance, that a four-digit number cannot be the year of publication if it occurs in a position where a majority of the references does not include numbers.

In more detail, RefParse works as follows: In a first phase, it finds the data elements that are easy to identify based on morphological clues, so-called *base elements*. Second, it disambiguates and verifies the positions of the base elements in the reference style at hand by comparing the references in a list to each other. Third, it uses the base elements as auxiliary delimiters to extract the remaining data elements. This process works in a total of six steps, explained in detail in Section 4.

Our evaluation on a large body of reference lists from the last 200 years shows that RefParse outperforms previous approaches for both contemporary and historic reference lists. This is due to the high degree of flexibility of RefParse regarding the reference style. In particular, it achieves around 94% word accuracy on average. Lexicons increases word accuracy further for several data elements. The latter makes RefParse very promising for community-backed construction of comprehensive bibliographies for entire scientific domains, like the Bibliography of Life effort currently undertaken as part of the ViBRANT project [17]: When users correct the parsed references, the lexicons can be extended at runtime to further increase accuracy.

Paper Outline. Section 2 describes the basic elements of bibliographic references. Section 3 discusses related work. Section 4 presents our generic approach to bibliographic reference parsing, which we evaluate in Section 5. Section 6 concludes.

2   Basics of References

This section describes the basic data elements of bibliographic references in our setting. We have compiled them from a sample of the documents we strive to extract detailed references from. The list of data elements is a subset of the one of BibTeX [13]. For practical reasons, we conflate several elements (see Table 1). Namely, these elements are as good as mutually exclusive in practice, and their instances exhibit very similar morphological features. The best criterion for their distinction is the other elements they occur with, so it is practical to treat them as a single meta element for extraction and disambiguate them once a reference is completely parsed. On the other hand, RefParse can handle a broader variety of data element values (especially person names) than covered by the BibTeX guidelines; this is necessary to handle existing data. Table 1 lists the data elements RefParse works with, together with examples and a mapping to BibTeX element names.

Table 1. Data elements and examples

Data Element (BibTeX names)	Examples / *Morphology or Remark*
Author (author)	A. U. van Thor, Jr. Thor, A. U. van, Jr. Thor, Jr., A. U. van van Thor, A. U. Jr. van Thor, Jr. A. U.
Title (title)	*no regularities, can contain anything, have any length*
Year of Publication (year)	*four-digit Arabic number from [1600, 2100]*
Publisher Name (publisher, institution, school)	Morgan Kaufman Harvard Univ. Press Univ. of Florida
Place of Publication (address)	*any city, possibly with state and/or country*
Periodical Name (journal)	Nature Science VLDB Journal Jour. Nat. Hist.
Part Designator (volume, number)	*(mostly Arabic) number, possibly with label (e.g. Vol.)*
Pagination (pages)	*two (mostly Arabic) numbers separated by dash*
Volume Title (booktitle)	*see title*
Editor (editor)	*see author*
Digital Identifier (url)	*any URL or DOI*
Content Information (note)	216 pp. 17 figures with illustrations

3   Related Work

The problem of parsing bibliographic reference strings into their individual data elements has received considerable attention recently. The approaches fall into two general categories: (1) Pattern based parsers like [12], and (2) parsers generated form training data by means of machine learning. The latter fall into three sub categories: (2a) parsers using statistical models [6, 8, 10, 16], (2b) parsers relying on knowledge bases created from training data [4, 5], and (2c) parsers based on sequence alignment algorithms [2, 9], which were originally developed for protein sequence alignment.

Pattern based parsers like ParaCite [12] use morphological clues to identify the data elements. They represent reference styles as meta patterns concatenated from patterns for individual data elements. Patterns are highly reliable for data elements with a distinctive morphology, e.g., years of publication, person names (authors, editors), part designators, or names of periodicals and publishers. However, the meta patterns have to be created manually at configuration time, rendering this approach incapable of self-adapting to unseen reference styles. RefParse uses patterns only to extract data elements with a distinctive morphology; it identifies the remaining data elements with other means, and it infers the reference style automatically at runtime.

Model based parsers identify the elements of references using statistical models, for instance Hidden Markov Models [8, 16], Conditional Random Fields [6, 14], or Finite State Transducers [10]. Their major weakness is the need for pre-parsed training data for supervised learning, which must cover all reference styles the parser is intended to process later on. While such data is readily available in sufficient amounts for many contemporary reference styles, this is not the case for the multitude of older styles used in historic publications. There, obtaining a sufficient number of representative training examples can become a major bottleneck and cost factor.

Knowledge based parsers [4, 5] create knowledge bases from their training data instead of statistical models. FLUX-CiM [4] divides reference into blocks at punctuation marks and then classifies these blocks by means of lookups in its knowledge base. Populating the knowledge base requires parsed references from the same scientific domain as the ones to parse, so to make sure that at least some percentage of the knowledge base lookups results in a match. Respective data cannot be assumed to be readily available for the reference lists found in historic publications. An additional problem with historic reference lists are abbreviated data elements, which (1) have internal dots that interfere with the blocking and (2) complicate knowledge base lookups. INFOMAP [5] creates a hierarchical, domain specific template base from its training data and then matches the references against these templates. Day has focused exclusively on references to journal articles in his evaluation, thus using a relatively homogeneous test set. The performance on heterogeneous reference lists like the bibliographies of real-world publications is unclear.

Despite its drawbacks as the main basis for parsing, the knowledge backed approach is promising for data elements whose instances likely occur multiple times. RefParse can use lexicons of names of persons, periodicals, publishers, and locations, facilitating recognition of known names. However, it does not depend on these lexicons and thus does not require training data. To handle abbreviations, lexicon lookups use a fuzzy matching mechanism.

Sequence alignment based parsers [2, 9] employ algorithms originally developed for protein sequences; the classes of individual words, numbers, and punctuation marks take the place of the base pairs. They rely on a large base of templates generated by means of supervised learning. Alignment matches the reference to parse against these templates and extracts the most similar one. The latter then specifies how to split the reference into its elements. Similar to model based techniques, the main drawback of sequence alignment parsers is their need for parsed training data. It has to include examples for all possible reference styles to make sure the parser has a respective template available. A further drawback is that only the best matching template is considered for each reference. When parsing a list of references that follow the same style, it seems more promising to select the template that is the best match from a global point of view, i.e., for all references.

RefParse uses alignment to identify the position and delimiting punctuation of data elements that occur in many references within a list, and whose possible instances allow morphology based extraction, namely the author list and the numeric data elements. As opposed to [2, 9], RefParse considers all possible element alignments for each reference in a list and then selects the globally best one. Further, it matches the alignments against one another, not against templates learned in a training phase.

4   The RefParse Algorithm

This section explains the RefParse algorithm in detail. First, we explain the rationale behind its design, and then we go through the parsing process step by step.

There is one **basic assumption** behind RefParse: All bibliographic references in the respective section of a publication are formatted in the same way. This applies both to the formatting of lists of person (author/editor) names and to the order of the individual data elements in the references. This assumption is reasonable: A thorough investigation of the Plazi Corpus (a corpus of over 1,000 real-word documents from the last 200 years, see Section 5.1) has not revealed any counterexamples.

The **basic idea** behind RefParse is inspired by experiences reported in related work. In particular, some data elements exhibit a highly distinctive morphology and thus are easy to identify by means of patterns; the most prominent example probably is the year of publication, a four-digit Arabic number. In addition, it occurs in many references. This makes it a reliable means of delimiting other data elements whose structure is less distinctive, e.g., author names and title.

However, experience also shows that such a delimitation scheme is problematic with references that comprise more than one four-digit number, e.g., as a part designator or as part of the title. The same applies to other numeric elements as well, namely pagination and part designators. The actual semantics of a number, i.e., which data element it belongs to, is often only clear in context, and the context may be confusing as well occasionally. However, this ambiguity rarely occurs across an entire reference list. And because all references in a list follow the same style, i.e., their data elements iave the same position, comparing possible arrangements of data elements across a eference list can resolve the ambiguous cases.

Capitalizing on this observation, RefParse works as follows: First, find the data :lements that are easy to identify based on morphological clues (Steps 1 and 2)elow); referred to as the **base elements**. Second, disambiguate and verify the posiions of the base elements in the reference style at hand, exploiting that all references n a list follow the same style (Steps 2 and 3). Third, use the base elements as uxiliary delimiters to extract the remaining data elements (Steps 4 to 6).

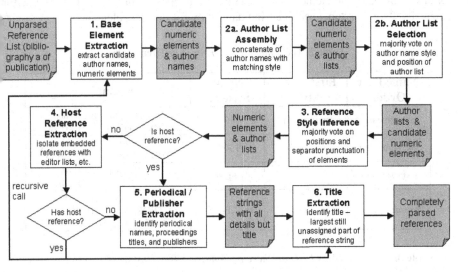

Fig. 1. The RefParse Algorithm

4.1 Parsing Steps

RefParse works in 6 steps, as visualized in Figure 1. It processes all references from the bibliography section of a publication together. The references go through the parsing process in parallel, i.e., one step runs on all references before proceeding to the next step. In the following explanations, the term *references* generally refers to the references that make up the bibliography section of a publication; *preceding reference* means the reference right above the one under investigation in such a list.

Step 1: Base Element Extraction. This step identifies possible instances of the base elements, i.e., years of publication, part designators, and pagination, as well as possible author names. The rationale is that these data elements have a distinctive morphological structure and at the same time occur in many references. The distinctive structure allows to reliably identify possible data element values by means of patterns. The high probability of occurrence ensures that there are several examples of the actual data element positions in the reference style in use. This is essential for the majority vote used in Step 3, to identify the positions and to deal with ambiguities. All patterns used in this step are designed to match any possible value of the data elements they represent, to avoid misses. The ambiguous cases occasionally resulting from this approach are tolerable in this phase; Step 3 deals with them later. Note that the patterns exclusively match individual data elements, not the reference as a whole.

RefParse relies on patterns for the numerical elements: years of publications are four digit Arabic numbers; paginations are either single Arabic numbers, or pairs of them with a dash in between; part designators are numbers, both Arabic and Roman, and sometimes single capital letters. If the interpretation of a number is ambiguous, all possible interpretations are considered. A four-digit number becomes a candidate for any of the numeric elements, for instance; numbers with preceding part designator labels like "vol." or "no." become candidates only for this data element.

Table 2. The name *Alex U. Thor* in different styles

Name Part Order	First Name Style	Name Style ID	Examples
last name last	written out	FL-N	Alex U. Thor Alex Thor Alex U. THOR Alex THOR
last name first	written out	LF-N	Thor, Alex U. Thor, Alex THOR, Alex U. THOR, Alex
last name last	initials	FL-I	A. U. Thor AU Thor A. U. THOR AU THOR
last name first	initials	LF-I	Thor, A. U. Thor, AU THOR, A. U. THOR, AU Thor AU THOR AU

Table 3. The names *Alex U. Thor, Steve E. Cond*, and *Tom Hird* listed in different styles

List Style	Example	Names Found in Step 1 (grouped by style)		
		LF-I	LF-N	FL-I
LF-I, FL-I	Thor, AU, SE Cond, T Hird	Thor, AU AU, SE Cond, T	AU, SE Cond Cond, T Hird	SE Cond T Hird
LF-I	Thor, AU, Cond, SE, Hird, T	Thor, AU Cond, SE Hird, T	AU, Cond SE, Hird	-
LF-I, FL-I	Thor, AU, SE Cond, and T Hird	Thor, AU AU, SE	AU, SE Cond	SE Cond T Hird
LF-I	Thor, AU, Cond, SE, and Hird, T	Thor, AU Cond, SE Hird, T	AU, Cond SE, Hird	-

For author names, RefParse uses multiple different patterns, each reflecting an individual formatting style. In particular, the patterns distinguish the order of first and last name, and whether first names are written out or generally given as initials. It can happen that parts of the references match several name styles: The pattern matching *Alex THOR* in Table 2, for instance, will also match *Thor AU* because both strings have the same morphological structure. *Alex THOR* could also be short for *Tom H. O. R. Alex*, with *Alex* being the last name, and *AU* could be the last name in *Thor AU*. To avoid errors, parts matched by patterns for several name styles become candidates for all of them. The patterns are designed to match both first names and last names consisting of multiple parts, with infixes, hyphenated double names, etc. Table 3 provides examples for the candidate names extracted from a given author list. Furthermore, there are patterns for author repetition marks (usually one or more dashes) that indicate that the authors are the same ones as in the preceding reference, and for the special string "*et al.*". The distinction of styles helps in Step 2, namely to only concatenate author names with matching styles.

In addition to the patterns, RefParse can use a lexicon of person names to look up the candidate author names; while increasing accuracy, however, this additional source of knowledge is not required for RefParse to work.

Step 2: Author List Assembly & Selection. This step identifies the list of author names in each reference, working in two stages: First, the *assembly stage* creates all possible author lists for each reference from the possible names found in Step 1. It concatenates names with matching style, i.e., the same order of first and last name and the same way of giving the first name. There is one exception with regard to the order of first and last name: It may be last name first for the first name of a list and first name first for all subsequent ones. Between two concatenated names, there has to be a separating punctuation mark and/or conjunction. Author repetition marks (see above) can be only the first part in a list, and "et al." can only be the last. Second, the *selection stage* chooses the most probable author list for each reference from the candidates created in the assembly stage. Its underlying assumption is that reference lists are consistent with regard to the style and positioning of the author lists, i.e., that the style of the author list is the same in each reference in a list, and that it has a similar position in each one. The actual selection works by means of a majority vote on the author list style: Each reference contributes one vote for each style it contains a candidate name list for. Finally, this step selects the actual author list for each reference from the candidates, based on compliance with the identified style and its position in the reference. The latter helps distinguish author lists from editor lists in the middle of a reference, which usually follow the same style as the author lists.

Step 3: Reference Style Inference. This step identifies the actual positions the reference style in use assigns to the numerical data elements identified in Step 1 and to the author lists identified in Step 2, as well as their separating punctuation. This again works in two stages: The *assembly stage* generates all possible arrangements of the data elements identified so far. A four-digit number, for instance, is (1) a possible year of publication, (2) a possible part designator, and (3) neither of the two. The latter reflects the fact that numbers can also occur as part of publication titles. This stage also considers that some data element might not be present in a reference, even though Step 1 has found a candidate instance. The *selection stage* chooses the most likely arrangement of data elements for each reference. It scores all arrangements generated in the assembly stage based on two criteria: The number of data elements contained, and the number of references the arrangement is possible for. The former is to favor arrangements covering many data elements. The latter exploits that all references in a list follow the same (unknown) style, so the actual instances of their data elements occur in similar positions within the references. Finally, this step extracts the actual instances of the data elements by matching the positions of the candidate instances against the selected arrangement.

Step 4: Host Reference Extraction. This step identifies possible embedded references to volumes that consist of multiple data elements themselves and thus require parsing. Table 4 provides examples of such embedded references. This step mostly aims at references to books embedded in references to book chapters, but there also are other cases, e.g., special issues of periodicals. The most common case of embedded references is a list of editor names and a volume title, followed by the name of a publisher or a periodical. In any case, a preposition like "in" usually precedes the list of editor names, often separated with a colon.

Table 4. References with embedded references to volumes

Reference to	Example (embedded references in bold)
book chapter	Thor, A. U. 2011: The title of the chapter. In: **Itor, E. D. (Ed.): The book title. Publisher, Place**: 8-15.
proceedings paper	Thor, A. U. 2011: The paper title. In: **Newton, G. et al. (Eds.): Proceedings of JCDL 2011, Ottawa, ON, Canada**

RefParse handles the embedded references found in this step recursively, processing them through Steps 1-3, 5, and 6. The rationale is that the embedded references follow the same style, just as the top level ones they are embedded in. Note that technically any pair of a periodical name and a part designator also constitutes a reference to a volume. However, such pairs do not exhibit the complexity found in embedded references to books and thus do not require recursive handling.

Step 5: Periodical & Publisher Extraction. In the references Step 4 has not found an embedded reference in, this step identifies the names of periodicals, the titles of conference proceedings, and the names and locations of publishers. It considers only those parts of the references that are not assigned to any data element yet. These unassigned parts mostly consist of the above data elements and the title of the referenced works. The extraction of multiple data elements in one step is helpful because they have a very similar morphology, and most references include either the name and location of a publisher, or the name of a periodical; co-occurrences of both in one reference are extremely rare. Embedded references to proceedings given completely with editors and/or publisher (see Table 3 for an example) are extracted in Step 4. Standalone proceedings titles exhibit the same morphology as names of periodicals.

In general, all instances of the data elements sought in this step are in title case (i.e., all words start with a capital letter, usually except for stop words like determiners and prepositions), and RefParse exploits this morphological property to identify them. Furthermore, co-occurrences of abbreviated title case words and stop words are extremely rare: Names of periodicals are either written out in full, including intermediate stop words, or they are abbreviated, omitting the stop words. RefParse exploits this as well, for each reference in isolation because the use of abbreviations is often inconsistent throughout a reference list. If a reference contains an abbreviated and a non-abbreviated title case block, the former becomes the periodical name or publisher. The rationale behind this choice is that the titles of referenced works are rarely abbreviated, whereas abbreviations are relatively common for periodical names.

In addition to the patterns, RefParse can use a lexicon of periodical and publisher names in this step. However, this additional source of knowledge is not required for RefParse to work.

Step 6: Title Extraction. This final step identifies the title of the referenced works. While the title arguably is the most important data element in bibliographic references, it is also the one whose instances exhibit the least regularity. In many reference styles, particularly contemporary ones, titles are not even in title case. This is the reason to identify them last, basically as the longest continuous part of a reference that has remained unassigned in all previous steps.

5   Evaluation

In this section, we report on a thorough evaluation of RefParse. We use two **test data sets** for our experiments: The *Cora Corpus* [11] is an artificial contemporary test data set; it has been used in the evaluation of related work [8]. This corpus consists of 500 individual parsed references that are not organized in reference lists. We have prepared the references as described in [8], in particular breaking the author field down to individual authors and moving separators between the fields. The *Plazi Corpus* originates from a real-world document collection; it consists of nearly 25,000 bibliographic references extracted from over 1,000 biological documents hosted by Plazi [15]. These documents have been published over the last 200 years. They are written in five different languages (English, French, German, Portuguese, and Italian). Due to the long time span and the many different origins and languages, the reference styles vary widely and cover many of the idiosyncracies of historic bibliographic referencing.

For each of the two test data set, we have run three **experiments**:

1. Without domain knowledge, i.e., with empty lexicons. This facilitates comparison to pattern based parsers. The results are labeled "RefParse-g", "g" for "generic".
2. With domain knowledge, i.e., with lexicons of person names, publishers, and periodicals. We use repeated random sub sampling validation in this experiment, with 5 repetitions and a training data / test data split ratio of 50% / 50%, i.e., splitting the corpora in half and using the first half for training (filling the lexicons) and the second half for the actual evaluation. This facilitates comparison to learning based parsers. The results are labeled "RefParse-l", "l" for "lexicons".
3. Comparison runs with two web services backed by ParsCit [6] and FreeCite (also a model based parser), respectively. The results are labeled accordingly.

We assess experimental results at several levels, using the same **metrics** as [10]: The *word accuracy* measures the fraction of the words (actually numbers and punctuation marks, too) of a bibliographic reference string that are assigned to the correct data element. The *field accuracy* reflects how many of the data elements of a bibliographic reference string are identified correctly. The *instance accuracy* is the fraction of the bibliographic references for which all data elements are identified correctly.

5.1   Experimental Results

Experiments with Cora Corpus (upper half of Table 5). This experiment degenerates RefParse to a purely pattern based parser because it has to work with individual reference strings and thus has no repeated patterns to exploit. Despite this disadvantage, RefParse clearly outperforms both ParsCit and FreeCite. Only for the title and volume title, ParsCit and FreeCite are more accurate than RefParse. The results for RefParse are also in the range of the numbers reported for other data sets in related work. Note that the latter mostly measure parsing quality as precision, recall, and f-score; however, these figures are easily converted to accuracy.

Experiments with Plazi Corpus (lower half of Table 5). Even without lexicons, i.e., solely relying on the structure inference and morphological clues, RefParse clearly outperforms all previous approaches. The accuracy for the title and the journal name are considerably better than reported in [2] for the plain pattern based ParaCite. This clearly emphasizes the benefit of structure inference. With lexicons, accuracy increases for the respective fields and for other fields as well.

Table 5. Experimental results with the Cora Corpus and the Plazi Corpus

	RefParse-g	RefParse-l	ParsCit	FreeCite
	Cora Corpus			
Word / Token	91.5%	89.8%	83.0%	83.8%
Field:				
- Author / Editor	98.6% / 74.6%	98.6% / 78.6%	95.7% / 0%	95.7% / 0%
- Title	79.0%	74.5%	91.0%	91.0%
- Year of Publication	98.8%	99.1%	96.7%	96.7%
- Pagination	97.7%	97.0%	88.9%	1.6%
- Part Designators	96.0%	89.2%	66.7%	96.0%
- Volume Title	38.8%	38.6%	46.3%	50%
- Journal / Publisher	68.0%	61.6%	53.1%	54.2%
Instance	58.4%	52.1%	23.4%	12.2%
	Plazi Corpus			
Word / Token	94.3%	93.7%	78.9%	79.7%
Field:				
- Author / Editor	97.2% / 83.7%	97.7% / 81.0%	88.3% / 0%	88.0% / 0%
- Title	78.4%	78.5%	40.4%	32.4%
- Year of Publication	99.5%	99.5%	95.5%	89.7%
- Pagination	99.3%	99.3%	20.4%	0.3%
- Part Designators	97.7%	95.1%	42.0%	64.3%
- Volume Title	63.2%	52.5%	0.6%	0.3%
- Journal / Publisher	76.6%	75.5 %	54.3%	44.3%
Instance	69.9%	69.2%	65.6%	3.4%

Discussion. With or without lexicons, RefParse significantly outperforms existing approaches. Surprisingly, populating the lexicons of persons, publishers, and periodicals, respectively, increases accuracy only for the author field, while the effect on the other fields is the opposite. We figured this is because with or without out lexicons, RefParse correctly parses most non-pathologic cases, while some lexicon entries incur partial matches in abbreviations and thus prevent the whole data element value from being found. A test run with a lexicon of all the journal and publisher names occurring in the Plazi Corpus has confirmed this assumption: accuracy increases to 92.4% for the respective field, to 97.2% for words / tokens, and to 84.2% for instances.

Generally problematic are reference lists from low-quality OCR, in particular ones whose punctuation is recognized incorrectly or not at all. However, the latter cases pose extreme challenges for blocking based parsers as well, so this problem is not specific to RefParse. Specifically problematic for RefParse are reference lists that deviate from their style guide, mostly due to sloppy editing.

The zero result of ParsCit and FreeCite in the editor field is due to the fact that neither parser can handle editors – this data element is not part of their backing model.

6 Conclusions

The extraction of data elements like author names and titles from the bibliographic references of scientific publications is an essential task when compiling comprehensive bibliographies. Most existing techniques are only applicable for the highly standardized reference styles used in the last two to three decades, however. They cannot handle the wide variety of reference styles used in older publications.

To reliably extract details from lists of bibliographic references, we have presented RefParse. This algorithm integrates existing approaches and adds two distinctive novel features: (1) It handles lists of references together to exploit structural similarities, and (2) it infers the reference style at hand at runtime. Our evaluation shows that RefParse significantly outperforms existing parsers.

Acknowledgements. This research has received funding from the Seventh Framework Program of the European Union (FP7/2007-2013) under grant agreement no 261532 (ViBRANT). RefParse is in productive use as part of RefBank, the platform of ViBRANT to build its Bibliography of Life.

References

1. Biodiversity Heritage Library, http://www.biodiversitylibrary.org/
2. Chen, C.-C., Yang, K.-H., et al.: BibPro: A Citation Parser Based on Sequence Alignment Techniques. In: Proc. AINAW 2008, Okinawa, Japan (2008)
3. CiteseerX, http://citeseerx.ist.psu.edu/index
4. Cortez, E., da Silva, A.S., Goncalves, M.A., et al.: FLUX-CiM: flexible unsupervised extraction of citation metadata. In: Proc. JCDL 2007, Vancouver, BC, Canada (2007)

5. Day, M.-Y., Tsai, R.T.-H., et al.: Reference metadata extraction using a hierarchical knowledge representation framework. Decision Support Systems 43, 152–167 (2007)
6. Councill, I.G., Giles, C.L., Kan, M.-Y.: ParsCit: an open-source CRF reference string parsing package. In: Proceedings of LREC 2008, Marrakech, Morocco (2008)
7. GoogleScholar, http://scholar.google.com
8. Hetzner, E.: A simple method for citation metadata extraction using hidden markov models. In: Proceedings of JCDL 2008, Pittsburgh, PA, USA (2008)
9. Huang, I.-A., Ho, J.-M., Kao, H.-Y., Lin, W.-C.: Extracting Citation Metadata from Online Publication Lists Using BLAST. In: Dai, H., Srikant, R., Zhang, C. (eds.) PAKDD 2004. LNCS (LNAI), vol. 3056, pp. 539–548. Springer, Heidelberg (2004)
10. Krämer, M., Kaprykowsky, H., Keysers, D., Breuel, T.: Bibliographic meta-data extraction using probabilistic finite state transducers. In: Proc. ICDAR 2007, Curitiba, Brazil (2007)
11. McCallum, A., Nigam, K., Rennie, J., Seymore, K.: A machine learning approach to building domain-specific search engines. In: Proc. IJCAI 1999, Stockholm, Sweden (1999)
12. ParaCite, http://paracite.eprints.org/
13. Patashnik, O.: BibTeXing - the original manual. Proceedings of the IEEE, 77 (1988)
14. Peng, F., McCallum, A.: Accurate information extraction from research papers using conditional random fields. In: Proc. HTL/NAACL 2004, Boston, MA, USA (2004)
15. Plazi, http://plazi.org/
16. Takasu, A.: Bibliographic Attribute Extraction from Erroneous References Based on a Statistical Model. In: Proc. JCDL 2003, Houston, TX, USA (2003)
17. ViBRANT: Virtual Biodiversity Research and Access Network for Taxonomy, grant 261532 in EU FP7/2007-2013

Catching the Drift – Indexing Implicit Knowledge in Chemical Digital Libraries

Benjamin Köhncke[1], Sascha Tönnies[1], and Wolf-Tilo Balke[2]

[1] L3S Research Center; Hannover, Germany
[2] TU Braunschweig, Germany
{koehncke,toennies}@L3S.de, balke@ifis.cs.tu-bs.de

Abstract. In the domain of chemistry the information gathering process is highly focused on chemical entities. But due to synonyms and different entity representations the indexing of chemical documents is a challenging process. Considering the field of drug design, the task is even more complex. Domain experts from this field are usually not interested in any chemical entity itself, but in representatives of some chemical class showing a specific reaction behavior. For describing such a reaction behavior of chemical entities the most interesting parts are their functional groups. The restriction of each chemical class is somehow also related to the entities' reaction behavior, but further based on the chemist's implicit knowledge. In this paper we present an approach dealing with this implicit knowledge by clustering chemical entities based on their functional groups. However, since such clusters are generally too unspecific, containing chemical entities from different chemical classes, we further divide them into sub-clusters using fingerprint based similarity measures. We analyze several uncorrelated fingerprint/similarity measure combinations and show that the most similar entities with respect to a query entity can be found in the respective sub-cluster. Furthermore, we use our approach for document retrieval introducing a new similarity measure based on Wikipedia categories. Our evaluation shows that the sub-clustering leads to suitable results enabling sophisticated document retrieval in chemical digital libraries.

Keywords: chemical digital collections, document ranking, clustering.

1 Introduction

During the last years, the information gathering process and the way of searching for literature has changed. Beside text-based Web searches many data providers extended their services by offering user-centered, personalizable Web portals enabling searches over heterogeneous document collections and topical databases. However, the way of information seeking and offered data strongly depends on the scientific domain and the task at hand.

In the domain of chemistry information seeking is essentially focused on chemical entities. Whereas domain experts can easily identify chemical entities and classify them in the context of the document, it is currently a hard task to extract this information automatically. Chemical documents refer to entities usually using trivial names

P. Zaphiris et al. (Eds.): TPDL 2012, LNCS 7489, pp. 383–395, 2012.
© Springer-Verlag Berlin Heidelberg 2012

relying on the reader to figure out the contextual information. But also this does not help indexing: each chemical structure may have several different trivial names, often chosen with respect to the paper's context, e.g., pharmaceutical names, brand names, or terms from natural product chemistry. As always, the challenge for search engines using entity names is to discover all related synonyms and disambiguate terms based on the document context. In particular, practitioners as well as academic researchers are usually interested in finding all related documents to individual chemical entities. For both the search is basically recall-oriented because especially for synthesis procedures missing information about for instance existing patents or expected yields may lead to considerable financial losses.

Facing these problems, chemical information service providers offer specialized indexes. These indexes are built up by *manually* identifying all chemical structures from a document collection in structure databases. The resulting structure databases are accessed through graphical interfaces. By drawing a chemical structure a domain expert can thus formulate a query, which in turn will be parsed by the chemical query parser and matched against entities' fingerprints stored inside the structure database. The amount of manual work required for building up and maintaining such indexes results in high costs. Today, the most important provider is the Chemical Abstract Service (CAS) offering high quality data at a price of about 30,000 USD/year for a single user subscription. Obviously for the growing open access movement this type of indexing documents is not a viable option.

Focusing on the important field of drug design, the information gathering and indexing process is even more complex. A chemist from the area of synthetic chemistry is not only interested in a specific chemical entity, but in a representative of a chemical class adopting a specific role. Especially he is interested in entities having the same or similar characteristic chemical reactions. To assess if a chemical entity is relevant for his task in mind he uses his implicit knowledge about chemical classes and reaction behaviors. The characteristic reaction behavior of a chemical entity is defined by its functional groups. Functional groups are specific groups of atoms that will undergo the same or similar chemical reactions independent of the molecule they are part of. However, currently there is no knowledge base available allowing for this kind of automatic classification of chemical entities.

The goal of this paper is to open up the chemical knowledge stored in the Web for the field of drug design, respectively chemical synthesis. We will introduce an approach allowing a chemist from the field of drug design to search for documents containing chemical entities reflecting his implicit similarity perception. The first part in the introduced workflow is responsible for the automatic extraction of chemical entities from document collections and the index creation with synonym mark-up and disambiguation. We rely on our approach described in [1] to create for each document in the collection an extended index page including synonyms and different representations of all included chemical entities. In a next step, we use the entity's structure information included in the index pages to extract all functional groups. Entities having the same functional groups are grouped together into one cluster. Since for most entities this segmentation still does not fit to the chemist's implicit perception of chemical classes, the respective clusters are further refined into sub-clusters using fingerprint based similarity measures. We use the resulting clusters in a document retrieval scenario showing that the computed sub-clusters decrease the number of compared entities a lot while still including almost all relevant documents. Furthermore, we introduce a similarity measure dealing with the specific requirements of not only taking the simple occurrence of the query entity into account but reflecting the implicit similarity perception of the chemist.

The rest of the paper is organized as follows: In section 2 we will give an overview of the related work. Section 3 introduces our workflow for clustering chemical documents based on the chemist's implicit knowledge. Our evaluation is described in section 4. We will conclude with a summary and outlook to future work in section 5.

2 Related Work

Considering the domain of chemistry, information seeking is focused on chemical entities. In most cases a substructure search is performed using a graphical query interface. However, when searching for entities showing a specific reaction behavior, which is characterized by their functional groups, a substructure search is not sufficient since not all relevant entities will be found [2]. The approach described in [2] shows how chemical entities can be indexed to perform a functional groups search in a fast way. But, as we will show later it is still necessary to further decompose the resulting entity sets to meet the chemist's implicit knowledge. Another interesting approach is introduced in [3] where a small ontology for functional groups is built.

For document retrieval the first necessary step is the extraction of all chemical entities from the documents. For an automatic extraction the only open source chemical entity recognition tool currently available is the OSCAR framework [4], which can identify and extract multiple name variations of chemical entities. In combination with name-to-structure algorithms these entity names can be transformed into chemical structure information [5]. Besides structural information also several text-based formats have been developed for the internal digital representation and exchange of structures. Morgan [6] and Gluck [7] have developed algorithms to store two-dimensional atom-bond structural representations of chemical entities in a tabular form, so-called connection tables. In addition, linear notations have found widespread use. The early Wiswesser Line Notation (WLN) [8] or the later SMILES [9] are representations of chemical structures in the form of a linear string of alphanumeric symbols. The latest development is the InChI Code, an open standard for chemical structure description, by the IUPAC [10].

The idea of clustering data into groups of similar objects is widely used through almost all domains. [11] gives a comprehensible review of different clustering techniques used in data mining. Also the need for clustering chemical entities is discussed in several papers. An early approach described in [12] uses hierarchical clustering based on ring substituents derived from the WLN. The authors compare different clustering techniques based on two small datasets. In [13] an algorithm is presented, called *HierS*, which is also based on a hierarchical clustering method. The algorithm is unsupervised and uses explicit topological chemical graphs to construct hierarchical relationships between ring features of chemical compounds. An overview of different clustering methods used in computational chemistry can be found in [14].

In our scenario clusters are built using specific properties of chemical entities. All entities in one cluster have the same functional groups and have to be representatives of the same chemical class. For building sub-clusters we use a fingerprint-based similarity measure computed between each pair of entities. However, for similarity computation between chemical entities many different measures are available. The first necessary step for computing similarity is the transformation of a chemical substance into a fingerprint. Of course, the idea of measuring the similarity between two objects, each defined by a set of common attributes, is also discussed in many other domains, like e.g. biology [15]. The important point is that the used similarity coefficients are almost the same across all different application areas. Researchers from all domains have worked on finding the most meaningful measure. For chemical information systems the work done by Willett et.al, [16] and [17], gives overviews of the coefficients that have found widespread use. However, only a few comparative studies are available. Hubalek collected 43 similarity measures for the field of biology. After evaluating similarities, correlations, transformations of the value range and symmetry, 23 were excluded. The remaining measures were used for clustering fungi data resulting in five clusters of related coefficients [15]. For the domain of chemistry, Willet evaluated 13 similarity measures for binary fingerprint code [18]. Our approach described in [19] uses these measures and combine them with different fingerprint representations to identify correlation between them. Since many of them are uncorrelated we presented a personalized retrieval system for chemical documents using a feedback engine to find the best similarity measure for each individual chemist.

3   Clustering of Chemical Documents

The following scenario showcases the daily tasks of a researcher in the chemical domain. Assume our scientist is interested in anti-tuberculosis drugs, their pharmacological activities and synthesis. He may start by looking for information about *Isoniazid* and related drugs. *Isoniazid* is an organic compound and the treatment of choice for tuberculosis. Naturally our researcher is looking for experimental procedures for the synthesis of *Isoniazid*-like structures as he would like to minimize the side effects and the risk of resistance. In a first step, the chemist analyzes the structure of *Isoniazid* and identifies the parts of the molecule responsible for the specific reaction behavior. Furthermore, he implicitly knows that *Isoniazid* belongs to the chemical class of *hydrazines*. In particular, he is interested in chemical substances having the same reaction behavior and chemical class as *Isoniazid*. As starting point, our chemist will use *4-cynaopyridine* as it is already available in his laboratory. The question to solve is how to synthesize *4-cynaopyridine* to get a substance having the same functional properties as *Isoniazid*. Therefore, our chemist is searching for literature where the synthesis of *Isoniazid*-like structures is described. Furthermore, also the chemical entity *4-cynopyridine* should be included as educt. The first step is the search for entities with the same functional groups as *Isoniazid*: *hydrazine derivative, aromatic compound, carboxylic acid hydrazide, heterocyclic compound.* Furthermore, relevant entities must also belong to the class of *hydrazines.* Finally the result set is filtered for papers including the chemical entity *4-cynopyridine* as an educt. As final step, the chemist can now examine, if the reaction described in the papers can also be used for his chemical entity.

To allow for these types of searches in a document retrieval system it is necessary to detect the role of a chemical entity in the respective document. In addition, the even more important part and the main focus of this paper is the question of how to reflect the chemist's perception of chemical entities belonging to the same chemical class.

Our idea is to build clusters of chemical entities based on their functional groups. Each cluster describes a class of entities with similar reaction characteristics. Our approach comprises the following four steps, which are illustrated in Fig. 1.

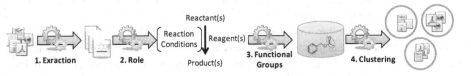

Fig. 1. Workflow Overview

Extraction of Chemical Entities: The first necessary step is to extract all chemical entities from the documents. We used our approach described in [1] to create an enriched index page for each document in the collection. The resulting index pages include further metadata, like synonyms, SMILES and InChI codes for all extracted chemical entities. The corresponding metadata has been extracted from a dumb of the PubChem database that includes all required information.

Role Identification: We have to determine the role of an entity within the document. Therefore, we analyzed our chemical document collection and identified 36 often used lexico-syntactic patterns. For a complete overview of the patterns see our website[1]. These patterns are in pseudo code, meaning that *[CHEMICAL]* is a placeholder for a recognized named entity, marked up by OSCAR. The placeholders are numbered consecutively and replaced by the respective role.

Calculation of Functional Groups: To determine the functional groups of a chemical entity the chemical structure of this entity is analyzed. There are different representations of chemical structures in the form of linear string notations. Usually the SMILES notation is used for functional groups analysis. We rely on the command line utility *checkmol*[2] to determine the functional groups of a chemical entity. Checkmol analyzes the input molecule for the presence of approximately 200 functional groups. We analyzed the output in a first short experiment with a group of domain experts and find out that checkmol simply recognize the presence of an *aromatic ring* but does not further investigate the dimension of contained aromatic rings. To

[1] http://www.13s.de/vifachem/pattern.pdf
[2] http://merian.pch.univie.ac.at/~nhaider/cheminf/cmmm.html

enhance the quality of the resulting clusters, we added an extra parsing step to check-mol's output, to *determine* the dimension of an aromatic ring, resulting in n/m-aromatic rings, where n stands for the number of contained aromatic rings and m for the number of connected ring groups.

Building (Sub-)Cluster: Each entity is located in one cluster and all other entities in that cluster have the same functional groups. These clusters were analyzed by a group of domain experts. The result was that some clusters are not specific because they contain entities from different chemical classes not fitting to the chemists' implicit perception. Therefore, we further decomposed such clusters into sub-clusters by computing fingerprint based similarity between the included entities. Fingerprints encode molecular structures in a series of binary digits (bits) where bits are set according to occurrences of particular structural features. For generating fingerprints, again the entity's unique SMILES representation is used. There are several types of chemical fingerprints focusing on different fragments of chemical entities [19].

As clustering algorithm we choose a partitioning method which constructs k partitions of the data. Each partition represents a cluster and satisfies the following requirements: each group must contain at least one object and each object must belong to exactly one group. One of the most famous algorithms from this group is the k-means clustering which we chose using the Weka[3] API.

3.1   Cluster Based Document Retrieval

Each document is associated to the clusters based on its included chemical entities. To search for documents, the query entity is assigned to the respective sub-cluster and all related documents are retrieved. Instead of just delivering all documents, the result set is ordered according to a similarity measure. In our scenario a document is relevant for a query term if some chemical entity in the document has the same functional properties, respectively the same chemical class, as the query entity. Therefore, we need a specific similarity measure not only taking the simple occurrence of the query term into account. We developed a measure which is based on the Wikipedia category information. In [20] we have shown that Wikipedia categories can be used to describe the content of chemical documents. The Wikipedia categories are structured in a taxonomic tree based on the relationships between them. Here, the idea is to retrieve for each document the associated categories based on the included chemical entities. Since Wikipedia includes information from many different domains it is not sensible to use the whole category tree for describing chemical entities. The evaluation in [20] has shown that only categories that are up to two nodes away from the query node are useful. We retrieve the respective categories for each query term and each document in the query's sub-cluster. The documents are ranked according to the following similarity measure:

$$swc(q_i, d_j) = \frac{cq_i d_j}{cq_i} \times \frac{cd_j}{ed_j}$$

---

[3] http://www.cs.waikato.ac.nz/ml/weka

where q_i is the query term and d_j the respective document. The *swc* measure consists of two parts. The first quotient divides the number of categories found for query term i in the respective document j ($cq_i d_j$) by the total number of categories found for query term i (cq_i). The second quotient divides the total number of categories for the document (cd_j) by the total number of chemical entities found in that document (ed_j).

4 Experiments

For our evaluation we used a dump of the PubChem database[4] containing around 31.5 million chemical entities. For each entity we determined the functional groups and created an inverted index with name and entity allocation. In addition, we used a collection of 2588 chemical documents from the journal Archive for Organic Chemistry (ARKIVOC) which is one of the most renowned open access sources for organic chemistry. For each of these documents we extracted the chemical entities and their roles within a reaction.

4.1 Clustering Based on Functional Groups

In the first experiment we want to gain first insights about the entities contained in the PubChem dump. We took the entities SMILES codes and used our extended version of the command-line tool *checkmol* to determine the respective functional groups, resulting in a set of functional groups for each entity. All entities containing exactly the same set of functional groups are grouped into one cluster. The resulting distribution of all 31.5 million entities is shown in Fig. 2 (left).

# Contained Entities	# Clusters
1	773092
$1 < x \leq 10$	816817
$10 < x \leq 100$	226147
$100 < x \leq 1000$	36535
$1000 < x \leq 10000$	3615
$10000 < x \leq 100000$	143
$100000 < x$	0

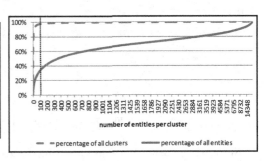

Fig. 2. Cluster sizes (left), Number of entities per cluster (right)

We did a survey with domain experts to analyze the clusters. The result is that clusters containing up to 100 chemical entities are still reasonable for domain experts meaning they correspond to the chemist's implicit knowledge. Therefore, we can discover that 97.84 % of the resulting clusters can already be used. But these clusters only contain around 30% of all chemical entities (see Fig. 2 right). Most of the entities

---

[4] http://pubchem.ncbi.nlm.nih.gov

(around 21 million) are located in the remaining 2.16 % of the clusters. Therefore, it is necessary to split them up into more meaningful clusters.

4.2   Building Meaningful Sub-Clusters

In this experiment we divide all clusters containing more than 100 chemical entities. For choosing a meaningful distance, respectively similarity function, we take the 5 different uncorrelated fingerprint/similarity measure combinations introduced in [19] into account. Each of these measures has different rankings with respect to the underlying fingerprint. To decide for a measure for the sub-cluster computation we evaluated for which of the measures the top-X ranked entities are in the same functional group cluster.

We randomly choose 100 clusters with more than 1000 entities per cluster. In addition, we choose 10 random queries and calculated the similarity between the query entity and all other entities from all clusters. For each entity we have six different fingerprint representations. The similarity is computed using the uncorrelated measures. Fig. 3 shows the average results based on the ten query entities.

The diagrams show that there are big differences between the different combinations. For the *substructure fingerprint* and the *Manhattan* distance always all top-100 entities are in the same functional groups cluster. For the top-1000 entities still around 800 are found in the same cluster. Since this combination retrieves the best results we decided to use it for sub-cluster computation.

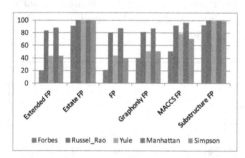 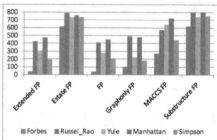

Fig. 3. Top-100 (left), Top-1000 (right)

Since we are using k-means clustering we also had to find a suitable k. The aim is that each entity in a cluster has the same chemical class. Therefore, we took a domain specific ontology including chemical classes as ground truth, the so-called ontology for chemical entities of biological interest (CheBI[5]). To the best of our knowledge CheBI is the only available ontology allowing for automatic classification of chemical entities. Since we are interested in decomposing the clusters with more than 100 entities (around 40000 clusters), we randomly took 2000 clusters (5%) out of this set. Since not all entities from our dataset are included in CheBI, we only choose clusters containing entities also included in the CheBI ontology. The idea is to take all entities

---

[5] http://www.ebi.ac.uk/chebi

from one cluster and assign the associated ontology classes to that cluster. Of course, it is not sensible to use all ontology nodes associated with one chemical entity. Nodes that are too general would lead to huge clusters that are again not meaningful. In [20] it was shown that only ontology nodes are meaningful that are at least three steps away from the entry node. Therefore, we only associated these classes with the respective cluster. We defined that the optimal segmentation is achieved, if all entities with different classes are in different sub-clusters. We manually built the respective sub-clusters and run the k-means algorithm varying the value for k. Our algorithm stops if k-means found the optimal solution, each entity is in one cluster for its own, or if no solution can be found. Evaluating the 2000 clusters we retrieved an optimal k for further splitting up the entities in the functional groups clusters in chemical classes of 4. Whereas CheBI includes for our dataset around 20000 chemical classes we were able to find more than 150000 classes for chemical entities. Please note, we cannot associate exact chemical class names to each cluster, but as we will see later, our results match the perception of the chemist's implicit knowledge of entities belonging to the same class.

4.3   Document Retrieval

During this experiment we evaluate the created clusters in a document retrieval scenario. Our document collection contains 2588 documents from the ARKIVOC journal. We generated enriched index pages using the approach described in [1] and extracted all contained chemical entities. Each document is associated to the functional groups clusters based on its contained entities. The last experiment has shown that the optimal segmentation of the entities in chemical classes is given for 4 sub-clusters. However, we will evaluate if this holds also for a document retrieval scenario.

First we have to randomly chose query entities and assess the relevance of each document for the respective query. Relevance can only be assessed manually by domain experts (in particular chemists), in what is a very expensive process. Therefore, we could not take the entire collection, but chose a subset of documents (still about 10% of the entire collection) for performing a precision/recall analysis. To choose a *representative* subset, we analyzed the number of occurrences of individual chemical entities in the document collection. It is not sensible to choose entities as query terms that either occur in almost every document or are extremely rare. We analyzed all entities occurring in less than 100 documents, but more than 20 documents. Furthermore, the entities should belong to functional groups clusters which have been further decomposed into sub-clusters using the fingerprint based similarity computations.

We retrieved all documents matching these queries and randomly chose a subset of 10%. From these documents we randomly selected a total of around 5% of the occurring entities resulting in 18 textual query terms. For the evaluation domain experts from the field of chemistry considered all retrieved documents with respect to each query and judged the relevance in a binary fashion. A document is marked as relevant if it contains entities having the same reaction behavior and belonging to the same chemical class as the query entity.

Now, we analyze if the sub-cluster decomposition is sensible meaning that all relevant documents for a query term are located in the same sub-cluster. Fig. 4 (left) shows the precision, recall and F-measure values for different values of k. The recall value is always around 93% meaning that some documents from other functional groups clusters were also marked as relevant. But if we are only using the functional groups clusters (k=1) we got a low precision value averaged over all queries of 42%. The precision value slightly increases up to 53% for k equals 12. According to the low precision values the classic F_1-Measure is on average only around 57%. The recall oriented F2-Measure results in an average of 68%.

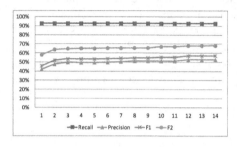 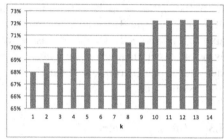

Fig. 4. Recall, Precision and F-Measures for varying k's (left), Mean Average Precision (MAP) for Wikipedia Categories Ranking and varying k's (right)

Please note we can further optimize the precision by using different k values for different queries, respectively different cluster sizes. For example, for query term 2 which is in a quite small cluster containing only 26 documents, already with k equals 2 we have a precision value of 60%. In contrast, for query term 15 the precision value for k equals 2 is 59% and for k equals 9 67%. Regarding all queries, the optimal value for k is varying between 1 and 12. But only 4 queries do not have their optimal precision value for k equals 12.

The last experiment has shown that we have an optimal segmentation for entities for k equals 4. For documents, k equals 4 is already good, but the precision value is slightly higher for k equals 12. We also tested higher values for k, but the precision did not increase anymore. Instead of delivering all documents in the sub-cluster randomly to the user, we also developed a similarity measure to rank the documents according to the query term. Fig. 4 (right) shows the mean average precision values for varying k's. Using the ranking we were able to reach a MAP value for k equals 12 of 72%. However, another interesting point is that even if the clusters include fewer documents, the recall value did not decrease. That means if a user is searching for documents with respect to a chemical entity with some characteristic reaction behavior and implicit knowledge of its chemical class, it is sufficient to find the cluster for this query entity and retrieve all included documents. If the query entity is located in a sub-cluster, it is not necessary to take chemical substances from other sub-clusters into account, even though they have the same functional groups. Our experiment has shown that almost all relevant documents are in the same sub-cluster as the query.

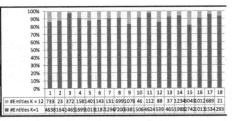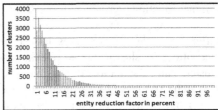

Fig. 5. Number of entities for k=1 and k=12 (left), Number of clusters including x percent of the entities for k=12 compared to k=1 (right)

Fig. 5 (left) compares the number of entities in the functional groups cluster (k=1) to the number of entities for 12 sub-clusters for each query. The figure shows that the cluster sizes decrease on around 90% on average over all queries. Since the recall does not decrease the conclusion is that the cluster quality is high and only irrelevant entities are located in other sub-clusters.

Fig. 5 (right) shows the number of clusters including a certain percentage of entities for k equals 12. For example, there are 3500 sub-clusters where the number of entities has been reduced to 3% of the number of entities for k equals 1. This observation is quite important considering a facetted-search scenario. For example, in the ViFaChem2 portal[6] the user has the possibility to decide for relevant chemical entities after submitting a query. Our evaluation has shown that this set of offered chemical entities can be highly decreased by only considering entities from the same sub-cluster leading to a more sophisticated search experience.

5   Conclusion and Future Work

We presented an approach, clustering chemical entities based on their functional groups to reveal the chemists' implicit knowledge. A manual analysis of the resulting clusters by domain experts has shown that a simple functional groups clustering is not enough since most clusters are too unspecific and do not fit to the chemists' perception of chemical classes. Therefore, we used fingerprint based similarity measures to further divide these clusters into sub-clusters. Even though, we cannot assign explicit class names to the resulting sub-clusters, our evaluation has shown that they reflect the chemists' perception of chemical classes.

Further on, we used the clusters for document retrieval. The documents are assigned to the clusters based on their contained chemical entities. We did a precision/recall analysis with a group of domain experts showing that almost all relevant documents (recall of 93%) are located in the respective sub-cluster of the query. Instead of just delivering all documents from the respective cluster, we also introduced a ranking measure based on Wikipedia categories to further enhance the precision.

Another important point is that, without losing any relevant documents, the number of entities in the sub-clusters is dramatically decreased about 90% compared to the

---

[6] www.chem.de

original functional groups clusters. This is quite important considering chemical Web portals using, e.g., facetted search, like for example the ViFaChem portal. It is a huge difference if the user retrieves 1000 or only 100 possible entries for the facets. Our evaluation has shown that all relevant chemical entities are located in the same sub-cluster. Thus, the number of possible hits is highly decreased resulting in a high quality search experience for the user.

For our future work we plan to investigate if our approach is also useful in other domains. Considering the field of biology, the search for information is also focused on entities, like e.g. genes or proteins. Just as in chemistry, a huge amount of these entities is available having specific characteristics, like e.g. phosphorylation sites that may be used in a clustering approach. We will see if this huge amount of data can also be confined without losing any relevant information in a document retrieval system.

References

1. Tönnies, S., Köhncke, B., Koepler, O., Balke, W.-T.: Exposing the Hidden Web for Chemical Digital Libraries. In: Proc. of the Joint Conf. on Digital Libraries, JCDL (2010)
2. Haider, N.: Functionality Pattern Matching as an Efficient Complementary Structure/Reaction Search Tool: An Open-Source Approach. Molecules 15(8) (2010)
3. Feldman, H.J., et al.: CO: A Chemical Ontology for Identification of Functional Groups and Semantic Comparison of Small Molecules. FEBS Letters 579(21) (2005)
4. Corbett, P., Murray-Rust, P.: High-Throughput Identification of Chemistry in Life Science Texts. In: Berthold, M., Glen, R.C., Fischer, I. (eds.) CompLife 2006. LNCS (LNBI), vol. 4216, pp. 107–118. Springer, Heidelberg (2006)
5. Townsend, J.A., et al.: Chemical Documents: Machine Understanding and Automated Information Extraction. Journal of Organic & Biomolecular Chemistry 2 (2004)
6. Morgan, H.L.: The Generation of a Unique Machine Description for Chemical Structures-A Technique Developed at Chemical Abstracts Service. Journal of Chemical Documentation 5(2) (1965)
7. Gluck, D.J.: A Chemical Structure Storage and Search System Developed at Du Pont. Journal of Chemical Documentation 5(1) (1965)
8. Smith, E., Baker, P., Wiswesser, W.: The Wiswesser Line-Formula Chemical Notation (WLN). Chemical Information Management (Cherry Hill, N.J.) 102(2) (1975)
9. Weininger, D.: SMILES, A Chemical Language and Information System. 1. Introduction to Methodology and Encoding Rules. Journal of Chemical Information and Modeling 28(1) (1988)
10. Stein, S.E., Heller, S.R., Tchekhovskoi, D.: An Open Standard For Chemical Structure Representation: The IUPAC Chemical Identifier. In: Proc. of the International Chemical Information Conference (2003)
11. Berkhin, P.: A Survey of Clustering Data Mining Techniques. Journal of Grouping Multidimensional Data (2006)
12. Adamson, G.W., Bawden, D.: Comparison of Hierarchical Cluster Analysis Techniques for Automatic Classification of Chemical Structures. Journal of Chemical Information and Modeling 21(4) (1981)
13. Wilkens, S.J., Janes, J., Su, A.I.: HierS: Hierarchical Scaffold Clustering Using Topological Chemical Graphs. Journal of Medicinal Chemistry 48(9) (2005)

14. Downs, G.M., Barnard, J.M.: Clustering Methods and their Uses in Computational Chemistry. Reviews in Computational Chemistry 18 (2002)
15. Hubálek, Z.: Coefficients of Association and Similarity, Based on Binary (Presence-Absence) Data: An Evaluation. Journal of Biological Reviews 57(4) (1982)
16. Willett, P., Barnard, J.M., Downs, G.M.: Chemical Similarity Searching. Journal of Chemical Information and Modeling 38(6) (1998)
17. Holliday, J., Hu, C., Willett, P.: Grouping of Coefficients for the Calculation of Intermolecular Similarity and Dissimilarity Using 2D Fragment Bit-Strings. Journal of Combinatorial Chemistry; High Throughput Screening 5(2) (2002)
18. Willett, P.: Similarity-based Approaches to Virtual Screening. Journal of Biochemical Society Transactions 31 (2003)
19. Tönnies, S., Köhncke, B., Balke, W.-T.: Taking Chemistry to the Task – Personalized Queries for Chemical Digital Libraries. In: Proc. of the Joint Conf. on Digital Libraries, JCDL (2011)
20. Köhncke, B., Balke, W.-T.: Using Wikipedia Categories for Compact Representations of Chemical Documents. In: Proc. of the Int. Conf. of Information and Knowledge Management, CIKM (2010)

Using Visual Cues for the Extraction of Web Image Semantic Information

Georgina Tryfou and Nicolas Tsapatsoulis

Cyprus University of Technology,
Department of Communication and Internet Studies,
Limassol, Cyprus

Abstract. Mining information for the images that currently exist in huge amounts on the web, has been a main scientific interest during the past years. Several methods have been exploited and web image information is extracted from textual sources such as image file names, anchor texts, existing keywords and, of course, surrounding text. However, the systems that attempt to mine information for images using surrounding text suffer from several problems, such as the inability to correctly assign all relevant text to an image and discard the irrelevant text as well. A novel method for extracting web image information is discussed in the present paper. The proposed system uses visual cues in order to cluster a web page into several regions and assign to each hosted image the text that most possibly refers to it. Three different approaches to the problem of text to image assignment are discussed and evaluated. The evaluation procedure indicates the advantages of using visual cues and two dimensional euclidean measures for extracting information for web images.

1 Introduction and Related Work

Currently, there are millions of images available on the web and the need for an efficient method for indexing and retrieving them has arouse. Web search engines share the objective to offer to the user an intuitive image search by minimizing the necessary human interaction for the optimization of the results. Currently, two main approaches exist in the literature for content extraction and representation of web images: (i) text-based and (ii) visual feature-based methods.

In this work we propose a text based method for indexing web images. There are several systems that use textual information in order to search for images on the web, for instance WebMARS [1], ARTISTIC [2] and the method described in [3] are some representative examples. These systems share one or more of the following drawbacks: the text in the web page is only partially processed; only a few words are considered as textual features; it is not clear how textual information is used to support image indexing and retrieval; term lists or taxonomies that are built in the set-up phase of the systems demand high user intervention [2]. Opposed to that, the proposed system processes the web page as a whole

P. Zaphiris et al. (Eds.): TPDL 2012, LNCS 7489, pp. 396–401, 2012.

ind attempts to assign each text block to the image it refers to. Moreover, any word or phrase is a possible keyword or representation for an image and no user intervention is necessary once the indexing model is built.

The paper is organized as follows: Section 2 presents the architecture of the proposed system while Section 3 presents the evaluation of it. Finally some conclusions and future perspectives are given in Section 5.

2   System Architecture

The general architecture of the proposed system is depicted in Fig. 1. As shown here, the system consists of two main parts, namely the "Visual Segmentation" and the "Text to Image Assignment" parts.

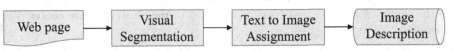

Fig. 1. The general architecture of the proposed system

2.1   Visual Segmentation

The content extraction of each web image is based on textual information that exists in the same web document and refers to this image. In order to determine to which image the various text parts of a web page refer to, we use the visual cues which are connected to the outline and the presentation of the hosting web page. In order to obtain this set of visual segments that form the web page, we use the Visual Based Page Segmentation (VIPS) algorithm [4], with the Permitted Degree of Coherence (pDOC) set to its maximum value.

2.2   Text to Image Assignment

Each text block found in the web page has to be assigned to an image block. After this assignment, the text blocks will form the image description. Fig. 2 shows how this assignment is done and each part is described in further details in the following few paragraphs.

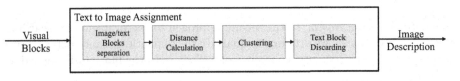

Fig. 2. The processing that takes place in the second part of the system: Text to Image Assignment

Image/Text Block Separation. The first task towards the assignment of each text block to the image block it refers to, is to determine whether a block contains an image or text. We use the HTML source code that corresponds to each block (as returned from the execution of the VIPS algorithm) in order to classify it into the correct category.

Distance Calculation. Once the blocks are separated into images and textual blocks, the Euclidean distance between every image/text block pair is calculated making use of the Cartesian coordinates, which are returned by the VIPS algorithm. The distance calculation in this case is not a trivial problem since the goal is to quantify the intuitive understanding of how close two visual blocks are. This quantification is done by calculating the Euclidean distance between the closest edges of each pair of blocks.

Clustering. After the distance calculation, the web page has to be clustered into regions. Each region is defined by a web image which is also considered to be the center of it. Each text block is assigned to the cluster, whose center it is closest to.

Text Block Discarding. Until here, each web page is separated into one or more clusters and all blocks of text that exist in the document are assigned to their closest image. However, it is possible that one or more blocks of text do not refer to a certain image. For this reason, it is necessary to discard one or more text blocks from the calculated clusters. In order to determine which blocks of text are irrelevant to the image they are connected to, we took into consideration 3 different approaches. In each approach we attempt to obtain a threshold that discards blocks based on euclidean or semantic distances. In the following paragraphs these methods are described.

Euclidean Distance Based. In this approach, a text block is discarded when its distance to the cluster center (*i. e.* corresponding web image) is bigger than a defined threshold t. In order to calculate this threshold, the distances that appear in a cluster are firstly normalized and using the normalized values, $\tilde{d}_i^c$, the threshold t is calculated as follows:

$$t_{ed} = t' + m_{ed} - s_{ed} \ , \tag{1}$$

where $t' = 0.1$ a static, predefined threshold, m_{ed} the mean value of the normalized euclidean distances found in the cluster c and s_{ed} the standard deviation of these distances.

Semantic Distance Based. In the second approach to the problem of discarding irrelevant text blocks, we attempt to quantify the semantic distance between two blocks of text. The semantic coherence of a block b_i to the block b_j is defined as follows:

$$C_{ij} = \frac{c_1 + c_2 + \ldots + c_N}{|b_i|} = \frac{\bigcup_{k=1}^{N} c_k}{b_i} \ , \tag{2}$$

where $c_1, c_2, \ldots, c_N$ the common sub-sequences of the blocks b_i and b_j. Similarly, we also calculate the semantic coherence C_{ji} of block b_j to block b_i. The harmonic mean of C_{ij} and C_{ji} expresses the semantic coherence of the two blocks, and is notated as sc. The semantic distance sd of the blocks is therefore calculated as follows:

$$sd = 1 - sc \ , \tag{3}$$

In every region of the web page only blocks that share high semantic coherence should exist. In every cluster, the semantic distance of each text block to the rest of the blocks is calculated and an adaptive threshold is defined as in (1) – this time using the mean value m_{sd} and standard deviation s_{sd} of the semantic distances:

$$t_{sd} = t' + m_{sd} + s_{sd} \ . \tag{4}$$

Vector Space Model Based. Semantic information, as obtained from the application of the Vector Space Model, is used in this approach, in order to discard text blocks from the initial clusters. Each text block may be considered as a document which is expressed as a vector of term weights. The corpus, from which the vocabulary is extracted, is the set of all the text blocks, *i.e.* the whole web page. The Vector Space Model, as described in [5], is employed for the current task. Each text block is considered as a query q and its similarity to the rest of the text blocks of the same cluster is calculated using the cosine similarity measure. The average similarity of each text block to the rest of the blocks in the same cluster, $\overline{sim}(q)$ is calculated and used for obtaining a measure of semantic distance, this time with the use of the Vector Space Model:

$$vd_q = 1 - \overline{sim}(q) \ . \tag{5}$$

As before, the goal is to allow in every region of the web page only blocks that have low semantic distance to co-exist. The adaptive threshold is again defined as in (1):

$$t_{vd} = t' + m_{vd} + s_{vd} \ , \tag{6}$$

where m_{vd} the mean value of Vector Space Model distances and s_{vd} the corresponding standard deviation.

3   Evaluation

3.1   Defined Measures

In order to evaluate the page segmentation and clustering results we use measures inspired from the traditional information retrieval measures *Precision, Recall* and *F-measure*.

Let A be the manually labelled region that the annotator determined to be connected with an image and $tb_1, tb_2, \ldots, tb_N$ the N regions that the proposed system assigned to this image. As shown in Fig. 3, there are three possible cases: (i) only a small part (sub-block) of tb_i is correctly identified (for instance in block

Fig. 3. The possible cases of textual blocks assigned to an image with annotation A. The grey regions correspond to the correctly identified parts of the blocks.

tb_1), (ii) the whole block tb_i is correctly identified (*e.g.* blocks $tb_2, \ldots tb_{N-1}$) and (iii) no part of the block tb_i is related to the image (block tb_N).

For all cases, we define $\tilde{tb}_i$ the correctly identified part of a block tb_i. Using the above notation we define the *content Precision*, C_p, of the system as follows:

$$C_p = \frac{\tilde{tb}_1 + \tilde{tb}_2 + \ldots + \tilde{tb}_N}{A} = \frac{\left| \bigcup\limits_{i=0}^{N} tb_i \bigcap A \right|}{|A|} . \tag{7}$$

The *content Recall*, C_r, is similarly defined as:

$$C_r = \frac{\tilde{tb}_1 + \tilde{tb}_2 + \ldots + \tilde{tb}_N}{tb_1 + tb_2 + \ldots + tb_N} = \frac{\left| \bigcup\limits_{i=0}^{N} tb_i \bigcap A \right|}{\left| \bigcup\limits_{i=0}^{N} tb_i \right|} . \tag{8}$$

Using equations 3 and 4 the *content F-measure*, C_f, is the harmonic mean of measures C_p and C_r:

$$C_f = 2\frac{C_p C_r}{C_p + C_r} . \tag{9}$$

The C_f measure is used for the evaluation of the proposed system.

3.2 Results

Using the above described evaluation measures, we evaluated a dataset that consists of 40 realistic web pages and a total of 131 images. We obtained three different sets of results, each one for the three different text block discarding methods. The results are presented in Table 1. It is clear that the use of the Euclidean distances is the best method to use and this may be due to the use of Euclidean coordinates for the initial clustering of the web page, as well as the lack

Table 1. The results for different Text Block Discarding Methods

Method	C_p	C_r	C_f
Euclidean Distance	0.8836	0.9656	0.8650
Semantic Distance	0.5143	0.6495	0.4736
Vector Space Model	0.5865	0.7656	0.5795

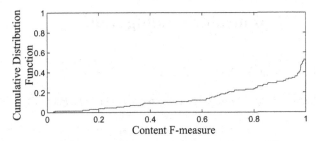

Fig. 4. Cumulative distribution function for the results based on the defined evaluation measures, and the use of Euclidean Distances for text block discarding

of a sophisticated way to locate the semantically relevant terms among various text blocks. In order to offer a better visualization of the results, the cumulative distribution function of them, obtained from the text to image assignment is presented in Fig. 4.

4   Conclusions

In this paper a system that extracts textual information for web images has been described. The method uses visual cues in order to identify the segments that form a web page and delivers a clustering of the contents of the web page in order to assign textual information to the existing images. Although the results so far are encouraging, it is critical for generalization purposes to also evaluate the system for a larger dataset, preferably annotated from more than one annotators. Moreover, the segmentation of the web page into visual blocks should be evaluated in a way that indicates the success of the VIPS algorithm in determining the segments that form a web page. Improving the calculation of semantic distances among text blocks with the use of a thesaurus and an ontology that describes the connections among different phrases is the direction of our future experiments. It is expected that a more appropriate measure for semantic distance calculations, and its combination to the euclidean distances will provide an even better assignment of text blocks to the corresponding images.

References

[1] Ortega-Binderberger, M., Mexico, A.: Webmars: A multimedia search engine for the world wide web (1999)
[2] Alexandre, L., Pereira, M., Madeira, S., Cordeiro, J., Dias, G.: Web image indexing: Combining image analysis with text processing. In: Proceedings of the 5th International Workshop on Image Analysis for Multimedia Interactive Services, WIAMIS 2004 (2004)
[3] Alcic, S., Conrad, S.: A clustering-based approach to web image context extraction. In: MMEDIA 2011 (2011)
[4] Cai, D., Yu, S., Wen, J.R., Ma, W.Y.: Vips: a vision based page segmentation algorithm. Technical report, Microsoft Research (2003)
[5] Salton, G., Wong, A., Yang, C.S.: A vector space model for automatic indexing. Commun. ACM 18(11), 613–620 (1975)

Malleable Finding Aids[*]

Scott R. Anderson[1] and Robert B. Allen[2]

[1] Ganser Library, Millersville University, Millersville, PA USA
sanderson@millersville.edu
[2] School of Information Management
Victoria University of Wellington, Wellington NZ
robert.allen@vuw.ac.nz

Abstract. We show a prototype implementation of a Wiki-based Malleable Finding Aid that provides features to support user engagement and we discuss the contribution of individual features such as graphical representations, a table of contents, interactive sorting of entries, and the possibility for user tagging. Finally, we explore the implications of Malleable Finding Aids for collections which are richly inter-linked and which support a fully social Archival Commons.

Keywords: Archives, Finding Aid, User Engagement, Wiki.

1   Towards Malleable Finding Aids

There have been increasing calls for an archival commons. Leveraging what users and archivists are already doing to enhance access or description in an online environment (and not as another process), is inherent in processes to bring it into the "archival box" of records contextualization [6]. Ian Anderson [2] identifies broad functionalities an archive can offer the user in a six-tiered model of archival functionality, ranging from the static poster to interactive elements. Anderson and Allen [4] describe several services which might be supported by an archival commons. These include: linked data; continuous (re)arrangement; tagging and "folksonomies"; names service; annotation and contribution with narrative tools; provenance of the narrative; reputation of the agents; recommendations and collaboration; collection-to-collection association; and visualization.

Strong finding aids would be key to an archival commons that engages users. Finding aids are the primary way for users to understand what is in an archival collection. They were originally developed to be used in conjunction with the materials they describe, not as stand-alone documents. The overview provided by collection-based finding aids is important to researchers as they search for suitable materials [5, 7]. However, finding aids are unfamiliar and, possibly, opaque to uninitiated users [13, 14]. Moreover, while traditional finding aids are helpful, they are generally inadequate to meet researcher

---

[*] This work was completed while both authors were at Drexel University.

P. Zaphiris et al. (Eds.): TPDL 2012, LNCS 7489, pp. 402–407, 2012.
© Springer-Verlag Berlin Heidelberg 2012

needs or preference in the electronic environment [3]. Even when they are online, finding aids are generally passive. They do not engage users. This passivity may make some basic tasks such as known-item searching nearly impossible [9]. Yet, a finding aid generated from an Encoded Archival Description (EAD) need not be passive. Gilliland-Swetland [8] describes the possibility of on-the-fly-faceting of a group of finding aids providing, for instance, teachers with the ability to note specific educational materials for instructional purposes, and cross EAD searching.

Given the difference between what actually exists (often unfamiliar and sparse finding aids), and what is desired by users (rich detail at the item level in a familiar presentation or environment), Yakel proposed that the reference mission for archivists should move beyond the simple provision of finding aids towards a new mission of active engagement with users, records, and systems, so that "clienteles can effectively use records". This mission should be accomplished in the user's time and place, without loss of context of what was (from the record perspective), while recognizing the many contexts of user need or intention (item tracking, summation, comparison, etc. of records). To this end, Light and Hyry [11] introduce the idea of colophons and annotations to capture the context of the work being done when processing materials within the context of the archive and subsequent use. Krause and Yakel [10] provide a test bed in which six interactive features are embedded in the Polar Bear Expedition Collections. Their Next Generation Finding Aid group implemented bookmarking, user commentary, link paths, browsing categories, searching, and (optional) user profiling. These are significant steps for making selected materials more malleable, accessible, and useful to a broader range of users. We develop a still more extensive set of services and base those services on a readily available development platform.

2   A Prototype Malleable Finding Aid

We aim to support users to more readily find items in a collection by the: (1) addition of descriptors to or about the collection; (2) annotation or addition of related objects that are better tied to general Web-based social or discovery services; and (3) facilitation of possible alternative arrangements of objects to develop new insights or meaning derived for a specific purpose or use. We implemented a prototype finding aid with these attributes in an off-the-shelf Wiki[1]. This prototype natively supports many interface features similar to those we have recommended for an archival commons [4]. For content we used an award-winning finding aid for the T.C.H. Jacobs collection[2] which was created and posted as a static Web page.

---

[1]   We used the Confluence Wiki which is a commercial application with liberal educational and not-for-profit licensing. http://www.atlassian.com/ software/confluence/

[2]   A finding aid for the T.C.H Jacobs Literary Papers at Pennsylvania State University was used as a example:
http://www.libraries.psu.edu/dam/psul/up/digital/findingaids/4995.htm

This Wiki-based Malleable Finding Aid implements the following services:

Key-Word Tags/Labels: The wiki platform has a built-in "labeling" function that allows users to contribute tags. In addition to appearing on the finding aid itself, the contributed tags are ingested into a "tag cloud" that aggregates all the tags in the defined space and can be applied to other wiki based pages (such as other finding aids in the same wiki space).

Table of Contents: The wiki application has a self-generating table of contents. In addition to providing navigational help with a familiar "look and feel", the table of contents adapts to changes in the section structure while also providing visual clues to the structure.

Brief Overview: This service implements a three-part section of the finding aid. First, a three-sentence narrative overview of the collection is created, drawing from the biographical and summary information in the collection. Second, there is a colored bar chart of the number of folders represented in each section. Third, some administrative information about the collection is included in a table.

Graphical Representations: In addition to the bar chart in the Brief Overview, there is a WorldCat-database-generated timeline with information about T.C.H. Jacobs's literary output linked from the "prolific author" phrase in the Brief Overview.

Maps: Another page shows an interactive Google map indicating where T.C.H. Jacobs was born.

Timeline: A basic timeline generated from information outside the collection itself is included as a graphical element through the "prolific author" link in the Brief Overview.

List Sorting: Traditional finding aids have a vertically-oriented and nested box/folder presentation which makes sorting or filtering difficult. The same information can be presented in a column-based table with sorting functionality. This approach compresses the same information into less space and is a well-known form for manipulation. In addition, collapsible sections can be introduced with links to larger pages of secondary information (bibliography, etc.).

Related Materials: A section of additional materials with three subsections is included in the finding aid and the table of contents.

Bibliography: The bibliography of works by T.C.H. Jacobs is partially wiki-enabled with functionalities such as sorting and linking.

Linking/Analytics Section: A section about the page provides common information, such as the inbound and outbound links spread throughout the finding aid, aggregated together in one place for quick viewing.

Comments: Comments are allowed at the bottom of the wiki-based finding aid. These are similar to commenting features which are commonly found on many social websites and services.

Page History/Recent Changes: The "page history" functionality is exposed via the "tools" drop-down icon on the wiki pages and allows users to make comparisons between page versions.

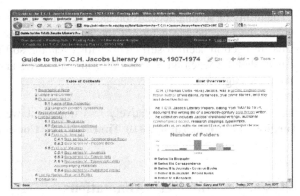

Fig. 1. Malleable Finding Aid page with table of contents, brief overview, links in, and graphical representation

3   Feedback about the Malleable Finding Aid

To compare the static Web-page version to an enhanced and malleable Wiki-based version, a 75-minute interview was conducted with a practicing academic librarian. This librarian is also a former special collections cataloguer with deep understanding of finding aids, cataloging practice, subject expertise in literature, and working knowledge of instructional pedagogy. In short, the user is especially well qualified to critique finding aids. The interview was conducted to highlight obvious advantages and weakness, than as a systematic user evaluation. The features of the wiki-based finding aid were explained and ratings of the utility of those features were collected. The features are listed below in rank order. Comments about the features made during the interview are also included.

#1 Table of Contents: The most useful feature was judged to be the interactive table of contents. The wiki table of contents was endorsed also because its purpose and functionality were instantly recognizable.

#2 Graphical Representations: The bar chart in the Brief Overview indicating folder volume within each series was deemed very valuable for its ability to convey at a glance where the bulk of the materials were within the various series.

#3 Brief Overview: This section was deemed very useful for providing a summary overview of some of the longer sections. The interviewee said it was like an executive summary adjacent to the table of contents and excelled at providing a basic context for the remainder of the document.

#4 Map: The interviewee said, "I frequently do mapping anyway" as a routine matter of trying to figure out where various geographic locations or entities are located either independently or in relation to one another.

#5 Linking: The interviewee said "Linking is natural" for aspects of this content. Not only is linking a very common activity, it provides all manner of "expected" functionality and navigational assistance.

#6 List Sorting: The utility of sorting depends on the size of the series and/or number of folders. Given the manageable nature of the individual series/folders in this collection, visual inspection seemed sufficient. However, the sorting utility was thought to be more helpful when the *entire* collection (all series and folders) was brought together in a single list.

The services which were judged to be relatively less important were **#7 Related Materials, #8 Tagging, #9 Bibliography, #10 Timeline, #11 Comments,** and **#12 Page History.**

4   Linked Resources and Social Annotations

The use of a Wiki as the platform for Malleable Finding Aids has advantages beyond natively providing interface features. It can be readily expanded; for example, each resource could be given a separate page. Moreover, multiple finding aids and their contents could be more readily inter-linked as a step toward modular digital libraries.

Thus, Malleable Finding Aids can be a step toward granular digital collections with rich conceptual models to represent the contents. In addition, the Wiki platform could support social annotation which is a key component of the archival commons. Zarro and Allen [15] explore the types of user annotations for individual items in a site with photos from LC. Ultimately, annotation may go beyond simple comments to weaving complex stories and positions. Indeed, those could be presented as trails in a graphical interface [1].

Interactive participation in traditional archival activities can address some weakly documented societal perspectives. The records or collections in an archive must be connected in both *strong* (well documented) and *weak* (less documented) societal settings to facilitate the ability to make full use of these collections by both those with in-depth knowledge and those with passing interest [12].

References

[1] Allen, R.B.: Using Information Visualization to Support Access of Archival Records. Journal of Archival Organization 3(1), 37–49 (2005)

[2] Anderson, I.G.: Necessary but not Sufficient: Modelling Online Archive Development in the UK. D-Lib. Magazine 14(1/2) (2008)

[3] Anderson, I.G.: Are you Being Served? Historians and the Search for Primary Sources. Archivaria 58, 81–129 (2004)

[4] Anderson, S.A., Allen, R.B.: Envisioning the Archival Commons. American Archivist 72(2), 383–400 (2009)

[5] Case, D.O.: The Collection and Use of Information by Some American Historians, a Study of Motives and Methods. Library Quarterly 61(1), 61–82 (1991)

[6] Duff, W., Fox, A.: 'You're a Guide Rather than an Expert': Archival Reference from an Archivist's Point of View. Journal of the Society of Archivists 27(2), 129–153 (2006)

[7] Duff, W., Harris, V.: Stories and Names: Archival Description as Narrating Records and Constructing Meanings. Archival Science 2(3-4), 263–285 (2002)

[8] Gilliland-Swetland, A.J.: Evaluation Design for Large-Scale, Collaborative Online Archives: Interim Report of the Online Archive of California Evaluation Project. Archives & Museum Informatics 12(3-4), 200 (1998)

[9] Gilliland-Swetland, A.J.: Popularizing the Finding Aid: Exploiting EAD to Enhance Online Discovery and Retrieval in Archival Information Systems by Diverse User Groups. Journal of Internet Cataloging 4(3), 207–208 (2002)

[10] Krause, M.G., Yakel, E.: Interaction in Virtual Archives: The Polar Bear Expedition Digital Collections Next Generation Finding Aid. American Archivist 70(2), 282–314 (2007)

[11] Light, M., Hyry, T.: Colophons and Annotations: New Directions for the Finding Aid. American Archivist 65(2), 216–400 (2002)

[12] Shilton, K., Srinivasan, R.: Participatory Appraisal and Arrangement for Multicultural Archival Collections. Archivaria 63, 87–101 (2007)

[13] Yakel, E.: Thinking Inside and Outside the Boxes: Archival Reference Services at the Turn of the Century. Archivaria 49, 140–160 (2000)

[14] Yakel, E.: Encoded Archival Description: Are Finding Aids Boundary Spanners or Barriers for Use? Journal of Archival Organization 2, 63–77 (2004)

[15] Zarro, M., Allen, R.B.: User-Contributed Descriptive Metadata for Libraries and Cultural Institutions. In: European Conference on Digital Libraries, pp. 46–54 (2010)

Improving Retrieval Results with Discipline-Specific Query Expansion

Thomas Lüke, Philipp Schaer, and Philipp Mayr

GESIS – Leibniz Institute for the Social Sciences,
Unter Sachsenhausen 6-8, 50667 Cologne, Germany
{thomas.lueke,philipp.schaer,philipp.mayr}@gesis.org

Abstract. Choosing the right terms to describe an information need is becoming more difficult as the amount of available information increases. Search-Term-Recommendation (STR) systems can help to overcome these problems. This paper evaluates the benefits that may be gained from the use of STRs in Query Expansion (QE). We create 17 STRs, 16 based on specific disciplines and one giving general recommendations, and compare the retrieval performance of these STRs. The main findings are: (1) QE with specific STRs leads to significantly better results than QE with a general STR, (2) QE with specific STRs selected by a heuristic mechanism of topic classification leads to better results than the general STR, however (3) selecting the best matching specific STR in an automatic way is a major challenge of this process.

Keywords: Term Suggestion, Information Retrieval, Thesaurus, Query Expansion, Digital Libraries, Search Term Recommendation.

1   Introduction

Users of scientific digital libraries have to deal with a constantly growing amount of accessible information. The challenge of expressing one's information need through the right terms has been described as "vocabulary problem" by Furnas et al. [1]. Specialized knowledge organization systems (KOS) like thesauri or classifications have been created to support the users in the search process and to provide a consistent way of expressing the information need. As a way to provide easy access to these tools methods like Entry Vocabulary Modules [6] or so-called Search-Term-Recommenders (STR) have been introduced. Such recommendation services are also in use in commercial end-user systems like Amazon or eBay.

A common approach in IR systems to improve retrieval results is the use of term based query expansion (QE). If semantically close concepts are added to an initial query term it often shows query results improvements. Typically language and terminology becomes more specific if a terminology in a subject discipline e.g. the social sciences is differentiated into sub disciplines. This has also been proven for suggestions given by STRs [7], however, effects on the application of QE have not been evaluated. Using a heuristic mechanism, our approach utilizes the phenomenon of

P. Zaphiris et al. (Eds.): TPDL 2012, LNCS 7489, pp. 408–413, 2012.

language specialization to recommend the most specific concepts from a controlled vocabulary through discipline-specific STRs. We conduct an empirical analysis in the domain of social sciences to evaluate retrieval performance in a standard IR evaluation environment using QE.

The paper is structured as follows: Section 2 gives a brief overview of previous findings in the area of search term recommendation, query expansion and IR evaluation. Section 3 describes the methods used in this paper to evaluate our setup. In Section 4 we present and discuss our evaluation. Section 5 summarizes the findings of this paper and presents ideas for future work.

2   Related Work

Hargittai [2] has shown that users need supporting mechanisms while expressing their information need through search queries. Such support may be provided by query recommendation mechanisms, which try to enrich the existing query with additional terms. This leads to better retrieval results or provides the searcher with a new viewpoint on the search as shown by Mutschke et al. [5].

Automatic Query Expansion mechanisms have been divided into two classes by Xu and Croft [8]: Expansion recommendations based on a global analysis of the entire document collection and recommendations based on the local subset of documents that were retrieved by using the unexpanded query (called pseudo-relevance feedback [4]). In their research the local approach outperformed the global one clearly. However, when the amount of non-relevant documents in the results to the original query increases, a so-called query drift may occur. Documents not relevant to the users' information need lead to mostly irrelevant expansion terms. If the query gets expanded with such terms, it drifts away from the original meaning and results in even less relevant documents. Mitra et. al [4] have proposed techniques to overcome query drift.

Additionally Petras [7] found that discipline-specific search term recommenders which are trained on sub disciplinary document corpora deliver more specific search term suggestions than general recommenders which are trained on an entire database.

3   Discipline-Specific Term Recommendation

In our approach we apply Petras' [7] idea of more specific STRs onto an an automatic query expansion setup. According to the findings in [4, 8] we expect retrieval results to improve. With real-world applications in mind and the intention to reduce the chance of query drift through non-relevant data sets we demand STRs to be created a-priori rather than on-demand as with pseudo-relevance feedback. To create document sets belonging to specific disciplines we use a hierarchically structured classification system. It is called "classification of social sciences" and part of the SOLIS[1] data set.

---

[1] http://www.gesis.org/en/services/research/solis-social-science-literature-information-system/

For example any class starting with 1 is connected to the entire field of *social sciences*, any class starting with 102 is connected to the sub-discipline of *sociology* and class 10209 is assigned to documents from the special field of *family sociology*. This structure allowed us to create STRs at the top level that cover every area of the classification system thereby allowing us to choose matching discipline-specific STRs for queries from various disciplines.

Table 1. Overview of the created discipline-specific (DS) STRs. Class describes the specific sub-discipline of the social sciences the STR was based on. #Docs and #CT shows the number of documents or controlled vocabulary terms in that given collection.

STR	Class	#Docs	#CT	STR	Class	#Docs	#CT
DS-1	Basic Research	26817	5642	DS-9	Economics	45217	6213
DS-2	Sociology	76342	7184	DS-10	Social Policy	26289	5586
DS-3	Demography	26298	5322	DS-11	Employment Research	61742	5610
DS-4	Ethnology	5409	3787	DS-12	Women's Studies	18116	5301
DS-5	Political Science	95536	6995	DS-13	Interdisciplinary Fields	38985	6454
DS-6	Education	18820	5199	DS-14	Humanities	53863	6703
DS-7	Psychology	24785	5725	DS-15	Legal Science	16330	5549
DS-8	Communications	37285	5893	DS-16	Natural Science	6083	4015
Global	Social sciences	383000	7781				

To test the effects of recommendation terms from different disciplines we use a standard IR evaluation environment. A set of pre-defined topics, each consisting of a title and a set of documents relevant to that query, is processed on the SOLR[2] search platform which uses *tf-idf* to rank results. The data sets that represent the basis for the different STRs are all created from the SOLIS dataset, a collection of more than 400,000 documents from various disciplines of the social sciences.

Based on the assumption of a specialized vocabulary in different scientific disciplines, 16 custom data sets (see DS-1 to DS-16 in Table 1) for different sub-disciplines of SOLIS are created. These 16 sets as well as the entire SOLIS data set (see Global in Table 1) are the basis for 17 STRs.

In addition SOLIS is indexed with the thesaurus TheSoz[3] which consists of almost 7800 descriptor terms. Our discipline-specific STRs reduce the vocabulary of TheSoz to about 5700 terms per data set (on average) with a trend of even smaller (and presumably more specific) vocabularies. Exact numbers can be found in Table 1.

Each STR is created to match arbitrary input terms to terms of the controlled vocabulary TheSoz. All documents of a data set are processed using a co-occurrence analysis of input terms (found in title and abstract of each document) and the subject-specific descriptor terms assigned to a document. In order to rank the suggested terms from the controlled vocabulary the logarithmically modified Jaccard similarity measure is used.

---

[2] http://lucene.apache.org/solr/
[3] http://thedatahub.org/dataset/gesis-thesoz

4   Evaluation

This section contains a description of our evaluation. In the first paragraph the setup is described. The second paragraph presents the main results and effects of discipline-specific expansion on retrieval performance measured through average precision values. In addition we give a single example of a query, going from the broader level to a detailed in-depth inspection of discipline-specific QE.

4.1   Evaluation Setup

In order to test the effects of discipline-specific STRs and a general STR within a query expansion scenario we choose the GIRT4 corpus [3], which is a subset of SOLIS. It is used in evaluation campaigns like CLEF or TREC. We use 100 of the CLEF topics ranging from years 2003 to 2006 (topic numbers were 76 to 175) and classify them through a heuristic approach based on the classification system mentioned above: All relevant documents for a given topic are put into groups based on their classification ID. The classification ID of the group that holds the most documents is assumed to be the topic's classification. Queries from these topics are created by removing stop words from the title of each topic. To test the performance of QE we expand the query with three different STRs:

- the general STR, based on the entire SOLIS data set (our baseline)
- the STR of the topic's class
- the STR that performs best for a given topic (out of the 16 discipline-specific STRs)

Every QE is performed automatically with the top 4 recommendations of a STR. We report mean average precision (MAP), rPrecision as well as p@5, p@10, p@20 and p@30. As additional comparison we include the results of a standard installation of SOLR without QE. Every query, whether expanded or not, is processed by this platform. We use Student's t-test to verify significance of the improvements.

4.2   Results

Table 2 shows the results of the QE with different STRs. The first observation which can be made is that the discipline-specific STRs always perform better than a general STR (and thereby also improve retrieval performance compared to an unexpanded query). The last line of Table 2 shows the maximum performance possible through the use of our 16 discipline-specific STRs. It is significantly better than a general STR in every case. However, the improvements for those STRs that are chosen based on the classification of topics did not always reach significance. Only the precision within the first 5 and 10 top ranked documents is significantly higher. Still a more precise classification of the original query is necessary to gain maximum benefit from our discipline-specific STRs. In addition to measuring the impact on average precision we further analyze the use of discipline-specific STRs by examining an individual query. According to Petras [6] the results of a QE could significantly improve if the "right"

Table 2. Evaluation results averaged over 100 topics for three different types of QE. Recommendations are based on a general STR (gSTR) which served as a baseline, a discipline-specific STR fitting the class of the topic (tSTR) and the discipline-specific STR performing best on each topic (bSTR). Confidence levels of significance are: $^{*}\alpha = .05$, $^{**}\alpha = .01$.

Exp. Type	MAP	rPrecison	p@5	p@10	p@20	p@30
gSTR (Base)	0.155	0.221	0.548	0.509	0.449	0.420
tSTR	0.159	0.224	0.578*	0.542**	0.460	0.424
bSTR	0.179**	0.233**	0.658**	0.601**	0.512**	0.463**

Table 3. Top 4 recommendations for the input terms "bilingual education" from three STRs

Recommendation	General	Topic-fitting	Best-performing
1	Multilingualism	Child	Multilingualism
2	Child	School	Speech
3	Speech	Multilingualism	Ethnic Group
4	Intercultural Education	Germany	Minority

terms are added to the query. We will see how different recommendations influence the results. Topic no. 131 has the title (and thus the query) "bilingual education". Table 3 shows the top 4 recommendations of each STR for this topic. While "multilingualism" is always a recommendation and "child" and "speech" appear twice, the rest of the terms appear only in one recommender. The general recommender proposes the most common terms while the two discipline-specific STRs propose more specific terms. In Table 4 we can see the effects of these different recommendations on retrieval precision. The unexpanded query performs satisfying but leaves room for improvement as it presents only 2 relevant documents within the top 5 and 3 within the top 10 documents (see p@5 and p@10). Expanding the query with terms from the general STR improves these results to 3 and 6 relevant documents in the top 5 or top 10 respectively. Using terms from the pre-defined, topic-fitting, discipline-specific STR 4 out of 5 documents within the top 5 are relevant. Finally, the best performing discipline-specific STR manages to expand the query in a way that all top 10 documents are relevant and even within the top 20 documents only 3 are not relevant.

Table 4. Evaluation results for topic 131 with three different types of QE Recommendations. Bold font indicates improvement

Exp. Type	AP	rPrecison	p@5	p@10	p@20	p@30
Solr	0.039	0.127	0.4	0.3	0.2	0.133
gSTR	**0.072**	**0.144**	**0.6**	**0.6**	**0.4**	**0.333**
tSTR	**0.076**	**0.161**	**0.8**	0.6	**0.45**	0.333
bSTR	**0.147**	0.161	**1**	**1**	**0.85**	**0.567**

5   Conclusion and Future Work

Our research shows that the use of discipline-specific Search-Term-Recommenders can improve the retrieval performance significantly if used as basis for an automated query expansion. However, it also becomes clear that choosing the best STR in an automated setting of query expansion is far from trivial. By doing an in-depth analysis of a single query we additionally demonstrate how discipline-specific term recommendations can improve the quality of search results for a user. This leads us to the conclusion that discipline-specific STRs can be a valuable addition to expert search platforms where users might not know how to optimally express their search.

In conclusion, STRs that are meant to assist users should be discipline-specific in order to recommend more specific terms. Still, it has to be determined how specific (or small) a data set may be while still producing reasonable results. To improve quality of QE it is essential to have a good algorithm for determining the specific discipline of the query. Besides having more specific recommendation another aspect of further research could be the use of additional metadata fields as it is common for users to enrich their search by explicitly specifying authors or other metadata fields (for further research on recommendations based on different types of metadata see the work by Schaer et al. in these proceedings). A STR providing recommendations of this kind could add additional benefits to a user's search, especially if it recommends e.g. the main authors of a specific discipline.

Acknowledgements. This work was partly funded by DFG, grant no. SU 647/5-2.

References

1. Furnas, G.W., et al.: The Vocabulary Problem in Human-System Communication. Communications of the ACM 30(11), 964–971 (1987)
2. Hargittai, E.: Hurdles to information seeking: Spelling and typographical mistakes during users' online behavior. Journal of the Association of Information Systems 7(1), 52–67 (2006)
3. Kluck, M.: The GIRT Data in the Evaluation of CLIR Systems – from 1997 Until 2003. In: Peters, C., Gonzalo, J., Braschler, M., Kluck, M., et al. (eds.) CLEF 2003. LNCS, vol. 3237, pp. 376–390. Springer, Heidelberg (2004)
4. Mitra, M., et al.: Improving automatic query expansion. In: Proceedings of the 21st Annual International ACM SIGIR Conference on Research and Development in Information Retrieval, pp. 206–214. ACM, Melbourne (1998)
5. Mutschke, P., et al.: Science models as value-added services for scholarly information systems. Scientometrics 89(1), 349–364 (2011)
6. Petras, V.: How one word can make all the difference - using subject metadata for automatic query expansion and reformulation. In: Working Notes for the CLEF 2005 Workshop, Vienna, Austria, September 21-23 (2005)
7. Petras, V.: Translating Dialects in Search: Mapping between Specialized Languages of Discourse and Documentary Languages. University of California (2006)
8. Xu, J., Croft, W.B.: Query expansion using local and global document analysis. In: Proceedings of the 19th Annual International ACM SIGIR Conference on Research and Development in Information Retrieval, pp. 4–11. ACM, Zurich (1996)

An Evaluation System for Digital Libraries

Alexander Nussbaumer[1], Eva-Catherine Hillemann[1],
Christina M. Steiner[1], and Dietrich Albert[1,2]

[1] Knowledge Management Institute, Graz University of Technology, Austria
{alexander.nussbaumer,eva.hillemann,
christina.steiner.dietrich.albert}@tugraz.at
[2] Department of Psychology, University of Graz, Austria
dietrich.albert@uni-graz.at

Abstract. Evaluation is an important task for digital libraries, because
it reveals relevant information about their quality. This paper presents a
conceptual and technical approach to support the systematic evaluation
of digital libraries in three ways and a system is presented that assists
during the entire evaluation process. First, it allows for formally modelling the evaluation goals and designing the evaluation process. Second,
it allows for data collection in a continuous and non-continuous, invasive
and non-invasive way. Third, it automatically creates reports based on
the defined evaluation models. On the basis of an example evaluation it
is outlined how the evaluation process can be designed and supported
with this system.

Keywords: evaluation, evaluation system, digital libraries, continuous
data collection, evaluation report.

1 Introduction

Evaluation is an important task in the context of digital libraries, because it
reveals relevant information about the quality of the technology for all stakeholders and decision makers [1]. Digital library evaluation involves conceptual
as well as technical tasks based on evaluation models [2,3]. This paper addresses
both the conceptual and the technical aspect and presents a combined approach
to evaluate digital libraries by the use of an evaluation system that has been
developed for this purpose.

Conducting evaluations is usually a time consuming task. Besides planning
and data collection, it requires a lot of time to analyse the collected evaluation
data. There are different types of software tools designed to support evaluation
tasks - tools for data collection (online survey tools for creating and adminis-
tering questionnaires, such as LimeSurvey[1]) or tools supporting the analysis of
collected data (for analysing qualitative or quantitative data, such as SPSS[2]).
Though these tools support the evaluation, there is still a high workload for the
evaluator due to the gap between data collection and analysis process. In order

---

[1] http://www.limesurvey.org/
[2] http://www-01.ibm.com/software/at/analytics/spss/

P. Zaphiris et al. (Eds.): TPDL 2012, LNCS 7489, pp. 414–419, 2012.

to address this situation, this paper presents an evaluation system that supports all steps of the evaluation. This initiative is made in the context of the European project CULTURA[3] which aims at building an adaptive system that allows for personalised access to and interaction with digital cultural collections.

2   Baseline and Problem Statement

Answering questions on technology evaluation can be considered as a process that critically examines a software or technology. It involves collecting and analysing information about a software's activities, characteristics, and outcomes. Its purpose is to make judgements about the benefit of a technology, to improve its effectiveness, and/or to inform programming decisions [4].

Typically, the evaluation process can be broken down into three key phases: (1) Planning, (2) collecting, and (3) analysing (see for example [5]). In the planning phase the evaluation questions are identified broken down into more detailed questions. Having decided on the "what" of an evaluation, the next step of planning is to decide on the "how". This is done by defining an evaluation plan or framework which contains the methodology in alignment with the evaluation questions and the operationalization of evaluation topics and aspects in terms of measurable indicators. In the collecting phase evaluation instruments are applied and the actual evaluation data are collected. Finally, in the third phase the collected data is analysed and aggregated. Analysis methods are chosen in alignment with the type of collected data. Evaluation results should be used in an effective way by feeding them back into the development process and/or disseminating them.

Several online (survey) tools can be found that support the evaluation process by enhancing and facilitating data collection and recording. Nevertheless, there are some limitations that prevent evaluators from using such tools during the whole evaluation process. One limitation is the statistical analysis of gathered data, which has to be done with external tools. In addition, statistical analysis is complicated by the fact that individual variables of collected data are not clearly related and mapped to the respective evaluation aspect. This has to be done manually. Furthermore, in most cases, online survey tools only allow to collect data in an explicit way after using the system. Such retrospective reports, however, might be biased and it would be use- and meaningful to collect evaluation data already during system usage.

3   Conceptual Approach

The general idea of the evaluation system presented in this paper is to support the evaluation in a systematic and most useful manner, in order to overcome (part of) the challenges and difficulties in the traditional evaluation process as outlined above. The goal is to conduct user-centered evaluation by capturing

---

[3] http://www.cultura-strep.eu/

evaluation data in order to identify the opinion of a user about a digital library. This section presents the key aspects of the conceptual approach.

The evaluation system supports evaluation in a systematic manner by explicitly modelling and representing the evaluation goal and the evaluator's expertise by formalising how and what should be evaluated. Therefore, an evaluation is based on an evaluation model that formally represents the evaluation approach. It consists of two parts, the quality model and the survey model. The quality model is an abstract model, that defines what should be evaluated. It defines evaluation aspects or parameters (such as ease of use or usefulness) which express the qualities of a system. These aspects are not directly measured, but items are defined that can be related to these aspects. In this way the quality of a system is measured indirectly by using different types of survey items as instruments for data collection (see Figure 1). The survey items are part of the survey model that defines how to collect data. Items might be concrete questions, but can also be specifications of specific usage data for monitoring the user's behaviour (see below). An item usually collects a user's opinion on a certain topic in terms of a numerical value, e.g. expressed on a rating scale. A set of items can be used as a questionnaire, but also as a collection of behavioural indicators.

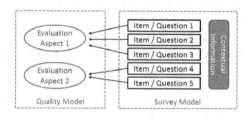

Fig. 1. Evaluation Model

Traditionally, the evaluation happens after a system has been used and consists in querying users, for example through questionnaires. However, in addition to questionnaires, the new approach also allows for collecting continuous evaluation data during the usage of a system. Small software snippets are integrated in the system to be evaluated, so that they can collect evaluation data continuously during the usage in an invasive way by asking the user with so-called jJudgets and in a non-invasive way by monitoring the user's behaviour with software sensors.

Three evaluation modes are supported by the evaluation system and its approach. Questionnaires capture the opinion on a system or digital library at a specific time and ask questions from a participant. They are filled out after the usage of a system, so they constitute a non-continuous and invasive way of collecting evaluation data. Judgets are little widgets integrated in the evaluated system and deliver evaluation data about the opinion on the digital library based on the user's explicit feedback. They allow for continuously evaluation providing data in an invasive way. With this approach a minimum of disruptiveness is assumed. Sensors are also integrated in the evaluated system, similar to udgets. However, they are not visible to the users, but monitor and log their usage behaviour and collect evaluation data in this way. In this way, evaluation is done continuously and non-invasively. A similar approach is known in the field of monitoring Contextual Attention Metadata (CAM), which is a structured and semantic way of capturing users' behaviour (see for example [6]).

An important feature of the evaluation system is the generation of automatic reports from the collected evaluation data on the basis of the underlying evaluation model. A report is made upon a survey model by aggregating all participants' data related to the respective survey model. The aggregation is conducted in the same way for questionnaires, sensor data, and Judget data. The report contains the raw data of each item and participant and the means for all items over all participants. Furthremore, it contains aggregated information according to the evaluation aspects of the quality model and an overall score. Reports are available for both evaluators and evaluated machines.

4 System Architecture and Implementation

The core part of the evaluation system is a Web service that exposes its functionality through both a Web Interface for the user and a REST interface for the evaluated system. Participants can access and fill out questionnaires using the Web interface and evaluators can get reports directly from the evaluation system. The REST interface is used by the evaluated system to send the collected data of the sensors and judgets to the Web service. The reports and overall scrores can also be retrived via the REST interface. The quality and survey models, as well as related evaluation data, are stored in a relational database that is part of this service. This service is implemented as an Apache Tomcat[4] Web application. An overview of the system architecture is given in Figure 2.

In order to create judgets and sensors and inject them in the evaluated system, example code snippets are available that can be used to create the actual judget or sensor. Judgets are usually implemented in JavaScript and send user feedback through AJAX to the evaluation system. Sensors also should send the information of the user behaviour to the evaluation system, independently how they are implemented (e.g. on the client or server side). In any case it depends on the evaluated system how exactly data is collected and how judgets and sensors are integrated.

5 Evaluation Process with the Evaluation System

This section describes the realisation of the three steps of the evaluation process as described in Section 2 with the support of the evaluation system. In contrast to traditional approaches of conducting evaluation, the presented evaluation system supports all steps of the evaluation in an integrated way.

---

[4] http://tomcat.apache.org/

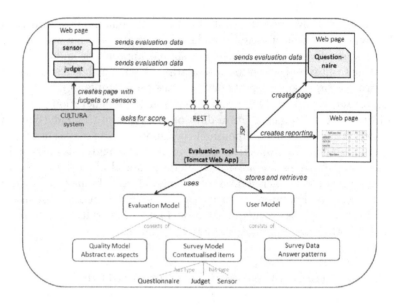

Fig. 2. System Architecture

First, the evaluator has to design the evaluation goal and decide what should be measured. For instance, investigating users' acceptance of a new digital library system might be an aim for an evaluation. This evaluation aim can be formally expressed in the quality model by defining the evaluation aspects. Then the evaluation method is defined describing the method of data collection (questionnaire, judget, sensor) in the survey model. The method of data collection depends on the evaluation mode describing if evaluation should be continuous or non-continuous, and in case of continuous evaluation, if it should be invasive or non-invasive. In the survey model the evaluation items are mapped to the evaluation qualities represented in the quality model. In case of a questionnaire or judgets items can be taken from standardised questionnaires, such as the System Usability Scale [7] or the User Acceptance Model [8].

In the second step the system collects the evaluation data from users implicitly (non-invasive) or explicitly (invasive) based on the defined underlying survey model.

In the last step the evaluation analysis is performed, which is also strongly supported by the evaluation system. The evaluator just has to navigate to the respective survey model and request the report. A Web page is created automatically that contains all raw data, aggregated data and scores (based on the definitions made in the survey model), as well as diagrams of aggregated data. This information is also available to the evaluated system through the REST Web interface.

6   Conclusion and Outlook

This paper presented an approach and system to support the evaluation of digital libraries. The key features of this approach include the formal design of the evaluation by defining evaluation models. Furthermore, the it provides the possibility to collect evaluation data not only through questionnaires after the usage of a system, but also during the system use in a invasive and non-invasive way. Finally, automatically generated reports based on the evaluation models reduce the work for the evaluator significantly.

Next step is to conduct a real evaluation in the context of the CULTURA project with the help of this evaluation system. The results of this evaluation will also be analysed in order to evaluate the applicability of the evaluation tool and its approach.

Acknowledgements. The work reported has been partially supported by the CULTURA project, as part of the Seventh Framework Programme of the European Commission, Area "Digital Libraries and Digital Preservation" (ICT-2009.4.1), grant agreement no. 269973.

References

1. Reeves, T.C., Apedoe, X., Woo, Y.: Evaluating digital libraries: a user-friendly guide. National Science Digital Library (2005)
2. Khoo, M., MacDonald, C.: An Organizational Model for Digital Library Evaluation. In: Gradmann, S., Borri, F., Meghini, C., Schuldt, H. (eds.) TPDL 2011. LNCS, vol. 6966, pp. 329–340. Springer, Heidelberg (2011)
3. Fuhr, N., Tsakonas, G., Aalberg, T., Agosti, M., Hansen, P., Kapidakis, S., Klas, C.P., Kovcs, L., Landoni, M., Micsik, A., Papatheodorou, C., Peters, C., Solvberg, I.: Evaluation of digital libraries. International Journal on Digital Libraries 8, 21–38 (2007)
4. Patton, M.: Qualitative evaluation and research methods. Sage, Thousand Oaks (1990)
5. Cook, J.: Evaluating learning technology ressources (2002), http://www.alt.ac.uk/docs/eln014.pdf
6. Schmitz, H.C., Scheffel, M., Friedrich, M., Jahn, M., Niemann, K., Wolpers, M.: CAMera for PLE. In: Cress, U., Dimitrova, V., Specht, M. (eds.) EC-TEL 2009. LNCS, vol. 5794, pp. 507–520. Springer, Heidelberg (2009)
7. Brooke, J.: Sus: a "quick and dirty" usability scale. In: Jordan, P., Thomas, B., Weerdmeester, B., McClelland, A.L. (eds.) SUS: a"quick and dirty" usability scale. Taylor and Francis, London
8. Thong, J., Hong, W., Tam, K.: Understanding user acceptance of digital libraries: what are the roles of interface characteristics, organizational context, and individual differences. International Journal of Human-Computer Studies 57, 215–242 (2002)

Enhancing Digital Libraries and Portals with Canonical Structures for Complex Objects

Scott Britell[1], Lois M.L. Delcambre[1], Lillian N. Cassel[2],
Edward A. Fox[3], and Richard Furuta[4]

[1] Dept. of Computer Science, Portland State University, Portland, OR
{britell,lmd}@cs.pdx.edu
[2] Dept. of Computing Sciences, Villanova University, Villanova, PA
lillian.cassel@villanova.edu
[3] Dept. of Computer Science, Virginia Tech, Blacksburg, VA
fox@vt.edu
[4] Dept. of Computer Science and Engineering, Texas A&M University,
College Station, TX
furuta@cse.tamu.edu

Abstract. Individual digital library resources are of interest in their own right, but, in some domains, resources can be part of (perhaps multiple) complex objects. We focus on domains with complex objects where a digital library user can benefit from seeing and browsing a resource in the context of its structure(s). We introduce canonical structures that can represent local digital library structures; the canonical structures allow us to provide sophisticated browsing/navigation aids in a generic way. We evaluate a means to transfer the structure of our resources to a digital library portal. We implement and evaluate approaches based on OAI-PMH and OAI-ORE using Dublin Core—with and without a custom namespace. We also transfer the canonical structure to a portal where our navigation widget is implemented.

1 Introduction

When a digital resource is composed of other digital resources, how should we describe it in a digital library or portal? As an example, should we represent a collection of books with the book as the first-class object? Or is a book a collection of individual chapters? We believe that showing fine-grained resources within the context of larger-grained resources that they are contained within is valuable. Thus, we want to enable viewing of structured resources (e.g., books, chapters, and questions) both independently and as part of complex object(s).

We focus on domain-specific digital libraries where commonly occurring structural patterns can be identified. To demonstrate our ideas, we use a digital library and a portal intended for educational materials and curricula. Such curricula typically have many resource types (e.g., unit, lesson, instructor guide, assessment, or instructional materials such as lectures and demonstrations). Resources such as a unit or lesson are quite naturally comprised of other resources; a digital

P. Zaphiris et al. (Eds.): TPDL 2012, LNCS 7489, pp. 420–425, 2012.

ibrary that directly supports that structure provides a more informative view
of the resource. The context(s) (i.e., structures) in which a given demonstration,
or example, is used can provide important information as to the suitability of
the demonstration for use in some new context. Students viewing a resource may
also benefit from seeing it within multiple contexts.

Our work has three goals: (1) describe the structure of complex resources
explicitly so that we can implement browsing and other widgets in a digital
library and portal generically, (2) transfer individual resources and the structure
of the complex objects from one source (e.g., digital library) to another (e.g.,
digital library portal), and (3) enable the use of generic widgets in the receiving
site.

2   Background

The Ensemble project[1] has built and is hosting a portal that provides access
to independent collections of resources that support education in computing.
The project also provides a place for public and private communities to work
and share materials, plus a set of detailed descriptions of technologies useful
in computing education. Ensemble collections are also shared to the National
Science Digital Library[2].

As one part of the Ensemble effort, we have built a collection called STEM-
Robotics[3] comprised of middle and high school robotics curricula. STEM-
Robotics is a publicly accessible digital library; it is not intended to be a learning
management system where students have individual accounts with their grades.

Our collaborators are providing content for STEMRobotics in a variety of
structures. Given the variation in structure across these courses while recognizing that they have structural patterns in common, we define *canonical structures*
and map the canonical structures to local structures. To exploit the canonical
structures along with the associated mappings to local structure, we have developed a set of generic widgets for the repository, called semantic widgets. One of
these, a general, hierarchical navigation widget, provides an easy-to-use browser
for all of the courses found in STEMRobotics. The general structure of the hierarchy is provided by the canonical structures while the specific titles and types
come from the local resources in the collection through their mappings. When
the canonical structures are mapped to a new course the navigation widget automatically works for that new course.

3   Bringing Complex Structure to a Portal

The process of transmitting complex structure from a source repository to the
Ensemble portal is shown in Figure 1. We transmit canonical structures using

---

[1] http://computingportal.org
[2] http://nsdl.org
[3] http://stemrobotics.cs.pdx.edu

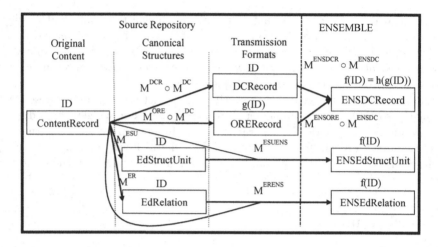

Fig. 1. The process of transmitting resource records from a source repository to the Ensemble portal is shown

two standard protocols for metadata transmission—the Open Archives Initiative Protocol for Metadata Harvesting (OAI-PMH) [4] and Object Re-use and Exchange (OAI-ORE) [12] in order to bring the complex structure within STEM-Robotics to the Ensemble portal. Both protocols provide mechanisms for transmitting metadata from standard namespaces such as Dublin Core (DC) [1] as well as metadata from custom namespaces. OAI-PMH transmits resource records as xml files whereas OAI-ORE transmits resources as aggregated collections represented as a set of RDF[4] triples.

A resource within the STEMRobotics repository is defined by a ContentRecord as shown in Figure 1. A ContentRecord holds the metadata associated with a resource, the ways in which a resource is related to other resources, and the content of the resource.

A Dublin Core record (DCRecord) consists of the metadata fields corresponding to our metadata and the fields that correspond to our relationships—the DC Relation field and the refinements of the Relation field such as IsPartOf and HasPart.

A resource transmitted via OAI-ORE (ORERecord) is represented by the DC record for that resource as well as a set of identifiers of aggregated resources that comprise that resource where the identifiers are either the resource itself or another ORERecord representing another aggregation.

The Ensemble portal stores the DC records(ENSDCRecord) for the resources in its partner collections. The records within Ensemble store the same information as the DC records transmitted from each partner collection with the addition of identifiers local to the portal. The Ensemble records also store the local record identifiers for the DC relation fields.

---

[4] http://www.w3.org/RDF/

We next define the two types of canonical structures. The first type (EdStructUnit) is defined as an identifier of a STEMRobotics resource and a set of identifiers of related resources, where the identifiers are those of the ContentRecords that have been mapped to the canonical structure. The second type of canonical structure (EdRelation) is defined as an identifier for a resource, its related resource, and the type of relationship between them.

The names of the resources and their ordinal position within relationships need not be specified in the canonical structure for STEMRobotics since it can be retrieved from the relationships in the structure in the original repository. But since the Ensemble portal only stores the metadata records for these resources, the original relationship types and ordinal positions must be stored in the canonical structures for Ensemble (ENSEdStructUnit and ENSEdRelation).

Mappings. Working from left to right in Figure 1, we start by defining the mapping from a ContentRecord in STEMRobotics to a DC record in Ensemble (ENSDCRecord) via OAI-PMH. This mapping requires an intermediate step that maps to a DCRecord for the OAI-PMH transmission (M^{DCR}). To achieve this we map the metadata of the ContentRecord to a DCRecord (M^{DC})—where we transform the STEMRobotics types to the types defined in the NSDL controlled vocabulary for types used by Ensemble—and we map the relationships from the content record to the DC HasPart relation refinement. This mapping transforms all of the possible relationships in the source repository to HasPart relationships. This transmits some of the complex structure (i.e., the HasPart links) in the case where the destination portal does not support canonical structures. The last part of the mapping to a record in Ensemble is the M^{ENSDC} and M^{ENSDCR} mappings that add Ensemble identifiers to the DCRecord in order to get an ENSDCRecord.

A resource from STEMRobotics transmitted to Ensemble via OAI-ORE, as shown in Figure 1, also first converts the metadata to a DCRecord, but then uses mapping M^{ORE} to transform all of the relationships of the resource into OAI-ORE aggregations. The mapping M^{OREENS} converts these aggregations to the Ensemble DC relation identifiers.

Canonical Structures. The mappings from canonical structures to resources in STEMRobotics are defined by M^{ER} and M^{ESU} where the identifiers of the related resources are mapped directly to the canonical structures. By composing the identifiers from these mappings with the identifiers of ContentRecords in STEMRobotics, widgets like the navigation menu can retrieve type and ordinal information from the relationships between ContentRecords.

M^{ESUENS} and M^{ERENS} map canonical structures from STEMRobotics to canonical structures in Ensemble where the source types and ordinal positions are added and the identifiers are mapped to identifiers in Ensemble. The canonical structure mappings are transmitted to Ensemble in OAI-PMH and OAI-ORE using a custom namespace.

Adapting Semantic Widgets for a Portal. Semantic widgets implemented in the STEMRobotics repository can exploit all of the local structure and content within the repository. But when we bring them to Ensemble, we must adapt them to the structures stored in the portal.

Widgets like the navigation browser display the source type and ordering transmitted to the portal. If this information does not exist, the adapted widgets will use the DC records in the portal. This is achieved by creating mappings from the canonical structures in Ensemble to the DC Relation and DC Type fields. The display of resources will reflect the hierarchy of the original resource but the ordering of resources within any level will be arbitrary, since DC records do not contain a specification for ordering of relationships.

4   Related Work

Our work on canonical structures [9] leverages years of research in the database community, notably the Higher Order Entity Relationship Model (HERM) [15] and work on data model patterns by Blaha [8] and Thalheim [15] as well as the Unified Modeling Language (UML) [6].

The IEEE Learning Object Metadata standard (LOM) [2] provides the ability to describe the structure within which an object resides, e.g. networked, hierarchical, or linear, as well as providing relations similar to those used by Dublin Core. But the LOM model falls short of expressing the semantics of our canonical structures since we cannot express ordering or arbitrary typed relationships. Extensions to LOM like those described above would allow functionality like we brought to Ensemble to be available in portals and search engines for LOM repositories similar to Ariadne [14]. Mapping canonical structures to LOM would also allow us to exchange complexly structured resources with e-learning systems that use LOM.

Similarly the Sharable Content Object Reference Model (SCORM) [5] can represent a complete course of educational materials. But SCORM does not capture all of the information necessary to represent resources of a course in a digital library. Projects like ASIDE [7] have worked to provide interoperability between e-learning systems and digital libraries by combining educational metadata in formats like LOM or SCORM with the Metadata Encoding and Transmission Standard (METS) [3].

METS provides a richer data model than that of Dublin Core, representing a structure map of each resource and its structure links. These structure links are often just in the form of hyperlinks within a digital object; thus the semantics of the relationships between resources are not captured explicitly as it is in canonical structures.

A number of projects, like Foresite [13], SCOPE [10], and LORE [11], have used OAI-ORE to represent complex digital objects for exchange purposes. By adopting published ontologies, these projects have shown how semantics can be added to OAI-ORE, similar to our proposed methods.

5   Conclusions

As repositories of user-generated materials become a larger part of the digital landscape, we have the opportunity to represent and make use of additional semantics that users add to these materials. In fact, the heterogeneity of structure in the STEMRobotics site motivated our interest in providing generic support for complex data models. By working within an application domain, we seek to identify and exploit commonly occurring structures by mapping them to our canonical structures.

Work in progress includes: mapping to additional standards, extending the set of canonical structures that we use, and significantly extending the set of enhanced widgets that can be built generically based on canonical structures. We are also working to define and use canonical structures in additional application domains.

References

1. Dublin Core Metadata Initiative, http://dublincore.org/
2. IEEE Final LOM Draft Standard,
   http://ltsc.ieee.org/wg12/20020612-Final-LOM-Draft.html
3. Metadata Encoding & Transmission Std.,
   http://www.loc.gov/standards/mets/
4. Open Archives Initiative Protocol for Metadata Harvesting,
   http://www.openarchives.org/pmh/
5. SCORM, http://www.adlnet.gov/capabilities/scorm
6. Unified Modeling Language, http://www.uml.org/
7. Arapi, P., Moumoutzis, N., Christodoulakis, S.: ASIDE: An Architecture for Supporting Interoperability between Digital Libraries and ELearning Applications. In: Proc. of the Intl. Conf. on Adv. Learning Tech., pp. 257–261. IEEE (2006)
8. Blaha, M.: Patterns of Data Modeling. CRC Press (June 2010)
9. Britell, S., Delcambre, L.M.L.: Mapping Semantic Widgets to Web-based, Domain-specific Collections. In: Conceptual Modeling ER 2012 (2012)
10. Cheung, K., Hunter, J., Lashtabeg, A., Drennan, J.: SCOPE: A Scientific Compound Object Publishing and Editing System. Intl. J. of Dig. Curation 3(2) (December 2008)
11. Gerber, A., Hunter, J., Buchanan, G., Masoodian, M., Cunningham, S.: Digital Libraries: Universal and Ubiquitous Access to Information. In: Buchanan, G., Masoodian, M., Cunningham, S.J. (eds.) ICADL 2008. LNCS, vol. 5362, pp. 246–255. Springer, Heidelberg (2008)
12. Lagoze, C., Van de Sompel, H., Nelson, M.L., Warner, S., Sanderson, R., Johnston, P.: Object Re-Use & Exchange: A Resource-Centric Approach. CoRR abs/0804.2 (2008)
13. Sanderson, R., Llewellyn, C., Jones, R.: Evaluation of OAI-ORE via large-scale information topology visualization. In: Proc. of the Joint Conf. on Digital Libraries, p. 441. ACM Press, New York (2009)
14. Ternier, S., Verbert, K., Parra, G., Vandeputte, B., Klerkx, J., Duval, E., Ordonez, V., Ochoa, X.: The Ariadne Infrastructure for Managing and Storing Metadata. IEEE Internet Computing 13(4), 18–25 (2009)
15. Thalheim, B.: Entity-relationship modeling: foundations of database technology, 1st edn. Springer, New York (2000)

Exploiting the Social and Semantic Web for Guided Web Archiving*

Thomas Risse[1], Stefan Dietze[1], Wim Peters[2], Katerina Doka[3],
Yannis Stavrakas[3], and Pierre Senellart[4]

[1] L3S Research Center, Hannover, Germany
{risse,dietze}@L3S.de
[2] University of Sheffield, UK
w.peters@dcs.shef.ac.uk
[3] IMIS / RC "ATHENA", Athens, Greece
katerina@cslab.ece.ntua.gr, yannis@imis.athena-innovation.gr
[4] Institut Mines–Télécom; Télécom ParisTech; CNRS LTCI, Paris, France
pierre.senellart@telecom-paristech.fr

Abstract. The constantly growing amount of Web content and the success of the Social Web lead to increasing needs for Web archiving. These needs go beyond the pure preservation of Web pages. Web archives are turning into "community memories" that aim at building a better understanding of the public view on, e.g., celebrities, court decisions, and other events. In this paper we present the ARCOMEM architecture that uses semantic information such as entities, topics, and events complemented with information from the social Web to guide a novel Web crawler. The resulting archives are automatically enriched with semantic meta-information to ease the access and allow retrieval based on conditions that involve high-level concepts.

Keywords: Web Archiving, Web Crawler, Text Analysis, Social Web.

1 Introduction

Given the ever increasing importance of the World Wide Web as a source of information, adequate *Web archiving* and *preservation* has become a cultural necessity in preserving knowledge. The report *Sustainable Economics for a Digital Planet* [1] states that "the first challenge for preservation arises when demand is diffuse or weakly articulated." This is especially the case for non-traditional digital publications, e.g., blogs, collaborative space, or digital lab books. The challenge with new forms of publications is that there can be a lack of alignment between what institutions see as worth preserving, what the owners see as of current value, and the incentive to preserve together with the rapidness at which decisions have to be made. For ephemeral publications such as the Web, this misalignment often results in irreparable loss. Given the deluge of digital information created and this situation of uncertainty, a first necessary step is to be

---

* This work is partly funded by the European Commission under ARCOMEM (ICT 270239).

P. Zaphiris et al. (Eds.): TPDL 2012, LNCS 7489, pp. 426–432, 2012.

able to respond quickly, even if in a preliminary fashion, by the timely creation of archives, with minimum overhead enabling more costly preservation actions further down the line. This is the challenge that the ARCOMEM[1] project is addressing.

A pivotal factor for enabling next-generation Web archives is crawling. Crawlers are complex programs that nevertheless implement a simple process: follow links and retrieve Web pages. In the ARCOMEM approach, however, crawling is much more complex, as it is enriched with functionality dealing with novel requirements. Instead of following a "collect-all" strategy, archival organizations are trying to build *community memories* that reflect the diversity of information people are interested in. Community memories largely revolve around *events* and the *entities* related to them such as persons, organizations, and locations. Thus, entities and events are natural candidates for focusing new types of content acquisition processes in preservation as well as for archive enrichment.

The crawler architecture we propose here is the basis for current implementation activities in the ARCOMEM project. Note that the system is only partially implemented at the moment, and we therefore do not present any evaluation.

The rest of the paper is structured as follows. Section 2 gives an overview of the overall architecture and the different processing phases. More details about the content and Social Web analysis as well as crawler guidance are presented in Section 3. We discuss the state of the art in Web archiving and related fields in Section 4. Finally, Section 5 gives conclusions and an outlook to future work.

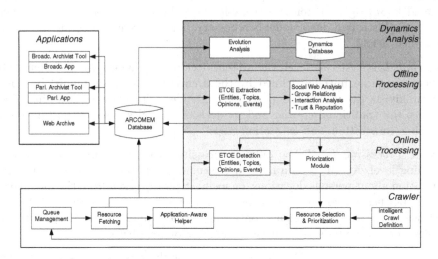

Fig. 1. Overall Architecture

---

[1] ARCOMEM – From Collect-All ARchives to COmmunity MEMories, http://www.arcomem.eu/

2   Approach and Architecture

The goal for the development of the ARCOMEM crawler architecture is to implement a socially aware and semantic-driven preservation model. This requires thorough analysis of the crawled Web page and its components. These components of a Web page are called *Web objects* and can be the title, a paragraph, an image or a video. Since a thorough analysis of all Web objects is time-consuming, the traditional way of Web crawling and archiving is no longer working. Therefore the ARCOMEM crawl principle is to start with a *semantically enhanced crawl specification* that extends traditional URL based seed lists with semantic information about entities, topics or events. This crawl specification is complemented by a small reference crawl to learn more about the crawl topic and intention of the archivist. The combination of the original crawl specification with the extracted information from the reference crawl is called the *intelligent crawl specification*. This specification, together with relatively simple semantic and social signals, is used to guide a broad crawl that is followed by a thorough analysis of the crawled content. Based on this analysis a semi-automatic selection of the content for the final archive is carried out.

The translation of these steps into the ARCOMEM system architecture foresees four processing levels: the *crawler level*, the *online processing level*, the *offline processing level*, and *dynamics analysis*, that revolve around the AR-COMEM database as depicted in Figure 1. The ARCOMEM database is the focal point for all components involved in crawling and content analysis. It stores all information from the crawl specification over the crawled content to the extracted knowledge. Therefore a scalable and efficient implementation together with a sophisticated data model is necessary. The different processing levels are described below.

Crawling Level: At this level, the system decides and fetches the relevant Web objects as those initially defined by the archivists, and later refined by both the archivists and the online processing modules. The crawling level includes, besides the traditional crawler and its decision modules, some important data cleaning, annotation, and extraction steps.

Online Processing Level: The online processing is tightly connected with the crawling level. At this level a number of semantic and social signals such as information about persons, locations, or social structure taken from the intelligent crawl specification are used to prioritize the crawler processing queue. Due to the near-real-time requirements, only time-efficient analysis can be performed, while complex analysis tasks are moved to the offline phase.

Offline Processing Level: At this level, most of the basic processing over the data takes place. The offline, fully-featured, versions of the entity, topics, opinions, and events analysis (ETOE analysis) and the analysis of the social contents operate over the cleansed data from the crawl that are stored in the ARCOMEM database. These processing tools perform linguistic, machine learning and NLP methods in order to provide a rich set of metadata annotations that are interlinked with the original data. The respective annotations are stored back in the

ARCOMEM database and are available for further processing and information mining. After all the relevant processing has taken place, the Web pages to be archived and preserved are selected in a semi-automatic way.

Dynamics Analysis Level: Finally, a more advanced processing step takes places. It operates on collections of Web objects that have been collected over time in order to register the evolution of various aspects identified by the ETOE and Web analysis components. As such, it produces aggregate results that pertain to a group archive of objects, rather than to particular instances.

3 Analysis for Crawl Guidance and Enrichment

We now describe in more detail the major analyses that are performed at all levels of the ARCOMEM architecture. We discuss over content analysis, analysis of the social Web, data enrichment, and crawler guidance itself. The discussion about dynamics analysis is post-poned to a later publication.

Content Analysis. The aim of this module is the extraction and detection of informational elements called ETOEs (Entities, Topics, Opinions, and Events) from Web pages (s. Section 2). The ETOE extraction takes place in the offline phase and processes a collection of Web pages. The results of the offline ETOE extractions are used to (1) get a better understanding of the crawl specification and (2) populate the ARCOMEM database with structured data about ETOEs and their occurrence in Web objects. In the online phase, single documents will be analyzed to determine their relevance to the crawl specification.

A crawl campaign is described by a crawl specification given by the archivist. This specification consists of, in addition to other parameters, a search string where the archivist specifies in their own words the semantic focus of the crawl campaign. The search string is a combination of entities, topics, and events, plus free terms. Since it will not always be possible to literally match the search string with the content of a Web page, it is important to learn from an initial set of pages how the search string will be represented on real pages. This analysis will be done in the offline phase since it requires a collection of Web pages and is computationally more expensive. The result of this analysis is used in the online phase to derive the relevance of a page with respect to the crawl specification.

Social Web Analysis. The aim of the Social Web analysis is to leverage the Social Web to contextualize content and information to be preserved, and to support the crawler guidance. In social networks users are discussing and reflecting about all kinds of topics, events and persons. By doing so, they regularly post links to other relevant Web pages or Social Web content. As these links are recommendations of individuals in the context of their social online activities they are highly relevant for preservation. However, since users are unknown and anonymous it is necessary to derive their reputation and trustworthiness in the social community during the Social Web analysis.

The results of the Social Web analysis can also be leveraged in the contextualization process to further enrich the Web objects, e.g., if the object is tweeted by many nature experts it may be a good candidate for nature topics. Furthermore, the similarity and overlap between the provided objects and objects already seen before is established in order to interlink those that are discussing the same event, activity, or entity, improving the contextualization of the involved Web objects.

Data Enrichment and Consolidation. Data extracted via dynamics analysis and content analysis is heterogeneous. For instance, during one particular cycle, the text analysis component might detect an entity from the term "Ireland", while during later cycles, entities based on the term "Republic of Ireland" or the German term "Irland" might be extracted. These would all be classified as entities of type *arco:Location* and correctly stored in the ARCOMEM data store as separate entities described according to the ARCOMEM RDF schema. Data enrichment and consolidation follows three aims: (a) enrich existing entities with related publicly available knowledge; (b) disambiguation and (c) identify data correlations such as the ones above. (a), (b) and (c) exploit publicly available data from the Linked Open Data cloud[2] which offers a vast amount of data of both domain-specific and domain-independent nature.

Crawler Guidance. As shown on the bottom part of Figure 1, the crawler used in the ARCOMEM project includes a number of functionalities that are not found in traditional Web crawlers. First, we replace the traditional crawl definition by an *intelligent crawl definition*, which allows the specification of relevance scores and the referencing of the particular kinds of Web applications and ETOEs that define the scope of the archiving task. *Queue management* functions similarly as in a traditional architecture, but the classical page fetching module is replaced by some more elaborate *resource fetching* component able to retrieve resources that are not just accessible by a simple HTTP GET request (but by a succession of such requests, or by a POST request, or by the use of an API), or individual Web objects inside a Web page (e.g., blog posts, individual comments, etc.).

After a resource (for instance a Web page) is fetched, an *application-aware helper* module is used in place of the usual link extraction function, to identify the Web application currently being crawled, decide on and categorize crawling actions (e.g., URL fetching, using an API) that can be performed on this particular Web application, and the kind of Web objects that can be extracted. This is a critical phase for using clues from the Social Web to crawl content, because, depending on the kind of Web application that is being crawled, the kind of relevant crawling actions and Web objects to be extracted vary dramatically.

Crawling actions thus obtained are sent for further analysis and ranking to online phase modules. They are then filtered and prioritized by a *resource selection & prioritization* module using both intelligent crawling definition and feedback from online analysis modules to prioritize the crawl. Semantic analysis can thus make an impact on crawl guidance: for example, if a topic relevant to the intelligent crawl specification is found in the anchor text of a link to an external Web site, this link may be prioritized over others on the same page.

---

[2] http://lod-cloud.net/

4   Related Work

Since 1996, several projects have pursued Web archiving (e.g., [2]). The Heritrix crawler [3], jointly developed by several Scandinavian national libraries and the Internet Archive through the International Internet Preservation Consortium (IIPC), is a mature and efficient tool for large-scale, archival-quality crawling. The method of choice for memory institutions is client-side archiving based on crawling. This method is derived from search engine crawl, and has been evolved by the archiving community to achieve a greater completeness of capture and a reduction of temporal coherence of crawls. These two requirements follow from the fact that, for Web archiving, crawlers are used to build collections and not only to index [4]. These issues were addressed in the European project LiWA (Living Web Archives)[3].

The task of crawl prioritization and focusing is the step in the crawl processing chain which combines the different analysis results and the crawl specification for filtering and ranking the URLs of a seed list. A number of strategies such as breadth-first, back link count and PageRank exist for this. PageRank and breadth-first are good strategies to crawl "important" content on the Web [5], but since these generic approaches do not cover specific information needs, focused or topical crawls have been developed [6]. However, these approaches have only a vague notion of topicality and do not address event-based crawling.

5   Conclusions and Future Work

In this paper we presented the approach we follow to develop a social and semantic aware Web crawler for creating Web archives as community memories that revolve around events and the entities related to them. The need to make decisions during the crawl process with only a limited amount of information raises a number of issues. The division into different processing phases allows us to separate the initial complex extraction of events and entities from their faster but shallower detection at crawl time. Furthermore, it allows in the offline phase to learn more about particular events and topics the archivist is interested in and to get more insights about trustful content on the Social Web.

The implementation of the presented architecture is underway. Parts of the system are built upon existing technologies while other, like the Social Web analysis, need to be developed from scratch. Also, a number of research questions need to be addressed. For example the typically limited set of reference pages and the limited time to detect topics, entities, and events during crawling are open issues. Also how the different extracted information, interaction patterns, etc., can be combined for prioritizing URLs is currently an open question.

References

1. Blue Ribbon Task Force on Sustainable Digital Preservation and Access: Sustainable economics for a digital planet, ensuring long-term access to digital information (2010), http://brtf.sdsc.edu/biblio/BRTF_Final_Report.pdf

---

[3] http://www.liwa-project.eu/

2. Arvidson, A., Lettenström, F.: The Kulturarw project – the Swedish royal Web archive. Electronic Library 16(2) (1998)
3. Mohr, G., Kimpton, M., Stack, M., Ranitovic, I.: Introduction to Heritrix, an archival quality Web crawler. In: 4th International Web Archiving Workshop (2004)
4. Masanès, J.: Web archiving. Springer (2006)
5. Baeza-Yates, R., Castillo, C., Marin, M., Rodriguez, A.: Crawling a country: better strategies than breadth-first for Web page ordering. In: 14th WWW Conf. (2005)
6. Menczer, F., Pant, G., Srinivasan, P.: Topical Web crawlers: Evaluating adaptive algorithms. ACM Trans. Internet Technol. 4, 378–419 (2004)

Query Expansion of Zero-Hit Subject Searches: Using a Thesaurus in Conjunction with NLP Techniques

Sarantos Kapidakis[1], Anna Mastora[1], and Manolis Peponakis[2]

[1] Laboratory on Digital Libraries & Electronic Publishing, Archives & Library Science Department, Ionian University, Corfu, Greece
{sarantos,mastora}@ionio.gr
[2] National Hellenic Research Foundation / National Documentation Centre, Athens, Greece
epepo@ekt.gr

Abstract. The focus of our study is zero-hit queries in keyword subject searches and the effort of increasing recall in these cases by reformulating and, then, expanding the initial queries using an external source of knowledge, namely a thesaurus. To this end, the objectives of this study are twofold. First, we perform the mapping of query terms to the thesaurus terms. Second, we use the matched terms to expand the user's initial query by taking advantage of the thesaurus relations and implementing natural language processing (NLP) techniques. We report on the overall procedure and elaborate on key points and considerations of each step of the process.

Keywords: Query expansion, Thesaurus, Zero-hit queries, NLP techniques.

1 Introduction

The focus of our study is zero-hit queries in keyword subject searches in an effort to increase recall by reformulating and, then, expanding the initial queries using an external source of knowledge, namely a thesaurus, and taking advantage of natural language processing (NLP) techniques. In case of zero-hit queries query expansion methods based on sets of retrieved results (implicit relevance feedback) cannot be implemented. Building on this fact, we chose to use a hand-made thesaurus to expand the initial queries taking advantage of the relations identified within a thesaurus' structure without letting the users interfere in the process.

In order to proceed to query expansion we first allocate an entry point within the knowledge base, i.e. match the initial queries to a term from the thesaurus. Exact string matching is unlikely to be successful in highly inflectional languages, like Greek, because of the various forms a word can take. Additionally, research has shown that typing errors are also responsible of delivering zero-hit queries [1]. To overcome the identified obstacles we used techniques for natural language processing, namely spelling, lemmatizing, removal of stop words, accent and case processing. The database and the thesaurus underwent the same processing where needed. Finally, we derived candidate expansion terms by considering the related, parallel, narrower and broader terms of the allocated entry point in the thesaurus moving one level towards each direction.

P. Zaphiris et al. (Eds.): TPDL 2012, LNCS 7489, pp. 433–438, 2012.

The remaining sections elaborate on the overall procedure and report on key points and considerations of each step of the process.

2   Aims and Objectives

The aim of the study is to improve the recall of user queries which returned zero hits by expanding them through multi-step implementation of natural language processing techniques and hand-made thesaurus browsing. To this end, the objectives of this study are twofold. First, we perform the matching of query terms to the thesaurus terms. Second, we use the mapped terms to expand the user's initial query terms by taking advantage of the thesaurus relations.

3   Related Work

In [2] is stated that the most critical language issue for retrieval effectiveness is the terms mismatch problem: the indexers and the users often do not use the same words. We say that this is undoubtedly one of the major reasons for a system's poor performance, especially as the outcome of subject searching [3, 4]. Agreeing with [5] we also suggest that the Subject facet is the only facet with semantic relations between its terms, making it the most suitable facet for our method and experimental setup.

In [6] is clarified that *query reformulation* involves either the restructuring of the original query or by adding new terms, while *query expansion* is limited to adding new terms to the original query. Three types of query expansion are discussed in literature: manual, automatic, and interactive (i.e., semiautomatic, user-mediated, or user-assisted). In [7] is stated that thesauri have been recognized as a useful source for enhancing search-term selection for query formulation and expansion. In their study they mention that in 50% of the searches where additional terms were suggested from the thesaurus, users stated that they had not been aware of the terms at the beginning of the search. In [8], about the performance comparison of thesaurus relationships in automatic versus interactive query expansion is concluded that synonyms and narrower terms are good candidates for automatic expansion, while related (associative) terms are better candidates for interactive expansion, leaving the report on broader terms rather equivocal.

In [9] is reported that the improvement in expansion increases when adding up to 20 terms, and reaches a plateau, then the improvement begins to decrease when more than 50 terms are added. The expanded queries, however, still perform better than the original queries. They also indicate that expanding a query with 30 to 40 top ranked terms seems to be the safest method with respect to the collection targeted in their evaluation. Additionally, the user may get confused if the system retrieves documents that do not contain the original query terms. But, using a thesaurus gives the user confidence and security that her needs are met, as reported in [10], while in [7] is mentioned that narrower and related terms taken together constitute approximately 50% of the query-expansion terms selected by users.

4 Methodology

The data analyzed in this paper was gathered from the log files produced during an in vitro experiment. More details about the experiment scenario can be found in [1].

We have to make clear at this point that the participants did not counteract with the thesaurus at any step of submitting their queries. Moreover, the subjects contained in the database were not formulated according to the thesaurus; they were free keywords provided by the cataloguers. The thesaurus was a tool used for our data post-processing.

4.1   The tools: Thesaurus, Speller and Lemmatizer

For our study we used the EuroVoc[1] thesaurus which is widely-used, multidisciplinary and available in 22 European languages, Greek being among them. We used a licensed version which is available in a variety of formats such as XML and SKOS/XML. The version we used contained 6,797 descriptors and 3,554 non-descriptors. In terms of the relationships identified in the thesaurus, we report on the *Equivalence* (UF and USE) and *Hierarchical* (Broader (BT) and Narrower (NT)).

For spell checking and correcting the queries we used Aspell, v.0.60.6[2]. Aspell is a utility program that connects to the Aspell library so that it can function as an ispell -a replacement, as an independent spell checker, as a test utility to test out Aspell library features, and as a utility for managing dictionaries used by the library. The Aspell library contains an interface allowing other programs direct access to its functions and therefore reducing the complex task of spell checking to simple library calls.

We also used the ilsp_nlp[3] tool, which lemmatizes Greek texts. The input is an XCES document with POS-tagged tokens and the output is an XCES document with lemmas assigned to each token. The tool is developed by the Greek Institute for Language and Speech Processing and is freely available through a web interface.

5   Query Matching and Expanding

The participants submitted 2,116 queries in total while 749 of them returned zero hits, which we further processed. Our first main task was to match the queries to the EuroVoc terms. Then we proceeded to expanding the allocated terms taking advantage of the thesaurus relations. Figure 1 depicts the basic steps of the process.

---

[1] Reproduced and adapted from the original language editions of the *Eurovoc Thesaurus (Edition 4.3)* © European Communities, 2008. //eurovoc.europa.eu/ (last accessed April 2012)
[2] Aspell, v.0.60.6. ©2000-2004 by Kevin Atkinson, //aspell.net/ (last accessed April 2012)
[3] ilsp_nlp, //nlp.ilsp.gr/soaplab2-axis/ (last accessed April 2012)

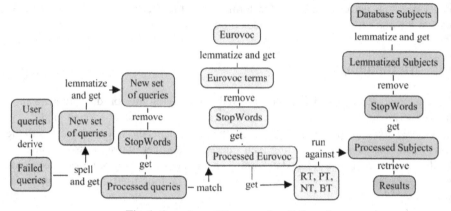

Fig. 1. Overview of the procedure followed

5.1   Matching Queries to EuroVoc Terms

In order to check if the query terms derived from the zero-hit queries were valid words we spell checked and corrected them with Aspell. The tool seems to work adequately, even for named entities, which is often a challenge. A decision we had to make concerned how many correction suggestions we would consider for each identified error. We decided, judging from a selective evaluation of the retrieved results, to use the first suggestion as the most likely to be correct.

During the execution of the search tasks, the participants submitted various forms of the word(s) representing their information need. Let us take for example the case of "Greenhouse effect". Within this nominal compound the word *Greenhouse* stands for "Θερμοκήπιο" *(Thermokēpio)*. We recorded the submission of the following forms of this word: *θερμοκήπιο (thermokēpio), θερμοκηπίου (thermokēpíou), θερμοκηπίων (thermokēpíōn), θερμοκήπια (thermokēpia),* one misspelled "*θερμοκηπείου*" (thermokēpeíou) and one truncated "*θερμοκήπ**" (thermokēp*). In order to eliminate the variant types of the words due to inflections we lemmatized the words.

A significant issue we also had to deal with was about accents. They play an important role in Greek language but we faced a rather intriguing situation. Most IR systems do not consider them while searching but tools for language processing do consider them for the analysis they perform because a change in the accent position may lead to a change to the word semantics. So, we have to carry accents along the processing and drop them at a convenient point. We followed a similar approach for capitalized words. Finally, we dropped several allocated stop words.

As far as the matching between terms is concerned, the ideal situation is met when all words in the query term match a EuroVoc term. For multi-word EuroVoc terms, this is highly unlikely. In this process, we accept as matching to EuroVoc terms all terms that contain at least one word from the user query. If the EuroVoc term contains additional words, we do not really broaden the query semantics very much, because in order to consider that a EuroVoc term, matching the user query, is contained in a record, all of its words must be present in the record. For example the user submitted

he term "organic food" and we located "organic law" and "organic farming" (be-
:ause of "organic") as well as "food production", "food fat", "pet food" (because of
food"). A record will match this term if it includes at least one EuroVoc term from
·ach of the two sets. Following these decisions, our pilot metrics showed that the
,346 words we imposed to matching with the EuroVoc, 810 of them gave us a posi-
ive match.

Truncation in the query terms is also considered during matching, but introduces
nore problems in the transformation procedures. A truncated word may seem miss-
)elled, and not be lemmatized, but should be left unchanged, as the replacement cor-
·ect word will in many cases fail to match the words that the original word did.

5.2   Expanding Queries

[he second step involves utilizing the thesaurus structure and relations in order to
)roduce the expansion terms. For each matching term, we currently follow all Re-
ated, Broader and Narrower relations for one level, and at the end we add the "Use
'or" relations for all the identified terms. All terms derived this way are accepted
ilternatives to the original word, and in order to be considered as matching a bibliographic record, all of its words are required to be present in this record. If more than
)ne query words are replaced by a set of EuroVoc terms, at least one term from each
;et must match the bibliographic record, as well as all words that did not lead to any
EuroVoc terms. Again, our pilot metrics show that for 810 mapped words to the the-
;aurus, we got 44,341 EuroVoc candidate-for-expansion terms in total.

Table 1. Improvement during transformation stages after database runs

State of Queries	Initial	Spelled	Lemmatized	Expanded
No of zero-hit queries	749	676	624	531

6   Conclusion and Future Considerations

Queries with zero hits are not a fruitful outcome for implementing relevance feedback
:echniques for query expansion. The use of techniques for language processing can be
·ffective towards improving recall rates when it comes to highly inflectional languag-
·s, like Greek. Lemmatizing appears to be a more appropriate process though more
:omputationally intensive. The improvement accomplished so far is shown in Table 1.
·or each subsequent step, the new set of queries derived from the previous step was
·e-run to the database. The overall conclusion is that by implementing the proposed
·ramework, 219 (29.07%) of the initial 749 failed queries were dealt with successfully, meaning that they now lead to returning results, instead of zero hits.

The spell checking and correcting seem to be of considerable assistance towards
the improvement of recall but requires extensive work on data interfacing and trans-
·ormations in order to derive quantitative results, since we had to combine tools (for
spelling, lemmatizing, thesaurus browsing and query matching) from various creators
having different interfaces and functions.

We do have to consider further options and take more meaningful decisions in selecting candidate expansion terms available in the thesaurus so as to deal with the problem of ambiguity and avoid any semantic drifts caused by arbitrary matches. Nominal compounds, named entities and acronyms are also part of our research in terms of properly recognizing them during our processing. Another consideration would be to deal with some exceptional cases, like when replacing a word, in order to avoid missing matches of it, it would be useful to also keep the original word together with the suggested alternatives, as if it was one more alternative. Finally, we plan to measure the exact recall rates at each stage of our method using a test collection.

Acknowledgment. This research has been co-financed by the European Union (European Social Fund – ESF) and Greek national funds through the Operational Program "Education and Lifelong Learning" of the National Strategic Reference Framework (NSRF) - Research Funding Program: Heracleitus II.

References

1. Mastora, A., Kapidakis, S., Monopoli, M.: Failed Queries: a Morpho-Syntactic Analysis Based on Transaction Log Files. In: First Workshop on Digital Information Management, Corfu, Greece, pp. 1–7 (2011),
   http://eprints.rclis.org/handle/10760/15845 (accessed April, 2012)
2. Carpineto, C., Romano, G.: A Survey of Automatic Query Expansion in Information Retrieval. ACM Comput. Surv. 44(1), 1:1–1:50 (2012)
3. Lau, E.P., Goh, D.H.-L.: In Search of Query Patterns: a Case Study of a University OPAC. Information Processing and Management: an International Journal 42(5), 1316–1329 (2006)
4. Villén-Rueda, L., et al.: The Use of OPAC in a Large Academic Library: A Transactional Log Analysis Study of Subject Searching. The Journal of Academic Librarianship 33(3), 327–337 (2007)
5. Hollink, L., Malaisé, V., Schreiber, G.: Thesaurus enrichment for query expansion in audiovisual archives. Multimedia Tools Appl. 49(1), 235–257 (2010)
6. Selvaretnam, B., Belkhatir, M.: Natural language technology and query expansion: issues, state-of-the-art and perspectives. Journal of Intelligent Information Systems (2011),
   http://dx.doi.org/10.1007/s10844-011-0174-3 (accessed April 2012)
7. Shiri, A., Revie, C.: Query expansion behavior within a thesaurus-enhanced search environment: A user-centered evaluation. Journal of the American Society for Information Science and Technology 57(4), 462–478 (2006)
8. Greenberg, J.: Optimal query expansion (QE) processing methods with semantically encoded structured thesauri terminology. Journal of the American Society for Information Science and Technology 52(6), 487–498 (2001)
9. Mandala, R., Tokunaga, T., Tanaka, H.: Query expansion using heterogeneous thesauri. Information Processing & Management 36, 361–378 (2000)
10. Fang, H.: A Re-examination of Query Expansion Using Lexical Resources. In: Proceedings of ACL 2008: HLT, pp. 139–147 (2008)

Towards Digital Repository Interoperability: The Document Indexing and Semantic Tagging Interface for Libraries (DISTIL)

Michael Khoo[1], Douglas Tudhope[2], Ceri Binding[2],
Eileen Abels[1], Xia Lin[1], and Diana Massam[3]

[1] The iSchool, Drexel University, 3141 Chestnut Street, Philadelphia, PA 19104, USA
{khoo,ega26,linx}@drexel.edu
[2] University of Glamorgan, Pontypridd, CF37 1DL, Wakes, UK
dstudhop@glam.ac.uk
[3] The University of Manchester, Oxford Road, M13 9PL
Diana.Massam@manchester.ac.uk

Abstract. The question of how to integrate diverse digital repositories into a unified information infrastructure, accessible and discoverable through simple interfaces, remains a central research issue for digital libraries. Many collections are described by specialized metadata, which currently has to be mapped and crosswalked to a standard format in order to be useful. However, this metadata work can be expensive and resource consuming. We describe work-in-progress with DISTIL (Document Indexing & Semantic Tagging Interface for Libraries) to support federated cross-collection search in humanities and the social sciences. DISTIL proposes to support interoperability by generating Dewey Decimal Classification 'tags' from individual metadata records. The resulting tags can then be used to support cross-collection browsing. We focus here on some of the initial pre-processing stages of the metadata workflow, which include cleaning and formatting metadata records, in order to extract terms that can then be used to generate the DDC tags. Some initial strategies for and issues with this workflow are described.

Keywords: dewey decimal classification, digital humanities, interoperability, metadata, social sciences, tagging.

1 Introduction

What is the future for digital libraries? Technically, it has been argued for a while that some of the basic challenges of placing collections online have been addressed [12]. Practically, instantiations of individual digital libraries have become increasingly sophisticated, feature-rich, and user-friendly (see for instance, Isaac Newton's notebooks at the Cambridge Digital Library [6]). At the same time, theoretical definitions of digital libraries continue to evolve, and have moved increasingly towards descriptions of complex socio-technical systems. (DELOS, for example, defines a digital library extensively as "as a tool at the centre of intellectual activity

P. Zaphiris et al. (Eds.): TPDL 2012, LNCS 7489, pp. 439–444, 2012.

having no logical, conceptual, physical, temporal or personal borders or barriers on information. It has moved from a content-centric system that simply organises and provides access to particular collections of data and information to a person-centric system that aims to provide interesting, novel, personalised experiences to users. Its main role has shifted from static storage and retrieval of information to facilitation of communication, collaboration and other forms of interaction among scientists, researchers or the general public on themes that are pertinent to the information stored in the Digital Library. Finally, it has moved from handling mostly centrally located text to synthesising distributed multimedia document collections, sensor data, mobile information and pervasive computing services" [7].)

As part of this theoretical evolution, there remains the important question of how to integrate diverse repositories into a single information infrastructure, accessible and discoverable through federated search. Some of the fundamental issues here, as have been known for a long time, are rooted in standardized metadata, but the associated work required to achieve this can be expensive and resource consuming [11]. There is often a gap between the amount of digital resources requiring description, and the number of catalogers who can generate such descriptions. Realizing the full potential of information infrastructures includes therefore addressing the problem of finding ways to make resources in heterogeneous collections easily discoverable and accessible. This poster describes work-in-progress with one such initiative, to develop a tool called DISTIL (Document Indexing & Semantic Tagging Interface for Libraries), that will support cross-collection search in humanities and the social sciences.

1.1   Interoperable Metadata for the Humanities and Social Sciences

Humanities and social sciences metadata in this area supports researchers, educators, students and others to share social science data and information, and to collaborate in research, education and learning, across large-scale information infrastructures [e.g. 1, 3, 5, 10, 16]. However, many such collections are described by specialized standards and formats, making federated search a difficult proposition. Further, there may be useful resources for humanities and social science scholars, which are to be found in science-based collections. Cross-disciplinary information retrieval will be built on standardized metadata, and methods for generating standardized metadata. Two recognized approaches here involve adopting a common metadata format at the start of a project, or the systematic crosswalking of existing metadata to a common format. Both of these solutions can be labour- and resource-intensive, and can achieve less than optimal success. There is therefore scope to explore the development of alternate methods for generating interoperable humanities and social science metadata.

The DISTIL tool represents an innovative method for generating Dewey Decimal Classification (DDC) tags for digital resources from metadata records and digital resource content, by performing text analyses of the metadata for each resource, and eventually the content of the resource itself. DDC was chosen as a target for several reasons: first, the tool is being built on top of the existing PERTAINS (PERsonlisation Tagging interface INformation in Services) tool being developed at the University of Glamorgan, which is a tool for generating DDC tags from keyword input [4, 15]; second, existing research at Drexel has been looking at building browsing interfaces on top of DDC record numbers. The existing PERTAINS tool is a Web-based vocabulary-based tag recommender service that takes user-provided keywords and generates corresponding DDC terms (effectively, 'tags'). PERTAINS currently provides an initial set of suggestions before the user enters a keyword. A suggestion is then made after user keyword is entered, based on the user keyword (plus using available 'context' from document title and metadata to help disambiguate between options). An 'Area of Interest filter' allowed the user to refine suggested tags by selecting (multiple) options from a broad classification.

The DISTIL tool will build on PERTAINS and will be embedded in a wider workflow that will identify, analyze and extract meaning from, and represent in one single interface, resources from multiple collections. At an early stage of the project, it was decided to develop the pre-processing workflow for DISTIL, and DISTIL itself, as two separate modules, in order to reduced the dependencies between the two operations and the overall workflow. A proposed outline of the workflow is therefore shown in Figure 1, in which each metadata record is pre-processed before being analysed by the DISTIL too, with the output of the DISTIL tool then being used to support federated searches across multiple collections. The work-in-progress described in this paper covers the preliminary work for stages 1 and 2.

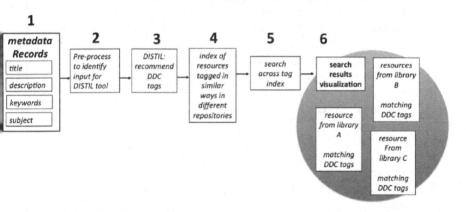

Fig. 1. High-level outline of proposed workflow

1.2   Pre-processing: Preliminary Research Questions and Findings

The partner repositories in the project are: The Internet Public Library, and Internet Librarian's Index (LII) (USA) [8]; Intute (UK) [9]; and the National Science Digital Library (NSDL) and Digital Library for Earth Systems Education (DLESE) (USA) [13]. Each repository uses a flavor of Dublin Core metadata, with variations in many of the qualified fields between each collection. An important initial research question for the project is therefore: How can a generalizable workflow for generating keywords from the existing metadata records in these different collections, keywords that will in turn be used by DISTIL to generate the DDC tags, be built?

The first step in the workflow outlined above is to select the metadata fields to include in the analysis. It was assumed initially, for heuristic purposes, that fields *title*, *description*, *subject(s)*, and any *keyword/topic* fields, would be used in the analysis, with no distinction being made between whether or not controlled vocabularies were applied. The following fields were therefore identified as being of use in the analysis:

- IPL/LII: *dc.description*; *dc.subject*; *dc.title*; *ipl.subject*
- NSDL: *dc.description*; *dc.subject*; *dc:subject xsi:type (optional)*; *dc.title*
- DLESE: *dc.description*; *dc.subject*; *dc:subject xsi:type (optional)*; *dc.title*
- Intute: *dc.description*; *dc.keywords*; *dc.title*

For each example, all XML markup was removed from these fields, leaving a body of text. The example in Figure 2 shows text extracted from an IPL/LII record for the Web site of the American Society for Tropical Medicine and Hygiene (http://www.astmh.org/).

```
The ASTMH represents scientists and clinicians
interested in research of tropical diseases and
education about control and prevention. The site
contains   membership    information,    tropical
medicine Questions and Answers, a newsletter, a
travel   clinic   directory,   and   information   for
medical   professionals,   including   funding   and
fellowship and overseas opportunities.
medicine
medical journals
virology
hygiene
global health
parasites
American   Society   of   Tropical   Medicine   and
Hygiene
Health and Medical Sciences Diseases, Disorders
and Syndromes
```

Fig. 2. Sample of text, extracted metadata record, for pre-processing, (c.f. Figure 1, stage 1)

Two basic types of analysis were then tested with the resulting texts, with the aim of identifying suitable methods for generating input keywords for DISTIL. In the first analysis, a frequency count was carried out on the text to identify frequently-occurring nouns and verbs (this analysis was carried out with the Nisus Pro software for Macintosh, which has several built-in basic text-parsing macros.) Following the frequency count, a hand analysis aggregated any appropriate stems; and a standard eviation-based cut-off point was then calculated, in order to identify the most requent nouns and verbs. In the second analysis, the cleaned text was submitted to an online text-parsing tool, the U.K. National Centre for Text Mining's TerMine [14], which identified compound noun phrases to be considered as inputs for DISTIL. A requency count of these noun phrases was then created from the TerMine output.

This process was applied to a small sample of records randomly selected from NSDL, DLESE, and Intute. Each of the two analyses produced differing results. On he one hand, the first (frequency) analysis produced a clear hierarchy of possible erms for submitting to DISTIL, in which a few terms occurred frequently, and many erms occurred infrequently. This made it easy to identify the most common terms; however, this analysis also tended to produce very general terms as the top results, which are not necessarily useful in producing more precise DDC tags. On the other hand, the second (TerMine noun phrase) analysis did produce more precise results, however, it also produced a less distinct hierarchy, with many individual noun phrases occurring once only, making it difficult to decide which precise result was the most significant.

The next goal in the analysis is to refine these initial steps, in order to balance the advantages offered by these two different approaches, and to generate useful terms for subsequent input to the DISTIL tool. The success of this initial stage will in turn be evaluated partly by asking indexing specialists to assess to what extent the DDC tags generated by DISTIL on the basis of the pre-processing, accurately describe the actual content of the resource that they are supposed to describe.

Table 1. Results of two different proposed text processing strategies (c.f. Figure 1, stage 2)

Frequency count: suggested terms and frequency	TerMine output
medic- tropical	american society global health medical journals medical professionals medical sciences diseases membership information overseas opportunities travel clinic directory tropical diseases tropical medicine tropical medicine questions

2   Conclusion

Realizing the potential of information infrastructures includes making resources in heterogeneous collections discoverable and accessible. We have introduced a tool, DISTIL (Document Indexing & Semantic Tagging Interface for Libraries), that aims to support federated cross-collection search in humanities and the social sciences. We have described some of the initial steps undertaken in order to analyse metadata records from a variety of digital libraries, and to extract key terms from those records for further processing. Issues identified during this initial work include maintaining a balance between methods that produce clearly-ranked lists of broad terms, and methods that produce ambiguously-ranked lists of more precise terms.

Acknowledgements. Current work in this project is funded by the National Endowment for the Humanities (NEH), United States, and by the Joint Information Systems Committee (JISC), UK. Previous work has been funded partly by JISC.

References

1. American Council of Learned Societies. Our Cultural Commonwealth (2006)
2. Audenaert, N., Furuta, R.: What Humanists Want: How Scholars Use Primary Source Documents. In: Proceedings of the 10th Annual Joint Conference on Digital Libraries (JCDL 2010), pp. 283–292 (2010)
3. Babeu, A.: Rome Wasn't Digitized in a Day. Building a Cyberinfrastructure for Digital Classics. Council on Library and Information Resources (2011)
4. Binding, C., Tudhope, D.: Terminology Web Services. Knowledge Organization 37(4), 287–298 (2010)
5. Burton, O.V., Appleford, S.: Cyberinfrastructure for the Humanities, Arts, and Social Sciences. EDUCAUSE Center for Applied Research, Research Bulletin 2009(1) (2009)
6. Cambridge Digital Library, The University of Cambridge. The Newton papers, http://cudl.lib.cam.ac.uk/collections/newton
7. Candela, L., Castelli, D., Ferro, N., Ioannidis, Y., Koutrika, G., Meghini, C., Pagano, P., Ross, S., Soergel, D., Agosti, M., Dobreva, V., Katifori, V., Schuldt, H.: The DELOS Digital Library Reference Model. Foundations for Digital Libraries. Version 0.98. N.p.: DELOS Network of Excellence on Digital Libraries (2007)
8. IPL: The Internet Public Library, http://www.ipl.org/
9. Intute, http://www.intute.ac.uk/
10. Kansa, E.C., Kansa, S.W., Watrall, E.: Archaeology 2.0: New approaches to communication and collaboration (2011)
11. Khoo, M., Hall, C.,, M.M.: A Sociotechnical Study of Crosswalking and Interoperability. In: 10th ACM/IEEE Joint Conference on Digital Libraries, Brisbane, Australia, June 21-25, pp. 361–364 (2010)
12. Lynch, C.: Where do we go from here? The next decade in digital libraries. D-Lib. Magazine 11(7/8) (July/August 2005)
13. NSDL: The National Science Digital Library, http://www.nsdl.org/
14. TerMine, http://www.nactem.ac.uk/software/termine/
15. Tudhope, D., Binding, C.: Faceted Thesauri. Axiomathes 18(2), 211–222 (2008)
16. Welshons, M. (ed.): Our Cultural Commonwealth. Report of the American Council of Learned Societies Commission on Cyberinfrastructure for the Humanities and Social Sciences. ACLS, New York (2006)

Aggregating Content for Europeana: A Workflow to Support Content Providers

Valentina Vassallo[1] and Marzia Piccininno[2]

[1] The Cyprus Institute - STARC
vassallo@cyi.ac.cy
[2] Ministero per i Beni e le Attività Culturali – ICCU
marzia.piccininno@beniculturali.it

Abstract. This document comes from the experiences of the authors as leaders of Work Packages about "Coordination of content" within the digital library projects to aggregate content to Europeana. In particular, it will focus on two projects, ATHENA and Linked Heritage (LH), with the definition of a workflow and a structured organization for content aggregation. The large amount of digital objects, coming from various European cultural institutions, has to be aggregated (at national level and/or for Europeana) creating good practices and implementing solutions to sustain the material aggregation in a long term perspective.

Keywords: Aggregation, Standards, Europeana.

1 Introduction

This paper comes from the experience of the two authors as leaders of Work Packages about the "Coordination of content" within the digital libraries projects to aggregate content for Europeana. In particular, it will be referred to ATHENA and LH projects, focusing on the importance of the creation of a workflow in content provision. It will support the Digital Humanists work in the difficult task of merging humanist tasks and technicalities. The large amount of digital and digitized objects owned by the European cultural institutions (museums, archives and libraries, research centres and preservation offices), implied the necessity of defining best practices and solutions for the sustainability on a long term perspective of the aggregated material. These methods imply the necessity of a standardized approach for mapping, aggregating and publishing the metadata for Europeana in the full respect of its standards and requirements.

1.1 Europeana, ATHENA and Linked Heritage projects

Europeana is the largest European digital library, which provides point of access to a large amount of databases and to over 23 million books, paintings, films, museum objects and archive records that were digitized all over Europe. It is a multi-annual strategy of the European Commission to foster the accessibility of digital cultural content, the promotion of standards agreed by all Member States, the research methods and techniques for an improved access to data and information culture. Europeana is also a huge source of information from the European cultural and scientific institutions and a collector of national and local aggregation experiences.

P. Zaphiris et al. (Eds.): TPDL 2012, LNCS 7489, pp. 445–454, 2012.

The first European projects funded by the European Commission for feeding Europeana date back to 2008: this core group (ATHENA, Europeana Local, European Film Gateway, etc.) needed to set a workflow for bringing the content into Europeana. Various repositories across Europe implementing the OAI-PMH protocol have been set up as well as portal for the metadata transmission, but only ATHENA approach and solutions are being reused in several other initiatives. Moreover, ATHENA set up a validation process of the work done by content providers and assured constant support training on tools, activities and procedures (mapping, IPR issues, etc.).

ATHENA (2008-2011)[1] was an eContentPlus Best Practice Network coordinated by the Ministry for Cultural Heritage and Activities (MiBAC, Italy) which brought together relevant stakeholders and content owners from museums and other cultural institutions all over Europe to evaluate and integrate specific tools, based on a common agreed set of standards and guidelines to create harmonised access to their content. ATHENA made available circa 4 million items to Europeana from 20 EU Member States plus Israel and Russia and elaborated standardized tools for the content aggregation now in use in other digitization activities.

LH[2] (2011-2013) is an initiative coordinated by the Central Institute for Union Catalogue of the Italian Libraries, depending on MiBAC that extends and implements the ATHENA results. It is a best practice network funded within FP7 that began in April 2011 and will run for 30 months; it will contribute new content to Europeana, from both public and private sector (mainly publishers), improve the quality of content in terms of richness of metadata, potential reuse and uniqueness, explore the potential of cultural Linked Open Data, and enable better search, retrieval and use of the content published in Europeana. Twenty-two countries are members of the LH consortium: culture ministries, government agencies, museums, libraries, and national aggregators, major research centers, publishers and small businesses, also involving organizations that contribute to Europeana for the first time with a contribution of 3 million records of a wide spectrum of cultural content.

2   Achieved Results

The need for a tool allowing the aggregation of content for Europeana, effective and performing but at the same time user friendly was the first requirement at beginning of the ATHENA project. This task was carried out by the National Technical University of Athens (NTUA) that elaborated a platform called MINT[3] which allows content providers to upload, map, validate a send data to Europeana in a single web environment (**Fig. 1**).

MINT implemented LIDO as intermediate harvesting schema which was developed within the ATHENA project for museum sector needs. It has been also successfully used in cross-domain contexts achieving a rapid and large application in many initiatives of the Europeana project group[4].

---

[1] www.athenaeurope.org

[2] http://www.linkedheritage.org/

[3] http://mint.image.ece.ntua.gr/

[4] http://network.icom.museum/cidoc/working-groups/data-harvesting-and-interchange/lido-community/use-of-lido.html

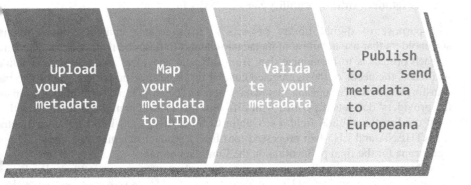

| Upload your metadata | Map your metadata to LIDO | Valida te your metadata | Publish to send metadata to Europeana |

Fig. 1. Metadata flow in MINT

The metadata ingestion service enables: the data uploading and mapping to LIDO by contributing partners; the transformations of the metadata records into LIDO records and the aggregation in the project repository; the validation of the content through the Europeana Content Checker[5]; the conversion of the stored data into ESE[6] and transmission to the Europeana ingestion office via OAI-PMH. The transmission of data to Europeana is technically managed by NTUA, however, the content providers have been asked by Europeana to check again the quality of the content once published online and to assess possible problems.

2.1   LIDO (Lightweight Information Describing Objects)

LIDO (Lightweight Information Describing Objects)[7] is based on CIDOC-CRM conceptual reference model. It comes from the integration between CDWA Lite and museumdat metadata schemas and it is based on SPECTRUM standard.

Specifically, LIDO (v.1.0) is organized in 7 areas of which 4 are descriptive and 3 are administrative (COBURN *et alii* 2010): Descriptive Information (Object Classification, Object Identification, Event, Relation); Administrative information: Rights Work, Record, Resource.

---

[5] Europeana Content Checker version 2 User Guide
http://pro.europeana.eu/documents/900548/ae5e78e8-ce78-424d-b360-5c01eddb3564

[6] ESE stands for Europeana Semantic Element, the Europeana data model. MINT will also support EDM (Europeana Data Model), the new application profile that will be implemented in the coming months.

[7] Its definition in an XML schema, together with the specification document, can be found at www.lido-schema.org. LIDO is the result of a collaborative effort of international stakeholders in the museum sector, starting in 2008, to create a common solution for contributing cultural heritage content to portals and other repositories of aggregated resources. Being an application of the CIDOC Conceptual Reference Model (CRM) it provides an explicit format to deliver museum's object information in a standardized way.

2.2   Metadata Interoperability Service (MINT)

The purpose of digital library projects is to aggregate content from different stakeholders that use a variety of metadata schemas to describe their content. The aim of each project is to harmonize data from different providers in a metadata schema that meets the needs of the aggregated content from the specific project, in order to be transformed in the Europeana metadata schema. The MINT platform allows to map the providers' datasets described with different metadata schemas (proprietary or the most used in the cultural field: i.e. Dublin Core, MARC, etc.) into LIDO, used both for ATHENA and LH, then processed into ESE metadata schema to be harvested by Europeana for the final publication on the European portal.

MINT is a server for the content ingestion based on open source software developed by the National Technological University of Athens (NTUA), a web platform designed to facilitate the cultural heritage content and metadata aggregation initiatives in Europe. The platform provides a management system both for user and organization, allowing the implementation and operation of different aggregation patterns (thematic or cross-domain, international, national or regional) and their access rights. The registered organizations can upload (via http, ftp, OAI-PMH) their metadata records in XML or CSV in order to manage, aggregate and publish their collections.

The metadata reference model (LIDO in our case) serves as aggregation schema which the other ingested schemas are aligned to (standard or proprietary). MINT is user-friendly: in fact, users can define their metadata crosswalks through a visual mappings editor for XSL language. Mapping is performed through easy drag-and-drop operations or inputs that are translated to the corresponding code (**Fig. 2**). The mapping editor displays both the input and the XSD target, thanks to the intuitive interface that provides access and navigation, data input schema, structure, documentation and restrictions of the output schema. These mappings can be applied to the ingested records, edited, downloaded and shared as templates among users of the platform. A preview interface shows to users the steps of aggregation, including the uploaded xml, the XSLT of their mapping, the transformed record into the target schema, the transformations from the target schema to other models of interest (for example ESE) and available html renderings of each xml record. Users can then transform their collections using complete and validated mappings, in order to publish them in available target schemas for the required aggregation and remediation steps. The process guarantees the possibility of updating, monitoring and ensuring a continuous enrichment and visibility of the contribution through the ingester tool and Europeana[8].

---

[8] The publication allows seeing data aggregated by the projects on Europeana's website, guaranteeing both the rights and use. In fact, each provider participating in the aggregation projects, must ensure the provision of mandatory elements required by Europeana, mainly constituted of a set of records containing information about the digital object, rights, the provider and the access to the digital resource. Furthermore, the publication in Europeana makes it possible to display the digital object in its original context that is the repository of the provider institution. This clearly allows greater access to the original data (in case of providers that give only mandatory elements) and guarantees the ownership right of intellectual data.

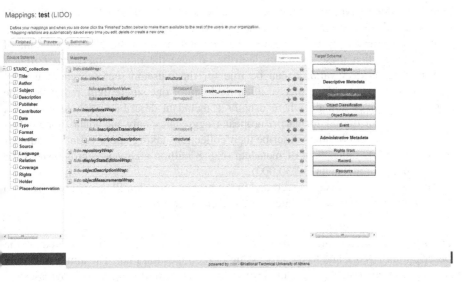

Fig. 2. The mapping procedure on the MINT tool

3   The Workflow in Providing Content

Taking into consideration the experiences of the described projects, the support to the aggregation process has been considered of paramount importance. We decided to create a workflow able to organize the content provision in order to support the providers work and facilitate their procedures and to get feedback about difficulties, probable improvements of the work chain, bugs, etc.

Following, the steps considered to elaborate a good procedure on the base of the projects' experiences.

Step 1 - Organising the content providers and investigation of the digital collections
The first step for taking the content into Europeana is assessing the digital collections that content providers described in the Description of Work. In fact, the experience of the past ATHENA project and of LH project currently, demonstrated that from the proposal writing to the approval and beginning of the project many things that affect the digital collection availability may happen: technical problems (re-engineering of the digital library or technical platform for the data management, metadata without digital objects, digitisation process in progress, etc.), managerial issues or copyright problems. This assessment can be easily done through a template living the operational details about each collection of digital items. This contribution is useful to analyse the problems described above and for the organisation of an ingestion plan. The content providers, answering to a survey, should provide the following information: Country, Data provider, Primary contact, Technical contact, Collection URL, Amount of metadata to be aggregated, Amount of digital objects (linked to the

metadata), Object types (image, text, sound, video), Description, Metadata used and Rights (details on the rights status of the collections).

It is fundamental to ask separately the amount of metadata and the amount of digital objects that these metadata describe because the ratio *1 metadata:1 digital object* is not always valid: it can be true for the museum sector because usually museum objects are unique pieces, often made up of one single item, rather than the library or archives domain where usually 1 metadata describes a whole book (i.e. several pages in digital format). Europeana aggregates only metadata; nonetheless sometimes content providers describe their collections taking into account the number of digital objects rather than of metadata (e.g. 8 million archival records instead the actual amount of metadata as 230,000).

The result of this survey is recorded in the project Ingestion Plan and contributes to the definition of the time schedule of the content aggregation.

Step 2 - The ingestion plan
The survey described brings to the revision of the list of content declared in the Description of Work and to the processing of the Ingestion Plan. The aim of this plan is to keep under control the content provision and monitor the ingestion progresses.

Step 3 – Training
Face-to-face training sessions with the projects content providers were organized to inform them on mapping to LIDO and the use of the ingester. The topics of the workshop are: General guidelines for the ingestion procedure to keep coordinators informed about the ingestion plan updates and changes; useful information about Europeana specifications; deadlines in the ingestion process; recruitment of the first providers ready to start mapping/upload content; launch of the help-desk support.

Step 4 - Training material
After the training workshop the coordinators prepared for all project partners a package of documentation: guidelines providing and aggregating content, use and mapping to LIDO, and guidelines about the use of the MINT tool. The documentation is divided in 6 main sections: General guidelines for providers, Useful information to aggregate content for Europeana, Guidelines on the use of the MINT tool, LIDO, Basic rules for mapping, Making the link with Europeana.

Step 5 - Frequently Asked Question
During the ATHENA and LH projects, the coordinators set up a help-desk service in order to face the providers' problems. In LH project the authors summarized the most important problems occurred to the providers. Therefore, the most frequent questions arrived to the help-desk have been re-elaborated and posted on the LH website[9], giving a further instrument to the providers.

Organising the work of the content providers is the first action to be carried out in order to allow them to get into the core of the project. The process of training content providers on the use of LIDO and MINT as aggregation tools lasts for the whole

---

[9] http://www.linkedheritage.eu/index.php?en/183/faq

duration of the project supporting content providers that may have different needs in content aggregation.

The training material will be also refined and increased in the future too.

3.1   Collecting Feedback from Content Providers

Within the experience carried out in ATHENA and LH projects, a methodology to gather feedbacks for the refinement and improvement of the ingestion process in order to support the contribution of large amounts of content to Europeana have been created. This includes various sources of information: Data report from the MINT ingester, for both overall and analytic views on the content ingestion; a help-desk service for keeping under control the partners' difficulties in the content aggregation process (mainly technical and mapping issues); the periodic interviews via e-mails and telephone/Skype calls mainly for collection management/content provision issues. The advancement of the state of the art of the ingestion is documented in the report menu of the MINT ingester, recorded in specific files elaborated for managing this information, traced in the request that the content providers make through the help-desk service.

MINT Data Report

MINT has data reporting functionalities and can provide additional features to make detailed reports both for the single content providers and for content coordinators: the partner can have a quick view on the amount of metadata he/she uploaded, while the coordinators (that have an account profile with more features) can supervise the work of all partners (**Fig. 3**). The "Data Report" menu offers the breakdown of the metadata uploaded in the tool country by country, and provider by provider under the umbrella of each country. This offers a quick look at the overall amount of registered organizations, registered users, metadata that have been uploaded and transformed to LIDO. As an example, FIG. 3 explains that at March 31, the LH aggregation tool has 75 registered users from 38 organisations, 68,909 metadata uploaded, 16,703 transformed into LIDO and 5,472 ready for the publication in Europeana. This means that the registered content providers made their mappings to LIDO, uploaded some of their metadata ("Total items"), transformed some of them into LIDO ("Total transformed Items") and validated them for the publication in Europeana ("Total Published Items").

The help-desk service

As soon as the ingestion phase started up, a help-desk should be created to support the content providers to overcome their problems and to keep content coordinators informed on the advancement of the activities. The help-desk service can be organized as a mailing list where interested people register and post their questions on various topics: mapping, software (bugs and information), deadlines, publication in Europeana. A group of experts answers for his/her interest field. To facilitate the distribution of the answers, each content provider is asked to specify the kind of problems in the e-mail subject: the person in charge replies and facilitates the work during the ingestion phase. In particular, four subjects have been considered: about the mapping procedure, the software, the ingestion and the work-flow. The service was tested in the ATHENA project and has proved to be an important support both in providing solutions both in sharing problems and experiences during all the ingestion process.

Fig. 3. MINT data report. On the left column: data report country by country and provider per provider. On the right column synthetic overview.

Periodic interviews

Periodically content coordinators should make a "tour de table" of the content providers via e-mails or telephone/Skype calls to gather direct feedbacks on the work they are carrying out; nonetheless, contacts outside this periodical check may also occur. The information is compared with that one given in the MINT data report (and eventually in the help-desk) and recorded in two files created for coordination purposes (not for a public consultation): the Ingestion Plan, describing the status and the availability of the collections to be ingested by each content provider; the "Monitoring" file which collects the personal history of each content provider (if their content provision differs from what has been declared in the original DoW, what are the criticisms and problems met, mapping or technical, if they still have to give an answer on some specific issue to coordinators, and any other business concerning this process. Every email exchange, every telephone/Skype call is here registered with the date, in order to keep under control the hundreds of messages that "Cordination of content" leaders have with content providers.

4   The Europeana Data Exchange Agreement (DEA) and Its Impact on the Metadata Aggregation

In September 2011 Europeana released a new licence that regulates the metadata provision with cultural institutions[10] and replaces the former two agreements for content providers and for aggregators. Unlike the elder agreements, DEA foresees that metadata descriptions (the texts only, not the thumbnails) are subjected to the CC0 licence, which effectively means releasing content as public domain[11] and allowing the commercial reuse of metadata; this implies the possibility for Europeana to publish the metadata as Linked Open Data (LOD).

The discussion about this licence begun before its release: as soon as it came into light, it created an immense uproar across the ATHENA community that was made up mainly of museums; the main concern of the European museums was that the commercial use of the metadata should have been explicitly excluded. For the museum communities the metadata of digital objects contains detailed descriptions that can be considered as small essays and moral rights lies on them. In short, they do not feel comfortable with the idea that their work may be reused by third parties also for commercial purposes.

The Data Exchange Agreement had a wide impact on the LH consortium: some partners subscribed it, some other did not yet because they are not convinced. To find a solution to this tangled matter and assure the content provision to Europeana, the LH management set up a task force of partners in order to present to the consortium practical ways to fulfil the project duties (that implies the DEA subscription) and to keep the integrity of their data. As a results, the task force elaborated a strategy that gives content providers 3 options for the metadata publication: they can send Europeana only the mandatory descriptive elements (basic set), a little wider description (intermediate set), or the full description (full set) if they feel that the CC0 license leaves them the control of their data. These three options implied also the updating of the MINT aggregation platform with a filter option in order to select the preferred one during the aggregation process.

5   Conclusions and Future Steps

The tools and procedures worked out and described in this paper have benefits for Europeana and for every initiative of aggregation of digital content. They support the content providers in every phase of their work for Europeana and in parallel allow content coordinators to monitor the evolutions of the activities. We can conclude that through the projects described and the methodology elaborated, the following results have been achieved.

LIDO: born within the museum sector, it has proven to be effective also for the material of other domains and it preserves the integrity of rich metadata. It is used by

---

[10] http://pro.europeana.eu/web/guest/data-exchange-agreement
[11] http://wiki.creativecommons.org/CC0

a large amount of European projects and national aggregators[12]. It is now an international standard with a devoted working group in CIDOC. Furthermore, as for the future, LIDO is expected to get even more used, in particular in national aggregator project environments (AUTERE, VAKKARI 2011).

MINT: it gathers the various steps that content providers should face to send their metadata to Europeana in a unique tool: mapping, aggregation, preview, publication in Europeana. It is user friendly and widely used in several project. The use of MINT allows also to aggregate metadata for Europeana (and any other aggregator) maintaining their integrity in a long term perspective and preserving the European portal from broken links.

WORKFLOW and FEEDBACK methodology: it is fundamental to assist content providers and keep under control the aggregation process; it also helps to build a strong community.

The main expected next outcome is the improvement of the multilingual terminology management functionalities of MINT that are being developed within LH but also in other European project using this ingestion software (Partage Plus, EuropeanaPhotography, etc.). This will allows cultural institutions to have in a unique environment tools both for the management of multilingual terminologies and mapping; by the other hand Europeana will benefit of metadata that are enriched and more effectively searchable within the portal.

References

1. Autere, R., Vakkari, M.: Towards Cross-Organizational Interoperability: The LIDO XML Schema as a National Level Integration Tool for the National Digital Library of Finland. In: Proceedings of the 15th International Conference on Theory and Practice of Digital Libraries: Research and Advanced Technology for Digital Libraries. Springer, Heidelberg (2011)
2. Coburn, E., Light, R., McKenna, G., Stein, R., Vitzthum, A.: LIDO - Lightweight Information Describing Objects Version 1.0 (2010)
3. McKenna, G., Rohde-Enslin, S., Stein, R.: Lightweight Information Describing Objects (LIDO: the international harvesting standard for museums), ATHENA project (2010)
4. McKenna, G., Stein, R.: Linked Data: Some preliminary results of the LH Project. In: Proceedings of the CIDOC 2011, Knowledge Management and Museums, Sibiu, Romania, September 4-9 (2011)
5. Niggemann, E., De Decker, J., Lévy, M.: The new renaissance. Report of the "comité des sages" on bringing Europe's cultural heritage online (2011), ISBN: 978-92-79-19272-2

---

[12]http://network.icom.museum/cidoc/working-groups/
data-harvesting-and-interchange/lido-community/use-of-lido.html

Diva: A Web-Based High-Resolution Digital Document Viewer

Andrew Hankinson[1], Wendy Liu[1], Laurent Pugin[2], and Ichiro Fujinaga[1]

[1] Centre for Interdisciplinary Research in Music Media and Technology,
Schulich School of Music, McGill University, Montreal, QC, Canada
{andrew.hankinson,wendy.liu}@mail.mcgill.ca,
ich@music.mcgill.ca
[2] RISM Switzerland, Berne, Switzerland
laurent.pugin@rism-ch.org

Abstract. This paper introduces the Diva (Document Image Viewer with Ajax) project. Diva is a multi-page image viewer, designed for web-based digital libraries to present documents in a web browser. Key features of Diva include: "lazily loading" only the parts of the document the user is viewing, the ability to "zoom" in and out for viewing high-resolution page images, support for Pyramid TIFF or multi-resolution JPEG 2000 images, a multi-page "grid" view for page images, and HTML5 canvas support for document image rotation and brightness/contrast control. We briefly discuss the history and motivation behind its development, provide an overview of how it compares to other document image viewers, illustrate the different components of Diva and how it works, and provide examples of how this may be used in a digital library context.

Keywords: Document images, image viewer, web applications.

1 Introduction

Document image collections, created by scanning or photographing document pages, are the primary means through which libraries and archives make multi-page physical items in their collections—books, magazines, newspapers, scores, sheet music—available online. For example, the HathiTrust, a consortium of research libraries and institutions digitizing their physical collections, claims to have digitized 10.2 million volumes, resulting in over 3.5 billion individual page images. JSTOR, an organization that is digitizing back-issues of scholarly journals, reports that they have over 300,000 issues available online, totalling almost 44 million pages. However, it is not just the large-scale digitization initiatives that are creating document image collections. Libraries and archives are digitizing items in their special collections, including rare items or items that help preserve and promote local interests (e.g., smaller newspapers, group newsletters, or genealogies).

While billions of images are being produced across the globe, there are still very few tools for quickly and efficiently browsing these document image collections.

P. Zaphiris et al. (Eds.): TPDL 2012, LNCS 7489, pp. 455–460, 2012.

Often institutions simply place their page images online in an "image gallery" format, providing static pages for each page image. Browsing these image galleries requires the user to constantly click "next" and "back" buttons to progress from one page to another, with each page image presented as a discrete entity. Furthermore, usually only a few image sizes are provided, one for "thumbnail" view, one for "web display," and one for "download."

Another prevalent option is to distribute the image files in the PDF or DjVu format. While this ultimately provides users with the ability to scroll through a document, they both require browser plugins or client-side applications to view the files. For PDF files, the complete file must also be downloaded before the user can view the first page. Items that are captured at very high resolution produce files that can be many gigabytes, making downloading the entire document a time- and bandwidth-consuming task.

Diva (Document Image Viewer with Ajax) was built to address a perceived need for a tool to help organizations display document image collections online. Document images are presented to the user as a single, scrollable entity on the page, much like a PDF viewer, but implemented using standard JavaScript, CSS, and HTML so no third-party plugins are required. Additionally, since it uses the IIP Image Server, a fast image server capable of serving high resolution images, users have the option of viewing page images at multiple resolutions. Smaller images allow users to quickly navigate the document, while larger (high-resolution) images allow users to see minute details on the images. Since Diva uses asynchronous HTTP requests, page images are only loaded when the user scrolls them into view. This means that a user never has to download a single multi-gigabyte file to access just a few pages.

To introduce Diva, we will first review some history behind its development. We will discuss some of the design considerations we formulated by examining current practices for displaying document image collections. We will present an overview of the current state of document image display, providing examples of other solutions. We will then describe how Diva serves page images, focusing on some optimizations we designed that are not present in other document image viewers. We will also introduce several new features we are developing that use the HTML5 canvas element, providing users with the ability to perform basic image manipulation in the browser.

2   Background

The Diva project began as a pilot project with the Swiss working group of the Répertoire International des Sources Musicale (RISM) project. RISM is an international initiative, founded in 1952, whose purpose is to identify and catalogue musical sources held in libraries and archives around the world. In 2008, the Swiss RISM working group received funding for an exploratory project in digitizing musical sources (prints and manuscripts) of Swiss composers held in libraries and monasteries across Switzerland. The goal of this project was to combine the extensive, research-quality metadata gathered over decades with high-quality images

)f the source itself, and publish these online in a freely accessible database. This will allow researchers to search, retrieve, and view the document images online.

We began by examining the current "state of the art" for displaying high-resolution images in a web browser. We conducted a survey of 24 digital libraries that contained document images, primarily focusing on their techniques used to display document images. Full details of this survey can be found in [1,2], but we will briefly summarize three of the functional requirements we formulated here; namely, preserve document integrity, display multiple page resolutions, and optimize page loading. One further consideration was to ensure that our software worked in modern web browsers with no dependencies on third-party plugins, such as Adobe Acrobat PDF Reader, DjVu, or Adobe Flash.

- **Preserve Document Integrity.** Many of the systems we surveyed displayed documents as collections of independent page images. As discussed earlier, a common method of document display is the "image gallery," with inefficient navigation methods for moving between adjacent pages. This does little to preserve the original document as an integral whole, rather than simply a collection of independent images.
- **Provide Multiple Page Resolutions.** For older materials, especially manuscripts, the ability to view details such as faint pencil markings or different ink colours is necessary. High-resolution images allow the user to identify small details, which is especially important in tasks such as reading cursive script or identifying faded or obscured portions of the image, while smaller low-resolution images give users the ability to quickly get an overview of the entire document.
- **Optimize Page Loading.** Presenting large page images, especially at higher resolutions, makes it difficult for users to quickly leaf through a document. Two common approaches are to allow users to download the high-resolution images one at a time, or have them download the entire document encapsulated within a container format like PDF. For ultra-high resolution documents, however, this download may involve hundreds of megabytes, and the user must wait until the entire document is downloaded before viewing it.

2.1 Existing Viewers

The Google Books project [3], perhaps the most well-known book digitization project, uses a viewer that presents items as a single scrollable entity embedded within the webpage. This allows users to quickly scroll and navigate through the entire work. Their viewer software, however, is not available as a separate component for libraries and archives to integrate into their own collections. The New York Times Document Viewer [4] provides an interface similar to the Google Books project. They have made their viewer available as an open-source project.

The Internet Archive BookReader [5], developed for the OpenLibrary project, displays document page images in either a scrolling frame or as a physical book metaphor, allowing users to navigate the book by "turning" the pages. Both the Google Books viewer and the Internet Archive BookReader serve full page images to the browser so that the entire image must be downloaded to view it. For high-resolution pages, this can be quite slow.

Other document viewers, such as the one provided by the World Digital Library (WDL) [6], use an approach where document page images are broken into smaller tiles. Only the portion of the image that is being viewed is served to the user, which allows the user to "zoom" in on high-resolution page scans. The WDL document viewer uses Microsoft's SeaDragon system [7] for serving high-resolution page scans.

3   How Diva Works

The three central components for Diva are: the JavaScript "front-end;" the IIP Image Server, a high-resolution image server capable of streaming large images efficiently; and the Diva data server, a server-side script that serves information about page layout to the JavaScript front-end.

To display a document (each document corresponds to an ordered list of image files in a directory), page images must be first processed and converted into multi-resolution image files using either the Pyramid TIFF [8] format or JPEG 2000 [9]. This creates multiple versions of the document page images in the same file, and segments each resolution into smaller square tiles.

The Diva data server, which is available as a PHP or Python script, is responsible for sending information, like image dimensions, to the browser about the entire document at each zoom level. The data server processes each file and caches the document measurements, such as the full width and height of the entire image collection at a given zoom level. This information is used to construct the container HTML elements required to display the entire document.

To optimize the page loading, we display only the pages, and the portions of the page, that the user is currently viewing. As the user scrolls through the document, the Diva.js JavaScript component manages the requests for new tiles from the IIP Image Server and fills in the tiles for the page images. If a user wishes to "zoom" in on a page to view details, or zoom out to navigate the document more quickly, the Diva.js front-end will manage all requests for that zoom level and the appropriate page tiles. As the user scrolls through the document images, pages are deleted from the browser's DOM (Document Object Model) tree once they are no longer being viewed, freeing up resources and reducing memory usage. This creates the illusion that the user is scrolling through the full document, while ensuring that the browser is only managing a small number of DOM objects. This results in a very efficient document viewing experience for long documents even on resource-constrained machines such as tablets and smartphones.

Diva is unique in that it incorporates many of the features available in separate document viewers in a single interface. The Google Books, NY Times, and Internet Archive software all present documents on a single page using asynchronous HTTP requests to load new pages. However, their software serves full, non-tiled page images. If a user requests a high-resolution view of the page, the viewer must download the entire page image to display it. Tiled page views, like those of the

WDL, provide the ability to serve only the portion of the high-resolution image the user is viewing instantly, without waiting for the entire page image to download. The primary advantage of Diva is that it incorporates both of these techniques in a single web-based document viewer, requiring no client-side browser plugins.

One unique feature of Diva is the ability to perform basic image manipulation on document page images using JavaScript image processing. We have integrated a page image view with the HTML5 Canvas element, allowing users to perform basic image manipulation on the page images such as adjusting angle of rotation, brightness, or contrast. This is especially useful for viewing older or degraded documents. Marginalia may be written at an angle, so page image rotation can aid users in reading these types of notes. Faded inks or faint pencil markings may be enhanced by using image brightness and contrast controls. We are planning further enhancements to allow users to perform more advanced image operations, like independent colour channel selection. All of this is performed using browser-native technologies, with no server-side image manipulation or plug-ins.

3.1   Performance

We have tested Diva on a number of document image collections, with the largest being the Salzinnes Antiphonal, a collection of 482 images, each approximately 230MB when processed as multi-resolution TIFF images for a total document size of 113GB. These document images are approximately 30 megapixels. This document is available for viewing at [11].

On a local area network connection (approximately 100 Megabits per second (Mbps)), downloading an entire document this size would take approximately 15 minutes. On slower connections, like domestic broadband, it would take significantly longer. With Diva, however, viewing these documents is almost instantaneous, since the page images are presented at a smaller size and the number of pages loaded into the browser at any given time is kept at a minimum. Users wishing to see details on a specific page can navigate to that page without downloading the entire document, and then zoom in on the details. This allows users to very quickly and easily navigate a high-resolution document without waiting minutes, or sometimes even hours, for the entire image collection to download to their local devices.

4   Conclusion

In this paper we have described Diva, a multi-page document image viewer capable of serving high-resolution document images, and facilitating fast and efficient navigation of document image collections. Diva is capable of serving high-resolution images in a web-browser without needing third-party plugins. Diva is available under the MIT open-source software license, allowing libraries and archives to implement and customize it for their own document image collections. The source code, documentation, and demo installations are all available at [12].

Through combining unified document presentation, highly optimized page loading, and the ability to view multiple page resolutions, we believe Diva presents significant advantages for the usability and navigability of document image collections in digital document libraries.

Acknowledgements. The authors would like to thank the Canadian Conservation Institute and Judy Dietz for providing access to the Salzinnes Antiphonal image collection. This work was funded by the Social Sciences and Humanities Research Council of Canada (SSHRC), the Swiss National Science Foundation, the Canadian Foundation for Innovation, and the Centre for Interdisciplinary Research in Music Media and Technology (CIRMMT) at McGill University.

References

1. Hankinson, A., Pugin, L., Fujinaga, I.: Interfaces for Document Representation in Digital Music Libraries. In: 10th International Conference on Music Information Retrieval (2009)
2. Hankinson, A., Liu, W., Pugin, L., Fujinaga, I.: Diva.js: A Continuous Document Viewing Interface. Code4Lib Journal 14 (2011),
   http://journal.code4lib.org/articles/5418
3. Vincent, L.: Google Book Search: Document Understanding on a Massive Scale. In: 9th International Conference on Document Analysis and Retrieval, pp. 819–823 (2007)
4. Maclean, A.: A New View: Introducing Doc Viewer 2.0.,
   http://open.blogs.nytimes.com/2010/03/27/a-new-view-introducing-doc-viewer-2-0/
5. Internet Archive: New BookReader!
   http://blog.archive.org/2010/12/10/2685/
6. World Digital Library, http://www.wdl.org
7. Seadragon Ajax,
   http://gallery.expression.microsoft.com/SeadragonAjax
8. TIFF, Pyramid,
   http://www.digitalpreservation.gov/formats/fdd/fdd000237.shtml
9. JPEG 2000, Part 1, Core Coding System (2000),
   http://www.digitalpreservation.gov/formats/fdd/fdd000138.shtml
10. VIPS, http://www.vips.ecs.soton.ac.uk/index.php?title=VIPS
11. Salzinnes Antiphonal, http://ddmal.music.mcgill.ca/salzinnes-cci
12. Diva.js, http://ddmal.music.mcgill.ca/diva

Collaborative Authoring of Walden's Paths

Yuanling Li[1], Paul Logasa Bogen II[2], Daniel Pogue[3],
Richard Furuta[1], Frank Shipman[1]

[1] Center for the Study of Digital Libraries
Department of Computer Science and Engineering
Texas A&M University College Station, TX
walden@csdl.tamu.edu
[2] Intelligent Computing Research Team
Computational Data Analytics Group Oak Ridge
National Laboratory Oak Ridge
TNbogenpl@ornl.gov
[3] User Experience and Interaction Design Team
Production Enhancement Halliburton Energy Services Houston, TX
daniel.pogue@halliburton.com

Abstract. This paper presents a prototype of an authoring tool to allow users to collaboratively build, annotate, manage, share and reuse collections of distributed resources from the World Wide Web. This extends on the Walden's Path project's work to help educators bring resources found on the World Wide Web into a linear contextualized structure. The introduction of collaborative authoring feature fosters collaborative learning activities through social interaction among participants, where participants can coauthor paths in groups. Besides, the prototype supports path sharing, branching and reusing; specifically, individual participant can contribute to the group with private collections of knowledge resources; paths completed by group can be shared among group members, such that participants can tailor, extend, reorder and/or replace nodes to have sub versions of shared paths for different information needs.

1   Introduction

It has been widely accepted that collaborative learning brings about a lot of improvements on human learning activities. For instance, Koschmann argued that human learning activities occur in social context where knowledge transfer takes place among social interactions [1]. Additionally, Scardamalia and Bereiter proposed that the central role of learning activities lie in facilitating learners' knowledge building process, where learners take the initiative to explore new knowledge [2]; therefore, a collaborative knowledge building environment promotes knowledge sharing, learning enthusiasm and critical thinking. Finally, Slavin claimed that learners from a learning group aiming at the same learning objective tend to be active in information exchange, creative thinking and take efforts on collective problem solving [3]. In order to facilitate this kind of progressive educational pedagogy new techniques are needed.

P. Zaphiris et al. (Eds.): TPDL 2012, LNCS 7489, pp. 461–467, 2012.
© Springer-Verlag Berlin Heidelberg 2012

1.1   Walden's Paths

Walden's Path [15] organizes resources on the World Wide Web into a linear ordered, contextualized data structure. A path is composed of ordered list of nodes that constitutes a resource URL, annotation and resource name. The World Wide Web could be viewed as a tremendously large collection of knowledge resources, with a range of media types like html, video, images, text and so forth. However, these resources are loosely linked into a highly complicated hypertext system; its dimension and size makes it more difficult to retrieve and organize to appropriately meet users' information requirements. The authoring of a path is the process of tailoring and filtering the resources from a large set of collection into a path with a specific topic. The path authoring of a path is a process of knowledge recreation, where each resource on the Web is evaluated, annotated and encapsulated into a node; the nodes are organized into a contextual knowledge artifact. Currently, authors of a path can reinterpret and comprehend the resources during the creation process and a path can be shared to other viewers who can traverse nodes in a path, where a learning process can be performed. However, to full embrace the knowledge building paradigm, the system needs to be redesigned with collaboration in mind. It is this goal that we have designed CoWPaths (Collaborative Walden's Paths) to address. Before delving further in to the motivation of our work we'd first present a quick overview of applicable work in pedagogy and in collaborative educational systems.

2   Prior Work

This need for new techniques to support knowledge building in many ways drives the introduction of computer technology to education and has brought profound impacts on the development of collaborative learning along with it. Specifically, computer supported collaborative learning improves the pattern of social interactions, knowledge construction and information sharing among learners. Researchers from different academic backgrounds have been extensively studying the computer supported collaborative learning from multiple perspectives, and found fruitful results. Liaw *et al.* found that students first take positive attitudes toward web based collaborative learning environment; then, individual learning performance improves as the group activities increases [4]; that is, active participation of high quality social interaction, such as opinion sharing, discussion, questioning and explanation, benefits the learning outcome of the group in general. Dewiyanti conducted observations and evaluations on learners' learning activities in web based collaborative environment with asynchronous communication provided [5]; her research found that group awareness among members is an vital prerequisite for successful collaborative activities where participants needed to be kept up to date with the status of group and peer activities in order to self-coordinate future group activities [5]. In the process of social interactions, participants can perform knowledge transfer through the communication cycle of questioning, explanation, interpretation, elaboration and evaluation. According to Balester *et al.*, the foundation of knowledge transfer is the differing levels of prior knowledge among participants and the mutual understanding of the reciprocal benefits of this interaction [6]. Additionally, Warshuerer showed that effective communication yields knowledge retention for longer periods [7]. Finally, active contexualization of knowledge into discussion and critical thinking helps people in longerterm memorization [8, 9].

The emergence of computer supported collaborative learning has brought advancement of computer technology to the study of pedagogical approaches. The central research topic focuses on the collaborative construction and sharing of knowledge and communication technology supporting effective social interaction. According to Stahl [10], collaborative learning is a dynamic process [11], where knowledge is initially constructed individually, this is called personal knowledge building; then, knowledge is transferred and shared among group members with social interactions, such as questioning, brainstorming, interpretation, explication and evaluation. It is during this process of social interaction that collaborative knowledge construction occurs. The creation of knowledge artifacts serves to persist the group's knowledge and enhances the later knowledge acquisition by individuals. Currently, a wide variety of synchronous or asynchronous communication technologies are available for social interaction, such as videoconferencing, email, discussion, and instant messaging [12].

3   Use Cases

From our prior observations and the literature on collaborative learning, we have identified two major use cases, each with two sub-cases that provide a refinement on the more general case.

3.1   Group Authoring (Extension and Modification)

Easing group authoring has been a sought after feature of our users. Authoring as a group in the past was handled in one of three means. Some users would create a shared account that each member would have access to. This allowed them to author paths but without any record of who made which changes. The second means was by having group members clustering around a single machine and working together. Again no history was maintained and this required synchronicity. The only advantage this mode had over shared accounts was that changes were made with a consensus and conflict over changes was minimized. The third way in which users re-shaped Walden's Paths to allow collaboration was by utilizing the path import and export features. In this mode, users would export a path when their revision was made and then transfer the file, typically over email, to other group members who would make their iterations and share the results again. This allowed a history to be obtained through the email records and allowed asynchronous work but required users to leave the system and manage the collaboration themselves.

Instead we seek to allow users to perform these collaborative activities that they are already trying to perform with greater ease and without the drawbacks of some of the other means they are currently using. Authoring is essentially two sub-cases, regardless or if it is collaborative or not: extension and modification. Extension is the addition of new nodes to a path. This is the most basic authoring activity as even the original path creation is an act of extension from the null path (the path without any pages). For this case the scenario is imagined as follows:

> *Peter and Valentine are preparing a path on current foreign policy issues between the United States and Russia for a project in their political science class. Peter initiates the creation of the path on his computer by adding a series of news articles on the current round of arms reduction treaties. After completing the path, Peter shares the path to a group that he and Valentine have access to. Peter indicates that other group members can extend the path. Later, Valentine on her machine logs and sees a new path in the group; she views the path and decides to add in a number of articles about Putin's election. In the group interface, she selects edit and can then add new pages to the path. She cannot change the path, as Peter did not grant modification privileges. Finally, Peter logs back in and see that his path has been updated and views the history to see that Valentine added a number of pages to it.*

In this example, the users can add new pages but not modify existing ones; this leads in to the other sub-case, modification. Modification allows the alteration of existing pages created by others. Similarly to the first subcase the second is imagined as the following:

> *Peter and Valentine's path has grown considerably and Valentine has called question to the order of the pages, the wording of some of the annotations and if one of the pages truly belongs. While Peter had been addressing her issues himself previously, he now decided that he trusts Valentine to let her edit the pages he has created. He then gives the group the modification privilege. The next time Valentine logs in to the system she notices that one of the articles is now pointing to a dead link. She finds an apt replacement and instead of sending the information to Peter in a comment on the path, she now has the ability to edit his pages in the editing interface.*

3.2   Future Reuse (Duplication and Adaption)

The second major use case is that of future reuse. In the past a path could be reused either in its entirety, by requesting an export from the author of the path, or via the duplication feature in Walden's Paths. However, duplication could only be done by the account that owned the path and the duplicated version could only be edited with that account. One group of users in the atmospheric science used a combination of a shared account and duplication to facilitate their reuse scenarios. These users were teaching assistants in a lower-level atmospheric sciences class. They used Walden's Paths to point students to self-study materials with some annotation to help guide their study. The class was offered in multiple different versions with altered curriculums and student populations. In order to tailor the canonical path that was authored collaboratively, each TA would copy the path and makes changes to suit her particular version of the class. When a new semester started the TAs would then repeat the same process of duplication and modification to make the path for the new semester. However all of these path versions were still reflecting an original that the new versions were not tied to. To help this kind of use we seek to allow duplication of paths between accounts and the adaptation of paths while maintaining the ties back to the original path. For this case we imagine another scenario building off the prior two:

Weeks after the assignment, Valentine and Peter are preparing to independently write research papers for their Political Science class. Valentine is writing about the interplay between President Obama and President-elect Putin. Peter is writing about how the arms negotiations and American foreign policy has affected the Russian election cycle. Both of them want to use their shared path as a basis for their research. Each of them seeks to focus and extend the path for their particular topic. Peter therefore turns on the adaption permission and both Peter and Valentine select to adapt the existing path from which they can make their changes while preserving the ties back to the original version which still exists unchanged. A third student in the class, Andrew, is writing his paper on US politics around Arms Control. Peter tells him of her path. Upon viewing her adaption, he sees the reference back to the original path which he decides he wants to use as the starting place for his own work. He asks Peter is he can have his own copy. Peter not wanting Andrew's path to be explicitly tied back to his, grants public duplication privileges. Andrew then can duplicate the path and edit it as it was his own.

One thing to note in this last use case is that privileges can be granted at multiple levels. Andrew's duplication was a public level permission while Valentine's adaption was at the group level. With these use cases in mind, we will now turn to our implementation of these scenarios.

4   Implementation

This section introduces the implementation and design for the collaborative authoring prototype. The goal of the tool is to provide a shared workspace that supports collaborative knowledge building, management, retrieval and sharing; at the meantime, asynchronous communication is provided in a form of contextual annotation and discussion. Local annotations allow users to annotate on any positions of the web pages in a path, which could be shared with, rated and commented by other users; therefore, local annotation feature provides an approach of social interaction and enhancement to knowledge building and information exchange. Group definition is proposed to facilitate collaborative authoring and social interaction; it includes four sections: group description, group knowledge repository, notification and group awareness visualization. Group description presents the group abstract of current group, introducing the group task and background information, group owner and group members. Personal notification customizes the most recent communication updates and pushes to each individual personal notification area to help maintaining individual active social interaction status. Specifically, new notifications include the following information: new annotations on knowledge page, unread discussion thread and reply, unread personal messages and group wise broadcasting message. Group awareness captures group status data, visualizes and delivers to group members. The data are extracted from two information sources, shared workspace status and group member activities. Group awareness includes group knowledge construction history and social interaction records. Specifically, all group-wise activities are recorded and visualized in a timeline ordered by timestamp sequence. These activities include page construction, modification, deletion, task modification, participant comments and annotations. From the activity visualization, members can keep track of the evolution of the shared workspace and maintain awareness of participants' recent activities and contributions. Besides, group awareness visualization achieves knowledge persistence by scripting interaction history. Knowledge transfer occurs in the course of social interactions among group members; therefore, longer knowledge retention can be achieved by scripting communication history for later viewing and reference.

5   Conclusions and Future Work

In summary, this paper presents a prototype design for a collaborative authoring tool for users to co-author Walden's Paths This prototype can be tailored as a collaborative learning tool for students to perform active collaborative learning in groups, where participants join in groups to accomplish group wise learning tasks. The collaborative creation of paths is a process of knowledge recreation, where participants perform knowledge exploration, knowledge filtering, contextualization, reinterpretation and evaluation. We believe that CoWPaths helps the active learning process by allowing member communication. Group members can initiate an asynchronous communication channel by creating annotations on resources created by other. This allows social interactions among group members, in which knowledge transfer is achieved. Users can also bookmark annotations to keep track and record valuable content in the course of collaborative authorship. A path can also be shared and reused by other users, including the extension or branching of the original path. Such customization allows for more flexible path creation and resource sharing.

By introducing Walden's Paths concept into the realm of computer supported collaborative learning, we open up further work. Our next step will be to evaluate the benefit of our approach for collaborative learning by observing students' learning behaviors in collaborative settings and identifying the differences between collaborative learning and independent learning.

Acknowledgements. Walden's Paths is being developed as part of the Ensemble Computing Pathways Project. This work has been funded in part by the National Science Foundation under grants DUE-0840715 and DUE-1044212.

This document was prepared by Oak Ridge National Laboratory, P.O. Box 2008, Oak Ridge, Tennessee 37831-6285; managed by UT-Battelle, LLC, for the US Department of Energy under contract number DE-AC05-00OR22725.

This manuscript has been authored by UT-Battelle, LLC, under contract DE-AC05-00OR22725 with the U.S. Department of Energy. The United States Government retains and the publisher, by accepting the article for publication, acknowledges that the United States Government retains a non-exclusive, paid-up, irrevocable, world-wide license to publish or reproduce the published form of this manuscript, or allow others to do so, for United States Government purposes.

References

1. Koschmann, T.: Paradigm shifts and instructional technology: an introduction. CSCL: Theory and Practice of an Emerging Paradigm (1996)
2. Scardamalia, M., Bereiter, C.: Knowledge building: Theory, pedagogy, and technology. In: Sawyer, K. (ed.) Cambridge Handbook of the Learning Sci., Cambridge UP (2006)
3. Slavin, R.E.: Cooperative Learning: Theory, Research, and Practice. Prentice-Hall (1989)
4. Liaw, S.S., Chen, G.D., Huang, H.M.: Users' attitudes toward Web-based collaborative learning systems for knowledge management. Computers & Ed. 50(3) (2008)
5. Dewiyanti, S., Brand-Gruwel, S., Jochems, W., Boers, N.J.: Students' experiences with collaborative learning in asynchronous Computer-Supported Collaborative Learning environments. Comp. in Human Behavior 23(1) (2007)
6. Stacey, E.: Collaborative Learning in an Online Environment. J. of Distance Ed. (1999)
7. Balester, V., Halasek, K., Peterson, N.: Sharing authority: Collaborative teaching in a computer-based writing course. Computers and Composition 9 (1992)
8. Warshauer, M.: Computer-Mediated Collaborative Learning: Theory and Practice. The Modern Language J 81(4), 4 (1997)
9. Gijlers, H., de Jong, T.: The relation between prior knowledge and students' collaborative discovery learning processes. J. Res. Sci. Teach. 42 (2005)
10. Del Soldato, T., du Boulay, B.: Implementation of motivational tactics in tutoring systems. Int. J. of Art. Int. in Ed. 6(4) (1996)
11. Stahl, G.: Meaning and Interpretation in Collaboration. Comp. Supp. for Collab. Learning (2003)
12. Stahl, G.: A Model of Collaborative Knowledge-Building. In: 4th Int. Conf. of the Learn. Sci. (2000)
13. Avouris, N., Komis, V., Margaritis, M., Fiotakis, G.: An Environment for Studying Collaborative Learning Activities. Ed. Tech. and Soc. 7(2) (2004)
14. Bogen, P.L., Pogue, D., Poursardar, F., Li, Y., Furuta, R., Shipman, F.: WPv4: A Reimagined Walden's Paths to Support Diverse User Communities. In: 15th Int. Conf on Theory and Prac. of Dig. Lib. (2011)
15. Feiner, S.: Seeing the Forest for the Trees: Hierarchical Displays of Hypertext Structures. In: ACM SIGOIS and IEEE-CS TC-OA 1988 Conf. on Office Info. Sys. (1988)
16. Calisir, F., Gurel, Z.: Influence of text structure and prior knowledge of the learner on reading comprehension, browsing and perceived control. Comp. in Human Beh. 19(2) (2003)
17. Zhang, Y.: The construction of mental models of information-rich web spaces: the development process and the impact of task complexity. SIGIR Forum 44(1) (2010)

Quantitative Analysis of Search Sessions Enhanced by Gaze Tracking with Dynamic Areas of Interest

Tuan Vu Tran and Norbert Fuhr

University of Duisburg-Essen
Information Engineering
Duisburg, Germany
vtran@is.inf.uni-due.de, norbert.fuhr@uni-due.de

Abstract. After presenting the ezDL experimental framework for the evaluation of user interfaces to digital library systems, we describe a new method for the quantitative analysis of user search sessions at a cognitive level. We combine system logs with gaze tracking, which is enhanced by a new framework for capturing dynamic areas of interest. This observation data is mapped onto a user action level. Then the user search process is modeled as a Markov-chain. The analysis not only allows for a better understanding of user behavior, but also points out possible system improvements.

1 Introduction

The analysis of user search behavior has a long tradition. The major focus of this research has been the development of cognitive models [5], which improved our understanding of the user's actions, but also has been considered in more recent designs for developing user-friendly systems.

On the other hand, quantitative analyses mostly have focused on system log analyses, but it is difficult to apply user-oriented criteria when regarding this data. More valuable information can be derived from observation data if we are able to relate this data to cognitive actions of the user, e.g. via intellectual mapping in the form of a post-hoc analysis (see e.g. [4]).

Log file studies have also been described by Jansen [6], Jones [8], Zhang and Kamps [11] who concluded that transaction log analysis can provide valuable insights into user-system interactions. Hoare et al. [4] developed a model of states and transitions that users traverse during a search session to understand user search processes. This model allows predictions of how changes to the interface might change user's behavior and thus enables comparisons between designs. Kumpulainen and Järvelin [10] modeled interaction between different information channels to optimize the process of task-based information access.

In recent years, gaze tracking has become a valuable tool for capturing implicit feedback from the users. Studies like [3], [7] and [1] analyze gaze-tracking data to understand the user search process.

P. Zaphiris et al. (Eds.): TPDL 2012, LNCS 7489, pp. 468–473, 2012.
© Springer-Verlag Berlin Heidelberg 2012

In the remainder of this paper, we first present the ezDL experimental framework for the evaluation of user interfaces to digital library (DL) systems. Then we describe a new quantitative analysis method based on logging and gaze tracking; by combining these two sources of evidence, we are able to perform an automatic mapping onto cognitive actions. As a major improvement over previous approaches using gaze tracking, we are able to consider dynamic areas of interest, i.e. display objects that appear only for a short period of time or change their position during the search session.

2 An Experimental Framework for DL User Interfaces

ezDL[1], the successor of Daffodil [9] is a configurable open source software for the design of advanced user interfaces for DLs and their evaluation. For the user, ezDL provides a desktop interface offering a configurable set of tools. Current tools comprise functions such as database selection, query formulation, result lists with sorting, grouping and clustering options, detail view, co-author graphs, search history, basket (clipboard) and a personal library for storing retrieved items in folders.

ezDL supports user-oriented experimentation at all levels of a DL: in the *user interface* existing tools can be replaced by variants with different designs or visualizations. At the *functional* level, new tools can be easily integrated into the system. ezDL can be connected with arbitrary *back-ends* via adapters and wrappers for retrieval systems (e.g. Solr), digital library systems or web-accessible digital libraries (currently, we have wrappers for about 10 libraries from computer science and medicine). Finally, at the *collection* level, experiments with new collections can be performed by running them with one of the back-ends currently supported.

Besides being a flexible, extensible system, ezDL comes with specific components for the user-oriented evaluation of digital library systems. The core ezDL system consists of the front-end (running on the user's computer), the back-end (running on a server) and the IR engine(s) managing the content. For running lab experiments, there is an experiment database that contains data for controlling the experiments and for collecting observation data. The former is used by an additional web application which schedules the user tasks (e.g. based on a latin square design) and starts the ezDL front-end, but also presents questionnaires to the users (e.g. before and after each task, as well as at the beginning and the end of an experimental session). As observation data, the experiment database collects the answers to the questionnaires as well as the logging data from the ezDL system, which can be configured according to the desired granularity. In addition, we also gather data from a gaze tracking tool.

The analysis of gaze tracking data is usually based on the definition of 'areas of interest' (AOI), which are rectangular areas on the screen like e.g. query interface, result items or details of a document. However, standard gaze-tracking software only allows for the monitoring of static AOIs. In contrast, users might

---

[1] www.ezdl.de

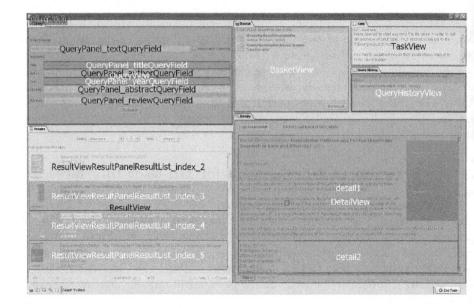

Fig. 1. User interface and areas of interest

like to resize the windows displayed on the screen or scroll through lists of items, and there also may be pop-up boxes. In order to be able to deal with these issues, we developed a framework called *AOILog*[2] which automatically keeps track of position, visibility and size of all user interface objects at any point in time. By combining this information with the gaze-tracking data, we always know the object the user is currently looking at. As an example, see the AOIs depicted in Figure 1; in the left lower window, each result item corresponds to one (dynamic) AOI, where the system keeps track of the items even during scrolling.

3   An Example Evaluation

For our experiments, we used a collection based on a crawl of 2.7 million records from the book database of the online bookseller Amazon.com. The data was crawled from February to March 2009 and indexed with Apache Solr 1.4. Besides the bibliographic data and the abstracts, a document record also contains the user reviews and a thumbnail of the book cover.

As test subjects, we recruited 12 students of computer science, cognitive and communication science and related fields. After an introduction into the system, users had to work on two tasks from each of the two categories, with a time limit of 15 minutes per task. The 'complex' tasks defined searches where users had to look into the user reviews of a book in order to judge about its relevance, while the 'narrow' tasks usually could be solved by reading the abstracts.

---

[2] *AOILog* is open source software based on Java; since components only have to register for using it, it can be easily integrated into any user interface based on Java Swing.

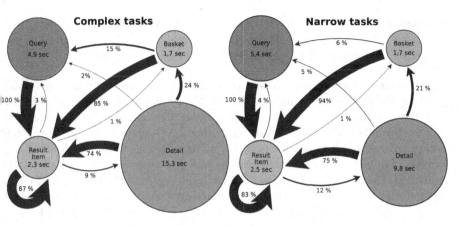

Fig. 2. Transition probabilities and user efforts

The user interface of our system consists of four major areas (see Figure 1): a query input field, a list of result items, a detail area showing all available data about the currently selected document, and a basket where users should place documents they deemed relevant. There are also two small areas showing the search history and the current working task. All windows are resizeable.

As mentioned above, our analysis is based on the combination of logging and gaze tracking data. For the latter, as in other studies, we focus on the so-called fixations, and also consider them only is they last for at least 80 ms, since this is the minimum time required for reading anything on the screen [2].

Corresponding to the four areas of the user interface, we can distinguish four types of user actions: formulating *queries*, looking at a *result item*, regarding document *details*, and looking at the *basket*. Thus, a one-to-one mapping of gaze tracking data and user actions seems to be straightforward. However, a closer analysis showed that this is too simplistic.

First, we noted that users were frequently looking back and forth between the details and the basket. This behavior was due to the fact that the book database often contains very similar entries (e.g. different editions of a book), thus forcing users to check if the current result item was substantially different from the books already placed in the basket. Occasionally, this also happened for items in the result list. In a similar way, when formulating a new query, users did not only look at the query field, but also checked the details of the current document as well as the items in the basket.

In order to deal with these problems, we separate between two levels, the gaze tracking level and the action level. Then we define a mapping from the former to the latter, which also considers the logging data. By default, the area the user looks at defines the current action, with the following exceptions:

– When the user looks from a result item to the basket and back, without moving the item to the basket, this is counted as part of regarding the item.
– The same holds for looking back and forth between details and the basket.

– Query formulation starts when the user's gaze wanders from the details or the basket to the query field for the first time, even when it returns to the basket/details several times before the query is actually submitted.

Applying these rules resulted in the transition probabilities and average times spend for the different actions displayed in Figure 2. On the left hand side, we show the Markov model for the complex tasks. One can see that users spend 4.9 s for formulating a query, after which they go through the result list. Each result item takes 2.3 s, and for only 9% of these items, users also look at the corresponding details, which takes another 15.3 s on average. In 24% of all details considered, the document is judged relevant and put into the basket, in 74% of the cases the document is considered not relevant and users go to the next result item, while with a probability of 2%, a new query is formulated. From the basket, users most often go to the next result item (85%), or they formulate a new query (15%).

Looking closer at this model, a number of observations can be made:

– Retrieval quality is surprisingly low—only about 3 % of the items in the result list are relevant. This is mainly due to the complexity of the retrieval tasks.
– After putting a document into the basket there is a probability of 15% that a new query is formulated. Most of the time, users go to the next result item, this may indicate that they hesitate to formulate new queries. A query expansion function could help here.
– Since only 24 % of the details regarded are judged as relevant, the quality of the result entries could be improved, in order to increase this rate (while the probability of the transition from result to details would decrease, thus leading to a reduction of the overall effort). Furthermore, we also see that click-through data is a poor indicator of relevance.

When we compared the statistics of complex vs. narrow tasks, we see that the number of queries (7.1 vs. 8.7) as well as the number of items placed in the basket (8.1 vs. 7.1) were about the same for the two task categories; for the complex tasks users looked at more items (88 vs. 78) but fewer details (17 vs. 25) than for the narrow tasks. However, users spend 50% more time per detail for the complex tasks (15.3 s vs. 9.8 s). This can be explained by the fact that complex tasks require the users to investigate more details about potential answers.

4   Conclusion

In this paper, we first presented the ezDL experimental framework for user-oriented evaluation of DL systems. Then we described a new quantitative analysis method which extends current methodologies for analyzing interactive IR by combining system logs with gaze tracking, which we enhanced by dynamic areas of interest. In order to relate this observation data to the user's cognitive actions, we separate between the gaze tracking and the action level, and ntroduce a corresponding mapping. Our example study points to possible system improvements, and shows that click-through rates are a poor indicator of relevance.

References

1. Buscher, G., Cutrell, E., Morris, M.R.: What do you see when you're surfing?: using eye tracking to predict salient regions of web pages. In: Proceedings of the 27th International Conference on Human Factors in Computing Systems, CHI 2009, pp. 21–30. ACM, New York (2009)
2. Buscher, G., Dengel, A., van Elst, L.: Eye movements as implicit relevance feedback. In: CHI 2008: Extended Abstracts on Human Factors in Computing Systems, pp. 2991–2996. ACM, New York (2008)
3. Guan, Z., Cutrell, E.: An eye tracking study of the effect of target rank on web search. In: CHI 2007: Proceedings of the SIGCHI Conference on Human Factors in Computing Systems, pp. 417–420. ACM, New York (2007)
4. Hoare, C., Sorensen, H.: Application of Session Analysis to Search Interface Design. In: Lalmas, M., Jose, J., Rauber, A., Sebastiani, F., Frommholz, I. (eds.) ECDL 2010. LNCS, vol. 6273, pp. 208–215. Springer, Heidelberg (2010)
5. Ingwersen, P., Järvelin, K.: The turn: integration of information seeking and retrieval in context. Springer-Verlag New York, Inc., Secaucus (2005)
6. Jansen, B.J.: Search log analysis: What it is, what's been done, how to do it. Library & Information Science Research 28(3) (2006)
7. Joachims, T., Granka, L., Pan, B., Hembrooke, H., Radlinski, F., Gay, G.: Evaluating the accuracy of implicit feedback from clicks and query reformulations in web search. ACM Trans. Inf. Syst. 25(2) (2007)
8. Jones, S., Cunningham, S.J., Mcnab, R., Boddie, S.: A transaction log analysis of a digital library. International Journal on Digital Libraries 3, 152–169 (2000)
9. Klas, C.-P., Fuhr, N., Schaefer, A.: Evaluating Strategic Support for Information Access in the DAFFODIL System. In: Heery, R., Lyon, L. (eds.) ECDL 2004. LNCS, vol. 3232, pp. 476–487. Springer, Heidelberg (2004)
10. Kumpulainen, S., Järvelin, K.: Information interaction in molecular medicine: integrated use of multiple channels. In: Proc. of the Third Symposium on Information Interaction in Context, IIiX 2010, pp. 95–104. ACM, New York (2010)
11. Zhang, J., Kamps, J.: A Search Log-Based Approach to Evaluation. In: Lalmas, M., Jose, J., Rauber, A., Sebastiani, F., Frommholz, I. (eds.) ECDL 2010. LNCS, vol. 6273, pp. 248–260. Springer, Heidelberg (2010)

Generating Content for Digital Libraries Using an Interactive Content Management System

Uros Damnjanovic and Sorin Hermon

The Cyprus Institute, STARC, Cyprus
{u.damnjanovic,s.heromon}@cyi.ac.cy.

Abstract. The goal of this paper is to present an interactive content management system for generating content of a digital library. The idea is to use interaction and data visualization techniques in the process of content generation, to check, understand and modify available information. We show the importance of interacting with data during the process of library creation and how this can lead to better quality of data. We present browsing functionalities for exploring relations within data that are used in the Human Sanctuary project. The set of developed tools can easily be extended and used for generating content in any other digital library project.

Keywords: Digital libraries, digital repositories, user interface, data visualization, human computer interaction.

1 Introduction

Ever since their inception, DL systems have had a very challenging objective: to provide organized content spaces and a suite of services that maximize synergies between content providers and content consumers, and boost the whole knowledge production lifecycle [1]. Providing means for using and understanding information is one of the most important features that distinguish digital libraries from simple digital repositories. Compared to various Internet sources, where relevance of information cannot always be known, DL's release users mental capacities from evaluating and collecting data, enabling them to focus on their work [2][3]. In our opinion interaction with content is one of the most important parts of the DL's since it is the place where information is received and new knowledge created [4]. Well-designed visual representation can replace cognitive calculations with simple perceptual inferences and improve comprehension, memory and decision-making [7][8]. When creating content for digital library it is not always clear what information should be used. Instead interaction with system may help content creators what information should be used in the library and how this information should be organized. EPrints, DSpace and Greenstone, are some examples of software's used for developing digital libraries [9][10]. These frameworks provide basic infrastructure for developing digital libraries, but in our opinion they lack interactive tools for accessing and using data.

We present in this paper, a content management system that we developed for generating and controlling content in digital libraries. We identified and developed

P. Zaphiris et al. (Eds.): TPDL 2012, LNCS 7489, pp. 474–479, 2012.
© Springer-Verlag Berlin Heidelberg 2012

umber of potentially useful interaction tasks that can give better control over data. In
ection 2 we present an ongoing project that is using our content management system,
nd describe the process of generating digital data. In section 3 we present
unctionalities for generating content and in section 4 we present tools for maintaining
nd interacting with data. In section 5 we conclude and discuss directions for future
vork.

2   Human Sanctuary Project

The goal of the project, done in collaboration with Israeli Museum, is to provide
ccess to digital documents that will bring the life of the Qumran community to a
vider public. The community of Qumran was behind the creation of the Dead Sea
Scrolls, the oldest remaining copies of the Bible and biblical documents. The
inderstanding of the community's life is important for researchers and general public
vho are interested in the history of religions and society. For this reason a 20-minute
eature movie that describes the most important aspects of the Qumran community is
nade by the museum. The movie is played at the museum, in addition to the
exhibition, and was well received with the audience. After success of the movie with
nuseum audience, next step was to make the movie available to a wider public by
outting it online together with innovative interactive methods that will be used to
ccess and explore the movie.

In order to present the movie, various criteria for interacting with the movie, and
or linking additional information to the movie are defined. The movie is manually
segmented into scenes, which will be used as one way of accessing information. Users
will be able to watch separate scenes, and access explanation on a scene level.
Another way of organizing the content is by topics. Users can use hierarchical topics
structure to explore the movie, on a topic level. Additionally, images, web pages and
ext documents will be used in combination with the movie, for explanation purposes.
Images or various objects related to the Dead Sea Scrolls, such as tools for writing
ind storing of the scrolls, images of scrolls and plans of the Qumran archaeological
site will be used to relate the content of the movie to the real world objects. Also links
o external web pages are used in places where additional explanation of some aspect
of the movie is already available online. For information that is not available, textual
lescriptions will be created and assigned to various segments of the movie.

3   Services for Generating Content

For the purpose of collecting and generating digital records that will be used in
combination with the movie we developed an interactive web based content
management tool, named Metadata Collector. The Metadata Collector provides basic
upload, edit, search and browsing functionalities for generating and accessing of the
content. In this project, Metadata Collector is not only used to upload information to
the server, but is also used to generate information. It has creative purpose, since at
the current stage of the project there is no definite data set that will be used. Instead

data is uploaded, evaluated and modified after interacting with the content, as we will show in the following section.

In order to link various external information with the movie, we defined metadata structure that assigns number of properties to every data record. All these properties will be used to build the information space, and to establish connections between stored data and the movie. The content upload process starts by selecting a type of the object being uploaded. At the moment the tool support images, audio, texts, pdf files, and links. At this stage users can fill in necessary form fields, and assign each object a number of time stamps, a number of topics, and a number of scenes. In this way the specific descriptive object is linked to the movie. Every object can then be described with key word labels and free text description. After each object is uploaded, users can browse the set of available records and edit objects if needed. It is also possible to upload more images at a time, and assign same metadata descriptions to all images, to save time and effort. Later individual images are accessed and modified. After content is being created there is a need for controlling and evaluating available information. This is done by interacting with data, using specially created tools for browsing and exploring the relations within data.

4   Interaction with Content for Control and Assessment

Since the process of generating content requires big mental effort, tools that help users manage their tasks are needed. From our experience on developing different digital libraries, the problem of content generation should not be seen only as a technical problem, that is, how to create digital objects. The problem of content generation also deals with the quality of created information and how it can be improved while interacting with data. Many times there is no way for users to understand how will the information they are creating be used in the final version of the library, and how will users interact with this information. This may lead to problems with end-users trying to understand and use this information in some practical manner. By being able to interact with data during the creation process, content creators may get better idea on what information to present, and how to present it. This means choosing the right content from the initial set of available information and presenting it in the best possible way. Our goal was to create tools that will help content creators to chose and organize data for a digital library by enabling "smart" interactions data.

Together with services for data upload and description creation, the core of our content management system are search and browsing functionalities. Users can access data in three levels of details. The first level is when only the basic information about the records is displayed in a grid-based layout for fast overview. A second level, displaying additional but not all information about the object, is accessed after user clicks on an item. Finally in the third level of data access users can get the full set of object descriptions. At any of this interaction stages users may select to edit object. In order to be able to understand relations between different items, additional tools for exploring various properties and relations between data were needed. As the work on data collection evolved, we identified number of possible interaction tasks that could help librarians in the process of content creation. One example of such interaction tool is an interactive data table where data records are displayed in rows, with data properties displayed in columns. Content creators needed a way to quickly inspect uploaded information with respect to specific metadata fields. Instead of browsing each individual items at a time, or browsing data in a data grid, where only small number of metadata properties may be displayed, information table gave content creators a chance to compare items on a single page, while having the overview of all properties. This proved to be very useful for understanding the current stage of the project and to get fast overview of the uploaded data. Without this tool, users would have to examine each item individually, and remember their properties, which was highly unpractical.

In order to focus on a specific data property like key word tags, we developed a tool that is showed in Fig 1. This tool helps users understand relations between images and tags. The screen is divided into three parts, tag cloud, image result part, and image details panel. A tag cloud shows all available tags, and displays them based on the appearance frequency. By clicking on a tag, images that are assigned with the selected tag are masked. Next, user can click on any image and explore its properties in details. Another way of using this tool is to select any metadata property by clicking on a specific field in the image description. This will again mask all images that are related to the selected image with respect to the selected criteria. By using this tool content creators have the chance to explore and understand relations between data with respect to their properties. It also helped to understand coverage of various topics with respect to the selected images.

Another tool that we have developed for interacting with content is shown in the Fig 2. This tool is used to interact with more properties at once. It uses hierarchical topic's tree, scene numbers and time intervals as queries and displays results in the image table. By selecting single criteria as a query, all other available properties are used to show relations between the query and given properties. For example if users select specific topic, set of images related to this topics is shown in the data table. Together with data table, criteria's that are not used in query such as scene number, and time intervals are used to highlight properties related to the chosen topic. Also user can investigate images in details, and edit them if needed.

As we already mentioned, the goal of our content management system was to help users create content, by helping them to upload and control data. At the initial stage of the project there was no precise idea on what information should be presented and how it should be structured. As the project progressed need for advanced tools emerged, tools that will give content creators a chance to better understand available information, and will give them sense on how this information will be used. By using these tools data was constantly improved either by introducing new information or by modifying the structure of the existing one. The metadata structure and thematic structure of the video were modified as a result of interactions with data. Interaction tasks that were used in our tools proved to be useful means of organizing and managing content. Generally content creators reacted positively to our system, emphasizing two main reasons for this. First, the process of uploading information and creating metadata descriptions was quite easy. They stated that not only it is easy to upload data, but also whole process was pleasurable experience, which is very important for them, since it helped them stayed focused and concentrated on their tasks for longer time periods. Another useful property of our system, according to content creators, was that it gave them good control over the content. They were able to explore available information from many different aspects and to easily make necessary changes. This proved to be very useful, especially when the process of generating content for a digital library can be seen as a creative process. This means that content creation task is not concerned only with creating of information in its digital form, but is also concerned with generating new knowledge from available information, and adding it to the system.

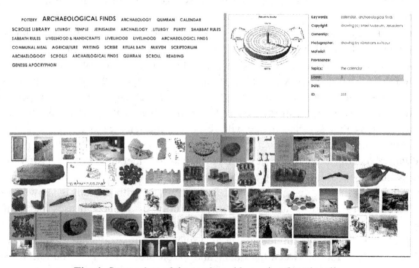

Fig. 1. Screenshot of the tag based browsing functionality

5   Conclusions and Future Work

We showed in this paper how interaction and interactive tasks can be used for better understanding of available information in the process of digital library creation. The process of creating content of the digital library should be incremental, and the tools for creating content should help librarians create, assess and modify available information. We presented our work on developing the digital library of the Human Sanctuary that needed carefully selected and generated content. Tools for interaction with the content proved to help librarians in their tasks and provided new ways of presenting information to end-users. After content is generated we will start developing tools that will be used by end users for using the data. This process will also be used as a final test of the available data, and will result in inclusion of new tools in the content management system.

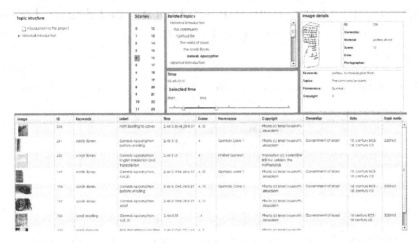

Fig. 2. Screenshot of the tool for exploring relations between topics, scenes and time intervals. Users can interact with data by selecting topic, scene or time interval. Set of relevant images is shown in the table.

References

1. Candela, L., Castelli, D., Fox, A.E., Ioannidis, Y.: On Digital Library Foundations. Int. J. Digit. Libr. 11, 37–39 (2010)
2. Becker, C., Kulovits, H., Guttenbrunner, M., Strodl, S., Rauber, A., Hofman, H.: Systematic Planning for Digital Preservation: Evaluating Potential Strategies and Building Preservation Plans. Int. J. Digit. Libr. 10, 133–157 (2007)
3. Liu, Z., Luo, L.: A Comparative Study of Digital Library Use: Factors, Perceived Influences and Satisfaction. The Journal of Academic Librarianship 37, 230–236 (2011)
4. Arms, W.Y., Aya, S., Dimitrev, P., Kot, B.J., Mitchell, R., Walle, L.: Building a Research Library for the History of the Web. In: Proceedings of the 6th ACM/IEEE-CS Joint Conference on Digital Libraries, pp. 95–102 (2006)
5. Gerken, J., Heilig, M., Jetter, H., Rexhausen, S., Demarmels, M., Werner, A., Reiterer, H.: Lessons Learned From the Design and Evaluation of Visual Information-Seeking Systems. Int. J. Digit. Libr. 10(2-3), 49–66 (2009)
6. Sastry, H., Manjunath, G., Venkatadri, M., Lokantha, C.R.: User Interface Design Framework for Digital Libraries. International Journal of Advanced Research in Computer Science 2, 505–508 (2011)
7. Chen, M., Ebert, D., Hagen, H., Laramee, R.S., Van Liere, R., Ma, K., Ribarsky, W., Scheuermann, G., Silver, D.: Data, Information, and Knowledge in Visualization. IEEE Comput. Graph., 12–19 (2009)
8. Karl, V.F., Sedig, K.: Interaction and the Epistemic Potential of Digital Libraries. Int. J. Digit. Libr. 11, 169–207 (2010)
9. Hitchcock, S., Carr, L., Jiao, Z., Bergmark, D., Hall, W., Lagoze, C., Harnad, S.: Developing Services for Open Eprint Archives: Globalisation, Integration and the Impact of Links. In: Proceedings of the Fifth ACM conference on Digital libraries, pp. 143–151 (2000)
10. Wolpert, A.: The Future of Electronic Data. Nature 420, 17–18 (2002)

Enhancing the Curation of Botanical Data Using Text Analysis Tools

Clare Llewellyn[1], Clare Grover[1], Jon Oberlander[1], and Elspeth Haston[2]

[1] University of Edinburgh, Edinburgh, United Kingdom
C.A.Llewellyn@sms.ed.ac.uk,
{Grover,Jon}@inf.ed.ac.uk
[2] Royal Botanic Gardens Edinburgh, Edinburgh, United Kingdom
E.Haston@rbge.ac.uk

Abstract. Automatic text analysis tools have significant potential to improve the productivity of those who organise large collections of data. However, to be effective, they have to be both technically efficient and provide a productive interaction with the user. Geographic referencing of historical botanical data is difficult, time consuming and relies heavily on the expertise of the curators. Botanical specimens that have poor quality labelling are often disregarded and the information is lost. This work highlights how the use of automated analysis methods can be used to assist in the curation of a botanical specimen library.

Keywords: text analysis, text mining, geographical location, assisted curation, botany.

1 Introduction

The aim of this work is to improve the interaction between users and automatic text analysis tools within a practical context. A tool has been created that allows users to curate botanical data by adding geographical locations. This is achieved via the user interacting with automatically generated locations extracted from textual records about botanical specimens. The user can correct or make additions to this generated output in order to specify the exact location where the botanical sample was collected.

To pursue the aim of creating a productive usable interface for textual analysis tools, a specific interface for such a tool has been created. The tool and interface are both applicable to many uses, but in order to evaluate it in detail, the focus is botanical science specimen data. The user group that this tool has been tested by are staff the Royal Botanic Gardens, Edinburgh (RBGE). The data curation experts at the RBGE expressed a need for the integration of a tool that extracts geographical locations from plant specimen data records into their current work flow. The demand for such a tool gives the opportunity to engage its likely users in evaluating the interface, therefore producing valid usability results.

P. Zaphiris et al. (Eds.): TPDL 2012, LNCS 7489, pp. 480–485, 2012.

2 Background

The Royal Botanic Garden of Edinburgh an internationally renowned centre of excellence for plant biodiversity research. The herbarium houses nearly three million specimens representing half to two thirds of the world's flora. It holds specimens collected from across the world and continues to receive approximately ten thousand new specimens each year [13].

Plant specimens are labelled with data that is relevant from their collection. Geographic referencing in this domain means converting the textual descriptions of where a plant was collected into machine readable geographic locations generally using a map based coordinate system. This is either done at the time when the plant is collected by GPS systems or retrofitted from textual descriptions [2, 9]. Historically, locations on plant specimens have been vague. Identifying and correcting plant specimens records that contain errors is time consuming and expensive for curators [9], therefore improvements to the speed and the accuracy of this process would be valuable. Currently, geographic referencing is conducted manually using resources such as gazetteers and maps to find the coordinates of the place names that have been identified in the plant specimen records by the curators. Tools have been built to assist in this process, including BioGeomancer [14] and GEOLocate [15], but these systems do not always fit well into the curation workflow [9]. Once the data has been geographically located it allows a botanical scientist to study environmental changes, particularly those concerning human impact and climate change.

Locations can be automatically generated through content analysis, natural language processing and text mining [3]. Widespread use of text analysis has not yet been achieved. For example in geographic referencing it is believed that the main barrier to uptake is that accuracy levels usually fall short of the expectations and needs of the user. It is proposed that this problem is rectifiable through the provision of interface extensions to existing text analysis tools to allow the user to correct and enhance automatically created output, thereby combining the efficiency of automatic processing with the accuracy of manual annotation [1]. Metrics for text analysis evaluation currently focus on comparisons with other text mining systems rather than evaluating the usefulness of the tool within a domain [1,10,12]. A study by Alex et al. in 2008 [1] found that the speed of curation can be increased by a third by assistance of text mining tools.

3 Prototype Tool

3.1 Data

Plant specimens have labels describing the collection details of that specimen. The text from these labels are stored in a database. The conversion of the label to a record is a manual process performed by the curators. This study focused on records from the United Kingdom and Ireland from 1747 to 2010. The total number of records processed was 43,060. It was found that, in total, 63.82% of records had some degree of geographical information, and could be geolocated.

3.2   Text Analysis Tools

The data from the database was processed using the Edinburgh Informatics information extraction tools which include LT-TTT2 and the Edinburgh Geoparser [6,7,11]. These are well established tools that process text and XML to identify place names and provide geographic coordinates for the locations. The Geoparser is made up of two main components; the Geotagger which provides place name recognition (identifies text strings as places) and the Georesolver which provides geographic referencing (looks up the names in a geographic gazetteer) [6]. The named entity recognition tool identifies word sequences as place-name entities and marks them up as XML elements. After initial tokenisation and part-of-speech tagging, it uses a rule-based method that takes into account information about part-of-speech, capitalisation, local context and lexicon look-up. The place-name entities recognised by this method are converted to gazetteer queries which are submitted to one of the Unlock or GeoNames gazetteer services [6,7,11].

3.3   The Tool

The data curation experts requested an automatic tool that extract geographical locations from plant specimen data records and could be integrated into their current work flow. The data was initially processed through text mining, database matching of similar fields and a National Grid Reference conversion to latitude and longitude. The information produced from this processing was stored in the database. The system is web based and the users interact with it through a webserver to query the database. The interface provides two views of the result, as a list of locations and as points on a map. The maps used are accessed through APIs - Google Maps and the National Library of Scotland's Ordinance Survey Maps.

4   Evaluation

An evaluation was conducted to test the hypothesis that a textual analysis tool can be used to improve, increase the speed or accuracy of the workflow of curators who are archiving plant specimen data. The current manual curation process was compared against a tool with textual analysis support. The evaluation was conducted in a manner adapted from a digital library evaluation framework [5]. Human computer interaction (HCI) within digital libraries has been studied extensively for the past ten years [8]. The tool was evaluated to ensure that it observed the basic HCI principles of a digital library such as obtaining correct results to a query quickly (precision and recall)[4]. It is important for the user to receive a manageable number of results so that they can see what the general content will be. Furh et al (2007)[5] provide an extensive framework for the evaluation of digital libraries which is adapted for this task. They suggest focusing on usability, usefulness (or relevance) and performance; therefore these were the areas evaluated in this work.

The evaluation was conducted with ten participants all of whom work at the RBGE. Each participant was asked to perform eight tasks. The data used for evaluation was data with known locations (a random sample from the 881 RBGE records that contained latitude and longitude values). This was then used to provide an accuracy measure for each task. A post task interview was used to provide qualitative information on the participants' opinion of the system.

Usability is measured by looking at the effectiveness, adaptability, enjoyability and learnability of the tool. Effectiveness was measured by how many tasks could be completed [4,5]. The results suggest that the tool performs slightly better than the traditional method. Adaptability was measured by whether they could adapt the experience to their own preferences. Comments on these features suggested that they were generally well liked. Enjoyability and learnability were measured through satisfaction scores for ease of use, visual appearance, contents, structure, error corrections and usefulness of help information. The tool scored highly in this category, the users liked the tool and found it easy to use. Participants found the tool easier to use than the traditional method.

Usefulness is evaluated through the relevance of provided content. If the content assist with the task defined in a satisfactory way [4,5] and whether the content provided led to participants accurately locating the samples. The text mining suggestions were not considered completely accurate, as many false positives were returned in order to include as many true positives as possible, but the text mining suggestions were still considered helpful. The users were willing to tolerate a degree of inaccuracy in the suggestions. Initially it was found that there was no significant difference between the accuracy of the two systems. However, there is a significant positive correlation between the tasks showing that some task was difficult with either method.

In order to look more closely at the accuracy achieved using the tool a further experiment was conducted. All participants in the test were asked to geo-locate all of the samples used in the initial evaluation using the tool. These locations were then clustered and the location which was the furthest from the others was left out, as was any location more than 25km away from the average point. Using an average location point from those left, a significant increase in accuracy was found when using the tool (p= 0.012, t=-2.742, df= 23). Thus the tasks may be difficult for specific individuals but when an average is taken over the whole group the result will be accurate.

Performance and efficiency of the tool was evaluated by assessing the efficient retrieval of information. It was measured by how much time it took to correctly complete tasks [4,5]. Each task was timed for each participant. In addition participants were asked to rate performance for each task. The performance, which was judged by speed of task completion, was better on average with the tool (see table 1). A paired sample t-test shows that the difference is not significant (p=0.539, t=-0.639 df=9).

Table 1. Average Speed for Tasks (in seconds)

	Minimum Time	Maximum Time	Mean Time	Standard Deviation
Tool	139.25	263.75	190.65	42.47748
Traditional	87.75	300.00	202.93	74.7781

A further experiment was conducted to investigate if the number of text mining locations offered had an effect on the time taken to complete the task to see if there was an optimal number of locations. Initially the total set of locations provided to the user was considered: every single latitude and longitude pair for every place name. Analysis indicates that it is possible that there is a positive effect of either offering very few or very many suggestions (below 2 and above 5). The total number of locations offered was contrasted with the number of unique locations. The tool often suggests a number of individual latitude and longitude locations for a single place name (as many location names are reused). A unique location is classified as a single place name (no matter how many suggestions are offered for that name). It was found that 2 unique locations may be beneficial, possibly because they are used to provide confirmation. With 1 location the task may take longer, as the user may need to consult other features. With more than 3 locations the task may take longer, as the locations may be contradictory.

5   Conclusions

The specific objectives of this work were to identify where text analysis tools can be used in the botanical curation workflow, to design and implement a prototype tool and to evaluate if this tool improves the ease, speed or accuracy of botanical curation. A tool was created that allowed users to interact with automatically generated geographical information in order to correct or make additions to the output. As requested by the data curation experts, the tool has been integrated into the current workflow.

In the evaluation it has been shown that, using the tool, average speeds are quicker. The tool scored highly for both usability and usefulness. The participants in the evaluation liked the tool and found it easy to use - they preferred it to the traditional method of using multiple data sources. The accuracy of the geographic location was compared between the tool and the traditional method. Initially, it was found that there was no significant difference in the accuracy of the two systems. When multiple participants used the tool to identify a location and values were clustered, leaving out the least similar location, a significant increase in accuracy is found. This shows that while the tasks may be difficult for individuals, higher accuracy can be gained from using locations from a number of individuals: they will collectively locate the specimen accurately. This suggests that this is an ideal tool for use with crowd sourcing. The analysis suggests that there may be an optimal number of total text mining suggestions and unique

ext mining locations that reduces the burden on the user and leads to more
fficient geographic location.

References

1. Alex, B., Grover, C., Haddow, B., Kabadjov, M., Klein, E., Matthews, M., Roebuck,
   S., Tobin, R., Wang, X.: Assisted curation: does text mining really help? In: Pacific
   Symposium on Biocomputing. Pacific Symposium on Biocomputing, pp. 556–567
   (2008)
2. Allen, W.H.: The Rise of the Botanical Database. BioScience 43(5), 274–279 (1993)
3. Auer, S., Lehmann, J., Hellmann, S.: LinkedGeoData: Adding a Spatial Dimension
   to the Web of Data. In: Bernstein, A., Karger, D.R., Heath, T., Feigenbaum, L.,
   Maynard, D., Motta, E., Thirunarayan, K. (eds.) ISWC 2009. LNCS, vol. 5823,
   pp. 731–746. Springer, Heidelberg (2009)
4. Blandford, A., Buchanan, G.: Usability of digital libraries: a source of creative
   tensions with technical developments (2003)
5. Fuhr, N., Tsakonas, G., Aalberg, T., Agosti, M., Hansen, P., Kapidakis, S., Klas,
   C.-P., et al.: Evaluation of digital libraries. International Journal on Digital Libraries 8(1), 21–38 (2007)
6. Grover, C., Givon, S., Tobin, R., Ball, J.: Named Entity Recognition for Digitised
   Historical Texts. In: Proceedings of the Sixth International Conference on Language
   Resources and Evaluation (LREC 2008) (2008)
7. Grover, C., Tobin, R.: Rule-Based Chunking and Reusability. In: Proceedings of
   the Fifth International Conference on Language Resources and Evaluation (LREC
   2006), Genoa, Italy (2006)
8. Jeng, J.: What Is Usability in the Context of the Digital Library and How Can It
   Be Measured? Information Technology and Libraries 24(2), 46–56 (2005)
9. Johnson, N.F.: Biodiversity Informatics. Annual Review of Entomology 52(1), 421–
   438 (2007)
10. Dietrich, R.-S., Kirsch, H., Couto, F.: Facts from Text Is Text Mining Ready to
    Deliver? PLoS Biol. 3(2) (February 15, 2005)
11. Richard, T., Grover, C., Byrne, K., Reid, J., Walsh, J.: Evaluation of georeferenc-
    ing. In: Proceedings of the 6th Workshop on Geographic Information Retrieval,
    GIR 2010, 7:17:8. ACM, New York (2010)
12. Winnenburg, R., Wchter, T., Plake, C., Doms, A., Schroeder, M.: Facts from text:
    can text mining help to scale-up high-quality manual curation of gene products
    with ontologies? Briefings in Bioinformatics 9(6), 466–478 (2008)
13. Edinburgh Royal Botanic Gardens website, http://www.rbge.org.uk/, (accessed
    March 29, 2012)
14. BioGeomancer, http://www.biogeomancer.org/, (accessed August 15, 2011)
15. GeoLocate, http://www.museum.tulane.edu/geolocate/, (accessed August 15,
    2011)

Ranking Distributed Knowledge Repositories

Robert Neumayer, Krisztian Balog, and Kjetil Nørvåg

Norwegian University of Science and Technology,
Department of Computer and Information Science, Trondheim, Norway
{robert.neumayer,krisztian.balog,kjetil.norvag}@idi.ntnu.no

Abstract. Increasingly many knowledge bases are published as Linked Data, driving the need for effective and efficient techniques for information access. Knowledge repositories are naturally organised around objects or entities and constitute a promising data source for entity-oriented search. There is a growing body of research on the subject, however, it is almost always (implicitly) assumed that a centralised index of all data is available. In this paper, we address the task of ranking distributed knowledge repositories—a vital component of federated search systems—and present two probabilistic methods based on generative language modeling techniques. We present a benchmarking testbed based on the test suites of the Semantic Search Challenge series to evaluate our approaches. In our experiments, we show that both our ranking approaches provide competitive performance and offer a viable alternative to centralised retrieval.

1 Introduction

In recent years the number of knowledge bases published as Linked Data has significantly increased. These range from general-purpose encyclopaediae like DBpedia or Freebase, to domain-specific databases, such as GeoNames for geographical entities. These knowledge bases are inherently organised around "entities" or "objects," such as persons, places, organisations, artifacts, etc. This, coupled with the fact that the most frequent types of queries in web search revolve around entities [10], lends significance to the *ad-hoc entity retrieval* task, defined as follows: "answering arbitrary information needs related to particular aspects of objects [entities], expressed in unconstrained natural language and resolved using a collection of structured data" [10]. The importance of search focused on entities is also witnessed by numerous tasks that have been featured at the TREC [15, 2, 3] and INEX [7] evaluation benchmarking campaigns.

Knowledge repositories are typically both heterogeneous and inherently distributed as they are located on disparate servers. Only in few cases it is possible to maintain a central index encompassing the contents of all individual data sources. Many sources can be covered only partially (for example, specific parts might be overlooked by spiders), while others may not be crawleable at all (due to authorisation settings prohibiting access). Instead of expending effort to crawl all data from these sources, one might pass the query to the search interface of multiple, suitable collections (usually distributed across several locations)—an approach known as *distributed information retrieval (DIR)*, also referred to as *federated search* [11]. For example, the query "painters of the gothic era" may be passed to a related collection, such as a digital library of a

P. Zaphiris et al. (Eds.): TPDL 2012, LNCS 7489, pp. 486–491, 2012.

museum, while for the query "San Antonio" collections containing information about the city, such as GeoNames or DBpedia, might be more appropriate. Of course, there are also queries for which multiple databases can contain answers. When querying distributed knowledge repositories (from now on *collections*) it is desirable to choose only those that (are likely to) contain relevant results. That is, we need to be able to rank individual collections with respect to a given query. This task, *collection ranking*, has received a lot of attention in the past in the DIR literature, but only in the context of traditional document search. We target the more general concept of entities, i.e., any digital object described in terms of ontology-based metadata from now *entity*). We focus on collection representation and ranking for ad-hoc entity search in a distributed environment. Building on prior DIR research we formulate two collection ranking strategies using a unified probabilistic retrieval framework based on language modeling techniques. According to one model (Collection-centric), each collection is represented as a term distribution computed over its contents. Our second model (Entity-centric) estimates the relevance of each individual entity within the collection and then aggregates these scores to determine the collection's relevance.

We introduce an experimental platform based on the data set and topics from the Semantic Search Challenge [9, 4]. We assume a cooperative environment, where collections provide information about their contents, such as their term statistics, and can implement the same retrieval function for ranking entities. We find that it is indeed necessary for collections to provide information about their term statistics; with these available, our models achieve very competitive performance. Assigning higher prior importance to larger collections, a reasonable heuristic, brings in further improvements.

In summary, this work makes the following contributions: (1) a unified generative modeling framework and two particular models for collection ranking, (2) a test set for evaluating the collection ranking task in Linked Data, and (3) an experimental comparison of our models using this data set.

2 Related Work

Distributed information retrieval (DIR) or *federated search*, is ad-hoc search in environments containing multiple, text databases [5]. DIR targets cases when documents cannot be copied into a centralised database. It involves three important sub-problems: I) *acquiring resource descriptions*, representing the content of each collection in some suitable form, (II) *resource selection*, selecting the most relevant collections, and, finally, (III) *result merging*. We restrict our attention to (I) and focus on both the representation and ranking of resources (collections).

Federated search techniques have recently been picked up by the digital library community too. MinervaDL, a digital library architecture for information retrieval in peer-to-peer (P2P) networks is presented in [16]. It differs from our work in three important aspects: they use a different architecture, do not consider retrieval effectiveness, and work with much smaller collections. Linked Data (LD) also bears increasing importance to digital libraries, as it can be beneficial to enriching metadata. For example, [14] suggest to automatically link FRBR works to the corresponding entity in LD.

3   Representing and Ranking Distributed Collections

In this section we present our approach for representing and ranking collections. We formulate this task in a generative probabilistic framework and rank collections based on their likelihood of containing entities relevant to an input query, $P(C|Q)$. Instead of estimating this probability directly, we apply Bayes' rule and rewrite it to $P(C|Q) \propto P(Q|C)P(C)$. Thus, the score of a collection is made up of two components: (1) *query generator* ($P(Q|C)$), that is, the probability of a query being generated by collection C; this can be interpreted as the collection's relevance to the query; (2) *collection prior* ($P(C)$), that is, the *a priori* probability of selecting collection C; this tells us how likely the collection is to contain the answer to any arbitrary query.

We propose two models for estimating the query generator by drawing upon existing strategies to collection ranking and formalise them within a language modeling framework. Our two approaches bear resemblance to the expert finding models of Balog et al. [1] and to the blog feed search models of Elsas et al. [8], but differ in the estimation of specific components. Let us remind ourselves that we assume a cooperative environment, in which collections can provide general term statistics and can implement the same retrieval function. Further, we assume that for each entity E, a document representing that entity, D_E, has already been created.

Collection-Centric Model. One of the simplest approaches to resource selection is to treat each collection as a single, large document [6, 13]. Once such a pseudo-document is generated for each collection, we can rank collections much like documents. In a language modeling setting this ranking is based on the probability of the collection generating the query. Formally:

$$P(Q|C) = \prod_{t \in Q} \left\{ (1 - \lambda_G)\left(\sum_{E \in C} P(t|D_E)P(E|C) \right) + \lambda_G P(t|G) \right\}^{n(t,Q)}, \quad (1)$$

where $n(t, Q)$ is the number of times term t is present in the query Q, $P(t|D_E)$ and $P(t|G)$ are maximum-likelihood estimates of the probability of observing term t given the entity and global (cross-collection) language models, respectively, and λ_G is the (global) smoothing parameter. We assume that all entities are equally important within a given collection, thus set $P(E|C) = 1/|C|$.

Entity-Centric Model. Instead of creating a direct term-based representation of collections, our second approach models and queries individual entities, then aggregates their relevance estimates:

$$P(Q|C) = \sum_{E \in C} P(E|C) \prod_{t \in Q} \left((1 - \lambda_G)P(t|\theta_E) + \lambda_G P(t|G) \right)^{n(t,Q)}, \quad (2)$$

where $P(t|\theta_E)$ is the probability of term t given the entity's language model and $P(t|G)$ is the global background language model. It is worth noting that this model employs smoothing on two levels: (1) on the entity's level, by smoothing the entity document with the collection (in estimating $P(t|\theta_E)$), and (2) on the collection level, by mixing with the global background model using coefficient λ_G.

This model resembles the ReDDE collection selection algorithm by Si and Callan 12]. The main difference is that we do not incorporate the collection size directly into he scoring formula, but accommodate it through the collection prior.

Collection Priors. To estimate the *a priori* probability of a collection, $P(C)$, we conider two alternatives. The simplest choice is to assume that all collections are equally mportant: $P(C) \propto 1$. We refer to this as the *uniform* prior. Intuitively, larger collecions are more likely to contain relevant entities to any information need. According to he *collection size* prior, we set the $P(C) \propto |C|$.

4 Experimental Setup

Our testbed is based on the 2010 and 2011 editions of the Semantic Search Chalenge [9, 4]. Queries there request specific named entities (e.g., "american embassy nairobi" or "martin luther king") from a collection of structured data, described as RDF. The data collection we use is the Billion Triple Challenge 2009 dataset.[1] Crawled during February/March 2009, it comprises about 1.14 billion RDF statements. For our experiments, we consider the top 100 second-level domains (measured in terms of the number of entities contained) as distributed knowledge repositories (i.e., our set of collections).

Queries originate from the Yahoo! Search Query Tiny Sample dataset.[2] Relevance judgments were obtained using Amazon's Mechanical Turk. We used all queries from 2010 and 2011, but filtered out those that did not have any relevant results from the top 100 domains that we considered as our collections; this left us with 136 queries in total. In our setting, a collection is considered relevant if it contains at least one entity that was judged relevant. For graded evaluation metrics, we set the relevance level of a collection to the number of relevant documents it contains.

We use standard IR evaluation metrics: Mean Average Precision (MAP), Mean Reciprocal Rank (MRR), Recall at 5 (R@5), and Normalised Discounted Cumulative Gain NDCG). Significance testing is done using a two-tailed paired t-test.

5 Experimental Evaluation

The main research question we seek to answer is the following: How to represent collecions for distributed entity retrieval? We address the following specific sub-questions: 1) What is the effect of taking global (cross-collection) term statistics into account? 2) Which of the Collection-centric (CC) and Entity-centric (EC) collection ranking approaches perform better? (3) What is the impact of collection priors?

Using Global Term Statistics. To investigate the potential benefits of having knowledge about the global (cross-collection) importance of query terms, we consider two settings: 1) not making use of using global term statistics; within our language modeling framework this is implemented by setting the λ_G (global) smoothing parameter to 0 in Eqs. 1

---

[1] http://vmlion25.deri.ie/

[2] http://webscope.sandbox.yahoo.com/catalog.php?datatype=1

Table 1. Collection ranking results. $^{\dagger}/^{\ddagger}$ denote significant differences at the 0.05/0.01 levels, respectively. Significance is tested for rows 2 vs. 1, 4 vs. 3, 5 vs. 3, and 6 vs. 4. Best scores within each block are typeset boldface.

Global term stat.	Coll. priors	Model	MAP	MRR	R@5	NDCG
No	No	Collection-centric	0.3048	0.4304	0.3552	**0.4873**
No	No	Entity-centric	**0.3710**‡	**0.4980**†	**0.4289**‡	0.4766
Yes	No	Collection-centric	**0.5149**	**0.8901**	0.5208	**0.8280**
Yes	No	Entity-centric	0.5134‡	0.8404†	**0.5262**	0.7967‡
Yes	Yes	Collection-centric	0.5368‡	0.9282†	0.4645†	0.8506‡
Yes	Yes	Entity-centric	**0.5494**‡	**0.9283**‡	**0.4817**†	**0.8564**‡

and 2, and (2) using global term statistics; we use Dirichlet smoothing and set λ_G proportional to the average collection length. Table 1 displays the results without and with global term statistics (rows 1-2 vs. rows 3-4). It is clear that using this information leads to substantial and significant improvements for both methods.

Collection-Centric vs. Entity-Centric Models. Based on the numbers in Table 1, if global term statistics are omitted, the Entity-centric model clearly outperforms the Collection-centric one (except for an insignificant degradation for NDCG). With global term statistics, however, their performance is much closer to each other, with CC actually performing better on MRR and NDCG. Interestingly, the two models have almost the same performance when averaged over all topics, but they generate significantly different results, i.e., on the level of individual topics, there are sometimes substantial differences between the two models, but these differences equal out on average.

Collection Priors. Our last set of experiments focuses on collection priors; the last two rows of Table 1 report the results. Priors improve for both collection selection methods on the precision-oriented metrics (MAP, MRR, NDCG), and especially help precision at the top rank (MRR). With MRR scores in the 0.9 range, there is not much room left for improvement at rank 1. Nevertheless, this is done at the expense of recall. All differences are significant.

6   Conclusions

To the best of our knowledge, ours is the first work to apply federated IR techniques in the context of entity search. In this paper, we presented two methods for collection ranking of distributed knowledge repositories. One approach scores an individual collection by collapsing all text associated with its entities into one pseudo-document. The other considers individual entities and aggregates their relevance scores on the collection level. Our experimental comparison of these two approaches showed that both deliver excellent performance. For both cases, we have shown the importance of having access to global term statistics and the benefits of incorporating collection priors. In future work, we plan to expand our work to non-cooperative environments.

References

[1] Balog, K., Azzopardi, L., de Rijke, M.: Formal models for expert finding in enterprise corpora. In: Proceedings of SIGIR, pp. 43–50 (2006)

[2] Balog, K., Soboroff, I., Thomas, P., Craswell, N., de Vries, A.P., Bailey, P.: Overview of the TREC 2008 enterprise track. In: The 17th Text Retrieval Conference Proceedings (TREC 2008). NIST (2009)

[3] Balog, K., de Vries, A.P., Serdyukov, P., Thomas, P., Westerveld, T.: Overview of the TREC 2009 entity track. In: Proceedings of the 18th Text REtrieval Conference (TREC 2009). NIST (February 2010)

[4] Blanco, R., Halpin, H., Herzig, D., Mika, P., Pound, J., Thompson, H., Duc, T.: Entity search evaluation over structured web data. In: 1st International Workshop on Entity-Oriented Search (EOS), pp. 65–71 (2011)

[5] Callan, J.: Distributed information retrieval. In: Advances in Information Retrieval, pp. 127–150. Kluwer Academic Publishers (2000)

[6] Callan, J.P., Lu, Z., Croft, W.B.: Searching distributed collections with inference networks. In: Proceedings of SIGIR, pp. 21–28 (1995)

[7] Demartini, G., de Vries, A., Iofciu, T., Zhu, J.: Overview of the INEX 2008 Entity Ranking Track. In: Geva, S., Kamps, J., Trotman, A. (eds.) INEX 2008. LNCS, vol. 5631, pp. 243–252. Springer, Heidelberg (2009)

[8] Elsas, J.L., Arguello, J., Callan, J., Carbonell, J.G.: Retrieval and feedback models for blog feed search. In: Proceedings of SIGIR, pp. 347–354 (2008)

[9] Halpin, H., Herzig, D.M., Mika, P., Blanco, R., Pound, J., Thompson, H.S., Tran, D.T.: Evaluating ad-hoc object retrieval. In: Proceedings of the International Workshop on Evaluation of Semantic Technologies, IWEST 2010 (2010)

[10] Pound, J., Mika, P., Zaragoza, H.: Ad-hoc object retrieval in the web of data. In: Proceedings of the 19th International Conference on World Wide Web, pp. 771–780 (2010)

[11] Shokouhi, M., Si, L.: Federated search. Foundations and Trends in Information Retrieval 5, 1–102 (2011)

[12] Si, L., Callan, J.: Relevant document distribution estimation method for resource selection. In: Proceedings of SIGIR, pp. 298–305 (2003)

[13] Si, L., Jin, R., Callan, J., Ogilvie, P.: A language modeling framework for resource selection and results merging. In: Proceedings of the Eleventh International Conference on Information and Knowledge Management, pp. 391–397 (2002)

[14] Takhirov, N., Duchateau, F., Aalberg, T.: Linking FRBR Entities to LOD through Semantic Matching. In: Gradmann, S., Borri, F., Meghini, C., Schuldt, H. (eds.) TPDL 2011. LNCS, vol. 6966, pp. 284–295. Springer, Heidelberg (2011)

[15] Voorhees, E.: Overview of the TREC 2004 question answering track. In: Proceedings of the 13th Text Retrieval Conference (TREC 2004) Gaithersburg, NIST Special Publication: SP 500–261 (2005)

[16] Zimmer, C., Tryfonopoulos, C., Weikum, G.: MinervaDL: An Architecture for Information Retrieval and Filtering in Distributed Digital Libraries. In: Kovács, L., Fuhr, N., Meghini, C. (eds.) ECDL 2007. LNCS, vol. 4675, pp. 148–160. Springer, Heidelberg (2007)

The CMDI MI Search Engine:
Access to Language Resources and Tools
Using Heterogeneous Metadata Schemas

Junte Zhang, Marc Kemps-Snijders, and Hans Bennis

Meertens Institute, Royal Netherlands Academy of Arts and Sciences

Abstract. The CLARIN Metadata Infrastructure (CMDI) provides a solution for access to different types of language resources and tools across Europe. Researchers have different research data and tools, which are large-scale and described differently with domain-specific metadata. In the context of the Search & Develop (S&D) project at the Meertens Institute within CLARIN, we present a system description of an advanced search engine that semantically converges differently structured metadata records based on CMDI for search and retrieval. It allows different groups of users – such as language researchers – to search across yet unexplored research data and locate relevant data for new insights, and find existing tools that could provide novel use cases.

1   Introduction

The Common Language Resources and Technology Infrastructure (CLARIN) initiative seeks to establish an integrated and interoperable research infrastructure of language resources and its technology.[1] Descriptive metadata is used to characterize large number of (legacy) data resources (collections) and tools (e.g. Web services) to facilitate their management and discovery. The Search & Develop (S&D) project within CLARIN in the Netherlands uses the Component MetaData Infrastructure (CMDI; [2]) to open up the sharing of resources and Web services for people and machines first within the collections of a single institution, then across institutions in the Netherlands and eventually across Europe as whole. This infrastructure enables new research methods in language research and stimulates the Digital Humanities, where new insights can be gained by combining and reusing resources from different institutions and domains, and existing tools can be more effectively found and reused based on new insights.

But how to search for data and services, which can be understood by both people from varying disciplines and machines? The challenge is that the data is heterogenous both in content and structure, and can be massive in amount. This flexibility encourages the exchange of resources and services, but at the same time is a challenge for access. Our aim is to develop a unified solution that can make language resources and research data and tools across collections and even institutions accessible and usable for advanced users like Humanities researchers based on metadata. We present our solution to this challenge with the CMDI Meertens Institute (MI) search engine.

---

[1] See http://www.clarin.eu/external/index.php?page=about-clarin

P. Zaphiris et al. (Eds.): TPDL 2012, LNCS 7489, pp. 492–495, 2012.

System Description of the CMDI Search Engine

າ this section we present the system description of the CMDI MI search engine, and ⁻e focus on our robust indexing method and user interface. This search engine is driven y CMDI metadata, which is used as the first step to gain access to resources.

.1   CMDI Files

. CMDI file in XML has a **<Header>**, **<Resources>**, and **<Components>**. The ɔrmer 2 are fixed, while the content and structure within **<Components>** is flexible ꞇnd can encapsulate any data in any structured form. An XML schema can be used ɔ make CMDI files coherent in structure for a (sub)collection and it contains refer-ꞇnces to ISOcat data categories (DC) stored in the Registry (DCR; [4]). The DCR was ҫstablished by the *ISO Technical Committee 37, Terminology and other language and .ontent resources* based on the ISO 12620:2009 standard. Because multiple elements ꞇay refer to the same DC, a certain degree of semantic interoperability can be achieved ꞇross different datasets. A specification using the DCR and projected for example in ꞇ XML schema is called a *profile* and can be (re)used for describing datasets.

We initially have indexed different CMDI datasets available across Europe and at ꞇe MI such as the Dutch Song Database, which have been converted from metadata of ҽgacy research datasets. At the MI, these are primarily Dutch language resources that .escribe variation in language and cultural data resulting from ethnographic research.

..2   Indexing Method

ʰhe CMDI MI search engine is driven by the Apache Solr enterprise search server that ꞇses the Lucene Java search library. This backend allows us to index the data efficiently ꞇnd implement state-of-the-art Information Retrieval (IR) features. We use the standard ҽlevance ranking based on the Vector Space Model. We have not yet enabled stop-word ҽmoval or conducted stemming, but this can be supported and fine-tuned in the future.

The following procedure is used to index the CMDI files, and can be used robustly to ꞇndex even more diversely structured CMDI files from more institutes in a loop. We use ꞇn XML schema parser to index on the XML element level by using XPath expressions ɔr focused access. XML elements are disambiguated with their parent.

- Export all CMDI files from the MySQL database, FTP, etc with a script.
- Create a list of the CMDI files that needs to be indexed.
- Use the XML Schema parser to link CMDI elements to their ISOcat DC and
  - run a Perl script based on all XML schemas and create a single Lucene indexing schema for all the CMDI files.
  - create an XSLT stylesheet β with a script and use β to convert CMDI elements to the Lucene indexing format with templates for each XPath. The same ele-ments with different XPaths are defined by setting the 'priority' attribute based on the XPath length. Elements without ISOcat DC are indexed as fulltext.
- Extract the ISOcat IDs as unique index field names and extract with the XML Schema parser the corresponding 'human-readable' (semantic) labels for metadata 'relabeling' in the user interface with a parallel array.

(a) Query autocompletion based on the result counts and the tags, and preserving the overview of the search trail and allow users to change it (in any order). The last used query is 'volksverhaal' (*Folktale*).

(b) The system projects location-based results on a map, zooms and aggregrates by clustering the markers. Users can directly click to go to the resources. The lists in 'collection' and 'CMDI profile' are kept in sync.

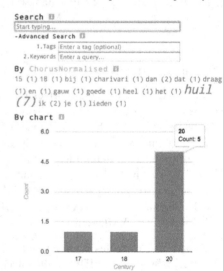

(c) Display of results with snippets and keywords in context within the last searched metadata label and the presentation of all used keywords in context given the fulltext. For each retrieved result in the list, there is a recommendation (when available) of related results based on the content similarity of the last used metadata label, or else the fulltext.

(d) Advanced Search allows user with (some) prior knowledge of the metadata schemas to select a tag, and then enter a query aided by the autocompletion feature based on that tag. Besides a tag cloud, the system uses bar and line charts to display aggregated time-based results (in tags 'Century', 'Year'). The results are narrowed down by clicking in the chart.

Fig. 1. The user interface of the CMDI MI search engine and its advanced features

Ve have automated this indexing method with a shell script in Linux, so that we can set
a cron job, for instance, as the CMDI files get revised or a collection gets extended with
a batch of new CMDI files. Our results show that this method is robust, as we can index
246,728 CMDI files from 18 different XML (metadata) schemas in a single stream. We
have indexed 150 different types of elements (based on the ISOcat DCs).

2.3   User Interface

For the user interface in Fig 1, we employ, modify, and extend the JavaScript library
AJAX Solr[2], which allows for faceted search that matches with the information seeking
behavior of Humanities users [1, 3]. It is setup with widgets, where the search box and
features of Fig. 1(a), 1(b) and 1(d) are located on the left and the result list and display
of Fig. 1(c) are presented on the right side. Users can express queries within the tags
and retrieve snippets within them for focused search, or else search in the fulltext.

The system supports serendipity with query autocompletion that directly includes the
hit counts, a tag cloud with keyword highlighting to give an overview on an aggregrated
level of the results to support query expansion, recommendation of similar results based
on content similarity, grouping of results by clicking on subject headings, display on an
aggregrated level of the results based on temporal (charts) and geographical (maps)
information, and different displays of the result types based on CMDI profile. It keeps
track of the cumulative search trail and allows users to revise it in any order.

3   Conclusion

We have presented the CMDI MI search engine, which includes the concise description
of CMDI metadata, the indexing method, and the user interface. The novelty is that
it is a common (semantic) gateway to diversely structured descriptions of language
resources with different metadata schemas for users such as Humanities researchers
with very specific and complex information (research) needs. It is a tool that provides
focused and interactive access to heterogeneous metadata of (legacy) language research
datasets and tools, supports serendipity, and provides new insights for further research
and development. It can be found and used at www.meertens.knaw.nl/cmdi/search.

References

1] Bates, M.J.: The design of browsing and berrypicking techniques for the online search inter-
    face. Online Review 13(5), 407–424 (1989)
2] Broeder, D., Kemps-Snijders, M., Uytvanck, D.V., Windhouwer, M., Withers, P., Wittenburg,
    P., Zinn, C.: A data category registry- and component-based metadata framework. In: LREC.
    European Language Resources Association, ELRA (2010)
3] Hearst, M.A., Karadi, C.: Cat-a-cone: an interactive interface for specifying searches and
    viewing retrieval results using a large category hierarchy. In: SIGIR, pp. 246–255. ACM,
    New York (1997)
4] Kemps-Snijders, M., Windhouwer, M., Wittenburg, P., Wright, S.E.: ISOcat: remodelling
    metadata for language resources. IJMSO 4(4), 261–276 (2009)

---

[2] See https://github.com/evolvingweb/ajax-solr

SIARD Archive Browser*

Arif Ur Rahman[1,2], Gabriel David[1,2], and Cristina Ribeiro[1,2]

[1] Departamento de Engenharia Informática–Faculdade de Engenharia,
Universidade do Porto
[2] INESC TEC
Rua Dr. Roberto Frias, 4200-465 Porto, Portugal
{badwanpk,gtd,mcr}@fe.up.pt

Abstract. SIARD Suite enables us to preserve a relational database in an open format. It migrates a relational database to SIARD format and preserves technical and contextual metadata along with the primary data ensuring long term accessibility.

This paper introduces a web application, the SIARD Archive Browser, which allows operations on the archive such as searching for a specific record, counting records in a table containing a keyword, sorting by a column and making joins. In many use cases, the application avoids the need to load a preserved database to a DBMS.

Keywords: SIARD, database archiving, database preservation.

1   Introduction

In many organizations relational databases are used as the main component of record-keeping systems. The amount and detail of information in these systems is growing very fast. However, at the same pace the concern also grows when the current hardware, operating systems, database management system (DBMS) and the actual applications become obsolete. There is a danger of losing the information as it may become unreadable.

The Software-Independent Archival of Relational Databases (SIARD) is a solution proposed by the Swiss Federal Archives to handle the problem [1]. An open format known as SIARD format was developed for archiving a relational database. Furthermore, a software package known as SIARD Suite was developed that facilitates the migration of a relational database to the SIARD format. It is based on open standards such as XML, SQL:1999, ZIP and UNICODE. It can convert a database originally in Oracle, Microsoft SQL Server or Microsoft Access to the SIARD format.

In a SIARD archive there are two folders named **header** and **content** [2]. In the **header** folder the file **metadata.xml** contains the structure of the archived database, information about constraints and the structure of the primary data. The **metadata.xml** file also includes the description of the database, schema, tables and columns of tables. The folder **contents** contains a folder **schema0** which

---

* This work is supported by FCT grant reference number SFRH/BD/45731/2008.

P. Zaphiris et al. (Eds.): TPDL 2012, LNCS 7489, pp. 496–499, 2012.

urther contains a folder for each table named `table0`, `table1` ... `table<n>`. Each older contains a file named `table<n>.xml` which contains the table data. The column names in each table are systematically numbered as `c1`, `c2` ... `c<n>`. The SIARD archive has a unique structure where the files in the `content` older depend on the metadata in the `header` folder. Because of this characteristic t is not possible to get meaning from the data in a SIARD archive while browsing t with existing XML readers. For example the files for table data may be opened ising Microsoft Excel but are presented without proper column names as these eside in another file. Similarly it is not possible to navigate to related tables by ollowing foreign keys without some Microsoft Excel programming. The SIARD Archive Browser (SAB) is proposed to facilitate users in browsing a SIARD archive.

SAB is a Java applet and uses XML parsers, namely the document object model (DOM) and the simple API for XML (SAX), for retrieving information rom a SIARD archive. It uses the `metadata.xml` file to retrieve any content rom an archived database. Besides browsing individual tables, SAB is able to ollow foreign keys in both directions thus enabling users to create table joins. The functionalities of the SAB are highlighted in the sequel.

2 Browser Functionalities

The basic principle for the development of the SAB is that it should be simple to use. Its main window is shown in Figure 1. It shows the names of all the tables n a SIARD archive in the combo box labeled "Select a Base Table". Choosing a table makes it the current focus (base table). Furthermore, the description explaining the contents of the selected base table is shown in the "Base Table Description" text area.

2.1 Browsing a Table

A user can select a table from the combo box labeled "Select a Table to Open" n Figure 1. The table opens in a new window as shown in Figure 2. The title bar of the window shows the table name. The description of the table is presented n the text area along with its total number of rows.

Users can filter a table by entering a string in the text box and then clicking the `Filter` button. Only filtered rows are displayed and the number of matching rows is shown under the check box. If no string is entered and the button is clicked, the total number of rows in the table is displayed. Filtering may be restricted to any single column using the check box. By default the first column s selected when the check box is ticked.

The columns of a table can be re-arranged in any order. Furthermore, it is possible to sort a table by clicking on the header of any column. If a column is a foreign key to another table, the name of the referenced table is shown in the tool tip for the cursor on the header of the column.

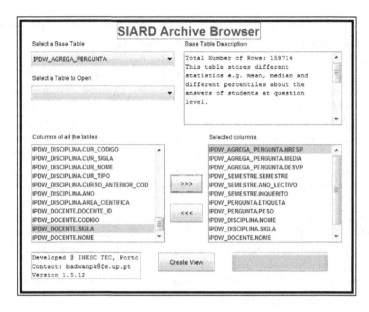

Fig. 1. SIARD Archive Browser

Users can navigate among tables using foreign keys in the base table as follows:

1. If the open table is the selected base table, clicking on a value in a foreign key column opens a new window showing the single row from the referenced table matching the clicked value as well as the description of the table.
2. If the open table is not the base table, then it is the target of at least one base table foreign key. Clicking a value in the referenced column opens a new window showing only the base table records matching that value. If there are multiple references to the same column, clicking on it opens a dialog to choose one of them to be navigated across. The records shown are those matching the clicked value on the chosen foreign key.

2.2   Making a Join

SAB makes it easy for users to make joins among tables related by foreign keys. When a base table is selected, the column names of the base table and column names of all the referenced tables (if any) are shown in the text area labeled "Columns of All the Tables" in Figure 1. Users can make a join by selecting the desired columns and then clicking the Create View button. The operation performed is a join between the base table and each of the tables with selected columns, using the corresponding foreign keys. The resulting view contains all the foreign keys and the selected columns. It can be browsed in the same manner as a table, i.e. re-arranging columns, sorting by a column, searching the whole table and searching by a specific column.

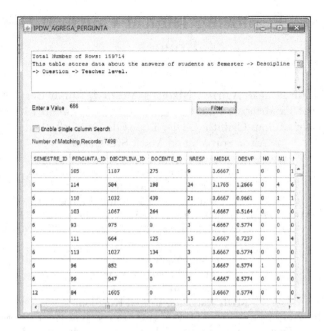

Fig. 2. Browsing a table

3 Conclusions and Future Work

The information in archived databases is not supposed to be needed very often. Sometimes it may be enough to perform some basic queries on the data which the SAB provides and thus avoid loading the archive to a DBMS. Whenever complex operations are needed, SIARD archives can be loaded to a DBMS which supports the SQL:1999 standard using the SIARD Suite.

Some functionalities which will be added in the future include support for BLOBs and CLOBs, improving the appearance of table data and addition of cumulative search facilities. Along with adding new functionalities some improvements will be made to make SAB more robust.

References

1. Heuscher, S., Järmann, S., Keller-Marxer, P., Möhle, F.: Providing authentic long-term archival access to complex relational data. In: Ensuring Long-Term Preservation and Adding Value to Scientific and Technical Data. European Space Agency (2004)
2. SFA. SIARD format description. Technical report, Swiss Federal Archives, Berne (September 2008)

PATHS – Exploring Digital Cultural Heritage Spaces

Mark Hall[2], Eneko Agirre[1], Nikolaos Aletras[2], Runar Bergheim[3],
Konstantinos Chandrinos[4], Paul Clough[2], Samuel Fernando[2], Kate Fernie[5],
Paula Goodale[2], Jillian Griffiths[5], Oier Lopez de Lacalle[6], Andrea de Polo[7],
Aitor Soroa[1], and Mark Stevenson[2]

[1] University of the Basque Country, Donostia, Spain
{e.agirre,a.soroa}@ehu.es
[2] Sheffield University, Sheffield, United Kingdom
{naletras1,p.d.clough,s.fernando,p.goodale,m.mhall,
m.stevenson}@sheffield.ac.uk
[3] Asplan Viak Internet AS, Norway
runarbe@gmail.com
[4] iSieve Technologies, Greece
chandrinos@gmail.com
[5] MDR Partners, United Kingdom
kfernie27@gmail.com,jillian.griffiths@mdrprojects.com
[6] Basque Foundation for Science / University of Edinburgh, Bilbao,
Spain / Edinburgh, United Kingdom
oier.lopezdelacalle@gmail.es
[7] Alinari, Florence, Italy
andrea@alinari.it

Abstract. Large amounts of digital cultural heritage (CH) information
have become available over the past years, requiring more powerful exploration systems than just a search box. The PATHS system aims to
provide an environment in which users can successfully explore a large,
unknown collection through two modalities: following existing paths to
learn about what is available and then freely exploring.

1 Introduction

Large amounts of digital cultural heritage (CH) information have become available over the past years, especially with the rise of aggregators such as Europeana[1], the European aggregator for museums, archives, libraries, and galleries.
These large collections present a challenge to the new user, primarily in discovering what items are present in the collection. In current systems support for item
discovery is mainly through the standard search paradigm [10], which is well
suited for CH professionals who are highly familiar with the collections, subject
areas, and have specific search goals. However, for new users who do not have a

---

[1] http://www.europeana.eu

P. Zaphiris et al. (Eds.): TPDL 2012, LNCS 7489, pp. 500–503, 2012.
© Springer-Verlag Berlin Heidelberg 2012

ood understanding of what is in the collections, what search keywords to use, nd have vague search goals, this method of access is unsatisfactory.

Alternative item discovery methodologies are required to introduce new users o digital CH collections [5,9]. Walden's Paths [8] pioneered the idea of providing sers with manually curated paths through the collection. Similarly [7] describe n automatic route-suggestion system for web-pages with the same aim of guid-ng novice users. To avoid relying on user input to create the initial paths a ariety of visualisation techniques [3] have been devised to provide the user with n overview over the collection [6]. Such visualisations provide improved access 11] as long as they ensure that similar items are grouped together [4]. Another enefit is that they promote serendipitous discovery of items [1].

2 The PATHS System

The PATHS system presented in this demonstration aims to provide a collection liscovery environment that combines the best aspects of these approaches into n integrated, seamless system. The system has two main modes: first, it allows he user to follow existing paths through the collection, a more focused variant of Walden's Paths [8]; second, it allows the user to easily branch off a path nd explore the collection on their own. The idea is to assist the novice user n transitioning from consuming existing paths, to exploring on their own, and inally creating paths for others to consume.

Fig. 1. The path-following interface consists of the item itself with its narrative (1), ath navigation (2), social features (3), and exploration starting points (4)

Figure 1 shows the user following an existing path. Such a path consists of a inear sequence of items taken from the collection, where each item consists of ts original meta-data plus a title and description (1) which the path author uses o create a narrative that the user can navigate along (2). In order to entice the ser away from the path and transition them into the exploration phase, similar tems and links to the item's topics and keywords are provided (4).

To support the exploration phase the system provides a number of different exploration modalities (Fig. 2), enabling the user to choose the exploration methodology that best suits their personal style and their current task or goal. Currently the system supports exploration via standard vocabularies (1), visual representations of hierarchical topic structures (2), and keyword clouds (3). Additionally, as in the path following interface, the provision of similar items provides a horizontal exploration network.

As the user explores the collection they can save the items, vocabulary entries, topics, or keywords to their workspace. The user can then form these into a path to share with their friends, colleagues, or everybody. Sharing and social interaction have become core ideas in the interaction with cultural heritage [2] and the PATHS system follows this trend by allowing the users to share, tag, and comment on anything they find while exploring the collection (Fig. 1 (3)).

Fig. 2. Exploration interface with the three primary exploration methodologies: standard vocabularies (1), visual topics (2), keyword clouds (3).

The PATHS system is built using a three-tier architecture consisting of a set of four component web-services that access a shared data storage and are tied together by the front-end user-interfaces (Fig. 3). The components provide acces to user-profiles, individual items, exploration visualisation, and path following and creating functionality that can be shared across a number of user-interfaces. The demo will focus on the web-based user-interface, but mobile apps and javascript components to enable integration into other sites are also under development.

Shared Data Storage			
User profile	Item access	Exploration visualisation	Path creation and following
Web-based Demo		Mobile Apps	JS Components

Fig. 3. The three-tier architecture used by the PATHS system

3 Demo Description

The PATHS demo will provide a guided tour to the PATHS system and demonstrate how a user can explore the collection, collect items and then form these into a path for others to follow. Participants can then explore the prototype system on their own, which will be provided at http://www.paths-project.eu. The demo will provide access to a sub-set of the data available in Europeana, covering approximately 1.7 million items from the English and Spanish collections.

Acknowledgements. The research leading to these results has received funding from the European Community's Seventh Framework Programme (FP7/2007-2013) under grant agreement n° 270082. We acknowledge the contribution of all project partners involved in PATHS (see: http://www.paths-project.eu).

References

1. Beale, R.: Supporting serendipity: Using ambient intelligence to augment user exploration for data mining and web browsing. International Journal of Human-Computer Studies 65(5), 421–433 (2007), Ambient intelligence: From interaction to insight
2. Bernstein, S.: Where do we go from here? continuing with web 2.0 at the brooklyn museum. In: Museums and the Web 2008: the International Conference for Culture and Heritage Online (2008)
3. Butavicius, M.A., Lee, M.D.: An empirical evaluation of four data visualization techniques for displaying short news text similarities. International Journal of Human-Computer Studies 65(11), 931–944 (2007)
4. Chen, C., Cribbin, T., Kuljis, J., Macredie, R.: Footprints of information foragers: behaviour semantics of visual exploration. International Journal of Human-Computer Studies 57(2), 139–163 (2002)
5. Geser, G.: Resource discovery - position paper: Putting the users first. Resource Discovery Technologies for the Heritage Sector 6, 7–12 (2004)
6. Hornbæk, K., Hertzum, M.: The notion of overview in information visualization. International Journal of Human-Computer Studies 69(7-8), 509–525 (2011)
7. Joachims, T., Freitag, D., Mitchell, T.: et al. Webwatcher: A tour guide for the world wide web. In: International Joint Conference on Artificial Intelligence., vol. 15, pp. 770–777. Lawrence Erlbaum Associates Ltd. (1997)
8. Shipman III, F., Furuta, R., Brenner, D., Chung, C., Hsieh, H.: Using paths in the classroom: experiences and adaptations. In: Proceedings of the Ninth ACM Conference on Hypertext and Hypermedia, pp. 267–270. ACM (1998)
9. Steemson, M.: Digicult experts seek out discovery technologies for cultural heritage. Resource Discovery Technologies for the Heritage Sector 6, 14–20 (2004)
10. Sutcliffe, A., Ennis, M.: Towards a cognitive theory of information retrieval. Interacting with Computers 10, 321–351 (1998)
11. Westerman, S.J., Cribbin, T.: Mapping semantic information in virtual space: dimensions, variance and individual differences. International Journal of Human-Computer Studies 53(5), 765–787 (2000)

FrbrVis: An Information Visualization Approach to Presenting FRBR Work Families

Tanja Merčun[1], Maja Žumer[1], and Trond Aalberg[2]

[1] University of Ljubljana, Ljubljana, Slovenia
[2] Norwegian University of Science and Technology, Trondheim, Norway
{tanja.mercun,maja.zumer}@ff.uni-lj.si,
trond.aalberg@idi.ntnu.no

Abstract. Although FRBR is becoming an important player in the bibliographic world, we have not seen many discussions or examples of how FRBR-based entities or relationships could best be displayed, explored or interacted with within a user interface. The paper presents a FrbrVis prototype as one possible approach to presenting FRBR-based bibliographic data using hierarchical information visualization structures and looks into how FRBR concepts have been implemented into an interactive user interface display.

Keywords: FRBR, Information Visualization, User Interface, Interaction.

1   Introduction

In the last few years, we have seen more frbrization projects as well as actual FRBR-inspired library information system implementations than before. However, the design and functionalities provided in these systems do not use the full potential of FRBR-based data and structures [1] and mostly only group records into work sets which are displayed in flat linear lists. With the Resource Description and Access (RDA) rules, FRBR is coming closer to its intended use in library catalogues and it is high time to start thinking about creating more innovative and effective systems for presenting FRBR-based data as the traditional approach is not capable of supporting all the possible relationships and entities provided by the FRBR model.

To our knowledge, no in depth studies have been performed so far on the choice of features, techniques, or elements for presentation of FRBR entities and relationships in user interfaces. But as Caryle stresses [2], user research on what kind of displays are most effective and what attributes most facilitate the use of catalogues are needed to guide the decision-making process of the new set of cataloguing rules and the design of online catalogue displays that incorporate FRBR. Our research presents a step in this direction by developing and testing a prototype that presents FRBR-based data using different hierarchical information visualization techniques. While such projects provide a lot of interesting topics that could be presented and discussed, this demo will focus on presenting some of the main characteristics of the FrbrVis prototype and its underlying information architecture.

P. Zaphiris et al. (Eds.): TPDL 2012, LNCS 7489, pp. 504–507, 2012.
© Springer-Verlag Berlin Heidelberg 2012

2   FrbrVis Prototype

FrbrVis prototype presents the result of our investigation into how FRBR-based re-cords could be presented and interacted with in library information systems. The process entailed a number of steps, starting with the identification of work family examples for fiction that would represent different levels of complexity within the bibliographic universe. For each work family, a selection between 40 and 100 records has been made manually in order to encompass as many scenarios and variations in entities, relationships, and attributes as possible. All the records were then manually edited and enriched (a more detailed report on the changes can be found in [3]).

While most frbrizations to date have been concerned with transforming existing re-cords as such, our study took an approach that primarily focused on creating as per-fect frbrization results as possible. By following the new RDA rules, manually adding relationships and other data that was not present in the original records, was inconsis-ently catalogued or was encoded in a way that did not support automatic processing, we were able to produce records that better captured the potential and the richness of the FRBR conceptual model. Using the FRBR tool [4] with rules adapted to this new frbrization approach, we have created frbrxml records and made sure the entities were correctly identified and all relationships were established.

In the next stage, FRBR structures and data were mapped to a user interface dis-play. Based on the three main relationships, we have divided the work record display using three tabs (Fig. 1): a) **versions** for displaying expressions and manifestations of the work, b) **related works** to show other works that are related to the work in ques-tion, and c) **works by/about the author** to present other works by the same author as well as works about that author.

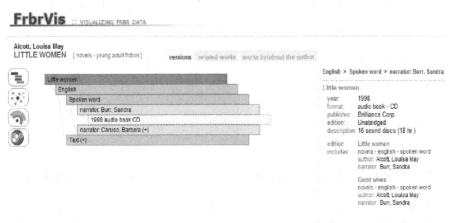

Fig. 1. FrbrVis prototype

Each of the three tab displays has been divided into two parts. On the left, the user is presented with a visual navigation feature which, using a hierarchical structure and aggregated categories, gives an overview of what is available in the collection, en-ables navigation between entities and the choice of level that most corresponds to the user's information need. By clicking on different categories, the user can drill down the hierarchy or he can decide to select a higher level category and browse through the results listed on the right side.

2.1   Information Architecture

The frbrized records were created based on the main FRBR entities and relationships. Implementing them to the user interface display, however, we have decided to aggregate these entities by attributes and relationship types instead of just creating lists of works, expressions and manifestations (see Fig. 2). Doing that allowed us to put the FRBR concepts in the centre of our design, but at the same time making it friendlier and more intuitive for the end user. The same approach has been suggested in [5] where the user testing results indicated that we should move beyond the simple transfer of work-expression-manifestation categories and "employ an in-depth hierarchical structure based on specifications of the bibliographic family".

For different versions of the work, the biggest change has been made on the expression level, where we have introduced three category levels: first the user could choose between languages, within a language between content types and then between different versions.

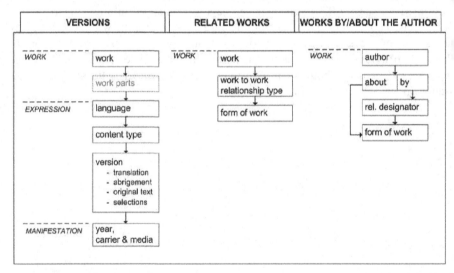

Fig. 2. Information architecture of FrbrVis display based on the FRBR conceptual model

To get an overview and to explore various related works, works have been collocated first by relationship type and then by form of work. Although the suggested relationships in the RDA are more specific, we have found the combination of more general relationships with the form of work much more useful for implementing into our hierarchically structured interactive display. The same has been done for works about and by the author, where collocation was made based on the main work-person relationships and the form of work.

.2   Information Visualization Techniques

To find a technique that would best accommodate the created hierarchical data structures, four different visualization techniques have been chosen for testing the implementation and interactive exploration of our work family data sets (Fig. 3): drop-down indented tree, sunburst , radial tree, and circular treemap technique.

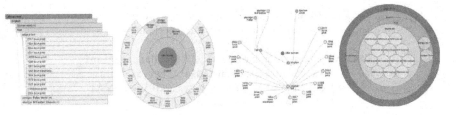

Fig. 3. Four information visualization techniques used in FrbrVis prototype

3   Conclusion

FrbrVis prototype has been developed to explore and test different presentation and interaction techniques that could be used for FRBR-based data, first through the development process itself and later through performance and subjective measures as well as observed user behaviour in a controlled experiment evaluation.

By creating the FrbrVis prototype, we have moved beyond a simple presentation of work-expression-manifestation and looked at how FRBR could help us make the most of our data by enabling features that we have not been able to provide to our users so far. However, the prerequisite is having bibliographic data that will support the creation of such systems and thinking about how we want our future information systems to look like and function is essential for better understanding and implementation of the changes in our cataloguing rules or practices as well as within frbrization projects.

References

1. McGrath, K., Bisko, L.: Identifying FRBR work-level data in bibliographic records for manifestations of moving images. The Code4Lib Journal 1(5) (2008)
2. Carlyle, A.: Understanding FRBR as a conceptual model. Library Resources and Technical Services 50(4), 264–273 (2006)
3. Aalberg, T., Merčun, T., Žumer, M.: Coding FRBR-Structured Bibliographic Information in MARC. In: Xing, C., Crestani, F., Rauber, A (eds.) ICADL 2011. LNCS, vol. 7001, pp. 128-137. Springer, Heidelberg (2011)
4. Aalberg, T.: A Process and Tool for the Conversion of MARC Records to a Normalized FRBR Implementation. In: Sugimoto, S., Hunter, J., Rauber, A., Morishima, A. (eds.) ICADL 2006. LNCS, vol. 4312, pp. 283–292. Springer, Heidelberg (2006)
5. Arastoopoor, S., Fattahi, R., Perirokh, M.: Developing user-centered displays for literary works in digital libraries: integrating bibliographic families, FRBR and users. In: Proceedings of International Conference of Asian Special Libraries (ICoASL 2011), pp. 83–91 (2011)

Metadata Enrichment Services
for the Europeana Digital Library

Giacomo Berardi[1], Andrea Esuli[1], Sergiu Gordea[2],
Diego Marcheggiani[1], and Fabrizio Sebastiani[1]

[1] Istituto di Scienza e Tecnologie dell'Informazione
Consiglio Nazionale delle Ricerche
56124 Pisa, IT
{Giacomo.Berardi,Andrea.Esuli,Diego.Marcheggiani,
Fabrizio.Sebastiani}@isti.cnr.it
[2] Austrian Institute of Technology, 1220 Vienna, AT
Sergiu.Gordea@ait.ac.at

Abstract. We demonstrate a metadata enrichment system for the Europeana digital library. The system allows different institutions which provide to Europeana pointers (in the form of metadata records - MRs) to their content to enrich their MRs by classifying them under a classification scheme of their choice, and to extract/highlight entities of significant interest within the MRs themselves. The use of a supervised learning metaphor allows each content provider (CP) to generate classifiers and extractors tailored to the CP's specific needs, thus allowing the tool to be effectively available to the multitude (2000+) of Europeana CPs.

1 Introduction

Europeana[1] is a digital library that acts as an aggregator of thousands of collections and archives of digitised printed and audiovisual material (including books, paintings, sculptures, movies, and artworks of different nature) [3]. Thousands of private and public institutions spread across the European Union are served by Europeana, ranging from major museums of international fame, to libraries of regional or domain-specific scope. Europeana users can thus search a virtual collection of millions of digital objects which altogether represent a significantly large window on Europe's cultural and scientific heritage.

Europeana does not contain the digitised objects themselves; it contains pointers to them, in the form of searchable metadata records (MRs). MRs are thus the objects around which most of the services that Europeana provides, including searching and browsing, revolve. MRs (compiled according to a format called *Europeana Semantic Elements*[2]) are fed to Europeana by each individual content provider (CP). Each such record describes (and links to) a digital object that resides in the CP's archive.

---

[1] http://Europeana.eu

[2] http://www.europeana.eu/schemas/ese/

P. Zaphiris et al. (Eds.): TPDL 2012, LNCS 7489, pp. 508–511, 2012.

We here demonstrate a service for the enrichment of Europeana MRs that has
een developed in the context of ASSETS, a project funded by the European
'ommission and aimed at developing new value-added, content-based services
ɔr Europeana. The MR is the *de facto* user's gateway to the digital object itself.
.nriching the semantics of a MR has thus a beneficial effect on the entire spec-
rum of the user's experience, including searching and browsing. The metadata
nrichment service that we describe here is to be installed on the Europeana
ɔortal, and will allow each contributing CP to enrich the semantics of its own
IRs prior to contributing them to Europeana. Different CPs are thus encour-
ged to use the same enrichment tool, thus allowing greater uniformity across
IRs of different provenance.

We view metadata enrichment as consisting of essentially two activities: *clas-
ification of MRs*, and *information extraction from MRs*. Classification consists
n the task of associating to a given MR one or more classes from a pre-specified
lassification scheme. Information extraction consists instead in the individua-
ion ("extraction") of substrings of text contained in the MR that instantiate one
.mong a set of prespecified concepts of interest. These two services are described
n Sections 2.1 and 2.2, respectively.

2 Architecture of the Metadata Enrichment System

2.1 The Metadata Classification Component

Classification refers to the task of associating to a MR one or more classes from
ι pre-specified classification scheme (i.e., a set of classes, possibly organized as a
axonomy). The chosen classification scheme can be domain-specific or general-
ɔurpose. For instance, the Accademia Nazionale di Santa Cecilia (an Italian
ultural institution active in the field of classical and contemporary music, and
ι Europeana CP) will typically be interested in adopting a music-specific classi-
ication scheme, while another institution of broader scope might want to adopt
ι general-purpose scheme such as the Library of Congress Subject Headings[3].

Setting up a classification system for Europeana is challenging, because of
he sheer diversity (a) of classification schemes that CPs might choose, and
b) of languages in which the MRs are going to be expressed in. Given this,
t would be implausible to provide a classification service based on manually
vritten classification rules, since this would place the burden of rule-writing on
he CPs themselves, who would then probably renounce using the service.

As a result, our classification service is based on supervised learning tech-
ιology: a learning algorithm learns, from a sample of manually preclassified
locuments that are provided to it, the characteristics that a given MR should
ιave in order to be associated to a given class [5]. This frees the CP from the
ɔurden of writing classification rules, and only requires it to provide a sample
ɔf manually classified MRs. In many cases these latter may already be available

---

[3] http://www.loc.gov/aba/cataloging/subject/weeklylists/

to the CP as a product of a classification activity that the CP has carried out in its daily operations.

As the supervised learning technology we have used the TREEBOOST system, a member of the family of "boosting"-based supervised learning algorithms that has shown state-of-the-art accuracy across a variety of datasets [2]. TREEBOOST allows the use of classification schemes organized either as a tree or as a directed acyclic graph. In order to cater for the multiplicity of languages in which the MRs can be expressed we use a completely language-independent preprocessing module which only consists of extracting the MRs from the records; stop word removal, stemming, and other types of linguistic analysis that are language-dependent are not used.

Classification accuracy results from experiments on many datasets of metadata records from Europeana CPs are presented in [1].

2.2   The Information Extraction Component

Information extraction refers to the task of identifying ("extracting"), in a given text, substrings that instantiate "concepts" belonging to a prespecified set [4]. Examples of "domain-independent" such concepts may be Person, Location, Organization; examples of domain-dependent concepts (e.g., for the domain of music) may instead be Director, Instrument, or Composer. Identifying instances of such concepts in a MR may be beneficial for browsing, and is ultimately a means of adding semantics to the MR and of enabling semantic search.

For reasons similar to the ones discussed for classification, it would be inappropriate to deploy an information extraction service based on manually written extraction rules. Again, we have chosen the supervised learning route, according to which the CP provides a general-purpose learning system with a set of texts in which the instances of the concepts of interest have been marked as such; from these annotated texts the system learns to extract the instances of the concepts.

As the supervised learning technology we have used an algorithm belonging to the family of *conditional random fields* (CRFs), which are nowadays considered state-of-the-art for addressing sequence learning tasks. Since syntactic analysis (particularly: POS tagging) is known to be beneficial in information extraction, we here do not avoid language-dependent processing. We first apply an automatic language recognizer to the record in order to determine the language it is written in, and then we submit the record to a POS tagging phase in the cases in which a POS tagger is available for the specified language.

Information extraction accuracy results from experiments on many datasets of metadata records from Europeana CPs are presented in [1], along with additional details about the preprocessing steps we have enacted.

2.3   The Ingestion Control Panel

The backend processing work performed by Europeana follows a complex process that includes operations related to customer relationship management, metadata

Fig. 1. The Ingestion Control Panel

‌arvesting, metadata processing (i.e. data normalization), thumbnail generation, ‌reation of submission and access information packages, etc. We extend the GUI ‌f the Europeana Ingestion Control Panel by integrating the invocation of en‌ichment services and supporting the following functionality (see Fig. 1):

- Enrichment model learning: by using a training set that suites their meta‌data, the content providers or Europeana are allowed to run the learn‌ing of enrichment models for metadata classification and/or information extraction;
- Enrichment by metadata classification: the classification of a collection can be performed by selecting an appropriate classification model. The user is also allowed to test the model on a particular collection object;
- Enrichment by information extraction: the extraction of the structured in‌formation from the object descriptions can be performed similarly to the classification by selecting an appropriate model and the collection or collec‌tion object to be enriched.

References

. Berardi, G., Esuli, A., Marcheggiani, D., Sebastiani, F., Gordea, S., Täckström, O.: Ingestion services: 2nd release. Deliverable D2.1.3, ASSETS Project ICT PSP 250527, Commission of the European Communities (2012)

. Esuli, A., Fagni, T., Sebastiani, F.: Boosting multi-label hierarchical text catego‌rization. Information Retrieval 11(4), 287–313 (2008)

. Purday, J.: Think culture: Europeana.eu from concept to construction. The Elec‌tronic Library (6), 919–937 (2009)

. Sarawagi, S.: Information extraction. Foundations and Trends in Databases 1(3), 261–377 (2008)

. Sebastiani, F.: Machine learning in automated text categorization. ACM Computing Surveys 34(1), 1–47 (2002)

Collaboratively Creating a Thematic Repository Using Interactive Table-Top Technology

Fernando Loizides, Christina Vasiliou, Andri Ioannou, and Panayiotis Zaphiris

Cyprus University of Technology
{fernando.loizides,christina.vasiliou,andri.i.ioannou,
panayiotis.zaphiris}@cut.ac.cy

Abstract. This paper reports on the design and development of a surface computing application in support of collaborative idea creation and
thematic categorisation. C.A.R.T (Collaborative Assisted Repository for
Tabletops) allows up to 4 users to simultaneously interact with virtual
objects, each containing a single concept, to create thematic categories.
Each object, which replicates a physical post-it on a multi-touch tabletop, is created by one of the team members either previous to the meeting
or during the initial stage. The application then encourages the exchange
of debate and conversation by presenting the ideas one at a time for users
to discuss and categorise. The resulting idea repository can be used for
roadmap creation as well as comparative studies using further participants. The application's main task is similar to that of card sorting and
affinity diagramming. We report on the functionality of the application
which was designed and developed following a user-centred approach.

Keywords: Digital Repositories, Idea Mapping, Tabletops.

1 Introduction

Surface computing and particularly multi-touch interactive tabletops have recently attracted attention in the fields of Digital Libraries, Human-Computer
Interaction and Educational Technology. A few empirical investigations have
demonstrated their affordances for collaborative brainstorming, yet a lot remains
to be done. The use of table-top computing for the creation and manipulation of
digital repositories remains under researched, largely due to the novelty of these
interactive surface devices. Our application, C.A.R.T, is specifically designed to
create repositories which contain ideas from a brainstorming session. The main
functionality, is the interactive and collaborative structuring of the repository
(underlying database) into thematic categories, rather than the input modality
of the ideas. We begin by describing the application architecture, followed by
the stages which a user undergoes to complete the task at hand.

2 System Overview

C.A.R.T can accommodate from 1 to 4 users simultaneously around the tabletop (See Figure 2 (left image)). It was created using Flash and Action Script

P. Zaphiris et al. (Eds.): TPDL 2012, LNCS 7489, pp. 512–516, 2012.

.0. Figure 1 provides an overview of the C.A.R.T architecture. We chose to use TouchMagix multi-touch tabletop (www.touchmagix.com/magixtable/) for its uitable hardware features. The tabletop accommodates 40 simultaneous touch etections for multiple user interactions. The 32 inch touch surface also requires o pressure for input recognition making it easier to use. Furthermore, we se-ected the use of this table-top over the alternative (Microsoft Surface 1) due to ts ambient light immunity. The application facilitates three stages; Idea gener-tion, Thematic Categorisation and Structured Repository Finalisation.

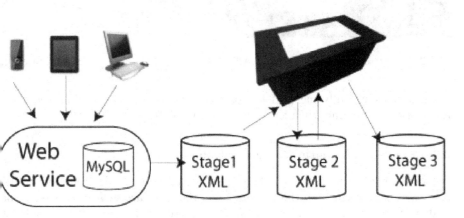

Fig. 1. C.A.R.T Architecture Overview

2.1 Stage 1: Idea Generation

Each collaborator generates new ideas for a designated amount of time. Ideas are typed into a web application (see Figure 2 (right image)) through the use of a mobile device (laptop, tablet, Smartphone) connected to the Internet. The need for the integration of mobile devices and a web application emerged from a constraint imposed by TouchMagix (also true for other platforms such as MS Surface); namely, that text entry can be done from one keyboard at a time. To resolve this problem, in the Beta version application, we developed four virtual keyboards on the tabletop (one for each user). However, users experienced diffi-culties typing extended ideas on the particular virtual keyboard; the keyboard interaction suffered from input latency and mistyping issues. We are aware that the benefits of an interactive tabletop do not depend on simultaneous input but perhaps lie in a more general quality of the form of input (i.e. touch) [3]. Thus, the use of mobile devices for input via a web application was considered as a practical solution to the problem by allowing collaborators to generate digital post-it notes at the same time. Indirect input devices are also more familiar to users and do not require much effort to use [2]. This set-up also allows for the participants to use their own devices (providing they are capable of wi-fi and web browsing) in order to input the notes into the system. Stage one can take place either during the initial part of the meeting, or alternatively, before the

514   F. Loizides et al.

Fig. 2. Left: Users interacting with C.A.R.T. Right: input system for ideas in Stage 1.

participants physically meet. If the participants have already submitted their notes prior to meeting, then the process begins at stage 2.

2.2   Stage 2: Thematic Categorisation

Before stage 2 begins, an XML file is created based on the ideas entered by the participants in stage 1. In the second stage, the ideas are presented one-by-one as digital post-it notes in the middle of the tabletop surface and become subject to discussion amongst the collaborators (see Figure 3 [left image]). "Orientation proves critical in how individuals comprehend information, how collaborators coordinate their actions, and how they mediate communication" [4]. We therefore chose to make each post-it appear in the direction of its creator who will explain his or her thought process. Participants can manipulate post-it notes to move them across the surface manipulation gestures such as rotation and zoom (two fingers used for these gestures), exploiting the abilities of multiple-touch interfaces. For each note, collaborators make an effort to categorize it in a thematic unit. In order to show the next note the previous note must be categorized. If there is controversy among the decision on how to categorize it, ideas can be placed in a Decide Later box to be revisited upon the categorization of other ideas. This is in-line with previous research into document triage, in which users create 'category piles', one of which is usually designated for documents which the user is uncertain about [1]. Thematic units can be created by any participant using a virtual keyboard and their title can be subsequently changed. Participants can drag and drop a post-it note over a thematic unit,for the two to be joined (see Figure 3 [right image]). At the current stage, participants do not have the ability to drag a note out of a thematic unit and reallocate it. At this stage participants cannot edit each others ideas, cannot generate new ideas, and post-its, and generated thematic units cannot be deleted. Once all the post-it notes are loaded on the screen and categorized by the participants. At the end of stage 2 a second XML file is created to store the latest ideas and structure.

Fig. 3. Left: stage 2, initial thematic categorisation of each post-it. Right: Stage 3, all interactions possible, such as deletion and adding of notes.

2.3 Stage 3: Structured Repository Finalisation

In this last stage, more flexibility is given to the participants to engage in collaborative decision making and reach a consensus on the thematic categories and taxonomy of ideas. In addition to the collaboration capabilities of Stage 2, participants can now: (a) Drag post-it notes and reallocate them in new thematic units, without any limitations until they reach a consensus, (b) Generate new post-it notes (c) Delete post-its or thematic units as needed by dragging them to the trash icon. Furthermore, ideas can be duplicated and placed in two categories if required. When all the participants are in agreement with any changes made, then the process ends and a third XML file is created showing the new structure (if any) of the entries.

3 Conclusions and Work in Progress

The process of creating digital libraries is enriched by technological advancements, collaborative abilities and novel input devices. In this paper we demonstrate one such technology in a prototype application using touch surfaces. The users collaborate in real time to produce a thematically structured database of brainstorming sessions using a process similar to affinity diagramming. The main aims of this application is encourage and stimulate discussion for the creation of databases of ideas that can enriched by collaboration between large groups of people. Our initial experiments included 3-4 individuals, while our future plans are to allow for significantly larger sets of participants to use the technology. Our future versions of the application will also allow for screen shots, photos from a web camera and video or voice messages to be stored.

References

1. Buchanan, G., Loizides, F.: Investigating Document Triage on Paper and Electronic Media. In: Kovács, L., Fuhr, N., Meghini, C. (eds.) ECDL 2007. LNCS, vol. 4675, pp. 416–427. Springer, Heidelberg (2007)

2. Ha, V., Inkpen, K.M., Whalen, T., Mandryk, R.L.: Direct intentions: The effects of input devices on collaboration around a tabletop display. In: Proceedings of the First IEEE International Workshop on Horizontal Interactive Human-Computer Systems, TABLETOP 2006, pp. 177–184. IEEE Computer Society (2006)
3. Harris, A., Rick, J., Bonnett, V., Yuill, N., Fleck, R., Marshall, P., Rogers, Y.: Around the table: are multiple-touch surfaces better than single-touch for children's collaborative interactions? In: Proceedings of the 9th International Conference on Computer Supported Collaborative Learning, CSCL 2009, vol. 1, pp. 335–344. International Society of the Learning Sciences (2009)
4. Kruger, R., Carpendale, S., Scott, S.D., Greenberg, S.: How people use orientation on tables: comprehension, coordination and communication. In: Proceedings of the 2003 International ACM SIGGROUP Conference on Supporting Group Work, GROUP 2003, pp. 369–378. ACM, New York (2003)

Author Index

Printed in the United States
by Baker & Taylor Publisher Services